Plant Ecology: Systems and Processes

Plant Ecology: Systems and Processes

Editor: Clive Koelling

Callisto Reference
www.callistoreference.com

Callisto Reference,
118-35 Queens Blvd., Suite 400,
Forest Hills, NY 11375, USA

Visit us on the World Wide Web at:
www.callistoreference.com

ISBN: 978-1-63239-796-6 (Hardback)

The publisher's policy is to use permanent paper from mills that operate a sustainable forestry policy. Furthermore, the publisher ensures that the text paper and cover boards used have met acceptable environmental accreditation standards.

Printed in China.

Cataloging-in-publication Data

Plant ecology : systems and processes / edited by Clive Koelling.
p. cm.
Includes bibliographical references and index.
ISBN 978-1-63239-796-6
1. Plant ecology. 2. Plant ecological genetics. 3. Plant ecophysiology. I. Koelling, Clive.
QK901 .P53 2017
581.7--dc23

Table of Contents

Preface

The study of distribution and abundance of plants focusing on the ecological factors is called plant ecology. The aim of this book is to present researches that have transformed this discipline and aided its advancement. This book is a valuable compilation of topics, ranging from the basic to the most complex advancements in the field of plant ecology. Some of the diverse topics covered in this book address the varied branches that fall under this discipline. Researches and case studies from across the world have been included in this text. It will prove to be immensely beneficial to students and researchers in this field.

The purpose of the book is to provide a glimpse into the dynamics and to present opinions and studies of some of the scientists engaged in the development of new ideas in the field from very different standpoints. This book will prove useful to students and researchers owing to its high content quality.

At the end, I would like to appreciate all the efforts made by the authors in completing their chapters professionally. I express my deepest gratitude to all of them for contributing to this book by sharing their valuable works. A special thanks to my family and friends for their constant support in this journey.

Editor

Cloning and Functional Characterization of a Vacuolar Na^+/H^+ Antiporter Gene from Mungbean (*VrNHX1*) and Its Ectopic Expression Enhanced Salt Tolerance in *Arabidopsis thaliana*

Sagarika Mishra[1], Hemasundar Alavilli[2], Byeong-ha Lee[2], Sanjib Kumar Panda[3,4], Lingaraj Sahoo[1]*

1 Department of Biotechnology, Indian Institute of Technology Guwahati, Guwahati, India, **2** Department of Life Science, Sogang University, Mapo-gu, Seoul, Korea, **3** Department of Life Sciences and Bioinformatics, Assam University, Silchar, India, **4** Department of Biochemistry & Molecular Biology, Noble Research Centre, Oklahoma State University, Stillwater, OK, United States of America

Abstract

Plant vacuolar NHX exchangers play a significant role in adaption to salt stress by compartmentalizing excess cytosolic Na^+ into vacuoles and maintaining cellular homeostasis and ionic equilibrium. We cloned an orthologue of the vacuolar Na^+/H^+ antiporter gene, *VrNHX1* from mungbean (*Vigna radiata*), an important Asiatic grain legume. The *VrNHX1* (Genbank Accession number JN656211.1) contains 2095 nucleotides with an open reading frame of 1629 nucleotides encoding a predicted protein of 542 amino acids with a deduced molecular mass of 59.6 kDa. The consensus amiloride binding motif (^{84}LFFIYLLPPI^{93}) was observed in the third putative transmembrane domain of *VrNHX1*. Bioinformatic and phylogenetic analysis clearly suggested that *VrNHX1* had high similarity to those of orthologs belonging to Class-I clade of plant NHX exchangers in leguminous crops. *VrNHX1* could be strongly induced by salt stress in mungbean as the expression in roots significantly increased in presence of 200 mM NaCl with concomitant accumulation of total [Na^+]. Induction of *VrNHX1* was also observed under cold and dehydration stress, indicating a possible cross talk between various abiotic stresses. Heterologous expression in salt sensitive yeast mutant AXT3 complemented for the loss of yeast vacuolar *NHX1* under NaCl, KCl and LiCl stress indicating that *VrNHX1* was the orthologue of *ScNHX1*. Further, AXT3 cells expressing *VrNHX1* survived under low pH environment and displayed vacuolar alkalinization analyzed using pH sensitive fluorescent dye BCECF-AM. The constitutive and stress inducible expression of *VrNHX1* resulted in enhanced salt tolerance in transgenic *Arabidopsis thaliana* lines. Our work suggested that *VrNHX1* was a salt tolerance determinant in mungbean.

Editor: Binying Fu, Institute of Crop Sciences, China

Funding: LS is grateful to Department of Biotechnology, Government of India for partial support from the grant (BT/PR10818/AGR/02/591/2008) to this work. BhL is grateful to Rural Development Administration, Republic of Korea for its support by Next-Generation BioGreen 21 Program (PJ009104). SM is grateful to MHRD for Research Fellowship. The funders had no role in study design, data collection and analysis, decision to publish, or preparation of the manuscript.

Competing Interests: The authors have declared that no competing interests exist.

* Email: ls@iitg.ernet.in

Introduction

Soil salinity poses increasing threat to plant growth and agricultural productivity worldwide [1]. More than 20% of the cultivated area and nearly half of the world's irrigated lands are adversely affected by salinity [2]. Enhanced crop production on salinity inflicted areas will rely on innovative agronomic practices coupled with use of genetically improved crop varieties [3]. In saline soils, Na^+ is the predominant toxic ion. Excess accumulation of Na^+ in cytosol is detrimental to many metabolic and physiological processes, vital for plant growth and productivity, as it causes ion imbalance, hyper osmotic stress, and oxidative damage to plants [4]. To cope with salinity stress, plants have evolved sophisticated mechanisms, including restricted uptake/ exclusion of Na^+ from cell, and compartmentalization of Na^+ into vacuoles. Na^+ efflux is catalyzed by a plasma membrane Na^+/H^+ antiporter (NHX) encoded by *SOS1* [5,6] while, a vacuolar Na^+/H^+ antiporter catalyzes the sequestration of Na^+ into vacuoles. Compartmentalization of Na^+ into vacuole not only provides an efficient mechanism to avert deleterious effects of Na^+ in cytoplasm, but also allows plant to use Na^+ as an osmoticum, for maintaining an osmotic potential for driving water into cell [4,7]. Vacuolar compartmentalization of Na^+ is a critical process in salt adaptation, which is conserved in both halophytes and glycophytes. Na^+ transport into vacuoles mediated through vacuolar Na^+/H^+ antiporter is an energy driven process involving H^+ transporting pumps such as H^+-ATPase and H^+-PPase [8]. The genes encoding for Na^+/H^+ antiporters (NHX) have been cloned from more than 60 plant species, including gymnosperms and dicotyledonous and monocotyledonous angiosperms. The expression of most *NHXs* was induced by NaCl treatment [9]. Overexpression of vacuolar NHX genes suppressed the salt sensitive phenotype of a yeast mutant defective for endosomal and vacuolar Na^+/H^+ antiporters and conferred salt tolerance in transgenic plants [10,11]. Several reports on improvement of salt tolerance through overexpression of vacuolar *NHXs* in agriculturally important but glycophytic crops implicate a pivotal function of the *NHXs* in intracellular compartmentalization of Na^+ and salt tolerance [3,12]. In legumes, *NHX1* has been reported in *Glycine max* [13], *Medicago sativa* [14], *Trifolium repens* [15], *Lotus tenuis* [16], *Caragana korshinskii* [17] and recently by our lab, in *Vigna unguiculata* (GenBank Acc. No. JN641304.2). However, no salt-tolerant genes including *NHX* yet reported from mungbean.

Mungbean (*Vigna radiata* L. Wilczek) is an important grain legume widely cultivated in south, east and south-east Asian countries for its protein rich grains. Salinity is recognized as major constraint in the production of mungbean [18,19]. Mungbean is moderately drought tolerant [20] and therefore, this distinctive character makes it a valuable tropical crop legume for studying the molecular tolerance mechanisms for various abiotic stresses including salinity. In this paper, we report the cloning and molecular characterization of *VrNHX1* antiporter from *V. radiata*, its expression pattern under various abiotic stresses like salt, dehydration and cold stress, functional complementation of *VrNHX1* in *Saccharomyces cerevisiae* salt sensitive mutant (AXT3) and finally, increased salt tolerance by constitutive and inducible expression of *VrNHX1* in transgenic *Arabidopsis thaliana*, highlighting the potent role of *VrNHX1* in salt tolerance mechanisms.

Materials and Methods

Plant Material and Stress Treatment

Mungbean (*Vigna radiata* L. Wilczek cv. K-851) seeds were surface sterilized with 0.2% mercuric chloride and rinsed three times with distilled water. The seeds were germinated in dark chamber for 2 days, transferred to Hoagland's nutrient medium, grown hydroponically in a controlled growth chamber at 25°C, 80% relative humidity with a 16 hr/8 hr photoperiod and photosynthetic flux intensity of 300 µmol m^{-2} s^{-1} for 14 days. For salt stress treatment, these two weeks old mungbean seedlings grown under hydroponic culture were transferred to 200 mM NaCl solution for 12 hrs and roots were harvested, frozen immediately, and stored at −80°C until further use.

Molecular cloning of *VrNHX1* cDNA by RACE approach

Total RNA was isolated from salt-treated roots of mungbean using AMBION RNAqueous Kit (Ambion, Carlsbad, CA, USA). One microgram of RNA was used for cDNA synthesis using Revert Aid First Strand cDNA Synthesis Kit (Thermo Fisher Scientific, Waltham, MA, USA). The cDNA was amplified with a pair of degenerate primers (Deg FP: 5′- TAT(A/T)ATATT-CAATGC(C/A)GGGTTTCA(G/A)GT(A/G) -3′ and Deg RP: 5′- GCATT(A/G)TGCCA(A/G)GT(A/G)TAATG(A/T)GA-CAT(A/G/C)AC -3′) designed from the conserved region of transmembrane domains of plant NHX antiporters submitted in NCBI database. The PCR condition was: 94°C for 3 min; 94°C for 30 sec, 52°C for 30 sec, 72°C for 30 sec with 30 cycles, and a final extension at 72°C for 10 min. Based on the resulting partial fragment, gene specific primers were designed for amplification of 5′- and 3′- untranslated regions of *VrNHX1*.

The 5′ RACE was performed using the 5′ RACE System for Rapid Amplification of cDNA Ends Kit, Version 2.0 (Invitrogen, Carlsbad, CA, USA). Briefly, five micrograms of RNA was used for first strand cDNA synthesis using a gene specific primer (GSP1: 5′- CTGCTTCTTTTTCACCTGAAACCCAGC -3′) and Superscript II reverse transcriptase (Invitrogen). cDNA was purified using SNAP column to remove unincorporated dNTPs and primer, that might interfere in the homopolymeric tailing of cDNA. Terminal transferase enzyme was used to add dCTPs to 3′ end of cDNA. The dc-tailed cDNA was amplified using abridged anchor primer (AAP: 5′- GGCCACGCGTCGACTAGTACGG-GIIGGGIIGGGIIG -3′) and gene specific primer (GSP2: 5′-ACCTGAAACCCAGCATTGAATAT-3′). The PCR condition was: 94°C for 3 min; 94°C for 30 sec, 55°C for 30 sec, 72°C for 30 sec with 30 cycles, and a final extension of 72°C for 10 min. Further, nested PCR was performed using abridged universal anchor primer (AUAP: 5′- GGCCACGCGTCGACTAGTAC -3′) and nested gene specific primer (GSP3: 5′- GGTATATGAA-GAAAAGATCTTC -3′) using the first PCR product as template. The PCR condition was: 94°C for 3 min; 94°C for 30 sec, 52°C for 30 sec, 72°C for 30 sec with 35 cycles, and a final extension of 72°C for 10 min.

The 3′ RACE was performed using 3′ RACE System for Rapid Amplification of cDNA Ends Kit, Version E (Invitrogen, Carlsbad, CA, USA). Five micrograms of RNA was used to synthesize cDNA using a dT-adapter primer (AP: 5′- GGCCACGCGTCGAC-TAGTAC$(T)_{17}$ -3′) and Superscript II reverse transcriptase (Invitrogen, Carlsbad, CA, USA). The first 3′-RACE-PCR was carried out using gene specific primer (GSP4: 5′- AGTGG-CATCCTCACTGTATTCTTTTGTG -3′) and abridged universal anchor primer (AUAP: 5′- GGCCACGCGTCGACTAG-TAC -3′). The PCR condition was: 94°C for 3 min; 94°C for 30 sec, 60°C for 30 sec, 72°C for 1 min and 30 sec with 30 cycles, and a final extension of 72°C for 10 min. The PCR product was diluted 10 times (1:10) and used as template for nested 3′ RACE-PCR. The nested 3′-RACE-PCR was carried out using gene specific primer (GSP5: 5′-GCTGTATATTGGAAGGCACTCT-3′) and abridged universal anchor primer (AUAP: 5′-GGCCACGCGTCGACTAGTAC -3′). The PCR condition was: 94°C for 3 min; 94°C for 30 sec, 55°C for 30 sec, 72°C for 1 min and 30 sec with 30 cycles, and a final extension of 72°C for 10 min. The above PCR products were cloned to TA cloning vector pTZR/T (Thermo Fisher Scientific, Waltham, MA, USA) sequenced and contiguous sequences aligned to obtain full length of *VrNHX1* cDNA.

Bioinformatic analysis of *VrNHX1*

Multiple sequence alignment and phylogenetic analysis were performed using Clustal W [21]. A phylogenetic tree was constructed using neighbor joining method and reliability of the tree was analyzed with bootstrap analysis with 500 replicates using MEGA4 (Molecular Evolutionary Genetics Analysis): Tree Explorer software [22]. Hydrophobicity plot and transmembrane domain prediction was performed using TMpred software [23]. Post-translational modification of *VrNHX1* was predicted by searching for conserved motifs of N- and O- glycosylation and N-myristoylation sites using ScanProsite [24].

Southern hybridization for *VrNHX1* copy number in mungbean genome

Twenty µg of genomic DNA was used for gene copy analysis of *VrNHX1* and digested with restriction endonucleases EcoRI and HindIII. Digested genomic DNA was electrophoretically fractionated on 0.8% agarose gel and blotted onto Zeta-Probe membrane (Bio-Rad, Hercules, CA, USA). The blot was hybridized with DIG-labeled 1.6 kb PCR product, corresponding to the coding region of *VrNHX1*. Southern hybridization was carried out using solution containing 50% formamide, 5 X SSC, 5 X Denhardt's solution, 0.05 M sodium phosphate pH 6.5, 0.1% SDS, 10% dextran sulfate, 0.1 mg/ml sheared denatured salmon-sperm

DNA and 20 ng/ml probe at 42°C for 18 hrs. Washing and detection was performed according to instructions of the DIG Labeling and Detection system (Roche Diagnostics, Mannheim, Germany).

Heterologous expression of *VrNHX1* in yeast mutant

Functional complementation assay was performed in yeast strains, W303-1B (*MATα ade2-1 can1-100 his3-11,15 leu2-3,112 trp1-1 ura3-1*) and AXT3 (*Δ ena1- 4::HIS3 Δnha1::-LEU2 Δnhx1::TRP1, ura3-1*). Yeast strains were grown in YPD (1% Yeast extract, 2% Peptone and 2% Glucose), YPGal (1% Yeast extract, 2% Peptone and 2% Galactose), SC (0.67% Yeast Nitrogen Base, 2% Glucose) and APGal synthetic minimal media (10 mM arginine, 8 mM phosphoric acid, 2 mM $MgSO_4$, 1 mM KCl, 0.2 mM $CaCl_2$, 2% Galactose, trace vitamins, and minerals; pH-4.0) supplemented with appropriate amino acids as indicated.

The CDS of *VrNHX1* was cloned into yeast expression vector pYES2.0 (Invitrogen, Carlsbad, CA, USA) with restriction sites of KpnI and BamHI. The yeast strains were transformed with pYES2.0 empty vector (labeled as AXTYES2.0 strain) or pYES*VrNHX1* recombinant plasmid (labeled as AXTVrNHX1 strain) by Lithium acetate method [25] and selected on SC ura^- medium.

For growth assay, precultured cells were grown till OD_{600} of 1.0, diluted to an OD_{600} of 0.006, and inoculated to liquid APGal ura^- synthetic minimal media supplemented with different concentrations of NaCl, KCl, and LiCl and grown at 30°C for 48 hrs. For complementation assay, saturated liquid cultures (OD_{600} 0.8) of each strain were serially diluted to 10, 100 and 1000 fold and spotted on APGal solid media supplemented with or without 50, 75 and 100 mM NaCl, 0.5 M KCl, 25 mM LiCl and and YPGal media supplemented with 50 μg/ml hygromycin. Plates were maintained at 30°C. Growth was monitored after 3 days.

Intracellular measurement of Na^+ and K^+ distribution in yeast mutant

Intracellular ion was extracted from yeast strains grown in liquid APGal media, pH 4.0 supplemented without or with 75 mM NaCl [26]. Briefly, cells were harvested at an OD_{600} of 0.3–0.4, centrifuged at 3000 g/3 min, washed twice in ice-cold 10 mM $MgCl_2$, 10 mM $CaCl_2$ and 1 mM HEPES buffer and resuspended in the same buffer. The relationship between cell density (Absorbance at OD_{600}) and yeast dry weight was determined. Total intracellular ion was determined by addition of HCl to a final concentration of 0.4% and incubated at 95°C for 20 min. After removal of cell debris the supernatant was measured for presence of Na^+ and K^+. Similarly, cells were grown and washed as above and resuspended in 2% cytochrome c, 18 μg/ml antimycin, 1 mM HEPES, 10 mM $MgSO_4$, 10 mM $CaCl_2$, and 5 mM 2-Deoxy D-Glucose solution. Cytochrome c selectively permeabilizes the plasma membrane. After 20 min incubation at room temperature, cells were washed thrice with the same solution without cytochrome c. Cytoplasmic ion content was determined by pooling the supernatants. The remaining vacuolar ions were extracted with addition of HCl in a final concentration of 0.4% and incubated at 95°C for 20 min. The Na^+ and K^+ distribution in the cytoplasmic and vacuolar fractions were measured in flame photometer (Systronics, MP, India).

Vacuolar pH estimation and fluorescence imaging

Yeast cells were grown in APGal medium (pH 5.0) to an OD_{600}: 0.25–0.3, pelleted, and washed with deionized distilled water. Further, the yeast cells were incubated with 50 μM 2′,7′-bis-(2-carboxyethyl)-5-(and-6)-carboxyfluorescein (BCECF-AM) (Molecular Probes, Eugene, Oregon) for 30 min, centrifuged, washed thrice and resuspended in APGal medium (pH 5.0) and immediately used for fluorescence measurement. Single emission fluorescence measurement at 490 nm excitation wavelength and absorbance at 600 nm were measured using LS 55 Fluorescence Spectrophotometer (Perkin Elmer, Waltham, MA, USA). The calibration curve for fluorescence intensities at different pH was obtained for each strain [27]. Briefly, the yeast strains (W303-1B, AXTYES2.0, AXTVrNHX1) were incubated in experimental medium containing 50 mM MES, 50 mM HEPES, 50 mM KCl, 50 mM NaCl, 0.2 M ammonium acetate, 10 mM NaN_3, 10 mM 2- deoxy glucose, 50 μM carbonyl cyanide m-chlorophenylhydrazone, titrated to five different pH values within the range of 4.0 to 8.0. Background subtracted I_{490} values were normalized to cell density for each strain, labeled as NI_{490} and plotted against pH values. For vacuolar pH estimation, experimental NI_{490} values corresponding to each strain was analyzed with the calibration curve specific for each strain.

For vacuolar pH imaging the yeast cells were grown, pelleted to be suspended in the same medium with 50 μM BCECF-AM pH specific dye as above. For fluorescence imaging, 100 μl of BCECF-loaded yeast suspension was plated onto glass cover slips precoated with concavalin-A (Sigma-Aldrich, St. Louis, MO, USA) and placed on glass slides. Fluorescence images were captured in Nikon eclipse Ti-U Fluorescence microscope (Nikon, Chiyoda, Tokyo, Japan).

Expression analysis of *VrNHX1* using semi-quantitative RT-PCR

Expression analysis under salt stress: Two different stages of growth in mungbean seedlings i.e. early and mid stage, were considered for expression analysis under salt stress (200 mM NaCl). Mungbean seedlings were germinated, grown in Hoagland's nutrient medium for five and ten days, in case of early and mid stage respectively, and transferred to 200 mM NaCl solution for salt stress assay. Leaves and roots of salt treated early and mid stage mungbean seedlings, were harvested at time intervals 0, 6, 12, 18, 24, and 48 hrs. Similarly, expression pattern for *VrNHX1* in response to different forms of abiotic stress such as salt (200 mM NaCl), dehydration (200 mM mannitol) and cold stress (4°C) was also studied at different time intervals (0, 6, 12, and 24 hrs) for mid-stage (10 days old) mungbean seedlings. Total RNA was extracted using RNeasy Plant Mini Kit (Qiagen, Venlo, Limburg, Netherlands) and reverse transcribed using Revert Aid First Strand cDNA Synthesis Kit. Semi-quantitative RT-PCR was performed using gene specific primers (RF: 5′- GTATTTCCACTGGCG-TAGTCATTTTGC -3′ and RR: 5′- GCATCATTCACAG-CACCCTCTCGG -3′). The PCR condition was: 94°C for 3 min; 94°C for 30 sec, 62°C for 30 sec, 72°C for 30 sec for 28 cycles, and a final extension at 72°C for 10 min. Housekeeping *VrTubulin-β* primers (FN: 5′- CTTGACTGCATCTGC-TATGTTCAG-3′ and RN: 5′-CCAGCTAATGCTCGGCA-TACTG -3′) were used as an internal control. The PCR condition was: 94°C for 3 min; 94°C for 30 sec, 58°C for 30 sec, 72°C for 30 sec for 28 cycles, and a final extension at 72°C for 10 min. Semi-quantitative RT-PCR was repeated three times. The PCR products were analyzed in 2% agarose gel stained with 10 mg/ml ethidium bromide.

Measurement of total ion content in salt stressed mungbean seedlings

Leaves and roots of untreated and salt-treated early and mid stage mungbean seedlings were harvested at different time intervals (0, 6, 12, 18, 24, 48 and 72 hrs). The samples were dried, digested with concentrated HNO_3 at 90°C for 1 hr and centrifuged at 12,000 rpm for 10 min [28]. The suspension was diluted with sterile milliQ water and analyzed for Na^+ and K^+ content in flame photometer.

Binary vector preparation and plant transformation

The 1.6 kb CDS of *VrNHX1* was cloned into standard plant binary vector pCAMBIA2301 (11.6 kb) flanked by cauliflower mosaic virus CaMV 35S promoter and terminator at *Pst*I restriction site. The resulting recombinant plant binary vector was labeled as pCAMBIA2301-35S::*VrNHX1* (13.9 kb). Further, a 0.898 kb promoter fragment of *AtRD29A* (DQ071887.1) was amplified from *A. thaliana* genomic DNA and cloned into *Eco*RI digested recombinant binary vector pCAMBIA2301-35S::*VrNHX1* (13.9 kb) by replacing the 0.4 kb 35S promoter fragment from 35SP::*VrNHX1*::35STer cassette. The resulting binary vector was named pCAMBIA2301-RD29A::*VrNHX1* (14.4 kb).

The recombinant plant binary vectors, pCAMBIA2301-35S::*VrNHX1* (13.9 kb) and pCAMBIA2301-RD29A::*VrNHX1* (14.4 kb) were transferred into *A. tumefaciens* GV3101 strain via electroporation at 1250 V with capacitance of 25 mF and resistance of 400 ohm. The constructs were used for transformation of *Arabidopsis thaliana* (ecotype Columbia) via floral dipping method [29]. The T_1 transgenic lines were screened on ½ MS medium (Duchefa, Haarlem, Netherlands) supplemented with 50 mg/l kanamycin (Duchefa, Haarlem, Netherlands). The transgenic selections were continued until T_3 generation to obtain homozygote transgenic lines with a single T-DNA locus (35S::*VrNHX1* or RD29A::*VrNHX1*).

RNA extraction and Real Time PCR of transgenic *Arabidopsis* lines

Total RNA was extracted from wild-type (WT) and T_3 independent 35S::*VrNHX1* and RD29A::*VrNHX1* transgenic lines using RNeasy Plant Mini Kit (Qiagen), quantified in Nanovue Plus Spectrophotometer (GE Healthcare, Little Chalfont, Buckinghamshire, UK) and cDNA was prepared using Revert Aid First Strand cDNA Synthesis Kit. The gene specific forward primer (VrRTF: 5′- TGATTCAATCCATCGACCAA-3′) and 35S poly-A reverse primer (TerparR: 5′-GCGAAACCC-TATAAGAACCCTAATTCC-3′) were used for amplification of a 0.283 kb fragment of *VrNHX1*::35S poly-A in transgenic *A. thaliana* plants. Housekeeping (UBQ1FP: 5′- AGAGCTGT-CAACTGCAGGAAGAA-3′ and UBQ1RP- 5′-ACAA-GAAAAACAAACCCTATCAAA GG) primers were used to amplify a 150 bp fragment of *AtUbiquitin* to be used as an internal control. Real time PCR was performed using USB VeriQuest SYBR Green qPCR Master Mix (2X) (Affymetrix, Santa Clara, CA, USA) and primers at a final concentration of 200 nM in 7500 Real-Time PCR System (Applied Biosystem, Foster City, California, USA) following the manufacturer's protocol. The experiment was repeated twice independently with three replicates. The expression values relative to the standard curve was calculated for each sample. The relative expression level of transgene *VrNHX1* in wild-type (WT) and transgenic 35S::*VrNHX1* and RD29A::*VrNHX1* *Arabidopsis* lines was estimated by normalizing *VrNHX1* expression values with respect to housekeeping *AtUBQ1* expression values in each case.

Salt tolerance assays of transgenic *Arabidopsis* lines

Wild-type (WT) and T_3 transgenic 35S::*VrNHX1* and RD29A::*VrNHX1* *Arabidopsis* seeds were germinated on ½ MS medium [30] in growth chamber maintained at 22°C and 60% relative humidity with a 16 hr/8 hr photoperiod under controlled conditions.

Studying germination efficiency under salt stress: The WT and T_3 transgenic 35S::*VrNHX1* and RD29A::*VrNHX1* lines were germinated on ½ MS medium supplemented with or without 150 mM NaCl and kept at 4°C for 3 days, prior to, transfer to growth chamber. The germination efficiency was studied after 10 days of salt stress.

Measurement of growth parameters under salt stress: The 4 days old germinated seedlings were transferred to ½ MS medium supplemented with or without 150 mM NaCl for 1 week and the difference in root length of wild-type WT and T_3 independent transgenic lines of *Arabidopsis* seedlings expressing *VrNHX1* was measured. Mean data was collected from ten replicates (n = 10) for wild-type (WT) and T_3 kanamycin selected transgenic *Arabidopsis* lines.

Measurement of physiological parameters under salt stress: The 10 days old germinated seedlings were transferred to ½ MS liquid medium supplemented with or without 200 mM NaCl for 5 days. For measurement of chlorophyll content, shoot samples were homogenized in 95% ethanol, lysate was centrifuged at 3,000 rpm for 10 min and absorbance was recorded for the extract at wavelength of 648 and 664 nm [31]. Lipid peroxidation was measured as the amount of malondialdehyde (MDA) determined by the thiobarbituric acid (TBA) reaction. Briefly, 0.2 g of fresh leaf samples were homogenized with 5 ml of 0.25% TBA containing 10% TCA (tricloroacetic acid). The homogenate was boiled for 30 min at 95°C and centrifuged at 10,000 g for 10 min Absorbance values were recorded at 532 nm and values corresponding to non-specific absorption at 600 nm were subtracted [32]. For colorimetric estimation of proline, leaf samples (0.5 g) were homogenized with 5.0 ml of sulfosalicylic acid (3%). 2 ml of homogenate was filtered through Whatman filter paper (No. 2) and incubated with 2 ml glacial acetic acid and 2 ml ninhydrin reagent at a ratio of 1:1:1 in boiling water bath at 100°C for 30 min. After cooling, 4 ml toluene was added to the reaction mixture, mixed vigorously and absorbance was measured at 520 nm [33]. Mean data was collected from three replicates (n = 3) for wild-type and T_3 kanamycin selected transgenic *Arabidopsis* lines.

Measurement of Na^+ and K^+ in transgenic *Arabidopsis* lines

The germinated seedlings were initially grown in ½ MS medium (0.5% agar) for 5 days and then subsequently transferred to soilrite and grown for 2 weeks. The WT and T_3 transgenic lines were subjected to salt stress for a period of 2 weeks by watering them with ½ MS nutrient liquid media supplemented with 250 mM NaCl. The whole plant was harvested for Na^+ and K^+ estimation using method described elsewhere [30]. Mean data was collected from three replicates (n = 3) for wild-type (WT) and T_3 kanamycin selected transgenic *Arabidopsis* lines.

Statistical analysis

Statistical comparison between the variances was determined by ANOVA (Analysis of variance) and significant differences between mean values were determined by Bonferroni analysis. Statistically significant mean values were denoted as "*" ($P \leq 0.05$).

Results

Isolation and in-silico analysis of *VrNHX1*

A *VrNHX1* cDNA of 2095 nucleotides in length (Genbank Accession number JN656211.1), with an open reading frame of 1,629 bp was obtained by RACE-PCR approach. It encodes a polypeptide of 542 amino acid residues with an estimated molecular mass 59.60 kDa and isoelectric point 6.76, predicted using ExPaSy bioinformatic tools for protein structure analysis (http://www.expasy.org/tools/). Multiple sequence alignment of deduced amino acid sequences of VrNHX1 revealed that it has 97.42% sequence identity with *Vigna unguiculata*, 92.25% with *Glycine max*, 88.48% with *Caragana korshinskii*, 87.27% with *Lotus tenuis*, 87.25% with *Trifolium repens*, 87.06% with *Medicago sativa*, and 86.72% with *Cicer arietinum* (Fig. 1 and S1). Phylogenetic relationship analysis performed using MEGA4 software indicated that VrNHX1 clustered into Class-I type IC-NHX legume *NHX* homologs, more closely to VuNHX1 and GmMHX1 (Fig. 1). The hydropathy plot of VrNHX1 protein predicted by TMpred software indicated highly hydrophobic N-terminal end with 11 putative transmembrane domains and a longer hydrophilic C-terminal end inside the vacuolar lumen (Fig. S2). The amiloride binding motif, ^{84}-LFFIYLLPPI-^{93}, a classic inhibitor of Na^+/H^+ antiporters [34] and also highly conserved among eukaryotic Na^+/H^+ exchangers, was detected in TM3 region (Fig. S1). The prediction of putative post-translational modification sites by ScanProsite software indicated presence of two potential *N*-glycosylation (ASN_glycosylation) sites, fifteen phosphorylation sites for protein kinase CK2 and protein kinase C, ten *N*-myristoylation sites, and one Leucine Zipper site (Table S1).

The Southern hybridization analysis revealed presence of single copy of *VrNHX1* in mungbean genome (Fig. 2). Two hybridization signals, one each for HindIII and EcoRI digested mungbean genome were detected, possibly due to the occurrence of a single *Hind*III site in *VrNHX1* (1.6 kb). Occurrence of a single *Eco*RI site in genome fragment of *VrNHX1* was accounted for getting two signals as probe lacked any *Eco*RI site.

Functional characterization of *VrNHX1* using salt sensitive yeast mutant

Previous work showed that heterologous expression of Na^+/H^+ antiporter genes in yeast mutant AXT3 could partly suppress its hypersensitivity to hygromycin and restore salt tolerance. The similar method was exploited to initially characterize the function of *VrNHX1*. The AXTVrNHX1 cells displayed enhanced Na^+, K^+ and Li^+ tolerance with statistically significant improvement in their survival at NaCl (75 and 100 mM) (Fig. 3 A) and 0.5 M KCl (Fig. 3 B), in contrast to AXTYES2.0 cells. Expression of *VrNHX1* in AXT3 cells under GAL1-inducible promoter restored salt tolerance upto 100 times dilution in 75 and 100 mM NaCl (Fig. 4 A), and better survival at 1000 times dilution range in 25 mM LiCl and 0.5 M KCl in AXTVrNHX1 cells on solid media (Fig. 4 B). *ScNHX1* has been suggested to ameliorate sensitivity of yeast cells by sequestering hygromycin-B, a cationic aminoglycoside antibiotic in vacuole. Therefore, yeast mutant lacking *NHX1* is more susceptible to hygromycin treatment [27]. *VrNHX1* expression showed suppression of hygromycin (50 µg/ml) sensitivity in AXTVrNHX1 cells (Fig. 4 C).

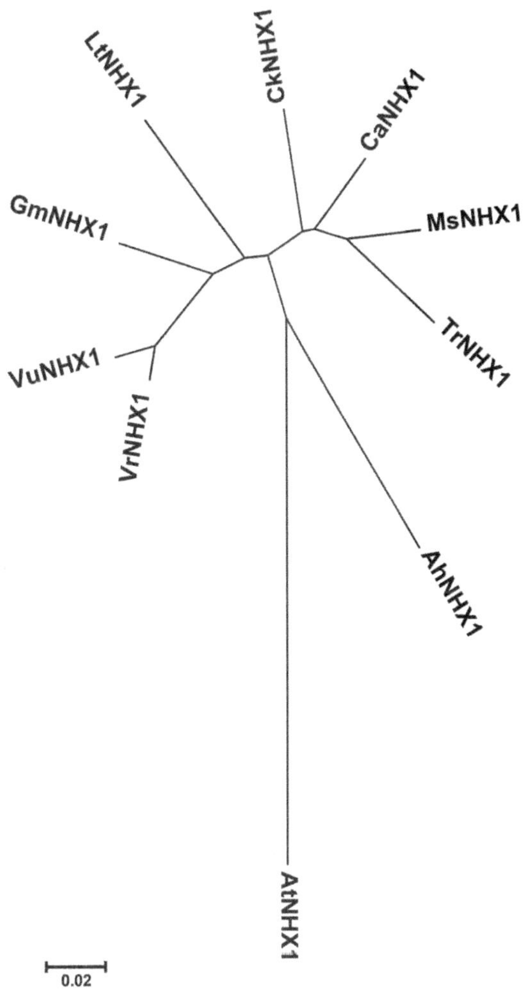

Figure 1. The phylogenetic tree for plant Na^+/H^+ antiporters was generated using MEGA4: Tree Explorer software. The evolutionary history was inferred using the neighbor-joining method and analyzed using bootstrap analysis with 500 replicates. Branches corresponding to partitions reproduced in less than 50% bootstrap replicates are collapsed. The tree is drawn to scale, with branch lengths in the same units as those of the evolutionary distances used to infer the phylogenetic tree. The evolutionary distances were computed using the Poisson correction method and are in the units of the number of amino acid substitutions per site. The GenBank Accession numbers for NHX proteins used are: VrNHX1 (AEO50758.1), VuNHX1 (AEO72079.2), GmNHX1 (AAY430061.1), CkNHX1 (ABG89337.1), MsNHX1 (AAS84487.1), CaNHX1 (ADL28385.1), TrNHX1 (ABV00895.1), LtNHX1 (ACE78322.1), AhNHX1 (ADK74832.1), AtNHX1 (NM_122597.2).

Na^+ and K^+ distribution in yeast mutants

The AXTYES2.0 cells displayed 2.3 times lower K^+ content than AXTVrNHX1 cells under normal condition owing to lack of yeast $Na^+/K^+/H^+$ antiporter activity (Fig. 5). Under salt stress, AXTVrNHX1 cells accumulated 2 times higher and 4.8 times lower vacuolar Na^+ content compared to AXTYES2.0 and W303-1B cells, respectively (Fig. 5). Similarly, vacuolar K^+ content observed for AXTVrNHX1 cells was 2.36 times higher than AXTYES2.0 cells. The cytoplsamic Na^+ content was higher in both the cell types as compared to W303-1B, due to the loss of NHA exchanger activity which cannot be solely compensated by *VrNHX1* complementation. However, cytoplasmic K^+ fractions measured were not statistically significant, though AXTVrNHX1

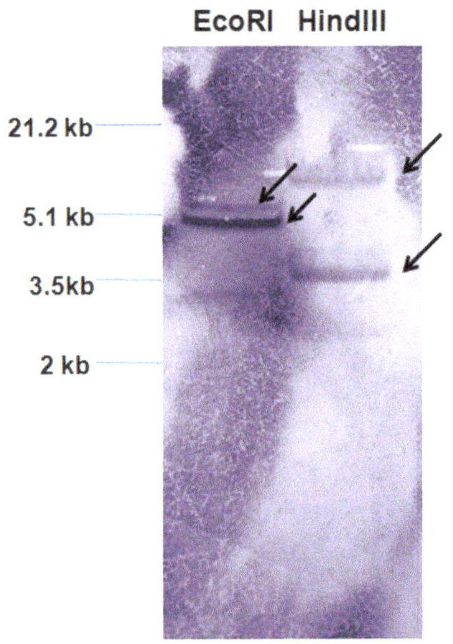

Figure 2. Copy number analysis of *VrNHX1* in mungbean genome. Mungbean genomic DNA (20 µg) was digested with EcoRI and HindIII, and hybridized with DIG-labeled probe corresponding to the CDS of *VrNHX1*. Hybridization signals are indicated as arrows.

cells exhibited higher K^+ values as compared to AXTYES2.0, indicating the improved ability of AXTVrNHX1 cells in maintaining a higher intracellular K^+/Na^+ ratio for ionic homeostasis. The total ion content in yeast cells was in accordance with distribution of Na^+ and K^+ in cytoplasm and vacuole.

Vacuolar pH estimation and imaging

2′,7′-bis-(2-carboxyethyl)-5-(and-6)-carboxyfluorescein (BCECF-AM), a widely used cell-permeant and pH-sensitive fluorescent indicator was used to measure the change in vacuolar pH of yeast mutant expressing *VrNHX1* grown under low pH environment. The study on the effect of low pH on growth efficiency of yeast cells showed that growth of AXTYES2.0 cells was highly affected with a 70.66% reduction in growth as compared to W303-1B. Moreover, AXTVrNHX1 mutant showed improved growth under acidic condition (Fig. S3). Vacuolar pH was estimated following calibration curve plotted for each strain (Fig. S4). An acidic vacuolar pH of 5.4 was observed for AXTYES2.0 cells whereas, a pH value 5.9 and 6.2 was recorded for AXTVrNHX1 and W303-1B cells, respectively in response to low pH stress condition (Fig. 6 A). Similarly, fluorescence images provided acidic vacuolar pH values for AXTYES2.0 cells and expression of *VrNHX1* alkalinized the vacuolar compartment (Fig. 6 B).

Expression pattern of *VrNHX1* under abiotic stress by Semi-quantitative RT-PCR

The expression of *VrNHX1* was studied by semi-quantitative RT-PCR, in roots and leaves of mungbean seedlings at early (five days old) and mid (ten days old) growth stages exposed to salt stress (200 mM NaCl) for different time interval (0, 6, 12, 18, 24 and 48 hrs). The results indicated that transcript levels of *VrNHX1* were induced by NaCl in both roots and shoots of early and mid stage mungbean seedlings, indicating the potent role of *VrNHX1* in salt tolerance mechanisms in mungbean. In case of early seedling stage, higher expression level of *VrNHX1* was observed in leaves at 12, 24, and 48 hrs and in roots after 6 hrs (Fig. 7 A). The differential expression of *VrNHX1* in roots and leaves was also observed in mid stage seedlings, with a significant accumulation observed at 48 hrs in leaves whereas, some basal level of *VrNHX1* transcript was observed in roots under normal condition which further increased steadily with salt stress treatment period (Fig. 7 A).

To determine whether the expression of *VrNHX1* was also induced by dehydration (200 mM Mannitol) and cold (4°C), mid-stage (10 days old) seedlings were given the respective stress treatments for different time intervals (0, 6, 12, and 24 hrs). The *VrNHX1* expression varied with salt, cold and drought stress. The accumulation of *VrNHX1* transcript under salt, cold and dehydration stress reached its peak at 24 hours (Fig. 7 B). The results indicated that osmotic and low temperature stress is involved in the up-regulation of *VrNHX1* in addition to an ion-specific signaling component in mungbean. The *VrNHX1* expression analysis revealed involvement of cross talk between salinity, low temperature and osmotic stress in mungbean.

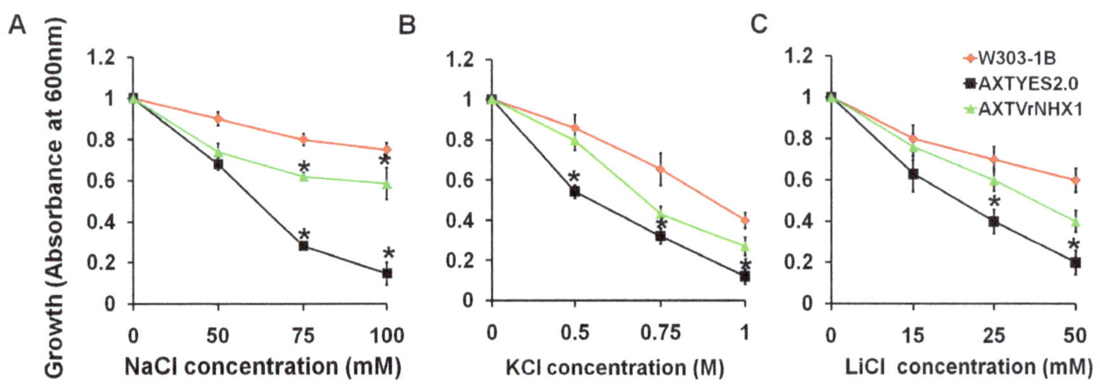

Figure 3. Cation sensitivity assay of transformed yeast strains (W303-1B, AXTYES2.0, AXTVrNHX1) under various concentrations of NaCl, KCl, and LiCl. Saturated seed cultures for each strain was diluted to an OD_{600} of 0.006 and inoculated to liquid APGal medium (pH 5.5) supplemented with or without various concentrations of (A) NaCl (0, 50, 75, 100 mM), B) KCl (0, 0.5, 0.75, 1.0 M), and (C) LiCl (0, 15, 20, 25 mM). Growth was observed at 30°C after 3 days and absorbance recorded at 600 nm. Data are means of 3 independent events (n = 3) and standard errors are plotted in the graph. Statistically significant values at P≤0.05 are indicated as "*", using Bonferroni analysis.

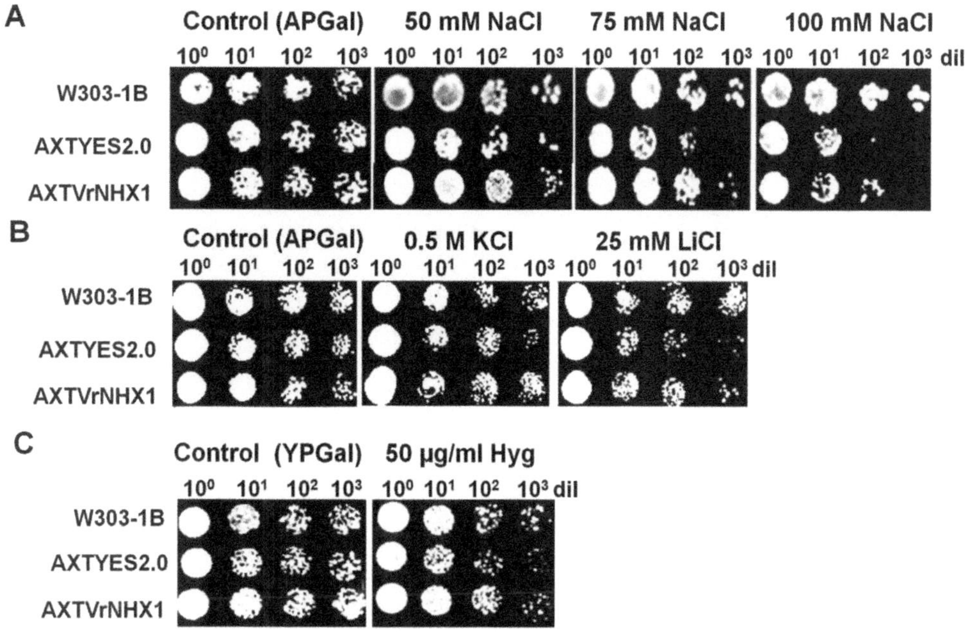

Figure 4. Heterologous expression of *VrNHX1* in yeast mutant. Wild type (W303-1B) strain was used as a control, *Δ ena1- 4 Δnha1 Δnhx1* mutant (AXT3) strain was transformed with null pYES2.0 (labeled as AXTYES2.0 strain) and pYES*VrNHX1* recombinant vector (labeled as AXTVrNHX1) were used for complementation assay. 10-fold serial dilutions of saturated seed cultures of each strain were spotted onto APGal media (pH-5.5) supplemented with or without (A) 50, 75 and 100 mM NaCl, (B) 25 mM LiCl, and (B) 0.5 M KCl. (C) Hygromycin sensitivity assay was performed by spotting 10-fold serial dilutions of saturated seed cultures of each strain onto YPGal media (pH- 5.5) supplemented with or without 50 µg/ml Hyg. The plates were incubated at 30°C for 3 days.

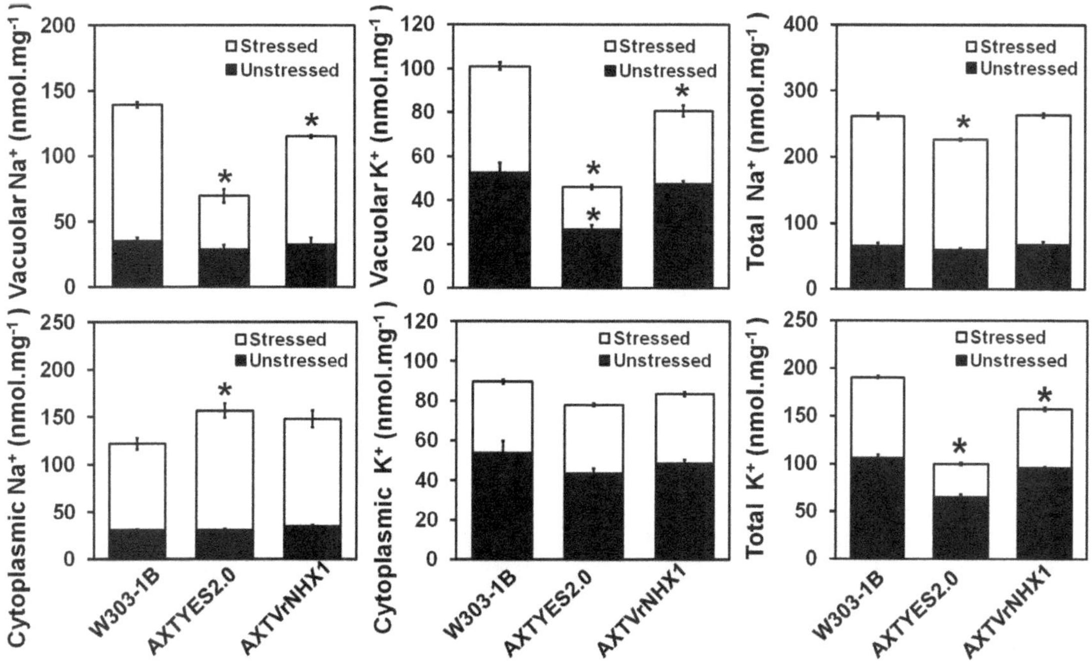

Figure 5. Total intracellular ion estimation in yeast strains W303-1B, AXTYES2.0 and AXTVrNHX1. Yeast cells were grown in APG medium (pH 4.0) with 1 mM KCl supplemented in presence (stressed) or absence of 75 mM NaCl (unstressed) and harvested at a cell density of 0.3. Total intracellular, vacuolar and cytoplasmic Na^+ and K^+ content was determined as described in the materials and methods section. Data are means of 3 independent events (n = 3) and standard errors are plotted in the graph. Statistically significant values at $P \leq 0.05$ are indicated as "*", using Bonferroni analysis.

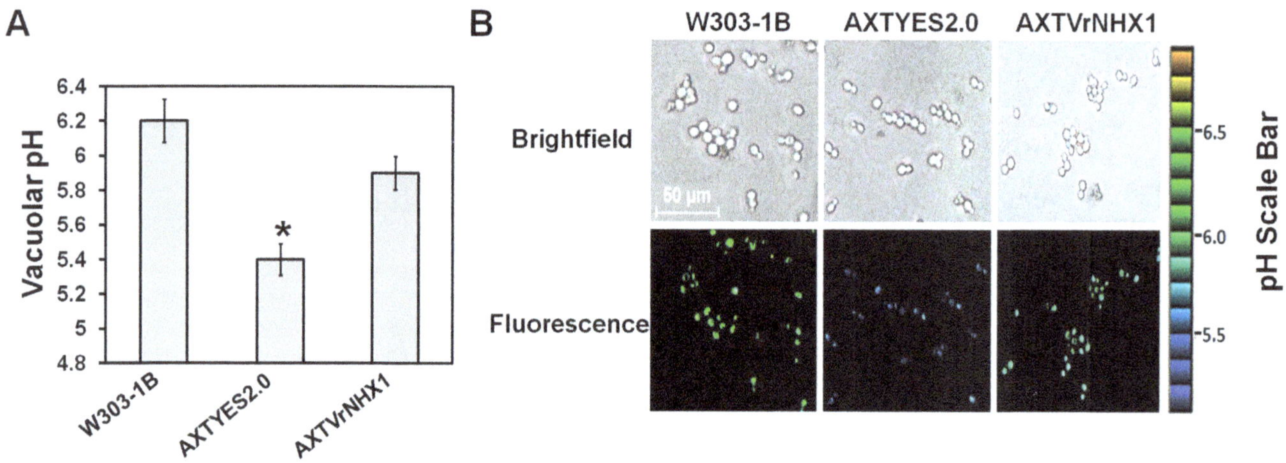

Figure 6. Measurement of vacuolar pH in yeast strains. (A) Vacuolar pH was measured for BCECF-AM loaded yeast strains W303-1B, AXTYES2.0 and AXTVrNHX1 as described in materials and methods following the calibration curve (Figure S4). Mean and SEs are plotted for three independent events (n = 3) in each case. Statistically significant values at P≤0.05 are indicated as "*", using Bonferroni analysis. (B) Accumulation of pH-sensitive fluorescent BCECF dye in yeast vacuoles was measured. The yeast strains were grown in APGal media (pH 5.0), resuspended in minimal medium with BCECF-AM dye for 30 min at 30°C. Yeast cells were visualized by Nikon eclipse Ti-U Fluorescence microscope (Nikon) at excitation wavelength of 440 nm. Bar scale: 50 μm.

Na^+ and K^+ measurement in salt stressed mungbean seedlings

The measurement of Na^+ and K^+ content in leaves and roots of untreated and salt-treated mungbean seedlings at different time intervals (0, 6, 12 18, 24, 48 and 72 hrs) showed that under salt stress, Na^+ accumulation increased in leaves/roots by 1.28/2.1, 1.1/2.3, 2.1/4.36, 4.8/4.3, 4.1/4.54 times whereas, K^+ accumulation decreased by 3.4/4.5, 1.6/1.78, 1.59/2.43, 2.2/3 and 2.1/3.5 times as compared to control condition at 6, 12, 18, 24 and 48 hrs, respectively in early stage mungbean seedlings (Fig. 8 A). Similarly, in mid stage seedlings, Na^+ accumulation in leaves/roots

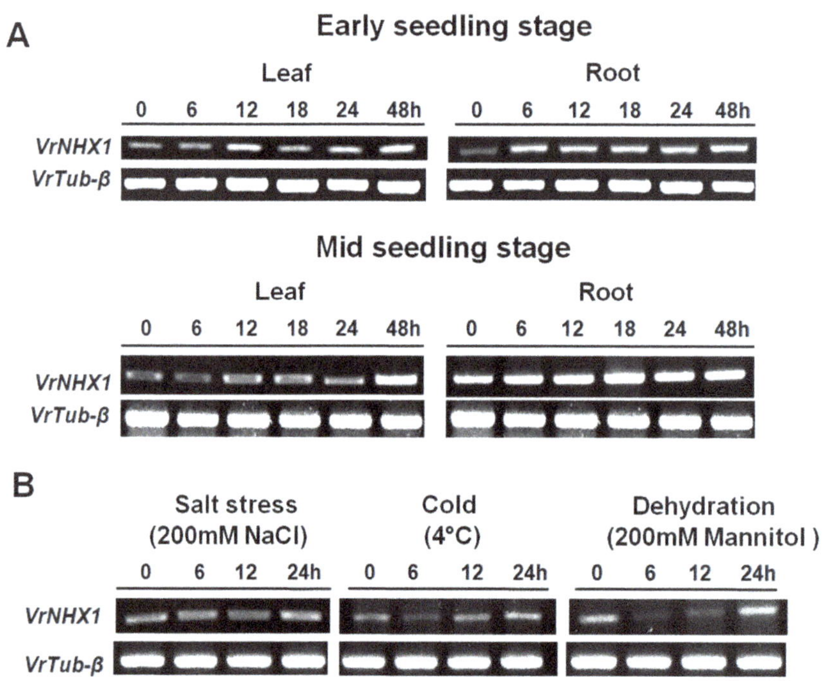

Figure 7. Expression analysis of *VrNHX1* in early and mid stage mungbean seedlings under various abiotic stresses. (A) Semi-quantitative RT-PCR for studying expression patterns of *VrNHX1* under salt stress was performed. Total RNA was isolated from leaves and roots of early (5 days) and mid stage mungbean seedlings (10 days) under 200 mM NaCl treatment at time intervals of 0, 6, 12, 18, 24, and 48 hrs. (B) Semi-quantitative RT-PCR for studying expression patterns of *VrNHX1* under different abiotic stress conditions such as salt, cold and dehydration stress was studied. Total RNA was isolated from mid stage mungbean seedlings under (A) 200 mM NaCl, (B) Cold (4°C), and (C) 200 mM Mannitol treatment at time intervals of 0, 6, 12, and 24 hrs. PCR fragments of 566 bp and 422 bp size corresponding to *VrNHX1* and *VrTubβ* were fractionated electrophoretically on 2% agarose gel stained with 10 mg/ml ethidium bromide.

also increased by 1.1/1.4, 1.4/2.4, 4/3.3, 4.5/3.5, 9.8/4.2 and 7.1/4.7 times whereas, K^+ accumulation decreased by 1.05/1.1, 1.03/1.66, 1.1/3.57, 1.34/3.2, 1.36/4.07, 1.77/4.03 times as compared to control condition at 6, 12, 18, 24, 48 and 72 hrs, respectively (Fig. 8 B). The overall higher accumulation of Na^+ (µmoles/g DW) in roots as opposed to leaves indicated the restriction of movement of toxic Na^+ to the aerial part of the plant as a plausible mechanism to confer salinity tolerance in mungbean.

Ectopic expression of *VrNHX1* resulted in enhanced salt tolerance in transgenic *Arabidopsis*

In order to characterize *VrNHX1* functionally *in planta*, T_3 homozygous *Arabidopsis* lines expressing *VrNHX1* under the control of constitutive CaMV35S promoter or a stress-responsive RD29A promoter were generated using the binary constructs pCAMBIA2301-35S::*VrNHX1* (Fig. S5 A) and pCAMBIA2301-RD29A::*VrNHX1* (Fig. S5 B), respectively, to study their performance under salt stress. The germination efficiency was studied in transgenic lines 1 (35S::*VrNHX1*) and 4 (RD29A::*VrNHX1*) after exposure to 150 mM NaCl stress for 10 days. Under normal condition, no difference was observed in WT and transgenic lines (Fig. 9 A). However, the transgenic lines exhibited better survival and germination efficiency than WT under salt stress (Fig. 9 A). Further, inhibition of root growth in WT and transgenic lines under salt stress (150 mM NaCl) was studied (Fig. 9 B). Transgenic lines 1 and 4 exhibited 2.65 and 3 times higher root length respectively, than WT (Fig. 9 C). The effect on physiological parameters was monitored in 10 days old wild-type (WT) and independent transgenic *Arabidopsis* lines expressing *VrNHX1* constitutively (Lines 1–3, 35S::*VrNHX1*) and inducibly (Lines 4–6, RD29A::*VrNHX1*) under 200 mM NaCl stress for 5 days, by analyzing the total chlorophyll, malondialdehyde (MDA) for lipid peroxidation and proline content. Under normal physiological condition, no qualitative and statistical difference was observed between wild-type and transgenic *Arabidopsis* lines (Fig. 10). However, under salt stress (200 mM NaCl), WT showed leaf senescence while transgenic *Arabidopsis* lines (Lines 1–3, 35S::*VrNHX1* and Lines 4–6, RD29A::*VrNHX1*) showed better growth and survival (Fig. 10 A). The transgenic lines showed higher chlorophyll (18–20 mg/ml) and proline (4.8–6 µmoles/g FW) content than WT (Fig. 10 B). The 35S::*VrNHX1* lines showed 1.35 times higher proline than RD29A::*VrNHX1* lines. A lower lipid peroxidation was detected in transgenic lines as WT showed 1.33 times higher malondialdehyde (MDA) content (Fig. 10 B).

Effect of salt stress was studied in mature WT and transgenic lines (Line 1, 35S::*VrNHX1* and Line 4, RD29A::*VrNHX1*). The transgenic lines displayed better survival efficiency while WT exhibited leaf senescence and growth inhibition upon salt stress (200 mM NaCl) (Fig. 11 A). Transgenic *Arabidopsis* 35S::*VrNHX1* plants displayed constitutively high expression of *VrNHX1* under both control (unstressed) and salt stress conditions, whereas RD29A::*VrNHX1* lines showed high induction of *VrNHX1* only after stress treatment with basal expression levels under normal conditions (Fig. 11 B). The total Na^+ and K^+ accumulated in transgenic lines was higher than WT. Further, transgenic 35S::*VrNHX1* and RD29A::*VrNHX1* lines exhibited 1.3 and 1.14 times higher Na^+/K^+ ratio, respectively, as compared to WT (Fig. 11 C, D).

Discussion

This is the first report on isolation and functional characterization of a vacuolar Na^+/H^+ antiporter (*VrNHX1*) from mungbean. Phylogenetic analysis and evolutionary relationship revealed that

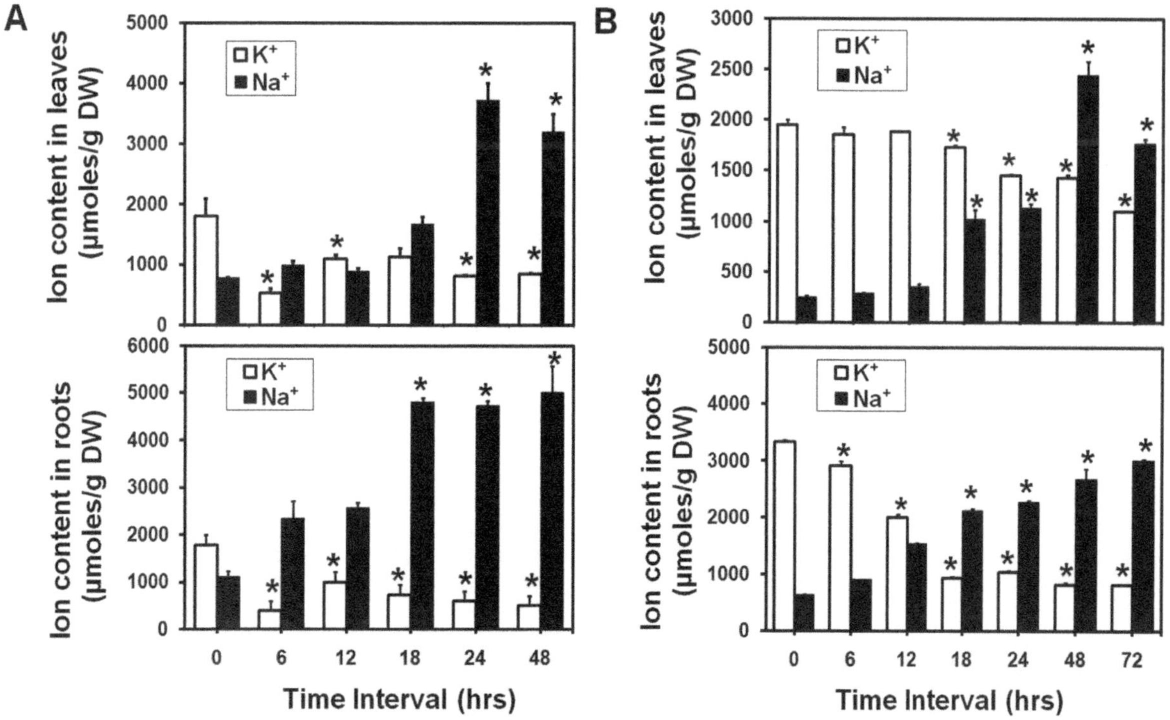

Figure 8. Total intracellular ion measurement in leaves and roots of early and mid stage mungbean seedlings. Na^+ and K^+ content in (A) leaves and (B) roots of unstressed and salt stressed mungbean seedlings harvested at time intervals of 0, 6, 12, 24, 48, and 72 hrs was measured using Flame Photometer. Values indicate means ± SE (n = 3). Statistically significant values at P≤0.05 are indicated as "*", using Bonferroni analysis.

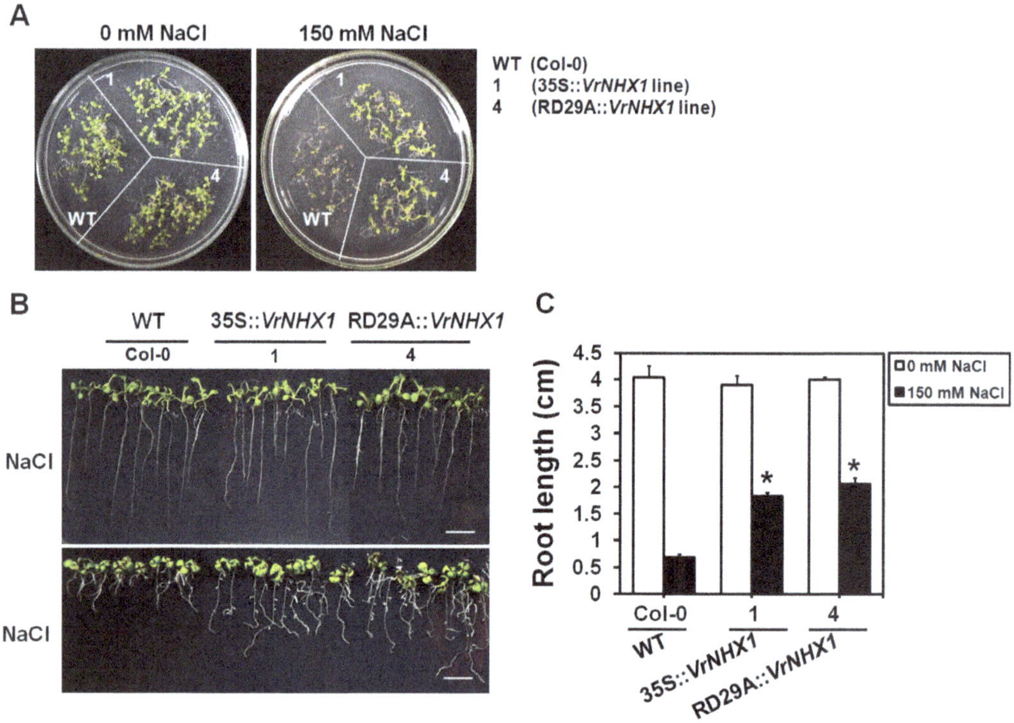

Figure 9. Effect of salt stress on germination efficiency and root growth of transgenic *Arabidopsis* lines. (A) The wildtype (WT, col-0) and transgenic (line 1, 35S::*VrNHX1* and line 4, RD29A::*VrNHX1*) seedlings were observed for germination score after 10 days exposure to salt stress (150 mM NaCl). (B) Root growth inhibition in wild type (WT, Col-0) and transgenic *Arabidopsis* (Line 1, 35S::*VrNHX1* and Line 4, RD29A::*VrNHX1*) plants upon salt stress (150 mM NaCl) was studied. The 4 days old germinated seedlings were transferred to 150 mM NaCl stress for a period of 7 days and (C) root length measured was plotted in graph. Values indicate means ± SE (n = 10). Statistically significant values at $P \leq 0.05$ are indicated as "*", using Bonferroni analysis.

VrNHX1 shared highest homology with reported legume Na^+/H^+ antiporters belonging to the Class-I type NHX exchanger group. The potential structural and functional similarity between yeast and plant endosomal Na^+/H^+ exchanger, serves as a valuable tool for validation of novel plant Na^+/H^+ exchangers for their role in salt tolerance [35,36]. Restored growth of AXTVrNHX1 cells in presence of high concentrations of Na^+, K^+, and Li^+ and suppression of hygromycin sensitivity indicated the functional complementation of *ScNHX1* by heterologous expression of *VrNHX1*. The Na^+ distribution pattern in vacuolar and cytoplasmic fractions of AXTVrNHX1 cells as compared to AXTYES2.0 cells, indicated the potent role of *VrNHX1* as a vacuolar Na^+/H^+ antiporter limited to vacuolar sequestration of alkali cations for establishing ion homeostasis. Similar findings were reported in functional complementation of *OsNHX1* in AXT3 mutant [37]. Moreover, *VrNHX1* expression in AXTVrNHX1 showed enhanced K^+ distribution within vacuolar fractions which was in accordance with the results obtained in heterologous expression of *AtNHX1* [10] and *TNHXS1* [38] in AXT3 mutant. It was also observed that cytoplasmic K^+ fractions were lower in AXTYES2.0 cells as compared with AXTVrNHX1 cells and W303-1B wild type cells. Alkalinization of endolytic compartments has been reported to be mediated by *ScNHX1* which serves as a leak pathway for H^+, thus, regulating the pH level for efficient survival against external acid stress [27,39]. In our studies, we observed that growth sensitivity of AXTVrNHX1 cells was lower than AXTYES2.0 cells under external acidic pH environment. Vacuolar acidification was reduced in AXTVrNHX1 cells under low pH indicating the role of *VrNHX1* in extrusion of excess H^+ by its ion specificity.

Differential regulation of Na^+ uptake, extrusion, compartmentalization, radial transport to stele, loading and unloading into xylem is responsible for the varied response of plants against salinity stress. Under salt stress, *VrNHX1* expression was induced in both leaves and roots of mungbean seedlings with concomitant higher expression in roots than leaves in both early and mid stage seedlings. This result was in accordance with previous reports on expression of *ZmNHX1*, *AeNHX1*, *AlNHX1*, and *ThNHX1* [40–43] and contrary to reports of expression *OsNHX1*, *AgNHX1*, *SsNHX1*, *PeNHX1*, *MsNHX1*, *TrNHX1*, *ZjNHX1*, *ZxNHX*, and *DmNHX1* [17,44–51] which had higher expression in leaves/shoots. The expression pattern of *VrNHX1* under various abiotic stress conditions in mungbean revealed gradual increase in expression under salt stress (200 mM NaCl) after 24 hrs, cold stress (4°C) at 12 hrs and dehydration stress (200 mM mannitol) after 24 hrs. The result was contrary to the previous reports on expression pattern of *PeNHX1* and *ThNHX1* [42,48] under cold stress that showed decrease in the transcript accumulation. No change in expression pattern of *AtNHX1* under cold stress has been reported [52]. Up-regulation of *VrNHX1* under cold stress can be attributed to the other unknown functional mechanisms that still remain to be deciphered. However, involvement of *NHX1* in conferring freezing tolerance has been reported in transgenic *A. thaliana* overexpressing *SsNHX1*, although the exact mechanism has not been explained [53]. Water deficit and altered water potential along with ionic imbalance are known to be primary effects of salt stress [4,8]. We found under dehydration stress the expression pattern of *VrNHX1* in mungbean seedlings similar to previous reports on expression of *GmNHX1*, *ThNHX1* and *EgNHX1*

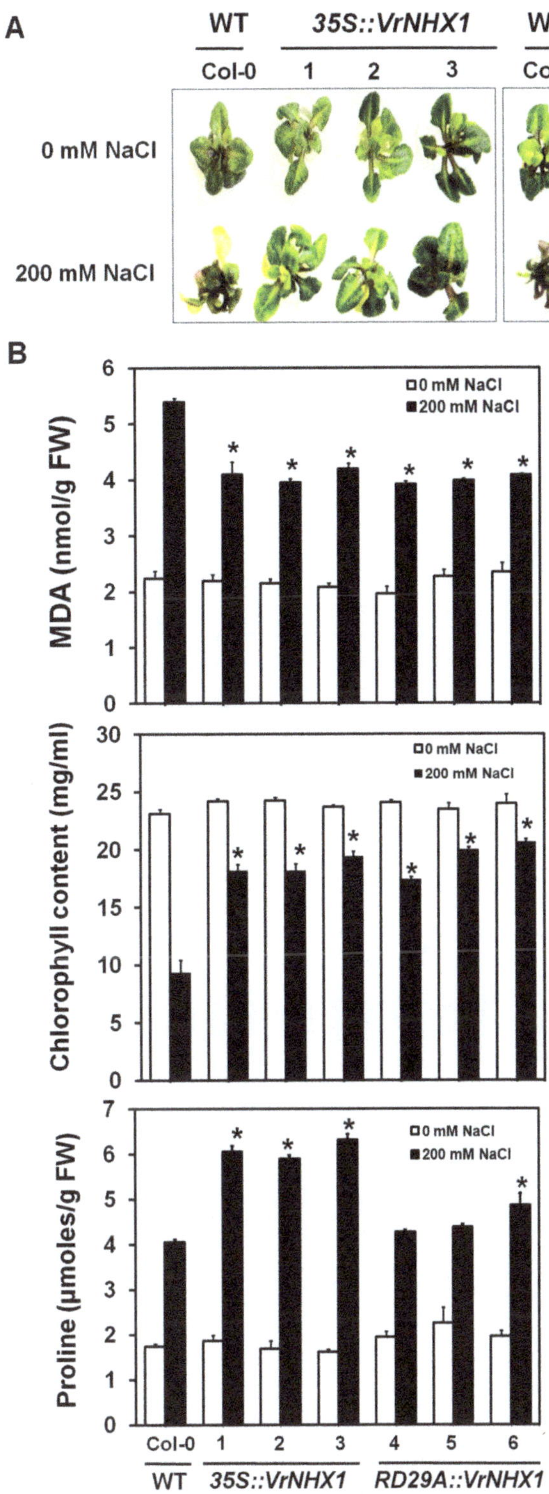

Figure 10. Studying the physiological changes in transgenic *Arabidopsis* lines under salt stress. (A) Effect of salt stress in wild type (WT, Col-0) and transgenic *Arabidopsis* lines expressing *VrNHX1* constitutively (Lines 1–3, 35S::*VrNHX1*) and inducibly (Lines 4–6, RD29A::*VrNHX1*). NaCl-induced morphological changes was visible in 10 days old WT and transgenic lines after exposure to 200 mM NaCl for 5 days. (B) Changes in chlorophyll, MDA and proline content were estimated and analyzed as explained in materials and methods section. Values indicate means ± SE (n = 3). Statistically significant values at P≤0.05 are indicated as "*", using Bonferroni analysis.

which displayed up-regulation under dehydration stress [13,42,54]. However, contrasting results have been reported for expression of *PeNHX1* and *AtNHX1* [48,52].

Physiological response under salt stress, indicated higher Na^+ accumulation in roots than shoots in early and mid stage mungbean seedlings, in contrast to the reports in *T. repens*,

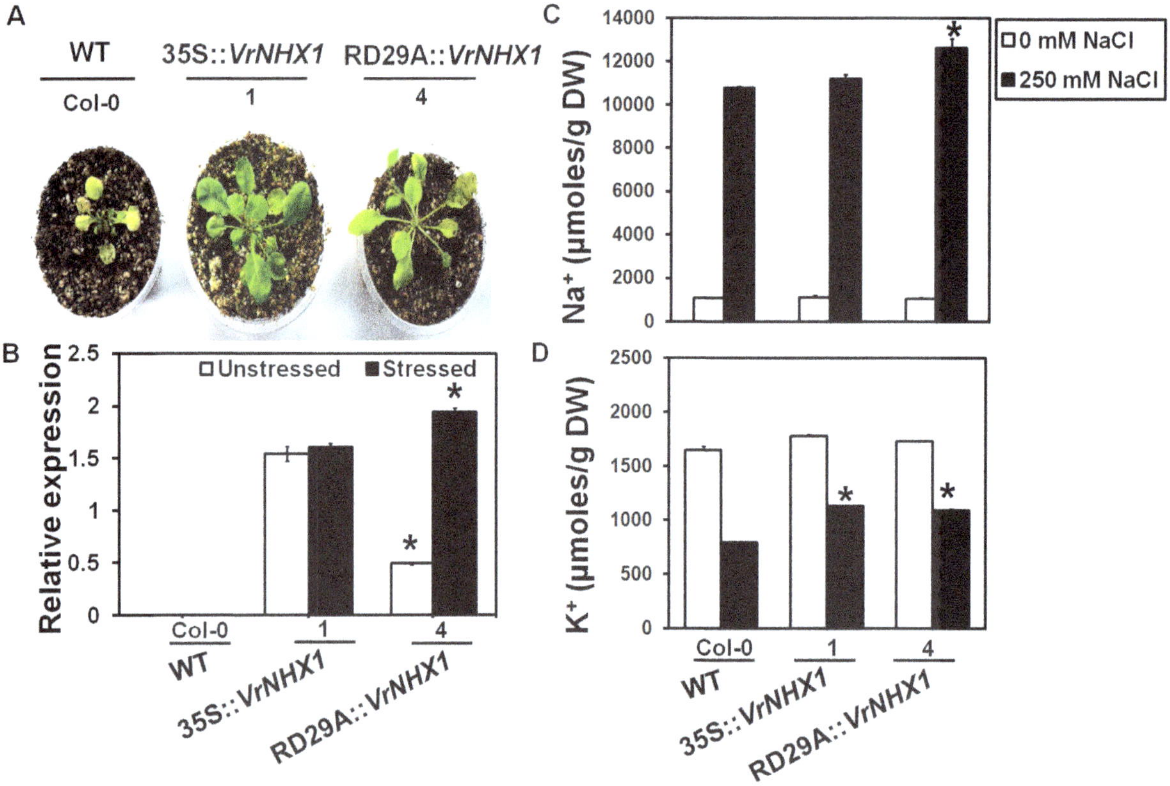

Figure 11. Salt tolerance assay in mature transgenic *Arabidopsis* lines under salt stress. (A) Effect of salt stress on wild type (WT, Col-0) and transgenic *Arabidopsis* lines expressing *VrNHX1* constitutively (Line 1, 35S::*VrNHX1*) and inducibly (Line 4, RD29A::*VrNHX1*) subjected to 250 mM NaCl treatment for 2 weeks (B) Relative transgene expression level of *VrNHX1* in transgenic *Arabidopsis* lines under unstressed and salt stressed conditions. No transgene expression was observed in WT. A 0.283 kb fragment of *VrNHX1*::35SployA and 0.150 kb fragment of *AtUBQ1* was amplified in quantitative RT-PCR analysis (C) Na^+ and (D) K^+ content (µmoles/g DW) was estimated in leaves of unstressed (0 mM NaCl) and salt stressed (250 mM NaCl) WT and transgenic lines, as described in materials and methods. Values indicate means ± SE (n = 3). Statistically significant values at P≤0.05 are indicated as "*", using Bonferroni analysis.

Z. japonica, *H. caspica*, *Z. xanthoxylum*, *D. morifolium* [17,49–51,55] that showed preferential accumulation of Na^+ in leaves/shoots. This indicated that higher K^+/Na^+ ratio is maintained in leaves owing to sequestration of higher Na^+ in root vacuoles thus, restricting their movement to the aerial part of plant. Combined together, increased *VrNHX1* transcript level coupled with higher sequestration of Na^+ in roots can be attributed as the tolerance mechanism of mungbean under salt stress.

Ectopic expression of *VrNHX1* conferred salt tolerance in transgenic *Arabidopsis* lines. Both, 35S::*VrNHX1* and RD29A::*VrNHX1* homozygous T_3 lines displayed better growth response in comparison to WT. Salt stress affects the photosynthetic system components including chlorophyll contents [56]. The reduction in chlorophyll content was less in transgenic lines (35S::*VrNHX1* and RD29::*VrNHX1*) as compared to WT. Lipid peroxidation is mediated by increase in accumulation of reactive oxygen species (ROS) under salinity stress [57]. Therefore, the extent of lipid peroxidation was measured using malionaldehyde (MDA), a by-product of lipid peroxidation. Transgenic lines showed lower extent of MDA generation as compared to WT indicating protection against membrane damage process. Metabolic response against salt stress, generally includes generation of proline, an osmoprotectant and compatible osmolyte, as a protective measure in plants [4]. Transgenic lines expressed higher proline content in response to salt stress. Proline is also known as a potent ROS scavenger [58] which might also be correlated with the lower levels of generation of ROS, thus rendering reduced lipid peroxidation in transgenic plants as compared to WT. Similar result was also reported for proline content in transgenic *Arabidopsis* lines overexpressing *DmNHX1* [51]. The regulation of K^+/Na^+ ratio to maintain K^+ homeostasis for proper cellular and enzymatic functioning is an essential mechanism against salinity stress in plants [59]. Our results demonstrated that the transgenic lines (35S::*VrNHX1* and RD29::*VrNHX1*) maintained a higher K^+/Na^+ ratio than WT plants under salt stress indicating effective tolerance in transgenic lines under salt stress. The phenotypical, physiological and expressional analysis using quantitative real-time PCR concluded that the transgenic RD29::*VrNHX1* line displayed comparable higher survival and growth than 35S::*VrNHX1* lines under salt stress and can be further exploited in crop plants.

The expression of *VrNHX1* under constitutive and inducible promoter enhanced salt tolerance in transgenic *Arabidopsis*. *AtNHX1* is one of the most effective genes in improving plant salt tolerance, however, it played a dominant role mainly in leaf. Our result suggested that *VrNHX1* might play an important role in the root resistance to Na^+ toxicity. Therefore, we could assume that overexpression of *VrNHX1* in crop plants might generate enhanced salt tolerance.

Supporting Information

Figure S1 Multiple sequence alignment was performed for amino acid sequences of plant NHX proteins using CLUSTAL W. The GenBank Accession numbers for NHX proteins are: VrNHX1 (AEO50758.1), *Vigna radiata*; VuNHX1 (AEO72079.2), *Vigna unguiculata*; GmNHX1 (AAY430061.1), *Glycine max*; CkNHX1 (ABG89337.1), *Caragana korshinskii*; MsNHX1 (AAS84487.1), *Medicago sativa*; CaNHX1 (ADL28385.1), *Cicer arietinum*; TrNHX1 (ABV00895.1), *Trifolium repens*; LtNHX1 (ACE78322.1), *Lotus tenuis*. "*" indicates identical amino acid (AA) residues. ":" indicates conservative AA substitutions and "." represents semi-conservative AA substitutions in the sequence alignment. The transmembrane region of VrNHX1 as indicated by TM 1–11 and conserved amiloride binding motif, ^{84}-LFFIYLLPPI-^{93}, a classic inhibitor of the Na^+/H^+ antiporters detected in TM3 region is also shown in the alignment.

Figure S2 Prediction of transmembrane helices of VrNHX1 (AEO50758.1).The hydropathy plot was generated using TMPred online software. The positive values indicate putative transmembrane domains as indicated as TM 1–11.

Figure S3 Growth measurement of yeast strains under low pH. Yeast strains were grown in synthetic medium APGal (pH 4.0) and absorbance was measured at 600 nm. The data shown above are normalized to growth under normal condition (APGal, pH 7.0). W303-1B:- Wild type strain, AXTYES2.0:- AXT3 mutant harboring null pYES2.0 plasmid, AXTVrNHX1:- AXT3 mutant harboring pYES*VrNHX1* recombinant plasmid. Data represent mean from three independent events (n = 3) and standard error plotted in the graph. Statistically significant values at $P \leq 0.05$ are indicated as "*", using Bonferroni analysis.

Figure S4 Calibration curve for pH sensitive BCECF fluorescent dye was plotted using standards ranging from pH 4.0–8.0. Yeast strains (W303-1B, AXTYES2.0, AXTVrNHX1) grown in APGal medium (pH 4.0) were loaded with BCECF dye as described in materials and methods, fluorescence intensity was measured at 440 and 490 nm, background values (measured with only cell extract and only BCECF dye) were subtracted and the ratio was plotted for each pH value. The data from the three yeast stains were pooled and mean ratio values were plotted with a fitted non-linear graph.

Figure S5 T-DNA region of pCAMBIA2301-35S::*VrNHX1* (13.9 kb) and pCAMBIA2301-RD29A::*VrNHX1* (14.4 kb). Restrcition enzyme PstI and EcoRI used for cloning 35SP::*VrNHX1*::35STer cassette (2.3 kb) and RD29A::*VrNHX1*::35STer cassette (2.8 kb) into plant binary vector pCAMBIA 2301 (11.6 kb) is also highlighted. Abbreviations: LB, left border; RB, right border; 35S Promoter, Cauliflower mosaic virus 35S promoter; RD29A promoter, Stress indicible AtRD29A promoter; CaMV 35S poly-A, Cauliflower mosaic virus 35S terminator; nos poly-A, nopaline transferase terminator; *nptII*, neomycin phosphotransferase; *intron-gus-A*, intron interrupted β-glucuronidase; *VrNHX1*, *Vigna radiata NHX1*.

Table S1 The putative post-translational modification sites predicted by ScanProsite software for VrNHX1.

Acknowledgments

We express our sincere thanks to Prof. Edward Blumwald and Dr. Olivier Cagnac for the yeast strains, W303 and AXT3 respectively. We also thank Dr. Luciana Loureiro Penha for providing vector for yeast expression analysis, Department of Civil Engineering, IIT Guwahati for ion estimation analysis using Flame Photometer.

Author Contributions

Conceived and designed the experiments: LS BhL SKP. Performed the experiments: SM HA. Analyzed the data: LS BhL SM. Contributed reagents/materials/analysis tools: LS SM HA BhL. Wrote the paper: SM LS. Supervised the study: LS BhL. Provided critical revision of the manuscript for important intellectual content: BhL.

References

1. Kronzucker HJ, Britto DT (2011) Sodium transport in plants: a critical review. New Phytol 189: 54–8.
2. Mahajan S, Tuteja N (2005) Cold, salinity and drought stresses: an overview. Arch Biochem Biophys 444: 139–158.
3. Schroeder JI, Delhaize E, Frommer WB, Guerinot ML, Harrison MJ, et al. (2013) Using membrane transporters to improve crops for sustainable food production. Nature 497: 60–66.
4. Hasegawa PM, Bressan RA, Zhu JK, Bohnert HJ (2000) Plant Cellular and Molecular Responses to High Salinity. Annu Rev Plant Physiol Plant Mol Bio 51: 463–99.
5. Shi H, Ishitani M, Kim C, Zhu JK (2000) The *Arabidopsis thaliana* salt tolerance gene *SOS1* encodes a putative Na^+/H^+ antiporter. Proc Natl Acad Sci USA 97: 6896–6901.
6. Shi H, Quintero FJ, Pardo JM, Zhu JK (2002) The Putative Plasma Membrane Na^+/H^+ Antiporter SOS1 Controls Long-Distance Na^+ Transport in Plants. Plant Cell 14: 465–477.
7. Blumwald E, Aharon GS, Apse MP (2000) Sodium transport in plant cells. Biochim Biophys Acta 1465: 140–151.
8. Blumwald E (2000) Sodium transport and salt tolerance in plants. Curr Opin Cell Biol 12: 431–434.
9. Pardo JM, Cubero B, Leidi EO, Quintero FJ (2006) Alkali cation exchangers: roles in cellular homeostasis and stress tolerance. J Exp Bot 57: 1181–1199.
10. Yokoi S, Quintero FJ, Cubero B, Ruiz MT, Bressan RA, et al. (2002) Differential expression and function of *Arabidopsis thaliana* NHX Na^+/H^+ antiporters in the salt stress response. Plant J 30: 529–539.
11. Zhang HX, Blumwald E (2001) Transgenic salt-tolerant tomato plants accumulate salt in foliage but not in fruit. Nat Biotechnol 19: 765–768.
12. Rodríguez-Rosales MP, Gálvez FJ, Huertas R, Aranda MN, Baghour M, et al. (2009) Plant NHX cation/proton antiporters. Plant Signal Behav 4:265–276.
13. Li WYF, Wong FL, Tsai SN, Phang TH, Shao G, et al. (2006) Tonoplast-located *GmCLC1* and *GmNHX1* from soybean enhance NaCl tolerance in transgenic bright yellow (BY)-2 cells. Plant Cell Environ 29: 1122–1137.
14. An BY, LuoY Li JR, Qiao WH, Zhang XS, et al. (2008) Expression of a vacuolar Na^+/H^+ antiporter gene of alfalfa enhances salinity tolerance in transgenic *Arabidopsis*. Acta Agron Sin 34: 557–564.
15. Tang R, Li C, Xu K, Du Y, Xia T, et al. (2010) Isolation, functional characterization, and expression pattern of a vacuolar Na^+/H^+ Antiporter gene *TrNHX1* from *Trifolium repens* L. Plant Mol Biol Rep 28: 102–111.
16. Teakle NL, Amtmann A, Real D, Colmer TD (2010) *Lotus tenuis* tolerates combined salinity and waterlogging: maintaining O_2 transport to roots and expression of an *NHX1*-like gene contribute to regulation of Na^+ transport. Physiol Plant 139: 358–374.
17. Yang DH, Song LY, Hu J, Yin WB, Li ZG, et al. (2012) Enhanced tolerance to NaCl and LiCl stresses by over-expressing *Caragana korshinskii* sodium/proton exchanger 1 (*CkNHX1*) and the hydrophilic C terminus is required for the activity of *CkNHX1* in *Atsos3-1* mutant and yeast. Biochem Biophys Res Commun 417: 732–737.
18. Hasanuzzaman M, Ali MR, Hossain M, Kuri S, Islam MS (2013) Evaluation of total phenolic content, free radical scavenging activity and phytochemical screening of different extracts of *Averrhoa bilimbi* (fruits). Int Curr Pharmaceut J 2: 92–96.
19. Jacoby B (1999) Mechanisms involved in salt tolerance of plants. In: Pessarakli M, editor. Plant and Crop Stress. pp. 97–123.

20. Nair RM, Yang RY, Easdown WJ, Thavarajah D, Thavarajah P, et al. (2013) Biofortification of mungbean (*Vigna radiata*) as a whole food to enhance human health. J Sci Food Agric 93: 1805–1813.
21. Thompson JD, Gibson TJ, Plewniak F, Jeanmougin F, Higgin DG (1997) The CLUSTAL_X windows interface: flexible strategies for multiple sequence alignment aided by quality analysis tools. Nucleic Acids Res 25: 4876–4882.
22. Tamura K, Dudley J, Nei M, Kumar S (2007) MEGA4: Molecular Evolutionary Genetics Analysis (MEGA) Software Version 4.0. Mol Biol Evol 24: 1596–1599.
23. Hofmann K, Stoffel W (1993) A database of membrane spanning proteins segments. Biol Chem 374: 166.
24. Gattiker A, Gasteiger E, Bairoch A (2002) ScanPROSITE: a reference implementation of a PROSITE scanning tool. Appl Bioinformatics 1: 107–108.
25. Gietz D, St. Jean A, Woods RA, Schiestl RH (1992) Improved method for high efficiency transformation of intact yeast cells. Nucleic Acids Res 20: 1425.
26. Venema K, Belver A, Marin-Manzano MC, Rodgriguez-Rosales MP, Donaire JP (2003) A novel intracellular K^+/H^+ antiporter related to Na^+/H^+ antiporters is important for K^+ ion homeostasis in plants. J Biol Chem 278: 22453–22459.
27. Brett CL, Tukaye DN, Mukherjee S, Rao R (2005) The yeast endosomal $Na^+(K^+)/H^+$ exchanger Nhx1 regulates cellular ph to control vesicle trafficking. Mol Biol Cell 16: 1396–1405.
28. Munns R, Wallace PA, Teakle NL, Colmer TD (2010) Measuring soluble ion concentrations (Na^+, K^+, Cl^-) in salt-treated plant. In: Sunkar R, editor. Plant Stress Tolerance: Methods in molecular biology. pp. 371–382.
29. Clough SJ, Bent AF (1998) Floral dip: a simplified method for Agrobacterium-mediated transformation of *Arabidopsis thaliana*. Plant J 16: 735–743.
30. Murashige T, Skoog F (1962) A revised medium for rapid growth and bio assays with tobacco tissue cultures. Physiol Plantarum 15: 473–497.
31. Lichtenthaler HK (1987) Chlorophyll fluorescence signatures of leaves during the autumnal chlorophyll breakdown. J Plant Physiol 131: 101–110.
32. Heath RL, Packer L (1968) Photoperoxidation in isolated chloroplasts. I. Kinetic and stoichiometry of fatty acid peroxidation. Arch Biochem Biophys 125: 189–198.
33. Bates LS, Waldren RP, Teare ID (1973) Rapid determination of free proline for water-stress studies. Plant Soil 39: 205–207.
34. Harris C, Fliegel L (1999) Amiloride and the Na^+/H^+ exchanger protein: mechanism and significance of inhibition of the Na^+/H^+ exchanger. Int J Mol Med 3: 315–321.
35. Darley CP, van Wuytswinkel OCM, van der Woude K, Mager WH, de Boer AH (2000) *Arabidopsis thaliana* and *Saccharomyces cerevisiae NHX1* genes encode amiloride sensitive electroneutral Na^+/H^+ exchangers. Biochem J 351: 241–249.
36. Quintero FJ, Blatt MR, Pardo JM (2000) Functional conservation between yeast and plant endosomal Na^+/H^+ antiporters. FEBS Lett 471: 224–228.
37. Kinclova-Zimmermannova O, Flegelova H, Sychrova H (2004) Rice Na^+/H^+-antiporter Nhx1 partially complements the alkali-metal-cation sensitivity of yeast strains lacking three sodium transporters. Folia Microbiol 49: 519–525.
38. Gouiaa S, Khoudi H, Leidi EO, Pardo JM, Masmoudi K (2012) Expression of wheat Na^+/H^+ antiporter *TNHXS1* and H^+-pyrophosphatase *TVP1* genes in tobacco from a bicistronic transcriptional unit improves salt tolerance. Plant Mol Biol 79:137–155.
39. Ali R, Brett CL, Mukherjee S, Rao R (2004) Inhibition of sodium/proton exchange by a Rab-GTPase- activating protein regulates endosomal traffic in yeast. J Biol Chem 279: 4498–4506.
40. Zhang GH, Su Q, An LJ, Wu S (2008) Characterization and expression of a vacuolar Na^+/H^+ antiporter gene from the monocot halophyte *Aeluropus littoralis*. Plant Physiol Biochem 46: 117–126.
41. Zorb C, Noll A, Karl S, Leib K, Yan F, et al. (2005) Molecular characterization of Na^+/H^+ antiporters (*ZmNHX*) of maize (*Zea mays* L.) and their expression under salt stress. J Plant Physiol 162: 55–65.
42. Wu C, Gao X, Kong X, Zhao Y, Zhang H (2009) Molecular Cloning and Functional Analysis of a Na^+/H^+ Antiiporter Gene *ThNHX1* from a Halophytic Plant *Thellungiella halophila*. Plant Mol Biol Rep 27: 1–12.
43. Qiao WH, Zhao XY, Li W, Luo Y, Zhang XS (2007) Overexpression of *AeNHX1*, a root-specific vacuolar Na^+/H^+ antiporter from *Agropyron elongatum*, confers salt tolerance to *Arabidopsis* and *Festuca* plants. Plant Cell Rep 26: 1663–1672.
44. Fukuda A, Nakamura A, Tagiri A, Tanaka H, Miyao A, et al. (2004) Function, intracellular localization and the importance in salt tolerance of a vacuolar Na^+/H^+ antiporter from rice. Plant Cell Physiol 45: 146–159.
45. Hamada A, Shono M, Xia T, Ohta M, Hayashi Y, et al. (2001) Isolation and characterization of a Na^+/H^+ antiporter gene from the halophyte *Atriplex gmelini*. Plant Mol Biol 46: 35–42.
46. Ma XL, Zhang Q, Shi HZ, Zhu JK, Zhao YX, et al. (2004) Molecular cloning and different expression of a vacuolar Na^+/H^+ antiporter gene in *Suada salsa* under salt stress. Biol Plantarum 48: 219–225.
47. An BY, LuoY Li JR, Qiao WH, Zhang XS, et al. (2008) Expression of a vacuolar Na^+/H^+ antiporter gene of alfalfa enhances salinity tolerance in transgenic *Arabidopsis*. Acta Agron Sin 34:557–564.
48. Rajagopal D, Agarwal P, Tyagi W, Singla-Pareek SL, Reddy MK, et al. (2007) Pennisetum *glaucum* Na^+/H^+ antiporter confers high level of salinity tolerance in transgenic *Brassica juncea*. Mol Breed 19: 137–151.
49. Du Y, Hei Q, Liu Y, Zhang H, Xu K, et al. (2010) Isolation and characterization of a putative vacuolar Na^+/H^+ Antiporter gene from *Zoysia japonica* L. J Plant Biol 53: 251–258.
50. Wu GQ, Xi JJ, Wang Q, Bao AK, Ma Q, et al. (2011) The *ZxNHX* gene encoding tonoplast Na^+/H^+ antiporter from the xerophytes *Zygophyllum xanthoxylum* plays important roles in response to salt and drought. J Plant Physiol 168: 758–767.
51. Zhang H, Liu Y, Xu Y, Chapman S, Love AJ, et al. (2012) A newly isolated Na^+/H^+ antiporter gene, *DmNHX1*, confers salt tolerance when expressed transiently in *Nicotiana benthamiana* or stably in *Arabidopsis thaliana*. Plant Cell Tiss Organ Cult 110: 189–200.
52. Shi H, Zhu JK (2002) Regulation of expression of the vacuolar Na^+/H^+ antiporter gene *AtNHX1* by salt stress and abscisic acid. Plant Mol Biol 50: 543–550.
53. Li J, Jiang G, Huang P, Ma J, Zhang F (2007) Overexpression of the Na^+/H^+ antiporter gene from *Suaeda salsa* confers cold and salt tolerance to transgenic *Arabidopsis thaliana*. Plant Cell Tiss Organ Cult 90: 41–48.
54. Baltierra Q, Castillo M, Gamboa MC, Rothhammer M, Krauskopf E (2012) Molecular characterization of a novel Na^+/H^+ antiporter cDNA from *Eucalyptus globules*. Biochem and Biophys Res Commun 430: 535–540.
55. Guan B, Hu YZ, Zeng YL, Wang Y, Zhang FC (2010) Molecular characterization and functional analysis of a vacuolar Na^+/H^+ antiporter gene (*HcNHX1*) from *Halostachys caspica*. Mol Biol Rep 38: 1889–1899.
56. Demetriou G, Neonaki C, Navakoudis E, Kotzabasis K (2007) Salt stress impact on the molecular structure and function of the photosynthetic apparatus—the protective role of polyamines. BBA Bioenergetics 1767: 272–280.
57. Bor M, Özdemir F, Türkan I (2003) The effect of salt stress on lipid peroxidation and antioxidants in leaves of sugar beet *Beta vulgaris* L. and wild beet *Beta maritima* L. Plant Sci 164: 77–84.
58. Szabados L, Savouré A (2010) Proline: a multifunctional amino acid. Trends Plant Sci 15: 89–97.
59. Maathuis FJ, Amtmann ANNA (1999) K^+ nutrition and Na^+ toxicity: the basis of cellular K^+/Na^+ ratios. Ann Bot 84: 123–133.

2

Comparative Transcriptional Profiling of Two Wheat Genotypes, with Contrasting Levels of Minerals in Grains, Shows Expression Differences during Grain Filling

Sudhir P. Singh[1]*, Raja Jeet[1], Jitendra Kumar[1], Vishnu Shukla[1], Rakesh Srivastava[2], Shrikant S. Mantri[1], Rakesh Tuli[1]*

1 National Agri-Food Biotechnology Institute, Department of Biotechnology (DBT), C-127, Industrial Area, Phase VIII, Mohali, 160071, India, **2** School of Biotechnology, Banaras Hindu University, Varanasi, India

Abstract

Wheat is one of the most important cereal crops in the world. To identify the candidate genes for mineral accumulation, it is important to examine differential transcriptome between wheat genotypes, with contrasting levels of minerals in grains. A transcriptional comparison of developing grains was carried out between two wheat genotypes- *Triticum aestivum* Cv. WL711 (low grain mineral), and *T. aestivum* L. IITR26 (high grain mineral), using Affymetrix GeneChip Wheat Genome Array. The study identified a total of 580 probe sets as differentially expressed (with *log2* fold change of ≥2 at p≤0.01) between the two genotypes, during grain filling. Transcripts with significant differences in induction or repression between the two genotypes included genes related to metal homeostasis, metal tolerance, lignin and flavonoid biosynthesis, amino acid and protein transport, vacuolar-sorting receptor, aquaporins, and stress responses. Meta-analysis revealed spatial and temporal signatures of a majority of the differentially regulated transcripts.

Editor: Pilar Hernandez, Institute for Sustainable Agriculture (IAS-CSIC), Spain

Funding: The authors have no support or funding to report.

Competing Interests: The authors have declared that no competing interests exist.

* Email: pratapsudhir17@gmail.com (SPS); rakeshtuli@hotmail.com (RT)

Introduction

Micronutrients play an important role in metabolism, such as in the production and functioning of enzymes, hormones and other substances. Therefore, an adequate intake of trace elements is necessary for proper growth and development [1]. Prevalence of mineral deficiency, especially of iron and zinc, has been estimated in more than two billion people worldwide (http://www.unicef.org/). In developing countries, every second pregnant woman and about 40% of preschool children are estimated to be affected by the mineral micronutrient deficiency (http://www.who.int/nutrition/topics/ida/en/). The extent of reliance on staple crop based diet is one of the major reasons for mineral micronutrient malnutrition in human beings, mainly in resource-poor countries [2]. Development of mineral enriched cereal grains through breeding or biotechnological interventions is important for addressing micronutrient insufficiency [3].

Bread wheat (*Triticum aestivum* L.) is a globally important cereal crop, accounting for 20% of the world's daily food supply [4]. Most high yielding wheat cultivars have low content of grain mineral nutrients [5,6]. The primitive wheat genotypes constitute a pool of significant variability for grain micronutrients, which can be utilized for breeding and/or genetically engineering wheat cultivars [7]. Landraces, the primitive cultivated genotypes, often exhibit elevated grain mineral concentration and biotic and abiotic stress tolerance [8,9]. For example, a primitive cultivar, *T. aestivum* L. IITR26 exhibits higher grain mineral concentration than the post-green revolution modern cultivar, *T. aestivum* Cv. WL711 [10,11,12]. We have earlier reported a distinct level of mineral distribution in maternal and filial grain tissues of the two wheat genotypes, IITR26 and WL711 [10,11]. IITR26 is more efficient in accumulating micronutrients (e.g. Fe, Zn and Mn) in grain tissues than WL711. Furthermore, comparatively higher occurrence of nutritionally important minerals was observed in the endosperm of IITR26 grains, than WL711 [10,11]. Differential expression profiling of genotypes with contrasting characters is an efficient tool for understanding the molecular basis of such phenotypic differences [13]. Therefore, comparative transcriptional profiling in the two wheat genotypes, IITR26 and WL711, during grain development was done in this study.

The three major grain developmental phases in wheat are 6–10 days after anthesis (DAA), 12–21 DAA and 28–42 DAA [14] (http://www.wheatbp.net). During grain filling, mineral nutrients are transported by the maternal tissues to filial grain tissues via endosperm cavity and transfer cells [15,16]. Grain growth rate has been recorded as the highest during 7th to 28th DAA and lowest during 28 DAA till maturity [17]. The grain developmental stages at, 14 and 28 DAA are major transition phases in wheat grain development with distinctive pattern of transcript abundance [14,18]. After 28 DAA, the grain starts to desiccate and gradually attains physiological maturity at about 42 DAA [14].

Microarray is a useful technique for accurate and high throughput gene expression analysis, though it is relatively less sensitive in detecting rare transcripts [19,20]. Microarray technology has been extensively utilized to examine transcriptome related to grain development in wheat [14,21,22,23,24,25,26,27]. We used Affymetrix GeneChip Wheat Genome Arrays to examine the transcriptome changes in developing grains of two genotypes, IITR26 and WL711, at 14 and 28 DAA. The study identifies several transcripts showing significant differential expression during grain filling, which could be candidate genes for mineral accumulation in grains at a higher level. The data set can be further examined for understanding the regulation of mineral accumulation and sequence polymorphism in the candidate genes.

Materials and Methods

Plant material

The two wheat genotypes, a wheat cultivar (*T. aestivum* Cv.WL711) and a landrace (*T. aestivum* L. IITR26), with contrasting levels of minerals in grain tissues [11], were grown in the experimental field of National Agri-Food Biotechnology Institute (NABI), SAS Nagar (Mohali) Punjab, India (310 m above sea level; Latitude 30° 47' North; Longitude 76° 41' East). The ears of each genotype were tagged at anthesis, in three replicates. The spikes were harvested at 14 and 28 DAA, frozen in liquid nitrogen immediately after harvesting and stored at −80°C for RNA extraction.

Mineral concentration analysis

About 25 seeds of 14 DAA and 28 DAA were lyophilized (VirTis, sentry 2.0, USA) and crushed in a clean mortar - pestle. Approximately 100 mg powder was used for acid digestion in the Microwave Reaction System (Mars 6, CEM Corporation, USA). Mineral concentration was estimated in the digested samples using inductively coupled plasma mass spectrometry (ICP-MS; 7700×AgilentTechnologies, Santa Clara, CA), following standard protocol.

RNA preparation

For RNA extraction, developing grains were extracted from the first and second floret of spikelet from the middle portion of the ear. Three grains were taken for RNA extraction in each sample. Total RNA was extracted using TRIzol (Invitrozen) and then purified with RNeasy Plant Mini Kit (Qiagen) following the manufacturer's protocol. On-column DNase (Qiagen) digestion was performed as instructed in the manual. RNA integrity was determined using the RNA 6000 Nano assay bioanalyzer (Agilent).

Microarray hybridization and data analysis

Double stranded cDNA synthesis, *in vitro* transcription to synthesize biotin labeled aRNA, purification and fragmentation of aRNA, and hybridization of arrays were performed according to Affymetrix technical manual. The Affymetrix GeneChip Wheat Genome Array contains 61,290 probe sets, representing about 25 K unigenes. The hybridized chips were washed, stained and scanned using the GeneChip scanner to generate the CEL files.

The CEL files were imported into GeneSpring GX v12 (Agilent Technologies). Signal intensities were recorded for all the 61,290 probe sets. The data has been deposited at NCBI (http://www.ncbi.nlm.nih.gov), with accession number GSE56032. The signal intensities were normalized by using Robust Multi-array Average (RMA) algorithm [28]. The Principal Component Analysis (PCA) in GeneSpring GX v12 established that the three biological replicates were located close to one another. The high correlation coefficient was observed among the three replicated samples, indicating less genetic background noise (Figure 1a). To correct the variability in the normalized expression values, the probe sets with coefficient of variation <50% were retained, and the rest were discarded. One-way analysis of variance (ANOVA, $p \leq 0.01$) was done with multiple testing correction algorithms using the Benjamini Hochberg approach with false discovery cut off, $q \leq 0.01$. Volcano plot with *t*-test unpaired computation was performed ($p \leq 0.01$), and the transcripts showing statistically significant expression difference of two fold (*log2*) or above were selected for further analysis (Figure 1b). A *log2* ratio = 1 indicates a change in the expression of two fold.

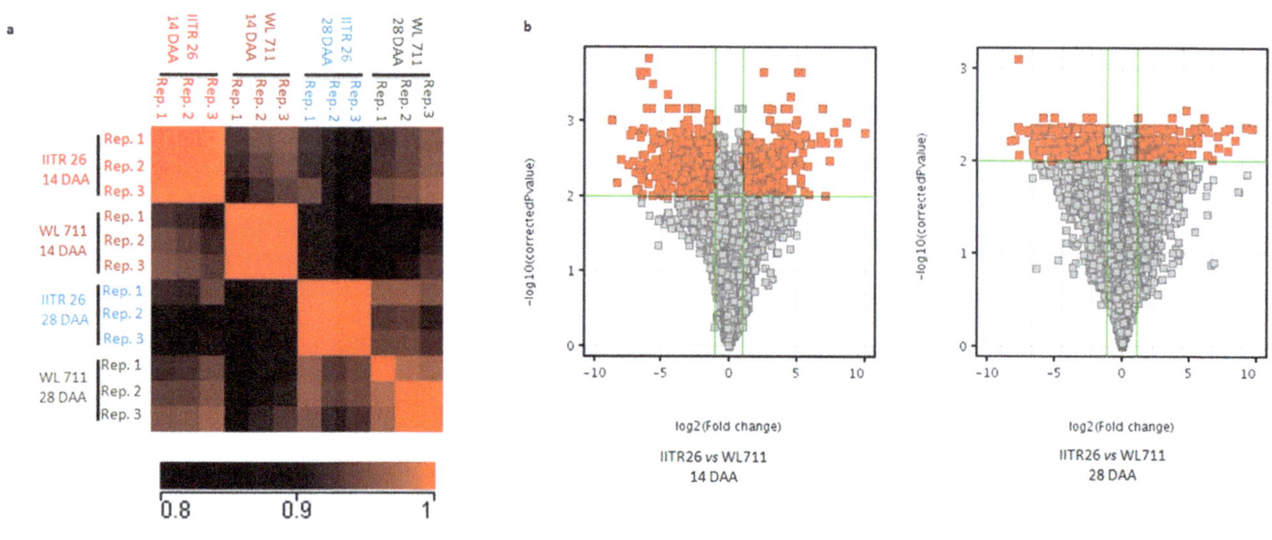

Figure 1. Differentially expressed transcripts at 14 and 28 DAA. (a) Correlation plot represents the pairwise correlation between biological replicates of the samples (b) Volcano plots represents the differentially expressed transcripts, satisfying the criteria of $p \leq 0.01$.

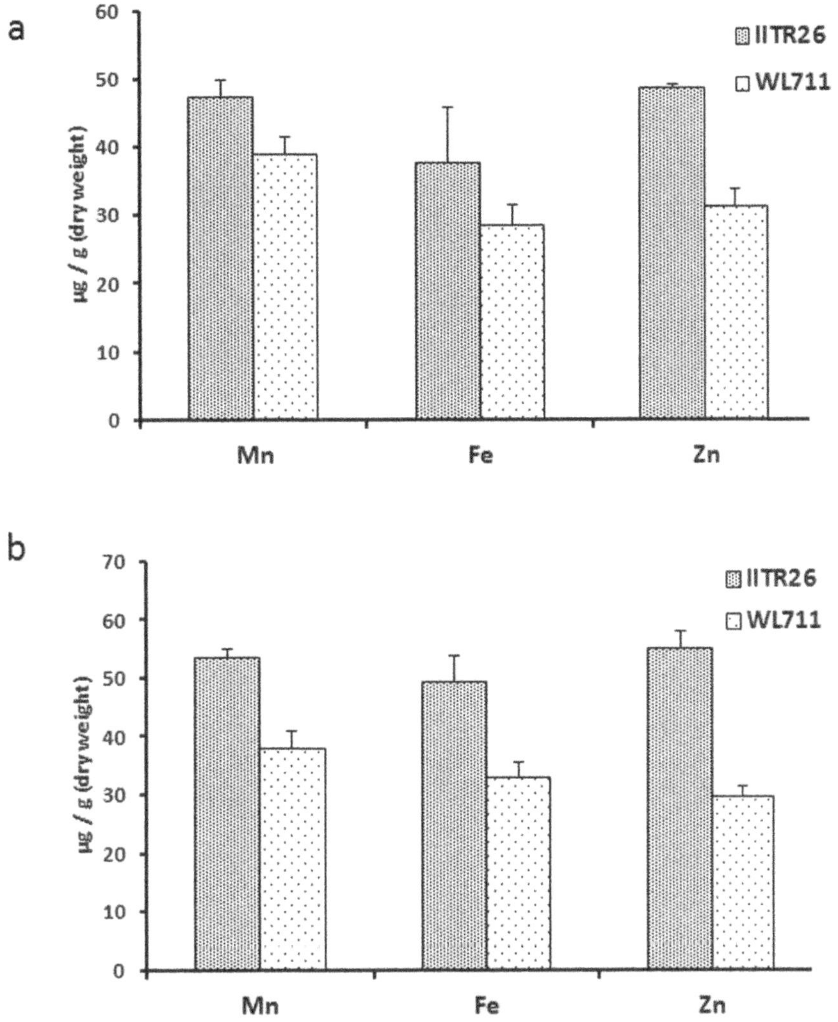

Figure 2. Concentration of micronutrients (Fe, Zn and Mn) in developing grains of IITR26 and WL711 of (a) 14 DAA and (b) 28 DAA.

Functional annotation and gene ontology analysis

Functional annotation of the differentially expressed probe sets (with *log2* fold change of ≥2 at 14 and/or 28 DAA) was done by employing HarvEST Assembly: XW Affymetrix Wheat 1 version 1.59 (www.harvest.ucr.edu). Blast2GO annotation of the differentially expressed transcripts was performed for each probe set query sequence to enrich the annotation and to generate the combined graph of biological process. Hierarchical clustered heat map was produced for the differentially expressed transcripts, using MEV (version 4.6.2) [29]. Enrichment of Gene Ontology (GO) terms in significantly differentially expressed genes (≥2 *log2* fold change) was evaluated using AgriGO analysis tool (http://bioinfo.cau.edu.cn/agriGO) with Fisher tests and Bonferroni multiple testing correction ($P<0.05$) [30].

In silico analysis of wheat transcripts

The differentially regulated transcripts were categorized by using MapMan software version 3.6.0RC1 (http://mapman.gabipd.org/web/guest/mapman). Spatial and temporal distribution pattern of the differential transcripts was examined at Genevestigator (https://www.genevestigator.com/gv/plant.jsp) using expression data of 1532 samples in *T. aestivum* microarray database of Affymetrix.

Quantitative real time PCR

Total RNA was extracted from the developing grains (14 DAA) of the two genotypes (IITR26 and WL711), in three biological replicates. First strand cDNA was synthesized by using SuperScript III First Strand Synthesis Kit (Life Technologies). Primers were designed using Primer Express software. Quantitative real time PCR was performed in three biological replicates using SYBR Green (Qiagen, USA) fluorescence dye and analyzed by 7500 Fast Real-Time PCR System (Applied Biosystems). The qRT-PCR was performed using gene specific primers (Table S1 in File S1), and analyzed as described previously [31]. The wheat 18S gene was used as the internal control to normalize the expression data.

Results and Discussion

Mineral concentration

The two genotypes, IITR26 and WL711, were analyzed for difference in mineral concentration in developing grains. The concentration of nutritionally important micronutrients, such as iron, zinc and manganese, was found significantly higher in the developing grains of IITR26, as compared to WL711 (Figure 2). This is in agreement with our previous reports which show a

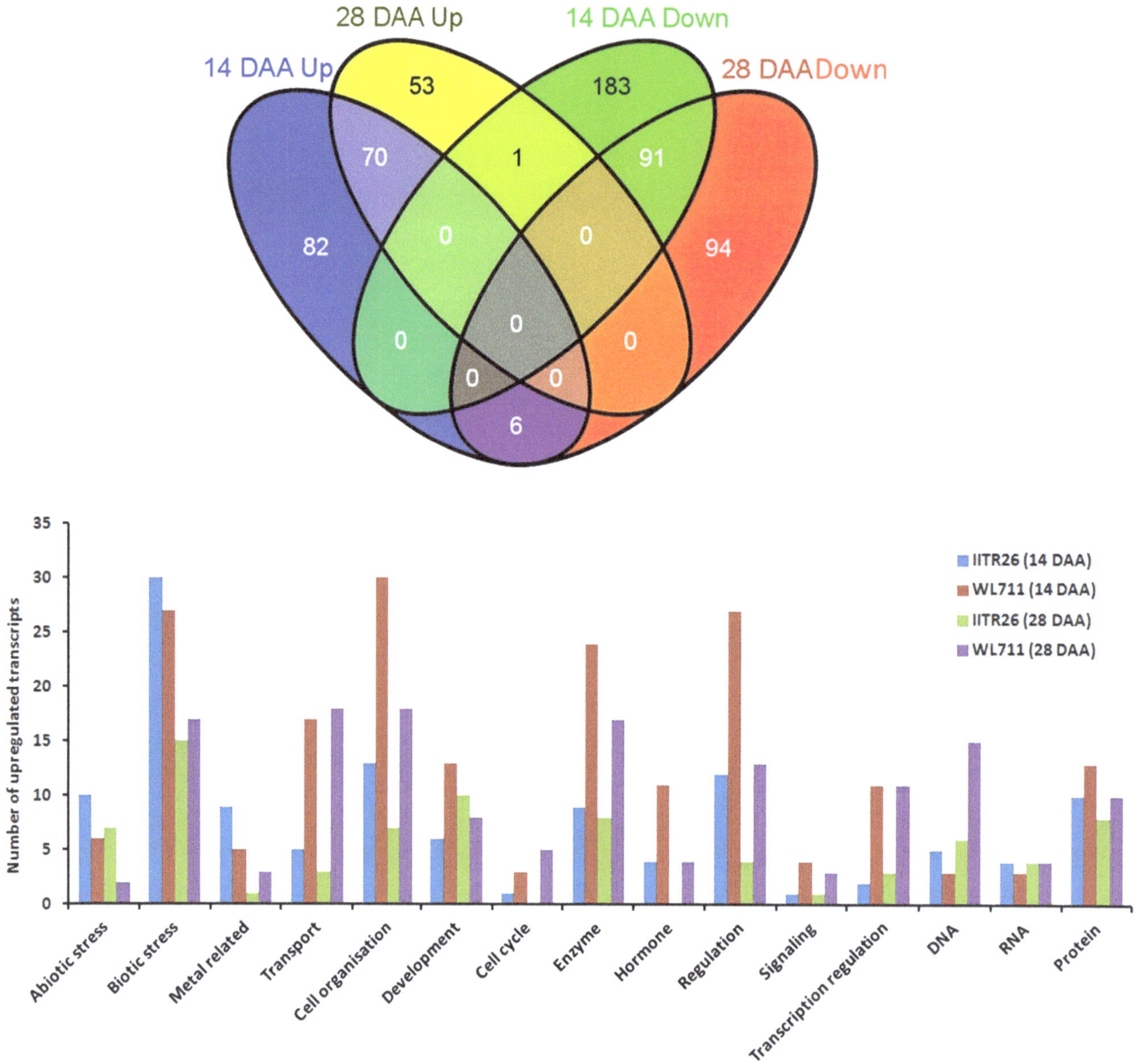

Figure 3. Differentially expressed transcripts with ≥2 *log2* fold change expression difference at p≤0.01, between IITR26 *vs.* WL711. (a) Venn diagram shows the total number of the differentially expressed transcripts and overlap at 14 and 28 DAA (b) Differentially regulated transcripts in biological and functional MapMan BINs.

comparatively higher level of accumulation of minerals in the grain tissues of IITR26 [10,11].

Comparative transcriptional profiling and data analysis

Transcriptome analysis of developing grain was compared between two wheat genotypes (IITR26 and WL711). A total of 433 and 315 probe sets (with *log2* fold change of ≥2 at p≤0.01) were identified as differentially expressed between IITR26 *vs* WL711 on 14 and 28 DAA, respectively (Figure 3a). Putative gene function was assigned to each probe set (Table S2 in File S1) using public databases as mentioned in materials and methods. A total of 149 differentially regulated probe sets could not be assigned to any function, out of them 56 were common between 14 and 28 DAA. Among these, the GO enrichment patterns show disproportionately represented transcripts of genes involved in the biological process of response to stress, polysaccharide metabolism, multi-organism processes, and the biological function of nutrient reservoir and hydrolase activities in IITR26 developing grains (Table 1). In the developing grains of WL711, the GO biological processes- lipid transport, DNA replication, DNA metabolic process, starch metabolic process, and the functions- lipid binding, enzyme regulation, hydrolase and peptidase inhibitor activities were over-represented (Table 2). The functionally annotated transcripts were categorized into 15 bins by using MapMan software (Figure 3b) to understand global changes in developing grains of the two contrasting genotypes. Hierarchically clustered heat maps were produced for the 580 differentially expressed transcripts (Figure S1 in File S1). Meta-analysis indicated spatial and temporal distribution of the differentially regulated probe sets, assigned to putative locations of 22 anatomical plant parts based

Table 1. Enrichment of GO terms in 282 genes up-regulated (≥2 *log2* fold) in developing grains of IITR26 during 14 and/or 28 DAA.

GO term	Ontology	Description	Contingency	P-value
GO:0044036	P	Cell wall macromolecule metabolic process	5, 85, 110, 34562	1.30E-05
GO:0051704	P	Multi-organism process	5, 366, 110, 34281	0.008
GO:0006950	P	Response to stress	12, 1726, 103, 34647	0.013
GO:0005976	P	Polysaccharide metabolic process	5, 455, 110, 34192	0.019
GO:0045735	F	Nutrient reservoir activity	5, 202, 110, 34445	0.00064
GO:0016798	F	Hydrolase activity, acting on glycosyl bonds	7, 430, 108, 34217	0.00065
GO:0004553	F	Hydrolase activity, hydrolyzing O-glycosyl compounds	5, 386, 110, 34261	0.0098

Key: P, biological process; F, molecular function.
Contingency denotes the number of genes in input list from the GO term, the number of genes on microarray from the GO term, the number of genes from input list not from the GO term, and number of genes on microarray not from the GO term.

on their expression potential in the previously reported 1532 *T. aestivum* microarray (Affymetrix) experiments listed in Genevestigator (https://www.genevestigator.com). Several transcripts were identified showing predominant expression in grain tissues, as compared to other plant organs (Figures 4, S2 in File S1).

Candidate genes involved in the transport and accumulation of minerals and other solutes

Transcripts related to metal homeostasis showed differential expression levels in the developing grains of the two genotypes, IITR26 and WL711 (Table 3). Metallothionein (MT) was up-regulated in IITR26 developing grains as compared to WL711. MT is a cystein rich protein having the capacity to bind with metals through its thiol group. It plays an essential role in metal ion homeostasis, transport and storage in the cell [32,33,34]. In the seed, its abundance has been reported in aleurone and embryo [35], whereas endosperm shows comparatively less expression. This coincides with aleurone and embryo being hot spots for mineral localization [10]. Elevated transcription of Manganese ion and Transition metal ion binding proteins was observed in IITR26, mainly during 14 DAA (Table 3). Transition metal binding protein, an orthologue of Farnesylated protein 3, is involved in executing metal ion homeostasis and, biotic and abiotic stress tolerance [34]. Its expression has been recorded mostly in seed tissues (Figure 4).

A transcript of the NAM-1 gene was found up-regulated in developing grains of IITR26 during grain filling. RNAi mediated suppression of NAM in wheat resulted in lower grain mineral and protein concentrations [36,37]. Further, expression of the probe set was found limited to endosperm in meta-analysis (Figure 4). Seed specific NAM-1 could be one of the candidate genes which play role in mineral accumulation in wheat, during grain filling.

Late embryogenesis abundant (LEA) proteins play crucial role in conferring tolerance to metal stress [38,39]. LEA proteins are suggested to be involved also in sequestering divalent metals, and micronutrient trafficking in phloem tissues [40,41]. Enhanced expression of LEA-12 could facilitate higher level of mineral accumulation in the grains of IITR26.

Elevated expression of lignin biosynthesis related genes have been observed, mainly during metal stress, in metal hyper-accumulator plants [42]. The up-regulated Cinnamoyl CoA reductase gene, an enzyme of the lignin branch biosynthetic pathway, presumably provides metal tolerance in developing grains of IITR26. Up-regulation of Flavonoid biosynthetic pathways genes has been notified under metal and biotic stresses [43]. Flavonoids, such as quercetin, have capacity of chelation of transitional metals [44]. Flavanone 3-hydroxylase has sites for putative iron binding [45]. Genevestigator analysis revealed the expression of Flavanone 3-hydroxylase and Dihydroflavonol-4-reductase genes largely in seed tissue (Figure 4). Their up-

Table 2. Enrichment of GO terms in 466 genes up-regulated (≥2 *log2* fold) in developing grains of WL711 during 14 and/or 28 DAA.

GO term	Ontology	Description	Contingency	P-value
GO:0006869	P	Lipid transport	5, 113, 210, 34534	0.00085
GO:0006260	P	DNA replication	6, 220, 209, 34427	0.0029
GO:0006259	P	DNA metabolic process	10, 625, 205, 34022	0.0064
GO:0009791	P	Post-embryonic development	6, 259, 209, 34388	0.0062
GO:0006073	P	Glucan/oligosaccharide/starch metabolic process	7, 352, 208, 34295	0.0072
GO:0030234	F	Enzyme regulator activity	8, 270, 207, 34377	0.00035
GO:0004553	F	Hydrolase activity, hydrolyzing O-glycosyl compounds	9, 386, 206, 34261	0.00083
GO:0004866	F	Endo-peptidase inhibitor activity	5, 121, 210, 34526	0.0011
GO:0008289	F	Lipid binding	6, 278, 209, 34369	0.0086

Contingency and key as described in Table 1.

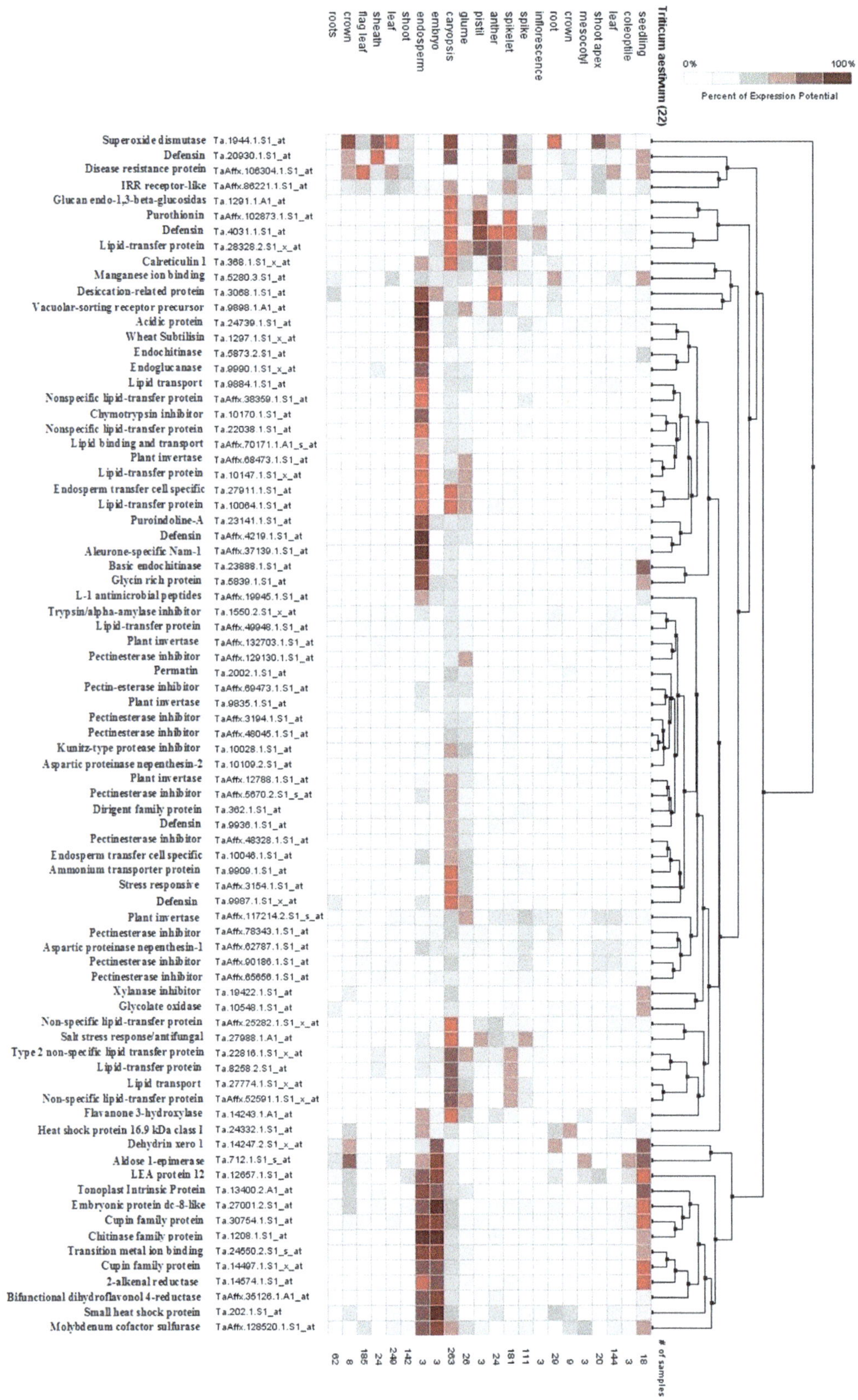

Figure 4. A heat-map of 170 transcripts listed in tables 3 and 4. The heat map shows the expression potential of the probe sets in 22 anatomical plant parts (Seedling, coleoptiles, leaf, shoot apex, mesocotyl, crown, root, inflorescence, spike, spikelet, anther, pistil, glume, caryopsis, embryo, endosperm, shoot, leaf, sheath, flag leaf, crown, roots), based on the previously reported 1532 experiments listed in Genevestigator (https://www.genevestigator.com). In the heat map, 'Crown' and 'Root' labeled at 6th and 7th column, respectively, represent the organs of seedlings.

Table 3. Differentially expressed transcripts involved in the transport and accumulation of minerals and other solute.

Probesets	Putative gene function	*Log2* Fold change (IITR26 *vs* WL711)		Corrected p-value
		14 DAA	28 DAA	
Ta.5280.3.S1_at	Manganese ion binding	2.01	0.14	1.65E-04
Ta.24550.2.S1_s_at	Transition metal ion binding	2.18	−0.52	2.20E-05
Ta.28347.1.S1_s_at	Metallothionein	3.56	2.39	1.86E-05
Ta.12657.1.S1_at	LEA protein 12	3.96	1.10	1.33E-05
Ta.6822.1.S1_at	Cinnamoyl-CoA reductase	2.54	0.60	4.83E-06
TaAffx.35126.1.A1_at	Bifunctional dihydroflavonol 4-reductase flavanone 4-reductase	3.71	0.73	1.30E-05
Ta.8881.1.S1_at	Dihydroflavonol-4-reductase	2.12	−0.29	1.57E-04
Ta.14243.1.A1_at	Flavanone 3-hydroxylase	2.35	−2.20	2.11E-04
TaAffx.37139.1.S1_at	Aleurone-specific Nam-1	2.00	0.15	1.10E-04
Ta.10548.1.S1_at	Glycolate oxidase	−3.13	−3.16	7.35E-05
Ta.368.1.S1_x_at	Calreticulin 1	−2.08	−0.99	6.33E-05
Ta.28543.1.A1_at	Protein phosphatase type 2c	−2.13	−1.92	4.75E-05
Ta.3862.1.S1_at	Probable protein phosphatase	−5.25	−1.13	6.69E-06
Ta.771.1.S1_at	Nicotianamine sythase 3	−0.19	−2.49	3.94E-05
Ta.556.2.A1_at	GDSL-like lipase/acylhydrolase	−7.16	−2.09	6.46E-06
Ta.8258.2.S1_at	Lipid-transfer protein (LTP) precursor	−5.09	−0.98	6.46E-06
Ta.28328.2.S1_x_at	Lipid-transfer protein (LTP) precursor	−4.89	−0.58	2.84E-05
Ta.27774.1.S1_x_at	Lipid transport	−3.71	−0.19	1.33E-05
Ta.9603.1.S1_s_at	Aquaporin NIP1	−3.37	−1.82	4.53E-05
Ta.1345.3.S1_x_at	Lipid-transfer protein (LTP) precursor	−3.21	−0.02	3.17E-05
Ta.9909.1.S1_at	Ammonium transporter protein	−2.05	−3.44	1.97E-04
Ta.11519.1.A1_at	Lipid transport	−3.21	−0.98	0.00128
Ta.22816.1.S1_x_at	Type 2 non-specific lipid transfer protein	−3.13	−2.07	4.83E-06
TaAffx.25282.1.S1_x_at	Non-specific lipid-transfer protein	−2.91	0.11	2.49E-05
Ta.7436.2.S1_at	Lipid-transfer protein (LTP) precursor	−2.86	−1.12	1.08E-04
Ta.10201.1.S1_x_at	Lipid-transfer protein (LTP) precursor	−2.84	−0.58	2.13E-05
TaAffx.70171.1.A1_s_at	Lipid binding and transport	−1.15	−6.94	3.11E-05
Ta.9884.1.S1_at	Lipid transport	−1.71	−6.81	6.77E-06
Ta.10147.1.S1_x_at	Lipid-transfer protein LTPL31	0.23	−6.00	1.93E-05
TaAffx.38359.1.S1_at	Nonspecific lipid-transfer protein	0.65	−5.64	6.01E-06
Ta.3605.2.S1_a_at	Lipid-transfer protein LTPL39	1.28	−5.43	1.05E-05
Ta.22038.1.S1_at	Nonspecific lipid-transfer protein	1.35	−5.28	1.33E-05
Ta.8141.3.S1_a_at	Lipid-transfer protein LTPL42	−0.41	−4.09	1.06E-04
Ta.14034.1.A1_at	Lipid-transfer protein LTPL128	1.06	−3.38	7.95E-05
Ta.23917.3.S1_x_at	Non-specific lipid-transfer protein	−2.95	−3.30	7.48E-05
Ta.13168.1.S1_a_at	Lipid-transfer protein LTPL38	−0.81	−3.27	1.87E-04
Ta.30659.1.S1_at	Aluminum-activated malate transporter	−1.52	−2.90	3.00E-04
Ta.25314.1.A1_s_at	Phosphoenolpyruvate phosphate translocator	−1.74	−2.73	3.89E-04
Ta.28055.2.S1_at	Lipid transport, lipid binding	−0.56	−2.47	4.83E-06
Ta.25583.1.S1_at	Lipid-transfer protein precursor	−1.07	−2.39	9.06E-06
Ta.10064.1.S1_at	Lipid-transfer protein precursor	−0.04	−2.13	4.88E-05
TaAffx.49948.1.S1_at	Lipid-transfer protein precursor	−2.44	−1.19	3.12E-05
Ta.28067.1.S1_x_at	Nonspecific lipid-transfer protein	−2.35	−3.10	4.64E-04
TaAffx.52591.1.S1_x_at	Non-specific lipid-transfer protein	−2.23	−0.24	2.13E-04
TaAffx.97181.1.S1_s_at	Lipid-transfer protein precursor	−2.06	−0.01	6.75E-04
Ta.13400.2.A1_at	Tonoplast Intrinsic Protein (TIP3)	2.98	0.30	1.08E-04
Ta.2895.1.S1_x_at	Plasma membrane intrinsic protein (PIP1)	3.41	1.97	0.01438
Ta.9898.1.A1_at	Vacuolar-sorting receptor precursor	5.21	3.20	1.20E-05

Table 3. Cont.

Probesets	Putative gene function	*Log2* Fold change (IITR26 *vs* WL711)		Corrected p-value
		14 DAA	28 DAA	
Ta.27916.1.A1_x_at	Transmembrane amino acid transporter	−3.41	0.34	1.23E-04
TaAffx.128497.1.S1_at	Amino acid transporter	0.56	2.28	4.31E-05
TaAffx.22704.1.S1_at	Amino acid transporter	2.68	0.39	0.0105
Ta.29726.2.A1_at	Protein transport protein (SecE/Sec61)	5.51	5.04	5.04E-06

regulation, during early grain filling stage in IITR26, could be helpful in favorably accumulating more transition metals [46].

In WL711 developing grains, transcripts for Protein phosphatase 2c-PP2c were significantly up-regulated (Table 3). PP2c is involved in the regulation of cation transporters and membrane polarization, influencing ion homeostasis in plant cells during stress and during phloem transport [47,48]. The Calreticulin-1 gene was up-regulated in WL711 developing grains, and its presence has been found mostly in seed tissues (Figure 4). It codes for multifunctional protein that binds Ca ions (as a second messenger) with low affinity, but high capacity, and its role has been demonstrated in drought and metal stress tolerance [49]. A probe set representing Glycolate oxidase gene, an important player in oxalate biosynthesis, was significantly up-regulated in the developing seeds of WL711 (Table 3). Oxylate has been notified to play important role in calcium regulation and metal detoxification, and its access reduces calcium bioavailability [50]. Nicotianamine (NA) is a metal chelator. NA is synthesized by trimerization of three molecules of S-adenosyl-Lmethionine by nicotianamine synthase (NAS). The role of NAS-3 has been reported in mineral homeostasis in developing organs in maize [51]. Induced expression of NAS3 has been recorded during excessive mineral, while expression is down regulated under mineral deficient condition [52]. The up-regulation of NAS-3, at later grain filling stage (Table 3), could be indicative of mineral sufficiency within the grain of WL711, as compared to IITR26.

In wheat, aleurone is a single cell layer storage tissue which accumulates minerals, phytic acid and proteins in globoids or protein storage vacuoles (PSV), within the cell [53]. High resolution X-ray micrograph of aleurone cell revealed the well defined edges of globoids, surrounded by a membrane, tonoplast [54]. Aquaporins, localized in the tonoplast of PSVs [55,56],

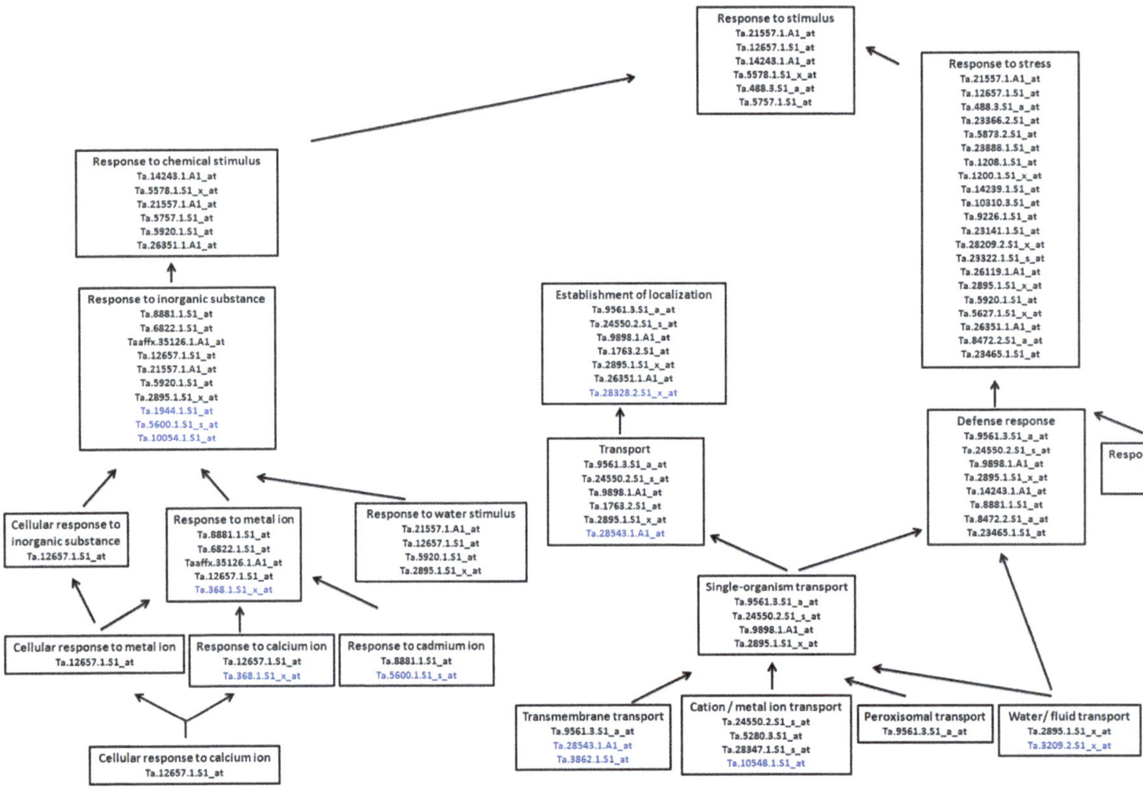

Figure 5. Distribution of differentially up-regulated (IITR26 *vs.* WL711; ≥2 *log2* fold change; p≤0.01) transcripts related to metal accumulation and stress response in IITR26 (probe sets in black) and WL711 (probe sets in blue) at 14 DAA. Data is extracted from Blast2GO results of combined graphs (Biological Processes). The annotation of the probe sets is given in tables S3.

Table 4. Differentially expressed transcripts representing metal tolerance and stress responses.

Probesets	Putative gene function	*Log2* Fold change (IITR26 *vs* WL711)		Corrected P-values
		14 DAA	28 DAA	
TaAffx.102873.1.S1_at	Purothionin	−7.33	−4.84	3.17E-05
Ta.20930.1.S1_at	Defensin	−7.22	−1.22	9.62E-05
Ta.9987.1.S1_x_at	Defensin Tk-AMP-D4	−4.03	−2.20	3.83E-04
Ta.9936.1.S1_at	Defensin Tk-AMP-D5	−3.33	−0.57	1.09E-05
Ta.4031.1.S1_at	Defensin	−2.69	−0.06	1.06E-04
TaAffx.4219.1.S1_at	Defensin	2.14	−0.53	0.001154
Ta.362.1.S1_at	Dirigent family protein	−6.75	−0.15	4.58E-05
Ta.7883.1.S1_x_at	Dirigent like protein	−3.76	0.00	5.30E-06
Ta.1944.1.S1_at	Superoxide dismutase	−6.61	−5.97	3.94E-06
Ta.5600.1.S1_s_at	Salt tolerance protein	−5.87	−0.31	7.86E-05
Ta.27988.1.A1_at	Salt stress response/antifungal	−2.41	−0.19	4.05E-04
Ta.27725.1.S1_at	Salt stress-induced hydrophobic peptide ESI3	2.77	0.30	7.10E-05
Ta.2002.1.S1_at	Permatin, thaumatin, osmotin 34	−5.64	−0.76	1.33E-05
Ta.5878.1.S1_at	Aldo/keto reductase family protein	−4.45	−5.09	1.22E-05
TaAffx.62787.1.S1_at	Aspartic proteinase nepenthesin-1	−3.52	−0.16	8.64E-05
Ta.10109.2.S1_at	Aspartic proteinase nepenthesin-2	−3.26	−2.00	3.25E-04
Ta.5072.1.S1_at	Wound induced protein	−3.26	−4.48	7.06E-06
Ta.19041.1.S1_at	Wound-responsive family protein	−2.74	−1.69	0.00163
TaAffx.3154.1.S1_at	Stress responsive	−3.25	−0.56	3.94E-06
Ta.9967.1.S1_at	Cytochrome P450	−2.88	−1.16	5.04E-06
Ta.3382.1.S1_at	Cytochrome P450	0.29	−2.42	1.79E-05
Ta.21115.3.A1_s_at	Cytochrome p450 like	1.05	−2.89	0.005749
Ta.3361.3.S1_a_at	Pathogenesis-related Bet v I family protein	−2.64	−3.78	9.06E-06
Ta.3094.2.S1_at	GDP-mannose 3,5-epimerase 1	−2.51	−1.75	2.35E-04
Ta.28866.1.S1_at	Patatin-like protein	−2.49	0.21	2.36E-05
Ta.21001.1.S1_at	Gamma-glutamyltranspeptidase 1 precursor	−2.49	−0.63	5.47E-05
TaAffx.56549.1.S1_at	Beta-glucosidase	−2.40	−1.91	9.84E-04
Ta.18082.1.S1_at	Beta-glucosidase	−2.13	0.04	4.09E-04
Ta.4683.1.S1_at	Beta-glucosidase, exo-beta-glucanse	0.17	−2.27	3.92E-05
Ta.2834.1.S1_at	Glucan endo-1,3-beta-glucosidase	−2.39	0.30	4.01E-04
TaAffx.73807.1.A1_at	Glucan endo-1,3-beta-glucosidase	−0.21	−3.08	3.82E-05
Ta.9958.1.S1_at	Glucan 1,3-beta-glucosidase	−4.69	−0.04	2.06E-05
Ta.1291.1.A1_at	Glucan endo-1,3-beta-glucosidase GIV	−3.86	−0.41	4.83E-06
Ta.9990.1.S1_x_at	Endoglucanase	−0.07	−4.17	1.65E-05
TaAffx.117214.2.S1_s_at	Plant invertase/pectin methylesterase inhibitor	−6.70	−0.06	2.12E-06
TaAffx.68473.1.S1_at	Plant invertase/pectin methylesterase inhibitor	−6.64	−2.54	2.22E-05
TaAffx.12788.1.S1_at	Plant invertase/pectin methylesterase inhibitor	−5.63	−0.13	5.09E-06
Ta.27567.1.S1_at	Plant invertase/pectin methylesterase inhibitor	−5.48	−0.24	1.30E-05
TaAffx.132703.1.S1_at	Plant invertase/pectin methylesterase inhibitor	−4.91	−0.06	3.06E-05
TaAffx.65656.1.S1_at	Pectinesterase inhibitor domain containing protein	−4.57	0.15	4.83E-06
TaAffx.78343.1.S1_at	Pectinesterase inhibitor domain	−3.97	0.05	1.54E-05
TaAffx.3194.1.S1_at	Pectinesterase inhibitor domain containing protein	−3.06	−0.73	6.77E-06
TaAffx.48045.1.S1_at	Pectinesterase inhibitor domain containing protein	−3.02	−2.26	0.001093
TaAffx.69473.1.S1_at	Pectin-esterase inhibitor	−3.00	−2.45	2.36E-04
TaAffx.90186.1.S1_at	Pectinesterase inhibitor	−2.60	−0.44	1.13E-04
Ta.9835.1.S1_at	Plant invertase/pectin methylesterase inhibitor	−2.52	−3.62	1.47E-05
TaAffx.48328.1.S1_at	Pectinesterase inhibitor domain containing protein	−2.02	−1.35	1.15E-04
TaAffx.5670.2.S1_s_at	Pectinesterase inhibitor	−2.11	−2.61	3.79E-04
TaAffx.129130.1.S1_at	Pectinesterase inhibitor domain containing protein	−5.28	−4.77	1.05E-05

Table 4. Cont.

Probesets	**Putative gene function**	***Log2* Fold change (IITR26 *vs* WL711)**		**Corrected P-values**
		14 DAA	**28 DAA**	
Ta.5969.3.S1_x_at	Heme-binding protein	−2.27	0.40	0.00568
TaAffx.86221.1.S1_at	IRR receptor-like serine threonine-protein kinase	−2.24	−0.31	0.00597
Ta.13256.2.S1_at	Thaumatin family domain containing protein	−2.12	−1.30	5.63E-05
Ta.10046.1.S1_at	Endosperm transfer cell specific PR9	−1.25	−7.16	5.09E-06
Ta.27911.1.S1_at	Endosperm transfer cell specific PR60	−0.17	−4.06	5.32E-05
Ta.1762.1.A1_at	Jasmonate-induced protein	−1.15	−2.75	4.99E-06
Ta.24739.1.S1_at	Acidic protein	−0.65	−4.21	7.63E-05
Ta.28539.1.A1_x_at	NAC domain-containing protein	−0.18	3.43	1.16E-05
Ta.5367.1.S1_s_at	NAC domain-containing protein 67	−0.66	3.28	1.19E-04
Ta.16815.1.S1_s_at	Tetratricopeptide repeat domain	0.01	2.01	2.20E-05
Ta.10170.1.S1_at	Chymotrypsin inhibitor WCI	0.06	−2.26	3.00E-05
Ta.14247.2.S1_x_at	Dehydrin xero 1	0.07	3.09	7.06E-06
Ta.8834.1.S1_x_at	Cysteine-rich repeat secretory protein 55 precursor	0.23	−2.63	3.34E-05
Ta.202.1.S1_at	Small heat shock protein	0.31	2.63	2.51E-04
Ta.712.1.S1_s_at	Aldose 1-epimerase	0.42	3.17	1.93E-05
TaAffx.10772.1.A1_at	Serine rich protein	0.46	2.79	5.80E-05
Ta.21508.1.A1_a_at	Oxidative Stress (OXS3)	0.61	2.02	6.07E-06
TaAffx.12591.1.S1_at	Disease resistance gene RAR1	2.05	1.49	1.94E-04
Ta.22179.1.A1_at	Disease resistance protein rga4	2.45	−0.44	0.007391
TaAffx.104814.1.S1_at	Disease resistance protein rga1-like	−2.62	−2.02	3.73E-05
TaAffx.106304.1.S1_at	Disease resistance protein NBS-LRR	0.14	2.05	1.78E-05
Ta.30674.1.S1_at	Peroxidase precursor	−2.38	0.88	1.36E-04
Ta.7262.1.A1_x_at	Peroxidase precursor	0.73	2.14	1.31E-04
Ta.9334.1.S1_s_at	Peroxidase precursor	1.32	−3.29	2.15E-04
Ta.488.3.S1_a_at	Peroxidase precursor	2.17	0.02	2.58E-04
Ta.23366.2.S1_at	Peroxidase precursor	2.98	7.14	6.85E-06
Ta.6572.2.S1_at	Peroxiredoxin Q	−2.22	−0.99	3.20E-05
Ta.9561.3.S1_a_at	Peroxisomes biogenesis proteins Peroxin Pex14	4.05	3.24	3.94E-06
Ta.28209.2.S1_x_at	BURP domain-containing protein RD22	2.07	0.95	0.00253
Ta.21557.1.A1_at	Senescence/dehydration associated protein	2.26	3.30	1.30E-05
Ta.6399.1.A1_at	Protease do-like 14	2.30	1.66	2.46E-04
TaAffx.128.1.S1_at	Chaperone protein dnaj 6-like	2.37	1.58	0.016101
Ta.27001.2.S1_at	Embryonic protein dc-8-like	2.46	0.25	2.12E-05
Ta.23322.1.S1_s_at	Thaumatin-like protein	2.56	1.88	2.55E-05
Ta.23465.1.S1_at	Ankyrin repeat family protein	2.57	1.81	0.009715
Ta.12273.1.A1_at	Ankyrin repeats	4.34	4.61	2.00E-05
Ta.8472.1.S1_at	Ankyrin repeat protein	4.21	3.81	1.40E-05
Ta.14574.1.S1_at	2-alkenal reductase	2.69	3.74	6.85E-06
Ta.14497.1.S1_x_at	Cupin family protein	2.72	−0.15	4.59E-05
Ta.30754.1.S1_at	Cupin family protein	2.99	−0.31	2.45E-05
Ta.26119.1.A1_at	3-isopropylmalate dehydrogenase	2.94	2.34	0.00429
Ta.3268.1.A1_at	F-box domain	2.05	1.86	0.009745
TaAffx.92919.1.A1_at	F-Box and Domain of Unknown Function Containing Proteins	2.12	0.55	5.93E-05
Ta.7304.1.A1_at	F-box domain containing protein	4.17	2.13	0.014434
TaAffx.64850.1.A1_at	F-box and FBD domain containing protein	4.56	4.80	3.94E-06
TaAffx.120564.1.A1_at	FDB-associated f-box protein at1g66310-like isoform x1	4.86	4.90	4.95E-06
Ta.5627.1.S1_x_at	VAMP protein	3.19	0.83	7.01E-05
Ta.5920.1.S1_at	ABA related (HVA22)	3.38	1.01	1.40E-05
Ta.10310.3.S1_at	ABA related	5.77	0.90	4.94E-05

Table 4. Cont.

Probesets	Putative gene function	*Log2* Fold change (IITR26 *vs* WL711)		Corrected P-values
		14 DAA	28 DAA	
Ta.5873.2.S1_at	Endochitinase	3.32	0.10	3.89E-05
Ta.23888.1.S1_at	Basic endochitinase	5.24	−0.16	1.68E-05
Ta.6952.1.S1_a_at	Chitinase	5.36	3.28	0.014148
Ta.1208.1.S1_at	Chitinase family protein precursor	8.00	2.02	3.94E-06
TaAffx.128520.1.S1_at	Molybdenum cofactor sulfurase family protein	3.43	0.30	2.59E-05
Ta.5257.2.S1_s_at	Low temperature and salt responsive protein	3.50	1.34	0.013306
Ta.3068.1.S1_at	Desiccation-related protein PCC13-62 precursor	3.56	0.18	6.99E-05
Ta.1550.2.S1_x_at	Trypsin/alpha-amylase inhibitor CMX1/CMX3	3.73	0.18	0.001797
Ta.1297.1.S1_x_at	Wheat Subtilisin	3.93	1.48	1.16E-04
Ta.20434.1.S1_at	Xylanase inhibitor	1.84	2.84	6.02E-04
Ta.19422.1.S1_at	Xylanase inhibitor	4.30	0.95	3.61E-05
Ta.1200.1.S1_x_at	Xylanase inhibitor	5.87	0.00	7.72E-05
Ta.26351.1.A1_at	Glucan synthase-like	4.46	0.33	7.77E-06
Ta.24332.1.S1_at	Heat shock protein 16.9 kDa class I	4.92	8.27	2.29E-04
Ta.20928.1.S1_at	Bowman-Birk inhibitor (BBI) gene	5.16	5.31	9.43E-06
Ta.9226.1.S1_at	Wheatwin	5.21	1.24	1.09E-05
Ta.10028.1.S1_at	Kunitz-type protease inhibitor	5.60	−0.71	3.92E-05
Ta.5839.1.S1_at	Glycin rich protein	7.06	−0.01	4.36E-05
TaAffx.19945.1.S1_at	L-1, precursor of antimicrobial peptides	7.36	1.59	3.76E-04
Ta.23141.1.S1_at	Puroindoline-A	9.99	7.51	5.30E-06

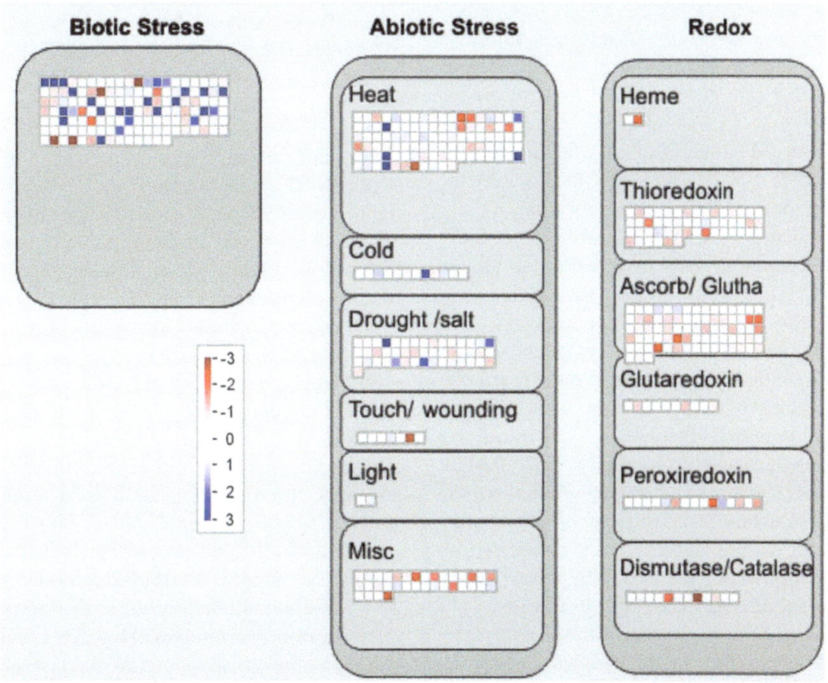

Figure 6. Differentially expressed transcripts, satisfying the criteria of p≤0.01 (IITR26 *vs.* WL711) at 14 DAA, identified in 'cellular response overview' using MapMan software version 3.6.0RC1. The *log2* fold change in the transcript levels were used for the analysis.

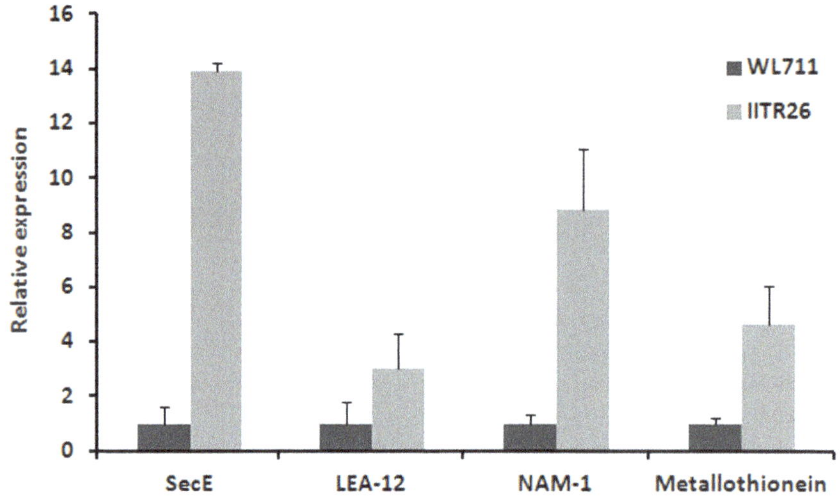

Figure 7. Quantitative RT-PCR analyses of a few candidate genes: Metallothionein, NAM-1, LEA-12, and Sec-E. Each bar indicates standard error in three biological replicates.

facilitate transmembrane transport of water and solutes, and provide cellular adaptation to metal stress by water homeostasis [57,58]. About three fold increased expression of aquaporin-Tonoplast intrinsic protein (TIP-3) was recorded in IITR26 developing grains (Table 3). Meta-analysis showed its abundance in seed tissues as compared to other organs (Figure 4). Some other aquaporins, Plasma membrane intrinsic protein (PIP-1) and Nodulin 26-like intrinsic protein (NIP-1), play an important role in osmo-adaptation and homeostosis of cells to facilitate transport of water and phloem derived nutrients to grain tissues [59,60]. However, NIP-1 has been notified as less efficient in transporting solute and ions [61]. Enhanced expression of TIP-3 and PIP-1 could promote higher level of mineral sequestration in IITR26 grains (Table 3). The aleurone globoids or PSVs store large number of organelle-specific proteins, synthesized by endosplasmic reticulum associated ribosomes and transported through vacuolar sorting [62]. Vacuolar sorting receptor-1 (VSR-1) is involved in sorting of proteins in PSV [63]. In meta-analysis, the predominance of VSR-1 was observed in endosperm (Figure 4). The enhanced expression of VSR-1 in IITR26 (Table 3) could be suggestive of higher storage activity [64] in IITR26 grains.

Several probe sets related to lipid transport and binding (LTP family protein) were significantly up-regulated in the developing grains of WL711 (Table 3). The expression of many of them has been recorded predominant in grain, as compared to other anatomical part (Figure 4). The enhanced expression of GDSL esterases/lipases, that catalyze hydrolysis of ester bonds- primarily in triglycerides, indicates relatively more active lipid metabolism [65] in WL711 developing grains. LTPs regulate lipid transport, homeostasis and lipid-mediated cellular processes [66]. Within grain, several LTPs have been observed up-regulated in aleurone cells around the embryo, scutellum and vascular bundle of the embryo, resulting maximum accumulation of lipids in embryo in grains [67,68]. Their role in providing protection to embryo from pathogen attack has been speculated [65,67,68]. Unlike WL711, transporters for amino acid and protein (SecE) [69] were expressed at higher level in IITR26 during grain filling (Table 3). This is in agreement with previous reports showing significant correlation between grain protein and micronutrient content [15,70,71]. The results also coincide with our observation of proportional association between elemental images of sulphur (possibly sulphur-rich peptides) and iron in cross-sections of the two contrasting genotypes [10].

Candidate genes representing stress responses and metal tolerance

Blast2GO results of combined graphs distributed the differentially expressed transcripts in different levels of biological processes, mapping the transcripts related with metal/cation transport, localization and responses etc. The results revealed transcripts representing stress response to metal ion activity in developing grains of IITR26, such as peroxins (Ta.9561.3.-S1_a_at), VSR-1 (Ta.9898.1.A1_at), PIP-1 (Ta.2895.1.S1_x_at), Ankyrin repeat protein (Ta.23465.1.S1_at) and Flavonoid (Ta.8881.1.S1_at) (Figure 5). Peroxisome biogenesis proteins (peroxins) are involved in docking cargo-receptor complexes at the peroxisomal membrane [72]. Several studies revealed essential role of peroxins in defense mechanisms conferring resistance against pathogen attack [73,74]. VSR-1 is a candidate gene controlling the defense network, and its up-regulation has been reported in metal tolerant and disease resistant plants [75,76]. Involvement of plasma membrane aquaporins (PIP-1) has been discussed in metal homeostasis and defense response in plants [77,78,79]. Flavonoids enhance protective abilities of the plant against biotic stresses [80,81]. Ankyrin repeat proteins have been reported to mediate metal tolerance and enhanced disease resistance in plants [82,83].

Several genes associated with abiotic and biotic stress were found differentially regulated in developing grains of both the genotypes (Table 4), presumably having coordination among the growth, maintenance and stress during seed development [84]. For example, the transcripts for β-purothionin, cysteine-rich proteins, and Defensin, up-regulated in WL711, are known to play significant role in protecting wheat grains from microbial pathogens [68,85]. In IITR26 developing grains, transcripts of Puroindolin-A (PINa), a cystine- and tryptophan-rich protein, was significantly predominant. However, the probe set for PINb did not pass the quality criteria. A relationship has been incurred between grain hardness and puroindoline content [86]. However, both the genotypes, IITR26 and WL711, exhibit similar grain hardness indices (~90) [12,87,88]. This indicates contribution of PINa in grain related traits, other than hardness. Role of PINa

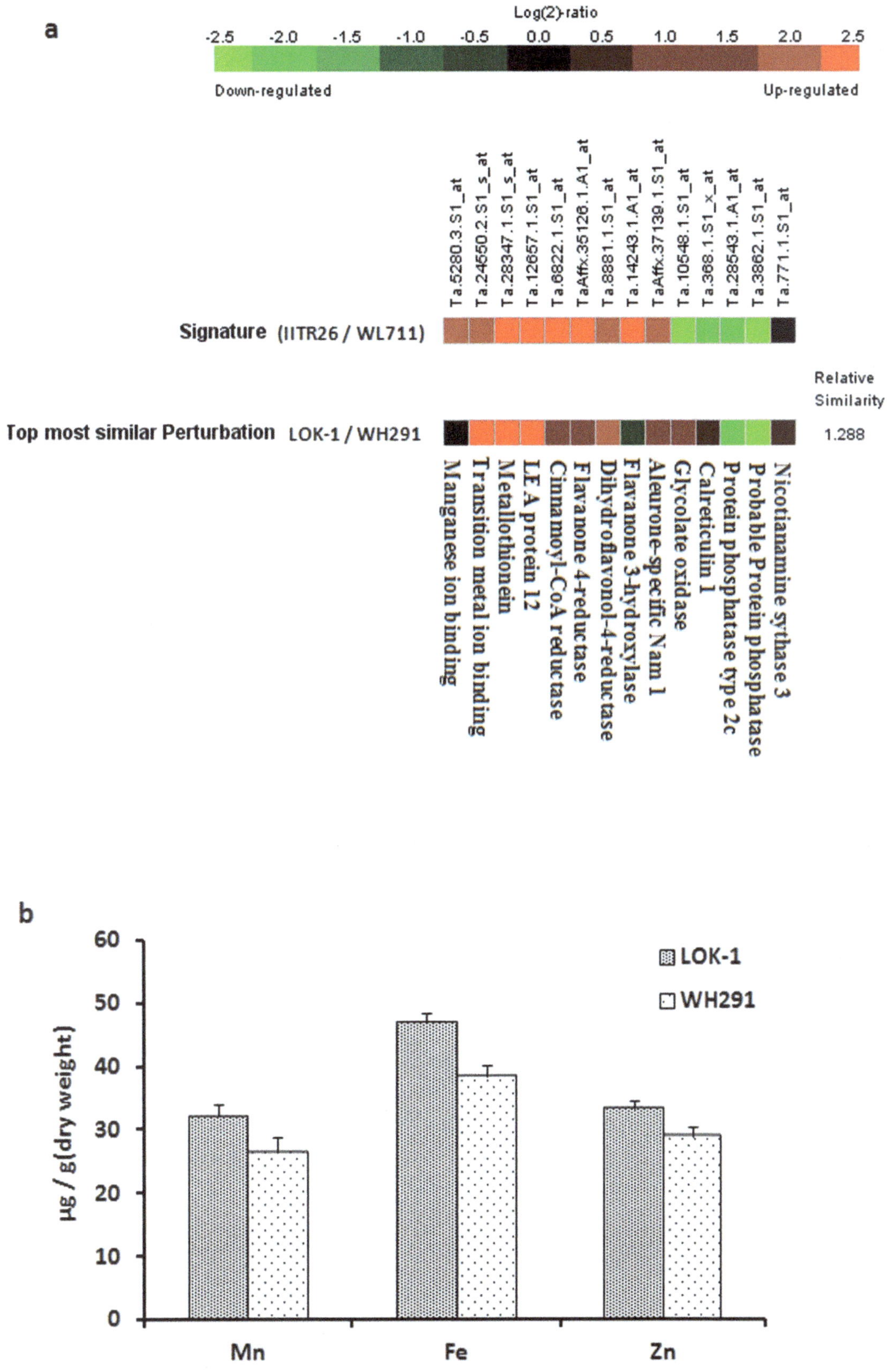

Figure 8. Similarity search meta-analysis and mineral concentration analysis. (a) The similarity search in Genevestigator, using the differentially expressed metal related transcripts in our data, revealed the top most perturbation comparing transcriptome between the developing grains of LOK-1 and WH291. (b) Concentration of micronutrients (Fe, Zn and Mn) in mature grains of LOK-1 and WH291.

protein has been established in plant defense as antimicrobial molecule [85]. Transcripts of many other antimicrobial molecules, Chitinase, Glycin rich proteins, Wheatwin, VAMP, Bowman-Birk and Kunitz-type protease inhibitor, were significantly up-regulated in the developing grains of IITR26 (Table 4). A positive correlation has been inferred between degree of biotic stress resistance and glucan synthase activity, on the contrary glucanase activity is negatively correlated [89]. The enhanced glucan synthase activity and reduced glucanase activity in IITR26 developing grains further corroborates enhanced resistance to pathogen attack in IITR26 (Table 4). Several fold reduction in superoxide dismutase activity in the developing grain of IITR26 (Table 4), could be indicative of less reactive oxygen species (ROS) formation [90]. Mampman analysis of cellular response overview reflected up-regulation of redox related transcripts in WL711 during grain filling (Figure 6). This is in agreement with the previous report of induced redox state in susceptible wheat as compared to the biotic stress tolerant line [91]. Meta-analysis determined dominant expression potential of several stress related transcripts in grain tissues (Figure 4), as compared to other plant organs, suggesting their role in seed development. The similarity search in Genevestigator, using the differentially expressed transcripts, revealed perturbations in which spikelet samples of wheat genotypes, with stress resistant (CS-7EL) and susceptible (CS) backgrounds (GEO accession GSE21386), have been compared (Figure S3 in File S1). The results coincide with the previous reports examining interactions between metal ion excess and biotic stress factors [92,93], and the responses to excess metal ions resemble to that of biotic and abiotic stresses [94,95]. Our results support the elemental defense hypothesis which suggests a crosstalk between metal accumulation and biotic and abiotic stress responses [96,97,98,99]. However, it needs to be investigated and validated in other wheat genotypes, with contrasting levels of minerals in grains, if the crosstalk substantiates the enhanced level of mineral sequestration in grains.

Pectin methylesterase, which catalyses the de-esterification of pectin, is regulated by pectin methylesterase inhibitor. Its role has been discussed in wounding, osmotic stress, senescence and seed development [100]. Significant up-regulation of several pectin methylesterase inhibitor related probe sets has been observed in WL711 (Table 4). Down regulation of pectin methylesterase inhibitor activity might inhibit de-esterification of pectin, which could result into reduced metal tolerance in the tissue [101].

The developing grains of IITR26 exhibited up-regulation of transcripts for F-box and NAC genes (Table 4). F-box proteins, apart from its role in seed development [102], provide tolerance during metal stress in plant tissues [103]. NAC proteins play a key role during stress signaling pathways and provide tolerance against various abiotic and biotic stresses [104]. Furthermore, the abscisic acid (ABA)/stress-induced proteins [105], such as HVA22 were expressed at higher level in IITR26 developing grains (Table 4).

The results are suggestive of IITR26 being comparatively more tolerant to stress during grain filling, which eventually could be a positive factor for metal accumulation.

Quantitative RT-PCR and similarity search meta-analysis

The relative fold change in expression of a few genes related to the transport and accumulation of metal was examined by real time PCR analysis (Figure 7). The expression of Metallothionein, NAM-1 and LEA-12, and protein, SecE, were found high in the developing grains of IITR26. The expression profile of these genes was in agreement with the microarray results. Hence, q-RT PCR results validated the microarray data, as reported in several published reports.

Similarity search meta-analysis was performed against 1532 experiments listed in Genevestigator, by using the metal related transcripts differentially expressed between IITR26 and WL711 (Table 3). The perturbation showing maximum similarity with our data comprised the transcriptional comparison between the developing grains of two wheat genotypes: LOK-1 and WH291 (Figure 8a). The mineral concentration analysis revealed relatively higher concentrations of micronutrients in mature grains of LOK-1 than WH291 (Figure 8b), which coincides with up-regulation of the metal related transcripts in early-stage developing grains (Figure 8a), supporting our data.

Conclusion

The microarray based comparative transcriptome profiling of developing grains of two genotypes contrasting for grain mineral concentration identified several differentially expressed genes. Prevalence of transcripts related to metal transport and accumulation, metal tolerance, vacuolar sorting receptor-1, aquaporins (TIP-3 and PIP-1), amino acid and protein transporters, lignin and flavonoid biosynthesis, and stress responses, was noticed in IITR26 during grain filling. For some of the differentially expressed probe sets, gene functions are yet to be determined (Table S2 in File S1). Thus, the differential expression analysis of microarray data revealed several candidate genes which may facilitate the elevated levels of minerals in the grains. However, the possibility of involvement of the differential transcripts in establishing other traits, for which the two genotypes differ, may not be denied. The differential expression data provided in this study would be useful for designing programs for functional genomics, exploration of sequence polymorphism in the candidate genes, and understanding other metabolic pathways common with mineral accumulation in wheat grains.

Supporting Information

File S1 Contains the following files: **Figure S1.** Hierarchically clustered heat-map for 580 differentially expressed transcripts (IITR26 *vs.* WL711; ≥2 *log2* fold change; p≤0.01) at 14 and 28 DAA. **Figure S2.** A heat map of differentially expressed transcripts (IITR26 vs. WL711; ≥2 *log2* fold change; p≤0.01) at (a) 14 and (b) 28 DAA. The rest of the details are as given in figure 4. **Figure S3.** The similarity search in Genevestigator, using the differentially expressed transcripts (IITR26 vs. WL711; ≥2 *log2* fold change; p≤ 0.01) at 14 DAA revealed perturbations (top 5) in which spikelet samples of wheat genotypes, with stress resistant (CS-7EL) and susceptible (CS) backgrounds (GEO accession GSE21386), have been compared. **Table S1.** Gene-specific primers used for qRT-PCR. **Table S2.** Table S1 Differentially regulated probe sets with ≥ 2 *log2* fold change expression difference at p≤0.01, between IITR26 vs. WL711, and their putative gene function during 14 and 28 DAA. **Table S3.** Annotation of the probe sets mentioned in Figure 5.

Acknowledgments

The authors thank Executive Director, National Agri-Food Biotechnology Institute (Department of Biotechnology) for facilitating the research work, and Prof. H. S. Dhaliwal of Eternal University for providing seeds. RJ and JK acknowledge ICMR and CSIR, respectively, for providing fellowship.

Author Contributions

Conceived and designed the experiments: SPS RT. Performed the experiments: SPS RJ JK. Analyzed the data: SPS VS RS SSM. Contributed reagents/materials/analysis tools: SPS SSM RT. Contributed to the writing of the manuscript: SPS RT.

References

1. Shenkin A (2006) Micronutrients in health and disease. Postgrad Med J 82: 559–567.
2. Sands DC, Morris CE, Dratz EA, Pilgeram A (2009) Elevating optimal human nutrition to a central goal of plant breeding and production of plant-based foods. Plant Sci 177: 377–389.
3. Nestel P, Bouis HE, Meenakshi JV, Pfeiffer W (2006) Biofortification of staple food crops. J Nutr 136: 1064–1067.
4. Reynolds M, Foulkes J, Furbank R, Griffiths S, King J, et al. (2012) Achieving yield gains in wheat. Plant Cell Environ 35: 1799–1823.
5. Shewry PR (2009) Wheat. J Exp Bot 60: 1537–1553.
6. Rawat N, Neelam K, Tiwari VK, Randhawa GS, Friebe B, et al. (2011) Development and molecular characterization of wheat-*Aegilops kotschyi* addition and substitution lines with high grain protein, iron, and zinc. Genome 54: 943–953.
7. Rawat N, Tiwari VK, Singh N, Randhawa GS, Singh K, et al. (2009) Evaluation and utilization of *Aegilops* and wild *Triticum* species for enhancing iron and zinc content in wheat. Genet Resour Crop Ev 56: 53–64.
8. Zeven AC (1998) Landraces: A review of definitions and classifications. Euphytica 104: 127–139.
9. Routray P, Basha O, Garg M, Singh NK, Dhaliwal HS (2007) Genetic diversity of landraces of wheat (*Triticum aestivum* L.) from hilly areas of Uttaranchal, India. Genet Resour Crop Ev 54: 1315–1326.
10. Singh SP, Vogel-Mikus K, Arcon I, Vavpetic P, Jeromel L, et al. (2013) Pattern of iron distribution in maternal and filial tissues in wheat grains with contrasting levels of iron. J Exp Bot 64: 3249–3260.
11. Singh SP, Vogel-Mikus K, Vavpetic P, Jeromel L, Pelicon P, et al. (2014) Spatial X-ray fluorescence micro-imaging of minerals in grain tissues of wheat and related genotypes. Planta 240: 277–289.
12. Tiwari VK, Neelam K, Rawat N, Dahiya J, Kumar R, et al. (2011) Characterization, evaluation and genetic diversity analysis of a set of landraces of wheat from high himalayan hills of Uttarakhand. In: Chibbar RN, Dexter J, editors. Wheat science dynamics: Challenges and opportunities. St. Paul, USA: AAAC International. 607–615.
13. Jiang SY, Ma A, Ramamoorthy R, Ramachandran S (2013) Genome-wide survey on genomic variation, expression divergence, and evolution in two contrasting rice genotypes under high salinity stress. Genome Biol Evol 5: 2032–2050.
14. Wan Y, Poole RL, Huttly AK, Toscano-Underwood C, Feeney K, et al. (2008) Transcriptome analysis of grain development in hexaploid wheat. BMC Genomics 9: 121.
15. Ozturk L, Yazici MA, Yucel C, Torun A, Cekic C, et al. (2006) Concentration and localization of zinc during seed development and germination in wheat. Physiol Plantarum128: 144–152.
16. Robert P, Jamme F, Barron C, Bouchet B, Saulnier L, et al. (2010) Change in wall composition of transfer and aleurone cells during wheat grain development. Planta 233: 393–406.
17. Houshmandfar A, Eradatmand AD (2011) Assessment of auxin and cytokinin levels at different grain type and position within a spike of wheat. Adv Environ Biol 5: 1058–1063.
18. Evers T, Millar S (2002) Cereal grain structure and development: some implications for quality. J Cereal Sci 36: 261–284.
19. Malone JH, Oliver B (2011) Microarrays, deep sequencing and the true measure of the transcriptome. BMC Biol 9: 34.
20. Xiao J, Jin X, Jia X, Wang H, Cao A, et al. (2013) Transcriptome-based discovery of pathways and genes related to resistance against *Fusarium* head blight in wheat landrace Wangshuibai. BMC Genomics 14: 197.
21. Clarke B, Rahman S (2005) A microarray analysis of wheat grain hardness. Theor Appl Genet 110: 1259–1267.
22. Wilson ID, Barker GL, Lu C, Coghill JA, Beswick RW, et al. (2005) Alteration of the embryo transcriptome of hexaploid winter wheat (*Triticum aestivum* cv. Mercia) during maturation and germination. Funct Integr Genomics 5: 144–154.
23. Laudencia-Chingcuanco DL, Stamova BS, You FM, Lazo GR, Beckles DM, et al. (2007) Transcriptional profiling of wheat caryopsis development using cDNA microarrays. Plant Mol Biol 63: 651–668.
24. Scossa F, Laudencia-Chingcuanco D, Anderson OD, Vensel WH, Lafiandra D, et al. (2008) Comparative proteomic and transcriptional profiling of a bread wheat cultivar and its derived transgenic line overexpressing a low molecular weight glutenin subunit gene in the endosperm. Proteomics 8: 2948–2966.
25. Yang X, Xu H, Li W, Li L, Sun J, et al. (2011) Screening and identification of seed-specific genes using digital differential display tools combined with microarray data from common wheat. BMC Genomics 12: 513.
26. Gregersen PL, Brinch-Pedersen H, Holm PB (2005) A microarray-based comparative analysis of gene expression profiles during grain development in transgenic and wild type wheat. Transgenic Res 14: 887–905.
27. Singh A, Mantri S, Sharma M, Chaudhury A, Tuli R, et al. (2014) Genome-wide transcriptome study in wheat identified candidate genes related to processing quality, majority of them showing interaction (quality x development) and having temporal and spatial distributions. BMC Genomics 15: 29.
28. Irizarry RA, Hobbs B, Collin F, Beazer-Barclay YD, Antonellis KJ, et al. (2003) Exploration, normalization, and summaries of high density oligonucleotide array probe level data. Biostatistics 4: 249–264.
29. Saeed AI, Bhagabati NK, Braisted JC, Liang W, Sharov V, et al. (2006) TM4 microarray software suite. Methods Enzymol 411: 134–193.
30. Du Z, Zhou X, Ling Y, Zhang Z, Su Z (2010) agriGO: a GO analysis toolkit for the agricultural community. Nucleic Acids Res 38: W64–70.
31. Singh SP, Pandey T, Srivastava R, Verma PC, Singh PK, et al. (2010) *BECLIN1* from *Arabidopsis thaliana* under the generic control of regulated expression systems, a strategy for developing male sterile plants. Plant Biotechnol J 8: 1005–1022.
32. Loebus J, Peroza EA, Bluthgen N, Fox T, Meyer-Klaucke W, et al. (2011) Protein and metal cluster structure of the wheat metallothionein domain gamma-E(c)-1: the second part of the puzzle. J Biol Inorg Chem 16: 683–694.
33. Walker EL, Waters BM (2011) The role of transition metal homeostasis in plant seed development. Curr Opin Plant Biol 14: 318–324.
34. de Abreu-Neto JB, Turchetto-Zolet AC, de Oliveira LF, Zanettini MH, Margis-Pinheiro M (2013) Heavy metal-associated isoprenylated plant protein (HIPP): characterization of a family of proteins exclusive to plants. FEBS J 280: 1604–1616.
35. Hegelund JN, Schiller M, Kichey T, Hansen TH, Pedas P, et al. (2012) Barley metallothioneins: MT3 and MT4 are localized in the grain aleurone layer and show differential zinc binding. Plant Physiol 159: 1125–1137.
36. Uauy C, Distelfeld A, Fahima T, Blechl A, Dubcovsky J (2006) A NAC Gene regulating senescence improves grain protein, zinc, and iron content in wheat. Science 314: 1298–1301.
37. Waters BM, Uauy C, Dubcovsky J, Grusak MA (2009) Wheat (*Triticum aestivum*) NAM proteins regulate the translocation of iron, zinc, and nitrogen compounds from vegetative tissues to grain. J Exp Bot 60: 4263–4274.
38. Tunnacliffe A, Wise MJ (2007) The continuing conundrum of the LEA proteins. Naturwissenschaften 94: 791–812.
39. Gao C, Wang C, Zheng L, Wang L, Wang Y (2012) A *LEA* gene regulates Cadmium tolerance by mediating physiological responses. Int J Mol Sci 13: 5468–5481.
40. Kruger C, Berkowitz O, Stephan UW, Hell R (2002) A metal-binding member of the late embryogenesis abundant protein family transports iron in the phloem of *Ricinus communis* L. J Biol Chem 277: 25062–25069.
41. Liu G, Xu H, Zhang L, Zheng Y (2011) Fe binding properties of two soybean (*Glycine max* L.) LEA4 proteins associated with antioxidant activity. Plant Cell Physiol 52: 994–1002.
42. van de Mortel JE, Almar Villanueva L, Schat H, Kwekkeboom J, Coughlan S, et al. (2006) Large expression differences in genes for iron and zinc homeostasis, stress response, and lignin biosynthesis distinguish roots of *Arabidopsis thaliana* and the related metal hyperaccumulator *Thlaspi caerulescens*. Plant Physiol 142: 1127–1147.
43. Hua C, Linling L, Shuiyuan C, Fuliang C, Feng X, et al. (2013) Molecular cloning and characterization of three genes encoding dihydroflavonol-4-reductase from *Ginkgo biloba* in anthocyanin biosynthetic pathway. PLoS One 8: e72017.
44. Leopoldini M, Russo N, Chiodo S, Toscano M (2006) Iron chelation by the powerful antioxidant flavonoid quercetin. J Agric Food Chem 54: 6343–6351.
45. Himi E, Maekawa M, Noda K (2011) Differential expression of three flavanone 3-hydroxylase genes in grains and coleoptiles of wheat. Int J Plant Genomics 2011: 369460.
46. Tolra R, Barcelo J, Poschenrieder C (2009) Constitutive and aluminium-induced patterns of phenolic compounds in two maize varieties differing in aluminium tolerance. J Inorg Biochem 103: 1486–1490.
47. Schweighofer A, Hirt H, Meskiene I (2004) Plant PP2C phosphatases: emerging functions in stress signaling. Trends Plant Sci 9: 236–243.
48. Flatman PW (2002) Regulation of Na-K-2Cl cotransport by phosphorylation and protein-protein interactions. BBA-Biomembranes 1566: 140–151.
49. Jia XY, Xu CY, Jing RL, Li RZ, Mao XG, et al. (2008) Molecular cloning and characterization of wheat calreticulin (CRT) gene involved in drought-stressed responses. J Exp Bot 59: 739–751.
50. Xu HW, Ji XM, He ZH, Shi WP, Zhu GH, et al. (2006) Oxalate accumulation and regulation is independent of glycolate oxidase in rice leaves. J Exp Bot 57: 1899–1908.
51. Zhou ML, Qi LP, Pang JF, Zhang Q, Lei Z, et al. (2013) Nicotianamine synthase gene family as central components in heavy metal and phytohormone response in maize. Funct Integr Genomics 13: 229–239.
52. Zhou X, Li S, Zhao Q, Liu X, Zhang S, et al. (2013) Genome-wide identification, classification and expression profiling of nicotianamine synthase (NAS) gene family in maize. BMC Genomics 14: 238.
53. Bohn L, Josefsen L, Meyer AS, Rasmussen SK (2007) Quantitative analysis of phytate globoids isolated from wheat bran and characterization of their sequential dephosphorylation by wheat phytase. J Agric Food Chem 55: 7547–7552.
54. Regvar M, Eichert D, Kaulich B, Gianoncelli A, Pongrac P, et al. (2011) New insights into globoids of protein storage vacuoles in wheat aleurone using synchrotron soft X-ray microscopy. J Exp Bot 62: 3929–3939.

55. Paris N, Stanley CM, Jones RL, Rogers JC (1996) Plant cells contain two functionally distinct vacuolar compartments. Cell 85: 563–572.
56. Takahashi H, Rai M, Kitagawa T, Morita S, Masumura T, et al. (2004) Differential localization of tonoplast intrinsic proteins on the membrane of protein body Type II and aleurone grain in rice seeds. Biosci Biotechnol Biochem 68: 1728–1736.
57. Li GW, Peng YH, Yu X, Zhang MH, Cai WM, et al. (2008) Transport functions and expression analysis of vacuolar membrane aquaporins in response to various stresses in rice. J Plant Physiol 165: 1879–1888.
58. Przedpelska-Wasowicz EM, Wierzbicka M (2011) Gating of aquaporins by heavy metals in *Allium cepa* L. epidermal cells. Protoplasma 248: 663–671.
59. Schuurmans JA, van Dongen JT, Rutjens BP, Boonman A, Pieterse CM, et al. (2003) Members of the aquaporin family in the developing pea seed coat include representatives of the PIP, TIP, and NIP subfamilies. Plant Mol Biol 53: 633–645.
60. Wallace IS, Choi WG, Roberts DM (2006) The structure, function and regulation of the nodulin 26-like intrinsic protein family of plant aquaglyceroporins. BBA-Biomembranes 1758: 1165–1175.
61. Zhou Y, Setz N, Niemietz C, Qu H, Offler CE, et al. (2007) Aquaporins and unloading of phloem-imported water in coats of developing bean seeds. Plant Cell Environ 30: 1566–1577.
62. Kang H, Kim SY, Song K, Sohn EJ, Lee Y, et al. (2012) Trafficking of vacuolar proteins: the crucial role of *Arabidopsis* vacuolar protein sorting 29 in recycling vacuolar sorting receptor. Plant Cell 24: 5058–5073.
63. Lee Y, Jang M, Song K, Kang H, Lee MH, et al. (2013) Functional identification of sorting receptors involved in trafficking of soluble lytic vacuolar proteins in vegetative cells of Arabidopsis. Plant Physiol 161: 121–133.
64. Fuji K, Shimada T, Takahashi H, Tamura K, Koumoto Y, et al. (2007) *Arabidopsis* vacuolar sorting mutants (green fluorescent seed) can be identified efficiently by secretion of vacuole-targeted green fluorescent protein in their seeds. Plant Cell 19: 597–609.
65. Chepyshko H, Lai CP, Huang LM, Liu JH, Shaw JF (2012) Multifunctionality and diversity of GDSL esterase/lipase gene family in rice (*Oryza sativa* L. *japonica*) genome: new insights from bioinformatics analysis. BMC Genomics 13: 309.
66. Lev S (2010) Non-vesicular lipid transport by lipid-transfer proteins and beyond. Nat Rev Mol Cell Biol 11: 739–750.
67. Boutrot F, Meynard D, Guiderdoni E, Joudrier P, Gautier MF (2007) The *Triticum aestivum* non-specific lipid transfer protein (*TaLtp*) gene family: comparative promoter activity of six *TaLtp* genes in transgenic rice. Planta 225: 843–862.
68. Kovalchuk N, Li M, Wittek F, Reid N, Singh R, et al. (2010) Defensin promoters as potential tools for engineering disease resistance in cereal grains. Plant Biotechnol J 8: 47–64.
69. Lycklama a Nijeholt JA, de Keyzer J, Prabudiansyah I, Driessen AJ (2013) Characterization of the supporting role of SecE in protein translocation. FEBS Lett 587: 3083–3088.
70. Kutman UB, Yildiz B, Ozturk L, Cakmak I (2010) Biofortification of durum wheat with zinc through soil and foliar applications of nitrogen. Cereal Chem 87: 1–9.
71. Cakmak I, Torun A, Millet E, Feldman M, Fahima T, et al. (2004) *Triticum dicoccoides*: An important genetic resource for increasing zinc and iron concentration in modern cultivated wheat. Soil Sci Plant Nutr 50: 1047–1054.
72. Monroe-Augustus M, Ramon NM, Ratzel SE, Lingard MJ, Christensen SE, et al. (2011) Matrix proteins are inefficiently imported into *Arabidopsis* peroxisomes lacking the receptor-docking peroxin PEX14. Plant Mol Biol 77: 1–15.
73. Reumann S, Babujee L, Ma C, Wienkoop S, Siemsen T, et al. (2007) Proteome analysis of *Arabidopsis* leaf peroxisomes reveals novel targeting peptides, metabolic pathways, and defense mechanisms. Plant Cell 19: 3170–3193.
74. Corpas FJ, Barroso JB (2013) Peroxynitrite (ONOO-) is endogenously produced in arabidopsis peroxisomes and is overproduced under cadmium stress. Ann Bot 113: 87–96.
75. Gupta S, Bhar A, Chatterjee M, Das S (2013) *Fusarium oxysporum* f.sp. *ciceri* race 1 induced redox state alterations are coupled to downstream defense signaling in root tissues of chickpea (*Cicer arietinum* L.). PLoS One 8: e73163.
76. Keinänen SI, Hassinen VH, Kärenlampi SO, Tervahauta AI (2007) Isolation of genes up-regulated by copper in a copper-tolerant birch (*Betula pendula*) clone. Tree Physiol 27: 1243–1252.
77. Gawwad MRA, Šutković J, Mataković L, Musrati M, Zhang L (2013) Functional interactome of Aquaporin 1 sub-family reveals new physiological functions in *Arabidopsis Thaliana* Network Biol. 3: 87–96.
78. Wintz H, Fox T, Wu YY, Feng V, Chen W, et al. (2003) Expression profiles of *Arabidopsis thaliana* in mineral deficiencies reveal novel transporters involved in metal homeostasis. J Biol Chem 278: 47644–47653.
79. Wu B, Steinbronn C, Alsterfjord M, Zeuthen T, Beitz E (2009) Concerted action of two cation filters in the aquaporin water channel. EMBO J 28: 2188–2194.
80. Kumar V, Nadda G, Kumar S, Yadav SK (2013) Transgenic tobacco overexpressing tea cDNA encoding dihydroflavonol 4-reductase and anthocyanidin reductase induces early flowering and provides biotic stress tolerance. PLoS One 8: e65535.
81. Sharma M, Chai C, Morohashi K, Grotewold E, Snook ME, et al. (2012) Expression of flavonoid 3′-hydroxylase is controlled by P1, the regulator of 3-deoxyflavonoid biosynthesis in maize. BMC Plant Biol 12: 196.
82. Cantu D, Yang B, Ruan R, Li K, Menzo V, et al. (2013) Comparative analysis of protein-protein interactions in the defense response of rice and wheat. BMC Genomics 14: 166.
83. Gao W, Li HY, Xiao S, Chye ML (2010) Protein interactors of acyl-CoA-binding protein ACBP2 mediate cadmium tolerance in Arabidopsis. Plant Signal Behav 5: 1025–1027.
84. McIntosh S, Watson L, Bundock P, Crawford A, White J, et al. (2007) SAGE of the developing wheat caryopsis. Plant Biotechnol J 5: 69–83.
85. Sanders MR, Clifton LA, Neylon C, Frazier RA, Green RJ (2013) Selected wheat seed defense proteins exhibit competitive binding to model microbial lipid interfaces. J Agr Food Chem 61: 6890–6900.
86. Cuesta S, Guzman C, Alvarez JB (2013) Allelic diversity and molecular characterization of puroindoline genes in five diploid species of the *Aegilops* genus. J Exp Bot 64: 5133–5143.
87. Sharma R, Rawat A, Misra BK, Nagarajan S (2012) Distribution of grain hardness in Indian wheat varieties and landraces. Wheat Information Service 114: 1–8.
88. Salunke R, Rawat N, Neelam K, Tiwari VK, Randhawa GS, et al. (2014) Effect of grain hardness on bioavailability of iron in wheat as determined using the coupled invitro digestion/Caco-2 model. LWT-Food Sci Technol: 1–6.
89. Frič F, Huttová J (1993) Glucanase, glucan synthase and chitinase activity in barley genotypes susceptible or resistant to *Erysiphe graminis* f.sp.*hordei*. Biol Plantarum 35: 95–101.
90. Alscher RG, Erturk N, Heath LS (2002) Role of superoxide dismutases (SODs) in controlling oxidative stress in plants. J Exp Bot 53: 1331–1341.
91. Reddy SK, Weng Y, Rudd JC, Akhunova A, Liu S (2013) Transcriptomics of induced defense responses to greenbug aphid feeding in near isogenic wheat lines. Plant Sci 212: 26–36.
92. Noret N, Meerts P, Tolra R, Poschenrieder C, Barcelo J, et al. (2005) Palatability of *Thlaspi caerulescens* for snails: influence of zinc and glucosinolates. New Phytol 165: 763–771.
93. Behmer ST, Lloyd CM, Raubenheimer D, Stewart-Clark J, Knight J, et al. (2005) Metal hyperaccumulation in plants: mechanisms of defence against insect herbivores. Funct Ecol 19: 55–66.
94. Vollenweider P, Gunthardt-Goerg MS (2005) Diagnosis of abiotic and biotic stress factors using the visible symptoms in foliage. Environ Pollut 137: 455–465.
95. Poschenrieder C, Tolra R, Barcelo J (2006) Can metals defend plants against biotic stress? Trends Plant Sci 11: 288–295.
96. Boyd RS (2007) The defense hypothesis of elemental hyperaccumulation: status, challenges and new directions. Plant Soil 293: 153–176.
97. Fones HN, Preston GM (2013) Trade-offs between metal hyperaccumulation and induced disease resistance in metal hyperaccumulator plants. Plant Pathol 62: 63–71.
98. Ueki S, Citovsky V (2002) The systemic movement of a tobamovirus is inhibited by a cadmium-ion-induced glycine-rich protein. Nat Cell Biol 4: 478–486.
99. Yao YA, Wang J, Ma X, Lutts S, Sun C, et al. (2012) Proteomic analysis of Mn-induced resistance to powdery mildew in grapevine. J Exp Bot 63: 5155–5170.
100. Hong MJ, Kim DY, Lee TG, Jeon WB, Seo YW (2010) Functional characterization of pectin methylesterase inhibitor (PMEI) in wheat. Genes Genet Syst 85: 97–106.
101. Weber M, Deinlein U, Fischer S, Rogowski M, Geimer S, et al. (2013) A mutation in the *Arabidopsis thaliana* cell wall biosynthesis gene pectin *methylesterase 3* as well as its aberrant expression cause hypersensitivity specifically to Zn. Plant J 76: 151–164.
102. Jain M, Nijhawan A, Arora R, Agarwal P, Ray S, et al. (2007) F-box proteins in rice. Genome-wide analysis, classification, temporal and spatial gene expression during panicle and seed development, and regulation by light and abiotic stress. Plant Physiol 143: 1467–1483.
103. Chen R, Guo W, Yin Y, Gong ZH (2014) A novel F-box protein CaF-box is involved in responses to plant hormones and abiotic stress in pepper (*Capsicum annuum* L.). Int J Mol Sci 15: 2413–2430.
104. Puranik S, Sahu PP, Srivastava PS, Prasad M (2012) NAC proteins: regulation and role in stress tolerance. Trends Plant Sci 17: 369–381.
105. Guo WJ, Ho TH (2008) An abscisic acid-induced protein, HVA22, inhibits gibberellin-mediated programmed cell death in cereal aleurone cells. Plant Physiol 147: 1710–1722.

3

Hyperspectral Detection of a Subsurface CO_2 Leak in the Presence of Water Stressed Vegetation

Gabriel J. Bellante[1]*, Scott L. Powell[1], Rick L. Lawrence[1], Kevin S. Repasky[2], Tracy Dougher[3]

1 Department of Land Resources and Environmental Sciences, Montana State University, Bozeman, Montana, United States of America, **2** Department of Electrical and Computer Engineering, Montana State University, Bozeman, Montana, United States of America, **3** Department of Plant Sciences and Plant Pathology, Montana State University, Bozeman, Montana, United States of America

Abstract

Remote sensing of vegetation stress has been posed as a possible large area monitoring tool for surface CO_2 leakage from geologic carbon sequestration (GCS) sites since vegetation is adversely affected by elevated CO_2 levels in soil. However, the extent to which remote sensing could be used for CO_2 leak detection depends on the spectral separability of the plant stress signal caused by various factors, including elevated soil CO_2 and water stress. This distinction is crucial to determining the seasonality and appropriateness of remote GCS site monitoring. A greenhouse experiment tested the degree to which plants stressed by elevated soil CO_2 could be distinguished from plants that were water stressed. A randomized block design assigned Alfalfa plants (Medicago sativa) to one of four possible treatment groups: 1) a CO_2 injection group; 2) a water stress group; 3) an interaction group that was subjected to both water stress and CO_2 injection; or 4) a group that received adequate water and no CO_2 injection. Single date classification trees were developed to identify individual spectral bands that were significant in distinguishing between CO_2 and water stress agents, in addition to a random forest classifier that was used to further understand and validate predictive accuracies. Overall peak classification accuracy was 90% (Kappa of 0.87) for the classification tree analysis and 83% (Kappa of 0.77) for the random forest classifier, demonstrating that vegetation stressed from an underground CO_2 leak could be accurately discerned from healthy vegetation and areas of co-occurring water stressed vegetation at certain times. Plants appear to hit a stress threshold, however, that would render detection of a CO_2 leak unlikely during severe drought conditions. Our findings suggest that early detection of a CO_2 leak with an aerial or ground-based hyperspectral imaging system is possible and could be an important GCS monitoring tool.

Editor: Rongrong Ji, Xiamen University, China

Funding: This research was funded by the U.S. Department of Energy and the National Energy Technology Laboratory through Award number: DE-FC26-05NT42587. Neither the United States Government nor any agency thereof, nor any of their employees, makes any warranty, express or implied, or assumes any legal liability or responsibility for the accuracy, completeness, or usefulness of any information, apparatus, product, or process disclosed, or represents that its use would not infringe privately owned rights. The views and opinions of authors expressed herein do not necessarily state or reflect those of the United States Government or any agency thereof. The funders had no role in study design, data collection and analysis, decision to publish, or preparation of the manuscript.

Competing Interests: The authors have declared that no competing interests exist.

* Email: gabriel.bellante@gmail.com

Introduction

Geologic Carbon Sequestration

Post-industrial atmospheric carbon dioxide (CO_2) concentration has risen from 280 ppm to over 380 ppm [1–6]. Atmospheric CO_2 absorbs and reemits radiation energy from the Earth's surface causing a warming of the surface environment [7]. Global warming could drastically reshape the Earth's climate and environment [8–12]. Geologic carbon sequestration (GCS) is a potential climate change mitigation strategy that captures point source CO_2 emissions from industrial sources and stores them in large sub-surface reservoirs. GCS, conceptually, has the potential to store gigatonnes (Gt) of CO_2, which would otherwise be released into the atmosphere, into specialized geologic formations for long-term storage [13–19].

There is a mandate to monitor GCS sites for CO_2 leakage to ensure the efficacy of this technology, given that it is on the brink of commercial-scale deployment [14,19,20–22]. Remote sensing is being investigated as a possible cost-effective, large-area monitoring method to detect surface CO_2 leaks at GCS sites [23–24]. A leak from a GCS site would not only compromise the viability of this technology as a climate mitigation strategy, but it could also threaten the safety of the surrounding environment and inhabitants at the surface. The 1986 eruption of CO_2 at Lake Nyos, Western Camaroon, killed over 1,700 people and demonstrated that a large pneumatic CO_2 explosion can have devastating consequences [25]. Natural sources of CO_2 leakage from the Long Valley Caldera in California have caused extensive forest mortality [26–28].

Elevated CO_2 levels in soil are known to cause anoxic conditions in plant roots [26,29], thereby interfering with plant respiration and inducing a stress response that could possibly be remotely sensed using aerial imagery [24]. Vegetated areas at GCS sites, presumably, could act as bellwethers to signal operational inefficiencies in hazardous CO_2 leak scenarios. Vegetation status and seasonality will determine when the remote detection of vegetation stress caused by elevated soil CO_2 would be feasible. Soil water availability is highly spatially variable, and water stressed vegetation could appear spectrally similar to vegetation stressed by elevated soil CO_2. Drought is another common environmental occurrence that can have lasting impacts on whole land regions and can be short-lived or persist for years. Regardless of duration and spatial extent, monitoring GCS sites with remote sensing data would require discrimination between water and CO_2 stress to accurately identify CO_2 leaks.

Remote Sensing as a Monitoring Technique for GCS Sites

A variety of monitoring techniques are being investigated for efficiency and accuracy at detecting a CO_2 leak from a subsurface storage reservoir [30]. Most methods tend to be expensive and resource intensive, while providing only limited spatial coverage. Remote sensing, alternatively, has the potential to provide near instantaneous monitoring over large swaths of land. Aerial imaging over GCS sites would be relatively inexpensive and might be performed with limited labor requirements. Regions of elevated soil CO_2, if accurately detected with remote sensing equipment, could be further investigated with instruments on the ground to properly diagnose the leak source. Early leak detection would allow site managers to take immediate remediation measures to prevent further CO_2 leakage, while minimizing safety risk and economic loss. GCS monitoring could tolerate false positives, within reason, whereas, a false negative resulting in a missed CO_2 leak could have serious consequences.

Hyperspectral Remote Sensing for Plant Stress Detection

Hyperspectral sensors collect nearly continuous spectral data across narrow channel widths throughout the visible and infrared portion of the electromagnetic (EM) spectrum (400–2500 nm). Spectral signatures or spectra can be used to detect subtle patterns in vegetation reflectance to better understand the physiological condition of plants. Plant pigment concentrations, leaf cellular structure, and leaf moisture content can be discerned with hyperspectral data to assess the overall health of vegetation and characterize plant stress [31–32].

Hyperspectral remote sensing has been used to detect and characterize numerous types of vegetation stressors within the visible and infrared portion of the EM spectrum. Hyperspectral data have been analyzed to model plant stress caused by elevated soil CO_2 [33], water stress [34–36], insect and pest invasion [37–40], heavy metal contamination [22], salinity stress [41], nutrient levels and crop status [42–43]; ozone exposure [44]; and natural gas leaks [45]. Certain spectral regions are known to be especially sensitive, because they relate to the specific chemical and physical responses of plants to stress.

Low reflectance in the visible portion of the EM spectrum (400–700 nm) is determined primarily by chlorophyll *a* and *b* pigment abundance and absorption, while high reflectance in the NIR (750–2500 nm) is governed by the structural, spongy mesophyll contained in plant leaves [31,46–49]. Chlorophyll absorption is highest in the visible blue (400–500 nm) and middle red (near 680 nm), therefore these regions are not sensitive to subtle changes in chlorophyll content, because the spectral signal is saturated [46,50]. Absorption by chlorophyll pigments is weakest in the visible green – orange (500–620 nm) and the far visible red (700 nm), therefore, these wavelengths are sensitive to stress and subtle changes in chlorophyll content [46,51]. Red edge indices are frequently used to detect stress at the boundary of the far red and NIR to detect changes in chlorophyll content. The red edge is a reflectance spike caused by steep transition from low reflectance in the far red to high reflectance in the NIR, which is used as reference because it is determined by the physical structure of plant leaves. Far red reflectance (near 700 nm) is known to increase as a plant exhibits a physiological stress response of reducing chlorophyll content to down regulate photosynthesis, thereby reflecting proportionally more potentially damaging photosynthetically active radiation compared to healthy vegetation, which can utilize all of that incoming energy. Vegetation stress caused by depleted oxygen concentrations around the roots from natural gas leaks and elevated soil CO_2 has been detected in the red edge region of the EM spectrum [23–24,33,45,52]. Water stress has also been found to cause reflectance changes related to chlorophyll pigment concentrations in the red edge [53–54].

Shifts of the red edge 5–10 nm towards shorter wavelengths, termed the "blue shift", has also been associated with a decline in chlorophyll absorption for stressed vegetation [32–33,45,52,55–58]. This phenomenon has been associated with the early detection of vegetation stress by hyperspectral remote sensing instruments. The sensitivity of hyperspectral instruments to detect vegetation stress early, and perhaps even before it is visible to the human eye, has distinct advantages over conventional monitoring methods, because land managers could then take quicker ameliorative action to minimize losses to valuable resources [37].

Vegetation reflectance in the green portion of the EM spectrum is associated with xanthophyll pigments in addition to chlorophyll. Xanthophyll pigments contained within plants' palisade mesophyll perform photoprotective roles that also influence photosynthetic radiation use efficiency and are sensitive to stress [42,59]. Vegetation reflectance within the green, therefore, will increase in response to stress to down regulate photosynthetic activity because of both xanthophyll pigment activity and decreases in chlorophyll content. The Photochemical Reflectance Index (PRI) is a narrow-band, hyperspectral index ratioing green reflectance and is known to be sensitive to xanthophyll pigment activity. The PRI has been used as an early indicator of water stress in plants caused by drought conditions [34–36,43].

Water stress responses in vegetation have also been detected in spectral water absorption features in the shortwave infrared (SWIR) portion of the EM spectrum. Reflectance within the 1300–2500 nm wavelengths has been associated with leaf water content and has been observed to increase as a primary response of plants to water stress [53]. Visible spectrum reflectance has also been shown to increase, but only as a secondary plant response to water stress, in which chlorophyll pigment concentration decreases as a mechanism of down regulating photosynthesis during prolonged dehydration in vegetation [53–54]. Spectral measurements of leaf water content in the SWIR, therefore, directly monitors for plant dehydration and would be a more immediate method of detecting a plant's response to water stress. The primary response of plants to drought within the water absorption bands also could possibly be spectrally distinct from other types of physiological stressors, including stress caused by elevated soil CO_2.

Our greenhouse experiment objectives were (1) to evaluate if elevated soil CO_2 stress can be detected in alfalfa plants using hyperspectral data, and if so how quickly; (2) to determine if CO_2 stress is spectrally distinguishable from water stress since water availability is likely to be spatially variable at GCS sites and water stressed vegetation will likely co-occur with CO_2 stressed vegetation during a CO_2 leak scenario; and (3) to determine if CO_2 stress is distinguishable during drought conditions when all vegetation is water stressed during a CO_2 leak scenario. It is still unknown whether the plant stress signal caused by elevated soil CO_2 is spectrally distinguishable from other forms of plant physiological stress. Water stress or drought conditions in particular, are common natural occurrences that might confound a CO_2 stress signal if both plant physiological stressors coincide. Remote CO_2 leak detection might therefore be problematic in the context of GCS monitoring. Although this experiment does not directly test the use of hyperspectral imaging for CO_2 leak detection in a real-world aerial monitoring context, understanding the spectral discernability between stress caused by elevated soil CO_2, water stress, and their interaction is important for determining the appropriate times and conditions in which remote sensing image acquisition should commence over GCS sites.

Methods

Data Collection

Alfalfa (*Medicago sativa*) seeds were planted on June 1, 2010, to obtain 40 mature plants for this greenhouse experiment. Each alfalfa seedling was transplanted into a 45 cm tall, 9.6-l tree pot (Stuewe & Sons, Inc.). These tree pots were modified in several ways prior to planting in order to facilitate CO_2 injection into each individual plant pot and to best emulate a CO_2 leak scenario from a GCS site. The drainage holes at the bottom of each pot were sealed using plastic wrap, silicone, and Gorilla tape to minimize injected CO_2 from readily escaping through the bottom of the pots. The pots were then filled with 1.0 l of pea-sized gravel placed on top of a layer of gauze cotton cloth that covered drain holes at the bottom of each pot. A 0.6-cm diameter tube was inserted into each plant pot above the bottom 5-cm layer of gravel. The tube terminated in a 5-cm long, ceramic air stone (Top Fin) intended to disperse the CO_2 gas radially, as opposed to being released from a single point source. The embedded air stone plumbing was topped with an additional l of gravel to assist in the dispersion of the CO_2 gas throughout each pot and to allow for adequate water drainage from the soil (Figure 1).

Soil was placed on top of another layer of gauze above the gravel to keep the media separate. The soil used in the pots was a mixture of equal parts loam, washed concrete sand, and Canadian Sphagnum peat moss. Additionally, AquaGro 2000 wetting agent was blended into the soil at a ratio of 0.5 kg/m^3 soil mix. The soil mix was pasteurized with aerated steam at 70°C for 60 minutes. A 10 cm long piece of 2.54 cm diameter PVC pipe was inserted 5 cm down into the surface of the soil of every plant to allow for immediate access for soil CO_2 concentration measurements with a Vaisala GMP221 probe to verify injection. The alfalfa plants were watered with approximately 300 ml of water at 0800 hours every Monday, Wednesday, and Friday. They were fertilized twice a week with a $^{1\ gm}/_{0.5\ l}$ dilution of NPK fertilizer (20-10-20) every Wednesday and Friday until CO_2 injection began. The greenhouse temperature averaged 21.25±1.60°C over the course of the experiment. Sixteen hours a day of supplemental artificial lighting was provided using 1000 watt metal halide growth bulbs (GE Multi-Vapor MVR 1000/C/U) from 0500 to 2100 hours. Greenhouse integral photosynthetic photon flux measured within each of the treatment blocks ranged from 9.41 to 15.79 mol/m^2/day on peak sunny days and 0.09 to 0.11 mol/m^2/day on cloudy days (Apogee Instruments Line quantum sensors).

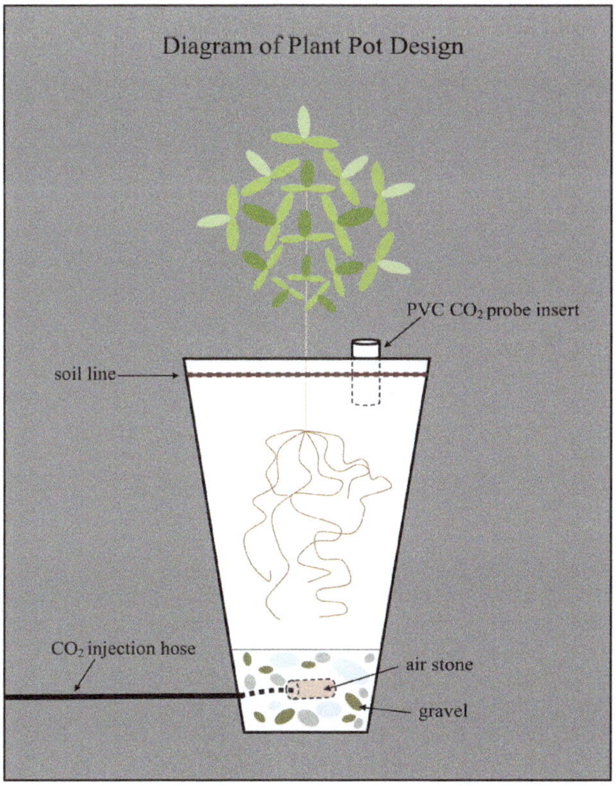

Figure 1. Diagram of a plant pot for the greenhouse experiment.

Five treatment blocks were developed, containing eight plants each (two replicates of each of four treatment groups per block). The four treatment groups included: (1) CO_2 injection (I); (2) water stress (WS); (3) an interaction group that was subjected to both water stress and CO_2 injection (WSI); and (4) adequate water and no CO_2 injection (C). Each treatment block contained two plants from each treatment group. CO_2 gas was delivered to the plant pots through a plumbing manifold that used equal lengths of tubing between each of the injected plants and a 2-stage pressure reducing regulator (Smith) at the CO_2 tank, which maintained constant pressure and did not require adjustment as the pressure within the CO_2 storage cylinder decreased over time. The gas was delivered from cylinders containing 50 lbs. liquid CO_2 at a constant pressure of 15 psig.

Alfalfa plants were randomly assigned to treatment blocks and to discrete positions within each block using a random number generator. Treatments within the blocks were randomly assigned using a stratified randomization developed to maintain equal lengths tubing between the injected plants (Figure 2). The plants that received water continued on the same watering regiment of 300 $^{ml}/_{day}$, while the water-stressed plants were not given any water in order to mimic drought conditions once CO_2 injection began.

Each alfalfa plant was scanned three times a week for three weeks, and all 40 alfalfa plants were sampled twice to obtain two spectral measurements per plant on each scanning date using an ASD Field Spec Pro 350. The ASD is a 16-bit spectrometer that has a spectral range of 350–2500 nm. The spectral resolution is 3 nm at 700 nm, 8.5 nm at 1400 nm, and 6.5 nm at 2100 nm. The sampling interval for the instrument is 1.4 nm at 350–1000 nm and 2 nm at 1000–2500 nm. Prior to interpolation there are a total of 1512 channels that are used in scan acquisition. The instrument takes a scan every 100 ms (ASD Inc.). A single sample consisted of two clipped alfalfa leaves, each containing three leaflets, from a randomly selected stem, determined by the roll of a 4-sided die. Each selected stem was clipped off at the bottom of the sixth node and spectral scans were taken from the four largest leaves on the cut stem. This sampling design intended to measure leaves from both the lower and upper canopy of each alfalfa plant to ensure that stress was not exhibited in one more than the other. The first sample consisted of the upper two biggest leaves and the second was taken from the lower two largest leaves of each cut stem. The leaves were dissected, meaning each leaflet was clipped from each sampled leaf (the leaves were left entire for ease and maximum coverage of the viewing spot later in the experiment,

Figure 2. Treatment block design. Single treatment block shown with plumbing and labeled treatments shown in white for each plant (left). Four of five treatment blocks are shown within the greenhouse (right).

when samples had leaves that were too desiccated and brittle to dissect). The three leaflets from the bottom leaf of the top sample were placed in parallel on a spectralon target overlapped by the three leaflets of the upper leaf from the same sample to achieve total leaf surface coverage of the ASD fiber optic sensor viewing spot, the region from which spectral information is collected. The process was repeated for the lower set of leaves from the same plant for the second sample. These protocols were implemented both to preserve the spatial relationship among the leaf samples, as well as to keep the orientation of the leaves relative to the plant consistent to best simulate what a remote sensing instrument from an airborne platform would view (Figure 3).

The fiber optic cable of the ASD was equipped with a plant probe accessory for leaf-level spectral measurements. The plant probe was 10 in. long and contained an internal, low-intensity halogen bulb, which produces little heat, for collecting spectral scans from vegetation. The viewing geometry of the mounted fiber optic cable created a 10 mm by 13 mm oval viewing spot from which spectral information was collected.

Initial spectral readings were taken for all 40 plants at 1000 hours on February 7, 2011, as a baseline before treatments were applied. CO_2 injection began at 1400 hours on February 7, 2011. A Vaisala soil probe was used to monitor CO_2 concentrations over time and to confirm successful injection into the individual plant pots. A Swagelok G2 model variable area flowmeter also was used

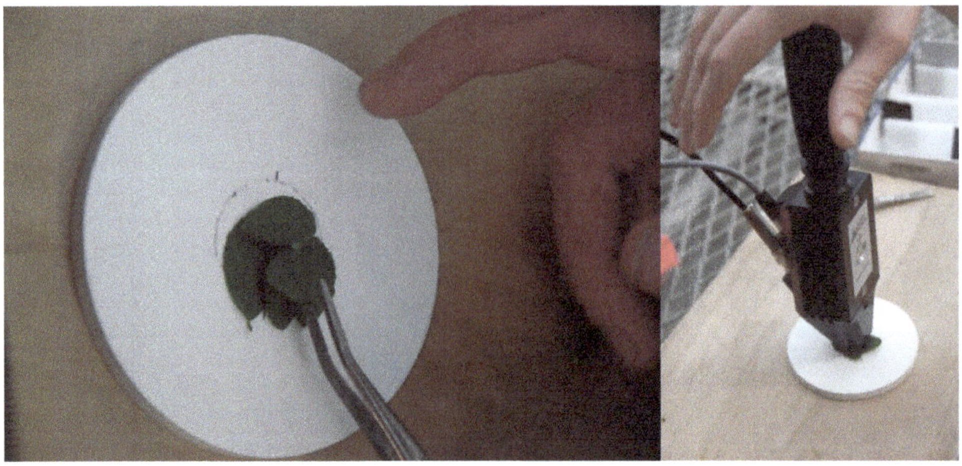

Figure 3. Sampling method. Two dissected alfalfa leaves on the spectralon target (left). The ASD fiber optic and plant probe assembly (right).

to measure CO_2 flux at the quick-connect union for each plant. Injection rates ranged from 0.25 to >5 l/hr for all injected plants. All 40 plants were spectrally scanned on Monday, Wednesday, and Friday for three weeks. Each plant was sampled and scanned on nine dates, except one plant, which was only scanned on eight dates because it was discovered on February 9 that a plant was not receiving CO_2 injection for some unknown reason. The plant therefore, was swapped out. A total of 18 scans were acquired for all plants, except for the exchanged plant, which had 16 scans. The injection was terminated on February 25, 2011, upon the completion of scanning all the plants earlier that day.

All spectral measurements were in reflectance and were a derived average of 25 individual scans. The instrument was calibrated to set gains and off-sets by an optimization process in the RS^3 software (ASD Inc.). A dark current was collected subsequently and the sensor was white referenced using a spectralon target. The ASD instrument was periodically recalibrated on each scanning date to ensure accurate and repeatable spectral measurements. A new white reference and dark current reading was made after the acquisition of both spectral scans for a single plant. The ASD was re-optimized upon the completion of acquiring data for an entire treatment block–eight plants.

Data Analysis

We developed a classification tree model for each individual scanning date to determine the timing and extent to which the different treatment groups were spectrally distinguishable. Classification tree analysis is a non-parametric statistical modeling method that has been successfully used to discriminate vegetation stress classes using hyperspectral data because it can utilize different band combinations to distinguish each class [37]. Classification trees use recursive, decision-based rules that can be interpreted by an analyst. This is important when trying to ascertain how and where the different plant physiological stressors are spectrally distinguished.

Although the ASD contained spectral data within the ultraviolet portion (350–400 nm) of the EM spectrum, these data were removed from further analysis because they were noisy and airborne hyperspectral sensors generally do not collect data at these wavelengths. The ASD measurements for the 400–2500 nm spectral range were output in an ascii format and exported into an Excel spreadsheet. TIBCO Spotfire S+ statistical software package was used to fit classification tree models to each scanning date's data using treatment group as the response variable and the ASD bands as explanatory (predictor) variables. Additionally, a factor variable indicating the relative position (upper or lower) of each leaf sample was included as a possible predictor variable in case sample location had an influence on treatment response. Cross-validation trees within S+ were used to prune the classification tree models for each date, so that they were unbiased and not over fit [60]. A standard 10-fold cross-validation was performed to determine the appropriate number of terminal nodes to lower deviance among samples, except that, given the small sample size, a script was written in S+ to take a random sample of scans stratified on treatment type (I, WS, WSI, or C) for each cross-validation tree to ensure that a balanced sample for each treatment group was withheld for validation purposes. This process was repeated five times, taking a different stratified random sample for each successive cross-validation tree and the plurality determined the appropriate size for pruning each classification tree. The spectral locations (wavelengths) used to distinguish between treatment groups for each of the binary nodes within each pruned classification tree model were examined to elucidate the spectral regions that best discerned between water stress and stress caused by elevated soil CO_2 in plants.

A random forest classification [61] also was performed on the individual scanning dates to evaluate the levels of prediction accuracy that could be achieved in discerning water and CO_2 stress agents. The algorithm is a "bagging" method that takes a bootstrap sample from the data observations to develop a classification tree by using a random subset of all possible explanatory variables (spectral bands) at each binary split [61]. This is done iteratively to form hundreds of classification trees (the forest) and then each observation is classified on the resultant plurality vote of the forest. Random forest models were derived with an ensemble of 500 classification trees with 45 randomly selected explanatory variables (of the possible 2100 spectral bands or the categorical variable indicating relative leaf position in the canopy) to be tried at each binary split within each tree. Variable importance plots from the random forest models were used to highlight the spectral regions used in distinguishing between the different treatment classes. Hyperspectral data have been modeled using random forest to greatly increase classification accuracies, as compared to other methods, as well as provide an unbiased, reliable internal estimate of accuracy for mapping land cover [62] and invasive plants [63]. A disadvantage of random forest is the inability to ascertain the relevance of individual explanatory variables and make meaningful interpretations of the model. Since hundreds of individual trees contribute to a random forest model, the classification results are essentially determined inside a "black box" that is effective for the purpose of prediction, but not for interpreting the decision-based rules that determine those predictions [64].

The lack of replicates in this study forced us to rely on internal measurements of predictive accuracy to evaluate spectral distinguishability. No data were withheld in the construction of the individual classification trees and the random forest classifier withheld approximately one third of each date's data as a bootstrap (out-of-bag) sample for internal validation within each of the 500 constructed trees. The random forest classifier has proved itself as a robust classification tool that contains an unbiased internal accuracy assessment that does not require a separately withheld dataset for validation [63]. Overall classification accuracies, Kappa statistics (a more robust estimate of classification accuracy because it takes into account chance agreement by differencing the observed accuracy from that of a total random assignment), and individual user's (mapping errors of commission) and producer's accuracies (mapping errors of omission) were used to assess predictive capabilities for the class distinctions through time [65].

Results

Examination of the individual sample spectral signatures for the different treatment groups illustrated the spectral regions that were critical in their distinction and accurate classification (Figure 4). Sample spectra for each of the treatment groups appeared to exhibit similar reflectance properties in the visible wavelengths, while subtle spectral deviations occurred in the near and SWIR peak reflectance features before treatments were applied, on February 07. The stressed sample spectra, in contrast, appeared to exhibit increased reflectance in the visible spectrum, especially within the visible green, as compared to the C class on February 21, at the height of spectral distinction among treatment groups. Water stressed samples exhibited increased reflectance in the SWIR as compared to the C and I classes. The WSI class predominantly exhibited higher reflectance in the visible and

SWIR regions compared to all other samples. The red edge spectral boundary shifted towards shorter wavelengths (the "blue shift") for stressed samples after four days of treatment application, however, visual symptoms of stress were also observed on this day. The blue shift was not uniquely expressed among stressed treatment groups.

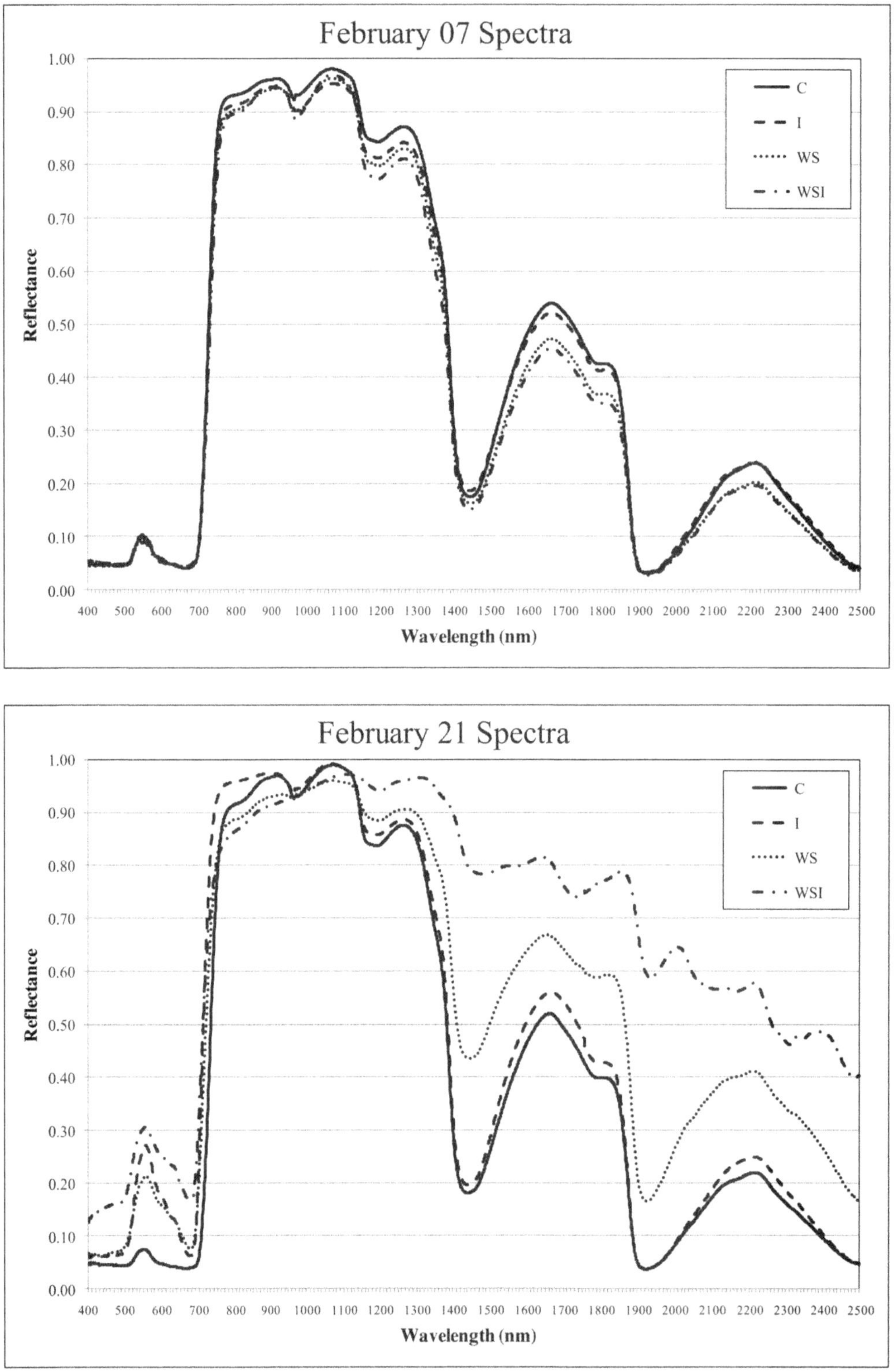

Figure 4. Sample reference spectra for each of the four treatment groups. Spectral signatures for February 07 (before treatment application) and February 21 (when spectral distinction was greatest between treatment groups).

Classification Tree Analysis

Cross-validation results determined that four terminal nodes were appropriate for February 21 and 23, otherwise the classification trees were pruned smaller to ensure they were not over fit (Table 1). Only the February 21 and 23 classification tree models, therefore, distinguished among all four treatment groups. A decision split or node was not justified for the classification tree models during the first week of the experiment. Not until February 14 were two or more terminal nodes warranted (Figure 5). The categorical variable indicating relative leaf sample position within each alfalfa canopy was never used as a splitting rule for any of the classification trees.

WS and WSI classes exhibited greater reflectance than the C and I classes within the water absorption bands near the 1400, 1900, and 2500 nm wavelengths. These distinctions in the SWIR infrared regions began on February 14 and persisted until the end of the experiment. Second level distinctions (decisions in trees with three or more terminal nodes) were made within the red edge at the 717 nm wavelength, where greater reflectance was exhibited by leaf samples from the I class as compared to the C class. The I class also exhibited greater reflectance in the visible green-visible orange at the 545, 621, and 541 nm wavelengths as compared to the C class. The WSI interaction group exhibited compound stress effects with greater reflectance as compared to the WS class in the water absorption bands at 1429 nm and 1901 nm.

Confusion early in the experiment, from February 14–18, occurred almost exclusively between (1) I and C classes, and (2) WSI and WS classes (Table 2). The C class was accurately distinguished by the second week of stress treatment application (user's and producer's accuracies ≥91%). Spectral differentiation between treatment groups was greatest on February 21 and February 23 (after two weeks of treatment application) with overall classification accuracies of 90% (Kappa of 0.87) and 81% (Kappa 0.75), respectively. Confusion during maximum spectral distinction primarily occurred (1) with I samples being classified as WS, and (2) WS being classified as WSI. Producer's accuracies for the I class were 80% and 55%, while the WSI class had perfect producer's accuracies (100%) on both February 21 and 23. Overall classification accuracy dropped to 66% (Kappa of 0.55) on the last scanning date, February 25.

Random Forest Analysis

The trend in accuracy of the random forest models throughout the time series was similar to that of the classification tree models (Figure 6). Out-of-bag accuracy predominantly exhibited a curvilinear trend with increasing out-of-bag accuracy from 25% (Kappa of 0) on February 7 (before treatments were applied) to peak classification accuracy of 83%, (Kappa of 0.77) on February 21. Subsequently, classification accuracy decreased. By February 25 out-of-bag accuracy had dropped to 65% (Kappa of 0.53). The C treatment group was best distinguished with user's and producer's accuracy ≥80% after one week of treatment application (Table 3). Most class confusion existed between WSI and WS treatment groups, once overall out-of-bag accuracy was ≥65% (February 14–25). I class producer's accuracy was ≥65% from February 14 onward. WSI class producer's accuracy was 90% for February 21 and 23, but was much lower on other dates.

The variable importance plots provided a visual display of the spectral regions that were the most influential in single date random forest model prediction (Figure 7). The variable importance plots were very noisy before there was spectral discernability among treatment groups. Spectral regions became more clearly defined as classification accuracies increased. The red edge region (700–750 nm) was shown to be the most important on February 14 and 16. On February 18, variable importance was predominantly placed on the spectral wavelengths centered on approximately 1400 nm. During maximum spectral distinction among treatment groups, variable importance was placed (1) in the visible green-yellow portion of the EM spectrum centered at approximately 550 nm; (2) the red edge (700–750 nm); and (3) the three water absorption features centered at approximately 1400, 1900, and 2500 nm. On February 25, the last scanning date, more emphasis was placed on the visible and NIR spectral regions rather than the water absorption features. The categorical variable indicating relative leaf sample position within each alfalfa canopy was never deemed important in any of the random forest classification models (variable importance<0.01).

Discussion

Our analyses demonstrated that hyperspectral spectrometry could distinguish between CO_2 stressed and healthy alfalfa leaves (I v. C). Furthermore, the spectral distinction of plant stress caused by elevated soil CO_2, water stress conditions, and their interaction was possible at certain times during the greenhouse experiment.

Table 1. Number of terminal nodes determined by the cross-validation results for the single date classification tree models.

Cross-Validation Tree Results						
Date	**x1**	**x2**	**x3**	**x4**	**x5**	**Plurality**
2/7	1	1	1	1	1	1
2/9	1	1	1	1	1	1
2/11	1	1	2	2	1	1
2/14	2	1	2	2	1	2
2/16	3	3	3	3	2	3
2/18	2	2	5	2	2	2
2/21	4	4	4	4	4	4
2/23	4	4	4	4	4	4
2/25	3	2	2	3	3	3

The plurality of the five cross-validation trials determined the appropriate size of the final trees.

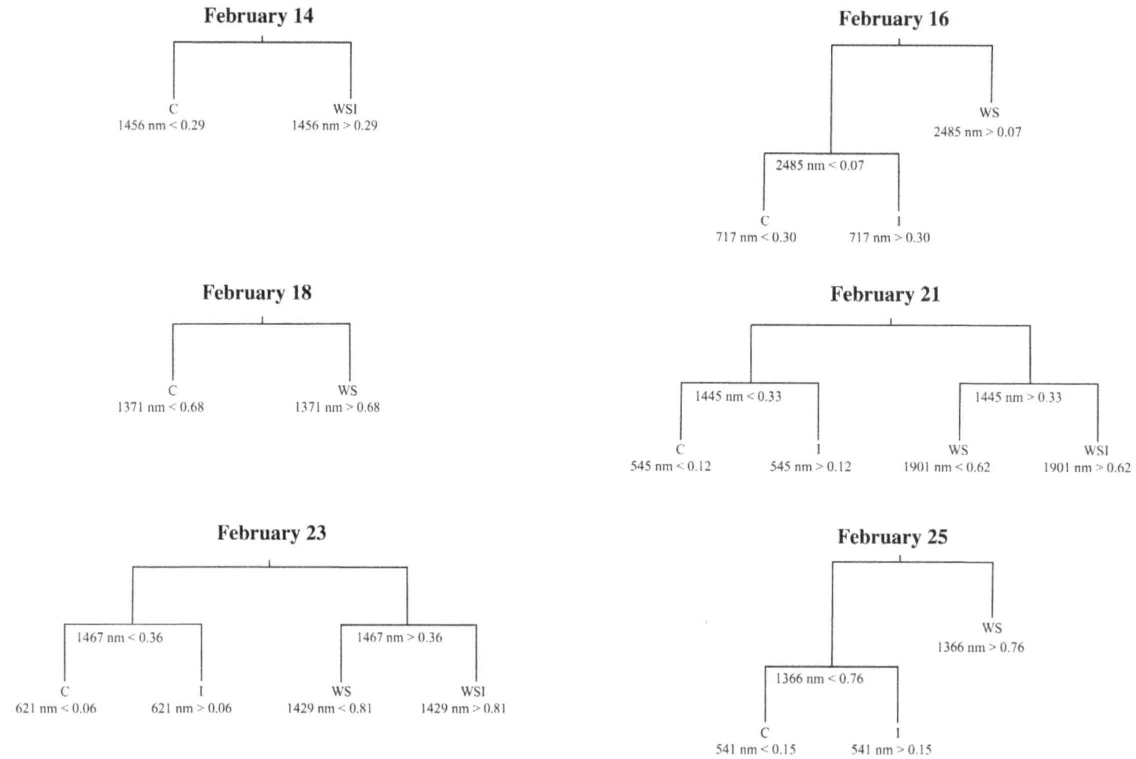

Figure 5. Pruned single date classification trees. Each tree shows the utilized spectral bands (by central band wavelength) and reflectance levels for each splitting rule. Splitting rules apply to the left branches of the tree. C = control class; I = CO_2 injection class; WS = water stress class; and WSI = water stress and CO_2 injection interaction class.

Those findings have significant implications for the use of hyperspectral imaging to monitor GCS sites for CO_2 leaks given that water stress is a prevalent environmental condition and could be a confounding factor. Remote sensing monitoring of CO_2 leaks amid a landscape of patchy, water stressed vegetation would most likely require differentiation of CO_2 and water stress (I v. WS), whereas during widespread drought, vegetation stressed by both CO_2 and water stress would need to be spectrally distinct from surrounding vegetation (WSI v. WS).

Confusion predominantly occurred between (1) I and C classes, (2) I and WS classes, and (3) WSI and WS classes, which could be problematic for CO_2 leak detection in a GCS monitoring context. Mapping errors of omission (low producer's accuracy) for CO_2 stressed vegetation (I and WSI) would be less acceptable from a hyperspectral monitoring standpoint, while mapping errors of commission (low user's accuracy) would be more tolerable, since a CO_2 leak would be less likely to go unidentified. High commission error and low omission error would result in the overestimation of CO_2 leaks, perhaps wasting labor resources for ground verification, however, CO_2 leaks would less likely be missed.

CO_2 Leak Detection When Water Is Not Limiting

Hyperspectral monitoring of GCS sites when water stressed vegetation is not present would consider the spectral discrimination between I and C classes. Spectral distinctions in the red edge, visible green, and visible orange regions were used to distinguish I from C classes. Rules discerned increased green reflectance within the I class multiple times near 550 nm, a spectral region sensitive to xanthophyll pigment activities associated with photosynthetic efficiency [42,59] and changes to chlorophyll content [46]. Additionally, increased reflectance at the 620 nm and 717 nm wavelengths distinguished I from C classes. These are also spectral regions that are known to be highly sensitive to chlorophyll *a* content and absorption [32]. The primary spectral response of the I class to elevated soil CO_2 was increased reflectance within the visible green–orange spectrum.

The I class was distinguished from all treatment groups with user's and producer's accuracies ≥70% beginning February 16, after ten days of CO_2 injection. Classification accuracies of the I class remained high until the end of the experiment, suggesting that a detectable stress signal caused by a CO_2 leak could persist for a substantial period of time. I and C classes were likely confounded in the experiment because some alfalfa plants received comparatively higher CO_2 injection rates due to minor differences in gas flow dynamics influenced by plant pot plumbing geometry and soil compaction. Severity and timing of stress caused by CO_2 injection therefore, varied. Healthy plants also appeared to be disproportionately targeted by a two-spotted spider mite infestation, which possibly caused a stress response that was most spectrally similar to stress caused by elevated soil CO_2 given that the insects caused a discoloration in the leaves most similar to the chlorosis observed in the alfalfa plants receiving CO_2 injection. The two-spotted spider mites were treated with numerous pesticides and biological controls, however, they probably remained an influential stress factor for healthier alfalfa plants. The results of the random forest classification demonstrated that stressed and non-stressed vegetation were accurately distinguished after one week of treatment application (C class accuracies ≥80%). Spectral differentiation among the stress classes was more difficult and time dependent, however.

Table 2. Internal classification accuracies and Kappa statistics for each single date classification tree model containing at least two terminal nodes.

February 14, 2011	C	I	WS	WSI	User's Accuracy	February 16, 2011	C	I	WS	WSI	User's Accuracy
C	20	20	6	3	41%	C	16	0	0	0	100%
I	0	0	0	0	na	I	4	19	0	0	83%
WS	0	0	0	0	na	WS	0	1	20	20	49%
WSI	0	0	14	17	55%	WSI	0	0	0	0	na
Producer's Accuracy	100%	0%	0%	85%	Overall Accuracy 46%	Producer's Accuracy	80%	95%	100%	0%	Overall Accuracy 69%
Kappa Statistic	0.28					Kappa Statistic	0.58				
February 18, 2011	**C**	**I**	**WS**	**WSI**	**User's Accuracy**	**February 21, 2011**	**C**	**I**	**WS**	**WSI**	**User's Accuracy**
C	20	20	0	0	50%	C	20	0	0	0	100%
I	0	0	0	0	na	I	0	16	0	0	100%
WS	0	0	20	20	50%	WS	0	4	16	0	80%
WSI	0	0	0	0	na	WSI	0	0	4	20	83%
Producer's Accuracy	100%	0%	100%	0%	Overall Accuracy 50%	Producer's Accuracy	100%	80%	80%	100%	Overall Accuracy 90%
Kappa Statistic	0.33					Kappa Statistic	0.87				
February 23, 2011	**C**	**I**	**WS**	**WSI**	**User's Accuracy**	**February 25, 2011**	**C**	**I**	**WS**	**WSI**	**User's Accuracy**
C	20	2	0	0	91%	C	20	1	0	0	95%
I	0	11	0	0	100%	I	0	13	0	0	100%
WS	0	6	14	0	70%	WS	0	6	20	20	43%
WSI	0	1	6	20	74%	WSI	0	0	0	0	na
Producer's Accuracy	100%	55%	70%	100%	Overall Accuracy 81%	Producer's Accuracy	100%	65%	100%	0%	Overall Accuracy 66%
Kappa Statistic	0.75					Kappa Statistic	0.55				

Treatment group columns represent the reference data while the rows represent the classified data.

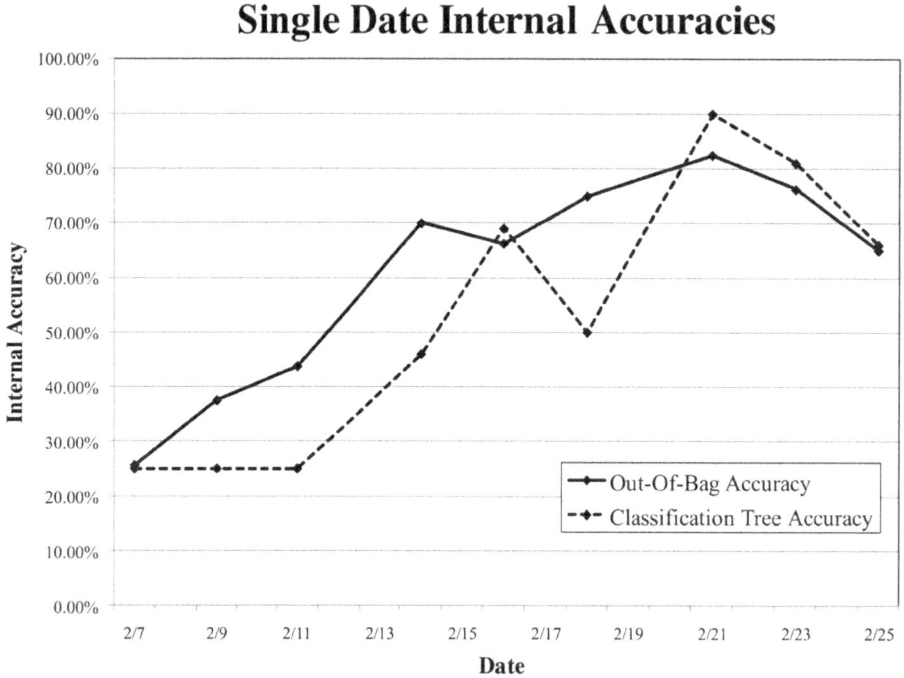

Figure 6. Overall internal accuracies for each random forest and classification tree model over the course of the experiment. Note that the February 7–11 classification trees did not contain ≥ two terminal nodes.

CO_2 Leak Detection When Soil Moisture Is Spatially Variable

Identification of a CO_2 leak would likely require water stress to be spectrally distinguished from stress caused by elevated soil CO_2 because soil moisture availability is highly variable across most landscapes [66–67]. Increased reflectance within the water absorption regions near 1400 and 2500 nm was the most important characteristic that distinguished water stressed classes (WSI and WS) from C and I classes. The primary spectral response of leaf dehydration occurs within the water absorption regions (near 1400, 1900, and 2500 nm) that are sensitive to leaf water content [53].

Patterns of variable importance for each of the random forest models identified discrete spectral regions that were particularly useful in discriminating among treatment groups. The red edge and the visible green-visible orange spectral regions were at least as important as the water absorption bands for distinguishing among the different classes. However, the "black box" nature of the random forest classifier did not allow for specific interpretation of the spectral distinctions made by the model [64]. It is possible that the variable importance within the visible and NIR wavelengths was related to distinguishing between the I and C classes given that these two treatment groups appeared to be spectrally similar in the SWIR region, while the water stress classes were likely separated by distinctions in the water absorption regions. Eventually increased reflectance in the visible, NIR, and SWIR regions can be exhibited by all stressed plants regardless of the cause, however [53].

Alfalfa plants of the I treatment group exhibited visual stress predominantly as chlorosis (yellowing of plant leaves due to lack of chlorophyll production), although some CO_2 injected plants' leaves did become desiccated by the end of the experiment, suggesting leaf water loss. Leaves of the WS treatment group generally exhibited a lack of vigor (leaves were droopy) early in the experiment, and eventually, they became dessicated and brittle as the water stress severity increased. Overall these classes were distinguished reasonably well. Only minor confusion occurred probably because of similar reflectance properties in the visible wavelengths due to eventual chlorophyll loss to water stressed samples and due to eventual leaf moisture loss, expressed as increased reflectance in the water absorption bands of I class samples [53].

CO_2 Leak Detection During Widespread Drought

Remote monitoring of GCS sites could require CO_2 leaks to be spectrally distinguished during drought conditions when all vegetation is subjected to water stress. Elevated soil CO_2 would thereby have to interact with water stress to cause a compound stress response in vegetation that was spectrally discernable. Greater reflectance in the water absorption bands near 1400 nm was used to distinguish WSI from WS classes, perhaps in response to diminished water-use-efficiency caused by comparatively faster leaf moisture loss or reduced water uptake at the root level. This was consistent with a compound stress interaction being exhibited by the WSI class. Treatment class accuracies indicated that alfalfa plants that went without water for one week exhibited water stress that became spectrally discernable on February 14. The confusion existed primarily *between* them and not with the other classes. Individual WS and WSI class accuracies, therefore, were low. This was consistent with a water stress signal that was more easily discerned because of the discrete water absorption features where leaf reflectance is directly related to leaf moisture content. Spectral distinction of the WSI class remained relatively convoluted with the WS class throughout the experiment. The WSI class was distinguished with high producer's accuracies (low omission error) on February 21 and 23, perhaps suggesting that vegetation could exhibit stress in response to a CO_2 leak that could be distinguished during drought conditions, albeit a narrow time window.

The plant leaves of the water stressed treatment groups became desiccated and brittle as the experiment progressed. These plants

Table 3. Out-of-bag accuracies and Kappa statistics for each single date random forest model.

February 7, 2011	**C**	**I**	**WS**	**WSI**	**User's Accuracy**	**February 9, 2011**	**C**	**I**	**WS**	**WSI**	**User's Accuracy**
C	5	2	3	5	33%	C	9	2	4	2	53%
I	3	4	6	2	27%	I	4	9	6	8	33%
WS	4	6	5	7	23%	WS	3	4	7	5	37%
WSI	8	6	6	6	23%	WSI	4	5	3	5	29%
Producer's Accuracy	25%	22%	25%	30%	Overall Accuracy 26%	Producer's Accuracy	45%	45%	35%	25%	Overall Accuracy 38%
Kappa Statistic	0					Kappa Statistic	0.17				
February 11, 2011	**C**	**I**	**WS**	**WSI**	**User's Accuracy**	**February 14, 2011**	**C**	**I**	**WS**	**WSI**	**User's Accuracy**
C	9	10	2	3	38%	C	17	2	2	0	81%
I	10	7	3	3	30%	I	2	13	1	2	72%
WS	0	2	10	5	59%	WS	1	2	12	4	63%
WSI	1	1	5	9	56%	WSI	0	3	5	14	64%
Producer's Accuracy	45%	35%	50%	45%	Overall Accuracy 44%	Producer's Accuracy	85%	65%	60%	70%	Overall Accuracy 70%
Kappa Statistic	0.25					Kappa Statistic	0.6				
February 16, 2011	**C**	**I**	**WS**	**WSI**	**User's Accuracy**	**February 18, 2011**	**C**	**I**	**WS**	**WSI**	**User's Accuracy**
C	16	4	0	0	80%	C	17	4	0	0	81%
I	4	15	1	1	71%	I	3	14	1	0	78%
WS	0	1	13	10	54%	WS	0	2	14	5	67%
WSI	0	0	6	9	60%	WSI	0	0	5	15	75%
Producer's Accuracy	80%	75%	65%	45%	Overall Accuracy 66%	Producer's Accuracy	85%	70%	70%	75%	Overall Accuracy 75%
Kappa Statistic	0.55					Kappa Statistic	0.67				
February 21, 2011	**C**	**I**	**WS**	**WSI**	**User's Accuracy**	**February 23, 2011**	**C**	**I**	**WS**	**WSI**	**User's Accuracy**
C	19	1	0	0	95%	C	19	2	0	0	90%
I	1	16	2	0	84%	I	1	14	3	0	78%
WS	0	3	13	2	72%	WS	0	4	10	2	63%
WSI	0	0	5	18	78%	WSI	0	0	7	18	72%
Producer's Accuracy	95%	80%	65%	90%	Overall Accuracy 83%	Producer's Accuracy	95%	70%	50%	90%	Overall Accuracy 76%
Kappa Statistic	0.77					Kappa Statistic	0.68				
February 25, 2011	**C**	**I**	**WS**	**WSI**	**User's Accuracy**						
C	19	2	1	0	86%						
I	1	15	2	0	83%						
WS	0	1	8	10	42%						

Table 3. Cont.

February 7, 2011	C	I	WS	WSI	User's Accuracy	February 9, 2011	C	I	WS	WSI	User's Accuracy
WSI	0	2	9	10	48%						
Producer's Accuracy	95%	75%	40%	50%	Overall Accuracy 65%						
Kappa Statistic	0.53										

Treatment group columns represent the reference data while the rows represent the classified data.

eventually began to die. Spectral distinction within the water absorption bands was probably difficult due to the extreme leaf moisture loss caused by severe drought by the end of the experiment. Little variable importance was placed on the water absorption features in the last classification date (February 25), where the water stressed treatment groups were poorly distinguished (accuracies≤50%), while the I and C treatment groups were classified with reasonable accuracy (accuracies ≥75%). This was evidence that the alfalfa plants' response to severe drought surpassed a stress threshold that rendered the CO_2 stress signal spectrally indistinguishable.

Pre-visual Plant Stress Detection

Pre-visual stress detection would be a critical attribute and an advantage to using hyperspectral remote sensing for early CO_2 leak detection at a GCS site. February 11 was the first day that subtle visual evidence of stress to some of the WSI and I treatment group plants was evident. Early visual symptoms of stress (i.e., chlorosis and languid leaves) were confined to older leaves located lower on the stems within the alfalfa plant canopies. Visible stress was most prominent within the WSI treatment group, which exhibited the compound effects of both CO_2 injection and lack of water. The accurate classification of the WSI samples did not occur until February 14, one week into the experiment, even though stress was noted visually three days before that time. The blue shift of the red edge towards shorter wavelengths has been associated with the pre-visual detection of plant stress [32,55,57], and was observed for some sample spectra by February 11. Blue shift spectral characteristics were similar in depth and slope for all of the stressed classes and, therefore, could not be used to distinguish among stress agents. This experiment provided no evidence that the pre-visual detection of plant stress was possible with hyperspectral data.

Summary

The primary conclusions that can be drawn from this greenhouse experiment are (1) that plant stress caused by elevated soil CO_2 was spectrally detected, probably in the visible spectrum, after approximately one week of CO_2 injection through the end of the experiment (random forest I class accuracies ≥65%). This is potentially, evidence that a spectral stress signal caused by elevated soil CO_2 could persist for a substantial period of time; (2) CO_2 and water stress were spectrally distinguishable; and (3) Elevated soil CO_2 appeared to cause a compound stress response detected in plants that were already water stressed, however, there was a relatively narrow time window when the WSI class was spectrally distinguished from the WS class. This was indicative of a time-dependent compound stress response caused by the interaction of elevated soil CO_2 and drought.

Although the results were variable between the individual classification tree and random forest classifier results, they both illustrated the same pattern of predictive success for plant stress detection. Differences between the two modeling strategies were likely due to the small sample size relative to the high dimensionality of the hyperspectral data. There were a total of 40 plants, 10 plants per class, as compared to the 1512 possible predictor variables. Depending on which samples were withheld for validation purposes during model development, variability did occur between outputs. Additionally, the random forest classifier uses a randomized node optimization which would introduce additional variability. Therefore, these differences were expected especially when using two distinctly different classification methods.

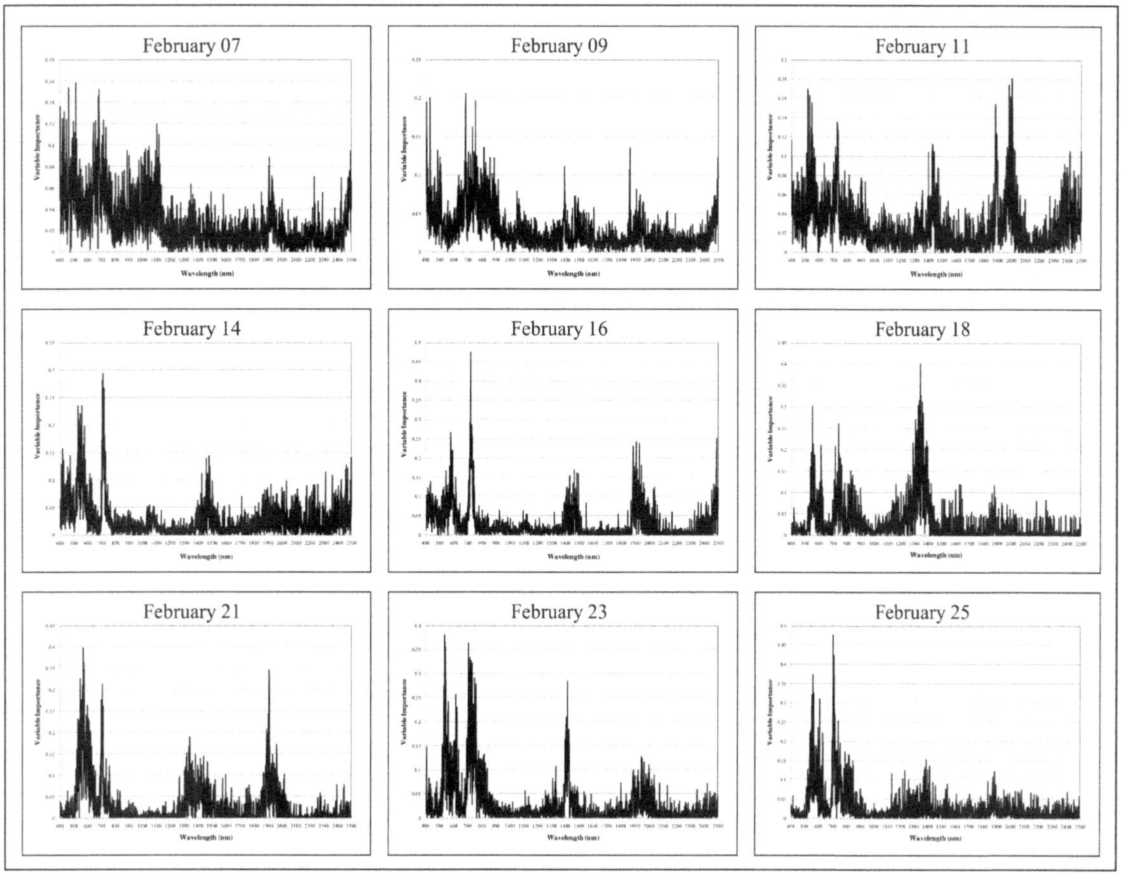

Figure 7. Variable importance plots for the single date random forest models. Note that the leaf position categorical variable is not included.

These results suggest that remote sensing may be used to monitor GCS sites for CO_2 leaks. Detection of a CO_2 leak when the availability of soil water is highly variable across space might be possible even if there are co-occurring patches of water stressed vegetation. Plants appear to hit a stress threshold, however, that renders spectral detection of a CO_2 leak unlikely during severe drought conditions. Regardless, this research demonstrated the necessity for remote sensing instruments to be spectrally sensitive to SWIR reflectance in order to accurately distinguish CO_2 and water stress.

Even though pre-visual stress detection was not possible at the leaf-scale, the early detection of stress caused by elevated soil CO_2 was achieved. Aerial hyperspectral imaging has been demonstrated to be effective for early stress detection at broader spatial scales [37] and the random forest classifier has been used to successfully map invasive plants in a natural setting [63]. Monitoring for CO_2 leaks at GCS sites will possibly require large area coverage, which will only be possible from an airborne platform. Hyperspectral instruments with high spatial resolution optics for canopy-level monitoring at altitude, therefore, might be important for the early detection of CO_2 leaks. The high dimensional nature of hyperspectral data will require robust data analysis methods that can detect a plant stress signature or signal within large datasets. Different data mining tools exist that could both reduce the dimensionality of hyperspectral data and discern the targeted CO_2 stress signature in an aerial monitoring context. For example, an orthogonal operator has been used to reduce the dimensionality of hyperspectral data while retaining the meaningful signature of interest [68–69]. Additionally, a hypergraph method that jointly relates the spatial context of image pixels to their associated spectral characteristics has been demonstrated to be effective in hyperspectral image classification and may be useful in detecting pockets of plant stress caused by elevated soil CO_2 [70]. Further exploration of different data analysis methods will be essential to optimize a workflow that would enable an analyst to efficiently assimilate large datasets and effectively detect a CO_2 leak in a GCS monitoring context.

Additional research is warranted to understand the extent to which elevated soil CO_2 stress could be detected at the canopy-level using airborne remote sensing, especially when water stress conditions exist. Spectral differentiation between CO_2 and water stressed vegetation would likely include spectral distinctions within the water absorption features. Spectral data acquired within the water absorption regions from an airborne remote sensing platform would likely be attenuated by atmospheric water vapor, perhaps further confounding CO_2 leak detection [47,54]. Snow cover and drought are going to limit the applicability of remote sensing data in temperate climates, therefore hyperspectral monitoring will need to be used in conjunction with other monitoring methods to obtain adequate temporal coverage of GCS sites. Improved understanding of vegetation spectral responses to stress caused by plant senescence will also be critical in determining the seasonal timing in which remote sensing data are appropriate for the detection of plant stress caused by an underground CO_2 leak.

Author Contributions

Conceived and designed the experiments: GB SP RL KR TD. Performed the experiments: GB SP RL KR TD. Analyzed the data: GB SP RL KR TD. Contributed reagents/materials/analysis tools: GB SP RL KR TD. Wrote the paper: GB SP RL KR TD.

References

1. Cuffey KM, Vimeux F (2001) Covariation of carbon dioxide and temperature from the Vostok ice core after deuterium-excess correction. Nature 412: 523–527.
2. Masarie K, Tans PT (1995) Extension and integration of atmosphere carbon dioxide data into a globally consistent measurement record. Journal of Geophysical Research 100: 11593–11610.
3. Monnin E, Indermühle A, Dällenbach A, Flückiger J, Stauffer B, et al. (2001) Atmospheric CO_2 concentrations over the last glacial termination. Science 291: 112–114.
4. Petit JR, Jouzel J, Raynaud D, Barkov NI, Barnola JM, et al. (1999) Climate and atmospheric history of the past 420,000 years from the Vostok ice core, Antarctica. Nature 399: 429–436.
5. Scripps Institute of Oceanography (2007) Monthly average carbon dioxide concentration. Available: http://scrippsco2.ucsd.edu/graphics_gallery/mauna_loa_record/mauna_loa_record.html. Accessed 2014 Sep 18.
6. Seigenthaler U, Stocker TF, Monnin E, Lüthi D, Schwander J, et al. (2005) Stable carbon cycle-climate relationship during the Late Pleistocene. Science 310: 1313–1317.
7. Intergovernmental Panel on Climate Change (IPCC) Fourth Assessment Report (2007) IPCC Fourth Assessment Report on Climate Change. Cambridge University Press, Cambridge, UK.
8. Parmeson C (2006) Ecological and evolutionary responses to recent climate change. Annual Review of Ecology, Evolution, and Systematics 37: 637–669.
9. Hansen J, Sato M, Kharecha P, Beerling D, Masson-Delmotte V (2008) Target atmospheric CO_2: where should humanity aim? Open Atmospheric Science Journal 2: 217–231.
10. Norby RJ, Luo Y (2006) Evaluating ecosystem responses to rising atmospheric CO_2 and global warming in a multi-factor world. New Phytologist 162: 281–293.
11. Shackleton NJ (2000) The 100,000-year ice-age cycle identified and found to lag temperature, carbon dioxide, and orbital eccentricity. Science 289: 1897–1902.
12. Vinnikov KY, Grody NC (2003) Global warming trend of mean tropospheric temperature observed by satellites. Science 302: 269–272.
13. Hepple RP (2002) Implications of surface seepage on the effectiveness of geologic storage of carbon dioxide as a climate change mitigation strategy. Lawrence Berkeley National Laboratory (LBNL), Paper LBNL-51267.
14. Hepple RP, Benson SM (2005) Geologic storage of carbon dioxide as a climate change mitigation strategy: performance requirements and the implications of surface seepage. Environmental Geology 47: 576–585.
15. Herzog HJ (2001) What future for carbon capture and sequestration? American Chemical Society 35: 148–153.
16. Intergovernmental Panel on Climate Change (IPCC) (2005) IPCC Special Report on Carbon Dioxide Capture and Storage. Cambridge University Press, Cambridge, UK.
17. Lawrence Berkeley National Laboratory (LBNL) (2000) An overview of geologic sequestration of CO_2. In: ENERGEX'2000: Proceedings of the 8th International Energy Forum, Las Vegas, NV.
18. Pacala S, Socolow R (2004) Stabilization wedges: solving the climate problem for the next 50 years with current technologies. Science 305: 968–972.
19. Wilson EJ, Friedmann SJ, Pollak MF (2007) Research for deployment: Incorporating risk, regulation, and liability for carbon capture and sequestration. Environmental Science & Technology 41: 5945–5952.
20. Cortis A, Oldenburg CM, Benson SM (2008) The role of optimality in characterizing CO_2 seepage from geologic carbon sequestration sites. International Journal of Greenhouse Gas Control 2: 640–652.
21. Oldenburg CM, Bryant SL, Nicot JP (2009) Certification framework based on effective trapping for geologic carbon sequestration. International Journal of Greenhouse Gas Control 3: 444–457.
22. Schuerger AC, Capelle GA, Di Benedetto JA, Mao CY, Thai CN, et al. (2003) Comparison of two hyperspectral imaging and two laser-induced fluorescence instruments for the detection of zinc stress and chlorophyll concentration in bahia grass (Paspalum notatum Flugge.). Remote Sensing of Environment 84: 572–588.
23. Keith CJ, Repasky KS, Lawrence RL, Jay SC, Carlsten JL (2009) Monitoring effects of a controlled subsurface carbon dioxide release on vegetation using a hyperspectral imager. International Journal of Greenhouse Gas Control 3: 626–632.
24. Male EJ, Pickles WL, Silver EA, Hoffmann GD, Lewicki JL, et al. (2010) Using hyperspectral plant signatures for CO_2 leak detection during the 2008 ZERT CO_2 sequestration field experiment in Bozeman, Montana. Environmental Earth Sciences 60: 251–261.
25. Kling GW, Evans WC, Tanyileke G, Kusakabe M, Ohba T, et al. (2005) Degassing Lakes Nyos and Monoun: Defusing certain disaster. Proceedings of the National Academy of Sciences 102: 14185–14190.
26. Bergfeld D, Evans WC, Howle JF, Farrar CD (2006) Carbon dioxide emissions from vegetation-kill zones around the resurgent dome of Long Valley caldera, eastern California, USA. Journal of Volcanology and Geothermal Research 152: 140–156.
27. Jong SM (1998) Imaging spectrometry for monitoring tree damage caused by volcanic activity in the Long Valley caldera, California. ITC Journal 1–10.
28. Pickles WL, Kasameyer PW, Martini BA, Potts DC, Silver EA (2001) Geobotanical remote sensing for geothermal exploration. Geothermal Resources Council Transactions 25: 307–312.
29. Maček I, Pfanz H, Francetič V, Batič F, Vodnik D (2005) Root respiration response to high CO2 concentrations in plants from natural CO2 springs. Environmental and Experimental Botany 54: 90–99.
30. Spangler LH, Dobeck LM, Repasky KS, Nehrir AR, Humphries SD, et al. (2009) A shallow subsurface controlled release facility in Bozeman, Montana, USA, for testing near surface CO2 detection techniques and transport models. Environmental Earth Science, http://dx.doi.org/10.1007/s12665-009-0400-402.
31. Goetz AFH, Vane G, Solomon JE, Rock BN (1985) Imaging spectrometry for Earth Remote Sensing. Science 228: 1147–1153.
32. Carter GA (1993) Responses of leaf reflectance to plant stress. American Journal of Botany 80: 239–243.
33. Noomen MF, Skidmore AK (2009) The effects of high soil CO_2 concentrations on leaf reflectance of maize plants. International Journal of Remote Sensing 30: 481–497.
34. Peguero-Pina JJ, Morales F, Flexas J, Gil-Pelegrin E, Moya I (2008) Photochemistry, remotely sensed physiological reflectance index and de-epoxidation state of the xanthophyll cycle in Quercus coccifera under intense drought. Oecologia 156: 1–11.
35. Penuelas J, Llusia J, Pinol J, Filella I (1997) Photochemical reflectance index and leaf photosynthetic radiation-use-efficiency assessment in Mediterranean trees. International Journal of Remote Sensing 18: 2863–2868.
36. Suarez L, Zarco-Tejada PJ, Sepulcre-Canto G, Perez-Priego O, Miller JR, et al. (2008) Assessing canopy PRI for water stress detection with diurnal airborne imagery. Remote Sensing of Environment 112: 560–575.
37. Lawrence RL, Labus M (2003) Early detection of Douglas-Fir Beetle infestation sith subcanopy resolution hyperspectral imagery. Westeren Journal of Applied Forestry 18: 202–206.
38. Pu R, Kelly M, Anderson GL, Gong P (2008) Using CASI hyperspectral imagery to detect mortality and vegetation stress associated with a new hardwood forest disease. Photogrammetric Engineering & Remote Sensing 74: 65–75.
39. Fitzgerald GJ, Maas SJ, Detar WR (2004) Spider mite detection and canopy component mapping in cotton using hyperspectral imagery and spectral mixture analysis. Precision Agriculture 5: 275–289.
40. Muhammed HH (2005) Hyperspectral crop reflectance data for characterizing and estimating fungal disease severity in wheat. Biosystems Engineering 91: 9–20.
41. Naumann JC, Young DR, Anderson JE (2009) Spatial variations in salinity stress across a coastal landscape using vegetation indices derived from hyperspectral imagery. Plant Ecology 202: 285–297.
42. Gamon JA, Serrano L, Surfus JS (1997) The photochemical reflectance index: an optical indicator of photosynthetic radiation use efficiency across species, functional types, and nutrient levels. Oecologia 112: 492–501.
43. Strachan IB, Pattey E, Boisvert JB (2002) Impact of nitrogen and environmental conditions on corn as detected by hyperspectral reflectance. Remote Sensing of Environment 80: 213–224.
44. Meroni M, Rossini M, Picchi V, Panigada C, Cogliati S, et al. (2008) Assessing steady-state fluorescence and PRI from hyperspectral proximal sensing as early indicators of plant stress: The case of ozone exposure. Sensors 8: 1740–1754.
45. Noomen MF, Smith KL, Colls JJ, Steven MD, Skidmore AK, et al. (2008) Hyperspectral indices for detecting changes in canopy reflectance as a result of underground natural gas leakage. International Journal of Remote Sensing 29: 5987–6008.
46. Carter GA, Knapp AK (2001) Leaf optical properties in higher plants: linking spectral characteristics to stress and chlorophyll concentration. American Journal of Botany 88: 677–684.
47. Knipling EB (1970) Physical and physiological basis for the reflectance of visible and near-infrared radiation from vegetation. Remote Sensing of Environment 1: 155–159.
48. Smith KW, Vogelmann TC, DeLucia EH, Bell DT, Shepherd KA (1997) Leaf form and photosynthesis. BioScience 47: 785–793.
49. Zarco-Tejada PJ, Miller JR, Mohammed GH, Noland TL (2000) Chlorophyll fluorescence effects on vegetation apparent reflectance: I. leaf-level measurements and model simulation. Remote Sensing of Environment 74: 582–595.
50. Lichtenhaler HK (1987) Chlorophylls and carotenoids: pigments of photosynthetic biomembranes. Methods in Enzymology 148: 350–382.

51. Buschmann C, Nagel E (1993) In vivo spectroscopy and internal optics of leaves as basis for remotes sensing of vegetation. International Journal of Remote Sensing 14: 711–722.
52. Smith KL, Steven MD, Colls JJ (2004) Use of hyperspectral derivative ratios in the red-edge region to identify plant stress responses to gas leaks. Remote Sensing of Environment 92: 207–217.
53. Carter GA (1991) Primary and secondary effects of water content on the spectral reflectance of leaves. American Journal of Botany 78: 916–924.
54. Tucker CJ, Elgin JH, McMurtrey JM (1980) Relationship of red and photographic infra-red spectral data to alfalfa agronomic values. International Journal of Remote Sensing 1: 69–75.
55. Carter GA, Miller RL (1994) Early detection of plant stress by digital imaging within narrow stress-sensitive wavebands. Remote Sensing of Environment 50: 295–302.
56. Smith KL, Steven MD, Colls JJ (2004) Spectral responses of pot-grown plants to displacement of soil oxygen. International Journal of Remote Sensing 25: 4395–4410.
57. Rock BN, Hoshizaki T, Miller JR (1988) Comparison of in situ and airborne spectral measurements of the blue shift associated with forest decline. Remote Sensing of Environment 24: 109–127.
58. Vogelmann JE, Rock BN, Moss DM (1993) Red edge spectral measurements from sugar maple leaves. International Journal of Remote Sensing 14: 1563–1575.
59. Gamon JA, Surfus JS (1999) Assessing leaf pigment content and activity with a reflectometer. New Phytologist 143: 105–117.
60. Venables WN, Ripley BD (1997) Modern Applied Statistics with S-Plus, Second Edition. Springer-Verlag, New York, NY 548 p.
61. Breiman L (2001) Random Forests. Machine Learning 45: 5–32.
62. Gislason PO, Benediktsson JA, Sveinsson JR (2006) Random Forest for land cover classification. Pattern Recognition Letters 27: 294–300.
63. Lawrence RL, Wood SD, Sheley RL (2006) Mapping invasive plants using hyperspectral imagery and Breiman Cutler classifications (RandomForest) Remote Sensing of Environment 100: 356–362.
64. Prasad AM, Iverson LR, Liaw A (2006) Newer classification and regression tree techniques: Bagging and random forests for ecological prediction. Ecosystems 9: 181–199.
65. Congalton RG, Green K (2009) Assessing the accuracy of remotely sensed data: principles and practices Second edition. Taylor and Francis Group.
66. Baier W (1969) Concepts of soil moisture availability and their effect on soil moisture estimates from a meteorological budget. Agricultural Meteorology 6: 165–178.
67. Denmead OT, Shaw RH (1962) Availability of soil water to plants as affected by soil moisture content and meteorological conditions. Agronomy Journal 54: 385–390.
68. Harsanyi JC, Chang C (1994) Hyperspectral image classification and dimensionality reduction: an orthogonal subspace projection approach. IEEE Transactions On Geoscience and Remote Sensing 32: 779–785.
69. Rodarmel C, Shan J (2002) Principal component analysis for hyperspectral image classification. Surveying and Land Information Science 62: 115–123.
70. Ji R, Gao Y, Hong R, Liu Q, Tao D, et al. (2014) Spectral-spatial constraint hyperspectral image classification. IEEE Transactions On Geoscience and Remote Sensing 52: 1811–1824.

4

The Endogenous Nitric Oxide Mediates Selenium-Induced Phytotoxicity by Promoting ROS Generation in *Brassica rapa*

Yi Chen[1,2☯], Hai-Zhen Mo[3☯], Liang-Bin Hu[3], You-Qin Li[2], Jian Chen[2*], Li-Fei Yang[1*]

1 College of Horticulture, Nanjing Agricultural University, Nanjing, China, 2 Institute of Food Quality and Safety, Jiangsu Academy of Agricultural Sciences, Nanjing, China, 3 Department of Food Science, Henan Institute of Science and Technology, Xinxiang, Henan Province, China

Abstract

Selenium (Se) is suggested as an emerging pollutant in agricultural environment because of the increasing anthropogenic release of Se, which in turn results in phytotoxicity. The most common consequence of Se-induced toxicity in plants is oxidative injury, but how Se induces reactive oxygen species (ROS) burst remains unclear. In this work, histofluorescent staining was applied to monitor the dynamics of ROS and nitric oxide (NO) in the root of *Brassica rapa* under Se(IV) stress. Se(IV)-induced faster accumulation of NO than ROS. Both NO and ROS accumulation were positively correlated with Se(IV)-induced inhibition of root growth. The NO accumulation was nitrate reductase (NR)- and nitric oxide synthase (NOS)-dependent while ROS accumulation was NADPH oxidase-dependent. The removal of NO by NR inhibitor, NOS inhibitor, and NO scavenger could alleviate Se(IV)-induced expression of *Br_Rbohs* coding for NADPH oxidase and the following ROS accumulation in roots, which further resulted in the amelioration of Se(IV)-induced oxidative injury and growth inhibition. Thus, we proposed that the endogenous NO played a toxic role in *B. rapa* under Se(IV) stress by triggering ROS burst. Such findings can be used to evaluate the toxic effects of Se contamination on crop plants.

Editor: Anil Kumar, University of Missouri-Kansas City, United States of America

Funding: This work was supported by National Natural Science Foundation of China (grant No. 31101537). The funders had no role in study design, data collection and analysis, decision to publish, or preparation of the manuscript.

Competing Interests: The authors have declared that no competing interests exist.

* Email: jacksonchen206@gmail.com (JC); lfy@njau.edu.cn (L-FY)

☯ These authors contributed equally to this work.

Introduction

Selenium (Se) is an essential and beneficial micronutrient for plants [1]. The important roles of Se in intrinsic physiological process has been confirmed by the identification of Se-containing proteins in plant cells [2,3]. The moderate supplement of Se is able to protect agricultural plants from multiple environmental stresses [4–10]. However, Se is becoming an emerging health hazards in a global scale because of the dramatically anthropogenic release of Se to the environment [11]. The excessive Se in agricultural environment poses potential threat to the growth of crop plants by causing phytotoxicity [12–14]. The lack of knowledge on the mechanism of Se-induced phytotoxicity limits the evaluation of the toxic effect of Se on crop plants. Several recent studies have suggested that the excessive Se could induce oxidative stress in plants by triggering the production of reactive oxygen species (ROS) [15–17]. However, how Se induces ROS generation and the subsequent oxidative injury in plants remains unclear. NADPH oxidase coding by *Rbohs* (*respiratory burst oxidative homologues*) has been suggested as a major source for ROS generation in plants under both biotic and abiotic stresses [18]. The *Rboh* family contains 10 annotated or putative *Rboh* genes (*RbohA-J*) in *Arabidopsis* genome [19]. The stimulated NADPH oxidase activity in plants has been associated with heavy metal stress [20–22], but the comprehensive regulation of different *Rboh* members by heavy metals (including Se) is hardly known.

Nitric oxide (NO) is a pivotal signalling molecule in regulating the key physiological processes during plant development and stress responses [23]. Nitrate reductase (NR) and nitric oxide synthase (NOS) are the two major enzymatic sources for the endogenous production of NO in plants [24]. *NOS* gene has not been identified from plants, but the activity of NOS has been successfully detected in plants [25,26]. There are other NO producing pathways including aerobic NO formation based on hydroxylamines or polyamines and anoxic NO formation based on deoxygenated hemeproteins, but the exact molecular mechanisms of these NO-producing pathways are still highly speculative [27]. The protective role of NO in plants against metal-induced oxidative stress has been well characterized [28–33], but recent studies have suggested that NO could also contribute to phytotoxicity induced by heavy metals, such as cadmium (Cd) [34–36], arsenic (As) [37], and aluminium (Al) [38]. NO is proposed to be a stimulator of hydrogen peroxide (H_2O_2) in Cd-induced cell death in Arabidopsis [34]. However, the mechanism of endogenous NO-mediated ROS generation and oxidative stress is currently not understood in plants under heavy metal stress.

In this study, the relationship between Se-induced phytotoxicity and NADPH oxidase-governed ROS generation was investigated in the root of *Brassica rapa*. The effect of Se stress on the production of endogenous NO from different sources was studied *in vivo* by fluorescent microscopy. To get deeper insights into the interplay between the endogenous NO and Se-induced oxidative stress in *B. rapa*, the involvement of the endogenous NO on the differential expression of *Rboh* genes, oxidative injury, and growth inhibition was investigated. All of these results are very important to help our understanding for the toxic role of NO in plants under Se stress, which will assist to the evaluation of the environmental risk of Se to crop growth.

Materials and Methods

Plant culture and treatment

Seeds of wild type *B. rapa* (LvLing) were sterilized with 1 % NaClO for 10 min, then rinsed several times with distilled water and germinated for 1 day in the dark on the floating plastic nets. After germination, young seedlings were transferred to Petri dish containing various treatment solution in a chamber with a photosynthetic active radiation of 200 μmol/m^2/s, a photoperiod of 12 h, and the temperature at 25±1°C.

Seedling roots were exposed to Na_2SeO_3 (sodium selenite) with different concentrations (0.03–0.46 mM) for various treatment time (0–72 h). DPI (diphenylene iodonium) at 10 μM, PY (pyridine) at 5 mM, and IMZ (imidazole) at 0.5 mM were applied as NADPH oxidase inhibitors [39,40]. Tungstate (Na_2WO_4) at 30 μM and L-NMMA (N^G-Monomethyl-L-arginine) at 200 μM were applied as NR inhibitor and NOS inhibitor, respectively [41,42]. The 0.1 mM of cPTIO [2-(4-carboxy-2-phenyl)-4,4,5,5-tetramethylinidazoline-1-oxyl-3-oxide] was applied as NO scavenger [43]. SNP (sodium nitroprusside) at 0.25 mM was applied as NO donor [43]. The treatment solution is composed of different chemicals mentioned above according to different experimental design. After treatment, the roots were washed with distilled water for physiological, histochemical, and biochemical analysis.

Histochemical analysis

Intracellular ROS was visualized using DCFH-DA (2′,7′-dichlorofluorescein diacetate) fluorescent probe described by Foreman et al [44]. The roots of seedlings were incubated in 1 μM of DCFH-DA at 25°C for 20 min. Then the roots were rinsed with distilled water for three times followed by the visualization (excitation 488 nm and emission 525 nm) with a fluorescence microscope (ECLIPSE, TE2000-S, Nikon).

Intracellular NO was visualized using DAF-FM DA (3-Amino, 4-aminomethyl-2′,7′- difluorescein, diacetate) fluorescent probe described by Guo et al [45]. The roots of seedlings were incubated in 20 mM of Hepes-NaOH (pH 7.5) buffer solution containing 15 μM of DAF-FM DA at 25°C for 15 min. Then the roots were rinsed with distilled water for three times followed by the visualization (excitation 490 nm and emission 525 nm) with a fluorescence microscope (ECLIPSE, TE2000-S, Nikon).

$O_2^{\cdot -}$ in shoots was visually detected by using NBT (nitro blue tetrazolium) described by Frahry and Schopfer [46]. The roots of seedlings after treatment were transferred to 10 mM Na-citrate buffer (pH 6.0) containing 6 mM NBT under light at 25°C for 20 min, and then the roots were rinsed with distilled water for three times, which allowed the dark blue insoluble formazan compound (by reaction of NBT with $O_2^{\cdot -}$) inside of roots to be clearly visualized and photographed.

Histochemical detection of lipid peroxidation was achieved by using Schiff's regent as described by Wang and Yang [28]. The roots of seedlings after treatment were incubated in Schiff's regent for 20 min. Then the stained roots were rinsed with a solution containing 0.5% (w/v) $K_2S_2O_5$ (prepared in 0.05 M of HCl) until the root colour became light red. After that, the roots were photographed using a digital camera.

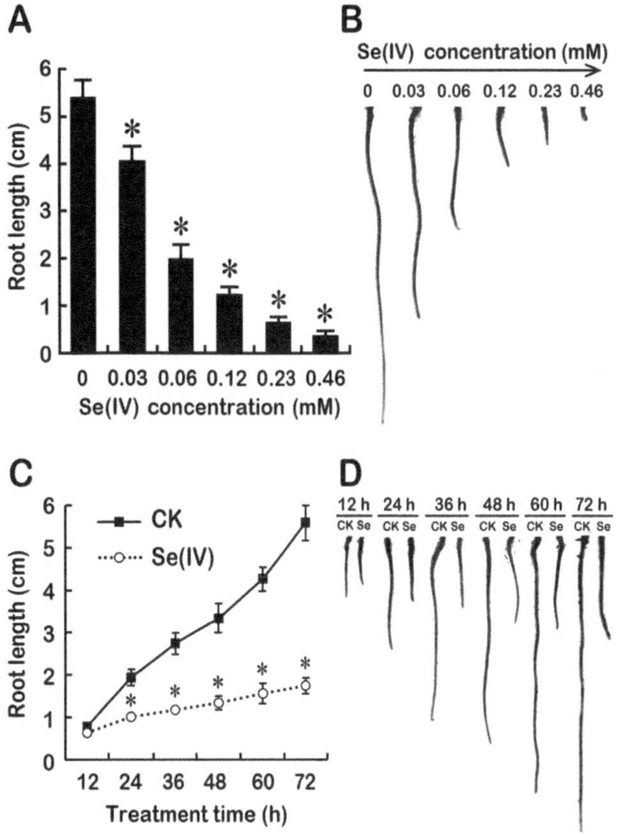

Figure 1. Effect of Se(IV) on the growth of *B. rapa* root. (A–B) The roots of seedlings were exposed to 0, 0.03, 0.06, 0.12, 0.23, and 0.46 mM of Se(IV) solution for 72 h. After that, the root length were measured (A). The images of roots were photographed (B). (C–D) The root length and root image were obtained when the roots of seedlings were exposed to 0.06 mM of Se(IV) solution for 12, 24, 36, 48, 60, and 72 h, respectively. *Asterisk* indicates that mean values of three replicates are significantly different between the treatments of Se(IV) and the control group (CK) ($P<0.05$).

Histochemical detection of loss of plasma membrane integrity was performed by using Evans blue as described by Yamamoto et al [47]. The roots of seedlings after treatment were incubated in 5 ml of 0.025% Evans blue solutions (w/v) for 20 min. After that, the roots were rinsed with distilled water for three times followed by photographed using a digital camera.

Screening and analysis of *Br_Rbohs* from the genome of *B. rapa*

The sequences of *AtRbohA-J* from *Arabidopsis* were used as baits for BLAST research in the genome of *B. rapa* from BRAD (*Brassica* database) (http://brassicadb.org/brad/index.php). The obtained sequences were retrieved and analyzed. The phylogenetic trees were constructed using the maximum likelihood method in MEGA 5.2. The multialignment of amino acid sequences was performed with ClustalX 2.0 and DNAMAN

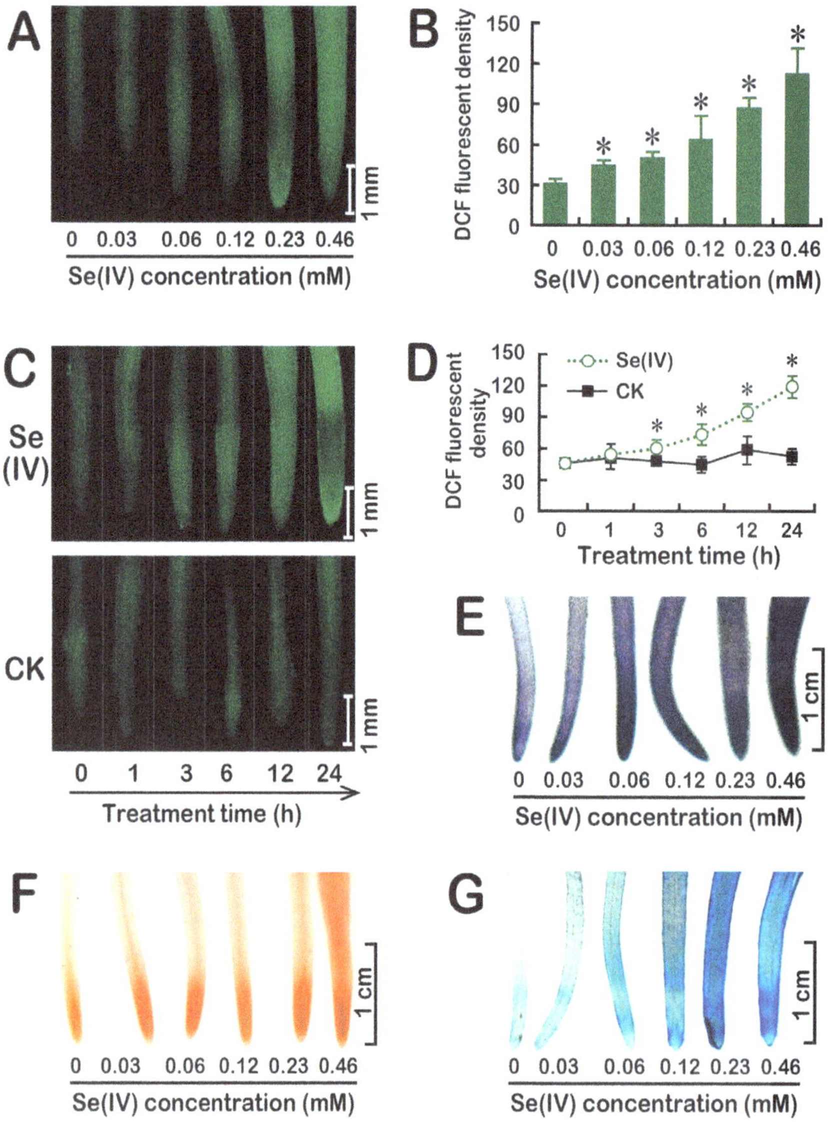

Figure 2. Effect of Se(IV) on endogenous total ROS (A–D), endogenous $O_2^{\cdot -}$ (E), lipid peroxidation (F), and the loss of plasma membrane integrity (G) in the roots of *B. rapa*. The roots of seedlings were exposed to 0, 0.03, 0.06, 0.12, 0.23, and 0.46 mM of Se(IV) solution for 48 h. Afterwards, the roots were loaded with DCFH-DA for 15 min and immediately photographed (A). The relative DCF fluorescent density in roots was estimated (B). (C–D) The image and density of DCF fluorescence were obtained when the roots of seedlings were exposed to 0.06 mM of Se(IV) solution for 0, 1, 3, 6, 12, and 24 h, respectively. The roots of seedlings were exposed to 0, 0.03, 0.06, 0.12, 0.23, and 0.46 mM of Se(IV) solution for 48 h. Afterwards, the roots were stained with NBT (E), Schiff's reagent (F), and Evan blue (G), respectively. *Asterisk* indicates that mean values of three replicates are significantly different between the treatments of Se(IV) and the control group (CK) ($P<0.05$).

5.2.2. Protein structure prediction was performed on SMART (http://smart.embl-heidelberg.de/).

The DNA sequences with the length of 2 kb were retrieved from the upstream of *Br_Rbohs* in *B. rapa* for promoter analysis. The sequence between the start of target gene and the end of its upstream gene was obtained for promoter analysis if the length of this sequence was less than 2 kb. The *cis*-elements in the retrieved promoter regions were analyzed using online tool PLACE (http://www.dna.affrc.go.jp/PLACE/signalscan.html).

Transcript analysis

Total RNA was extracted from root tissues using Trizol (Invitrogen) according to the manufacturer's instructions. Reverse transcription was performed at 42°C in 25 µl reaction mixture including 3 µg of RNA, 0.5 µg of oligo (dT) primers, 12.5 nmol of dNTPs, 20 units of RNase inhibitor and 200 units of M-MLV. The first cDNA was used as a template for polymerase chain amplification and to analyse the transcripts of genes by using real-time quantitative reverse transcription-polymerase chain reaction (qRT-PCR) (Applied Biosystems 7500 Fast Real-Time PCR System, LifeTechnologiesTM). The primers designed for the amplification of genes are listed in Table S1.

Determination of enzyme activities

About 0.05 g of fresh root tissue was homogenized in 1.5 ml of ice-cooled phosphate buffer (50 mM, pH 7.0, containing 1 mM

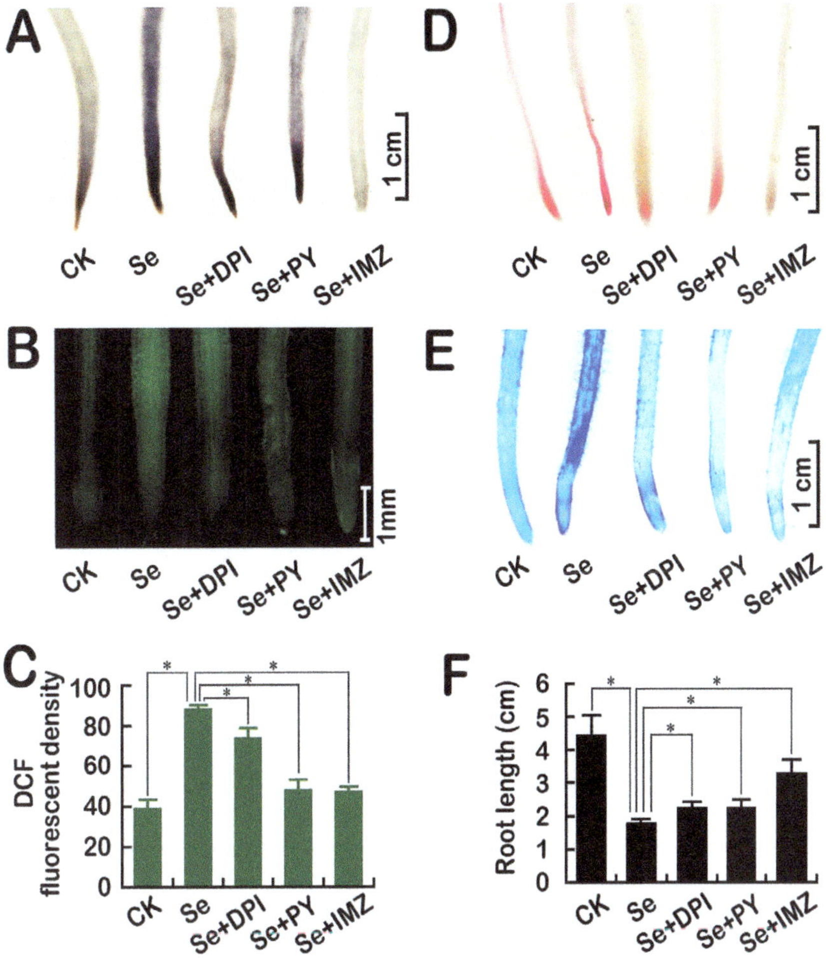

Figure 3. Effect of NADPH oxidase inhibitors (DPI, PY, and IMZ) on endogenous $O_2^{\cdot -}$ (A), endogenous total ROS (B–C), lipid peroxidation (D), the loss of plasma membrane integrity (E), and the growth of roots under Se(IV) stress. The roots of seedlings were exposed to Se(IV) (0.06 mM), Se(IV) (0.06 mM)+DPI (10 μM), Se(IV) (0.06 mM)+PY (5 mM), and Se(IV) (0.06 mM)+IMZ (0.5 mM) for 48 h. Afterwards, the roots were stained with NBT (A), DCFH-DA (B), Schiff's reagent (D), and Evan blue (E), respectively. The density of DCF fluorescence were estimated (C). The roots of seedlings were exposed to the above treatment solutions for 72 h. Afterwards, the root length were measured (F). *Asterisk* indicates that mean values of three replicates are significantly different between the different treatments ($P<0.05$).

ethylenediamine tetra-acetic acid (EDTA) and 1% w/v insoluble polyvinylpyrrolidone). The homogenate was centrifuged at 15,000 g for 10 min at 4°C. The supernatant was used as the crude extract for the assay of enzyme activities.

The activity of NR was determined as described by Xiong et al [48]. Enzyme extract was added to 0.4 mg/L of pre-warmed assay buffer containing 100 mM of HEPES-KOH (pH 7.5), 5 mM of KNO_3, and 0.25 mM of NADH. The mixed solution reacted at 30°C for 60 min, and then the reaction was stopped by adding Zn-acetate. The absorbance of produced nitrite was measured at 540 nm by adding 1 mg/L of 1% sulfanilamide in 3 M HCl plus 1 mg/L of 0.2% N-(1-naphthyl) ethylenediamine. A standard curve of NO_2^- in the range 0–2 μg was prepared to convert the measured absorbance to concentration.

The NOS activity was determined by NOS colorimetric kit as described by Li et al [49]. NOS activity was spectrophotometrically measured as the capacity of catalysing NO production from L-arginine, which is based on the oxidation of oxyhaemoglobin to methaemoglobin by NO. One unit was defined as generating 1 nmol of NO per minute at 37°C per microgram fresh weight.

Statistical analysis

Each result was presented as the mean ± standard deviation (SD) of at least three replicated measurements. The significant differences between treatments were statistically evaluated by SD and one-way analysis of variance (ANOVA) using SPSS 2.0. The data between two specific different treatments were compared statistically by ANOVA, followed by F-test if the ANOVA result is significant at $P<0.05$. For multiple comparison analysis, least significant difference test (LSD) was performed on all data following ANOVA tests to test for significant ($P<0.05$) differences among different treatments.

Results

Se(IV) induced growth inhibition and oxidative injury in the roots of *B. rapa*

Treatment with Se(IV) inhibited the root growth *B.rapa* in a dose-dependent manner. The roots were exposed to 0–0.46 mM of Se(IV) for 72 h. Compared to the control, the root length decreased by almost 50% at 0.06 mM Se(IV) level (Figure 1A, B).

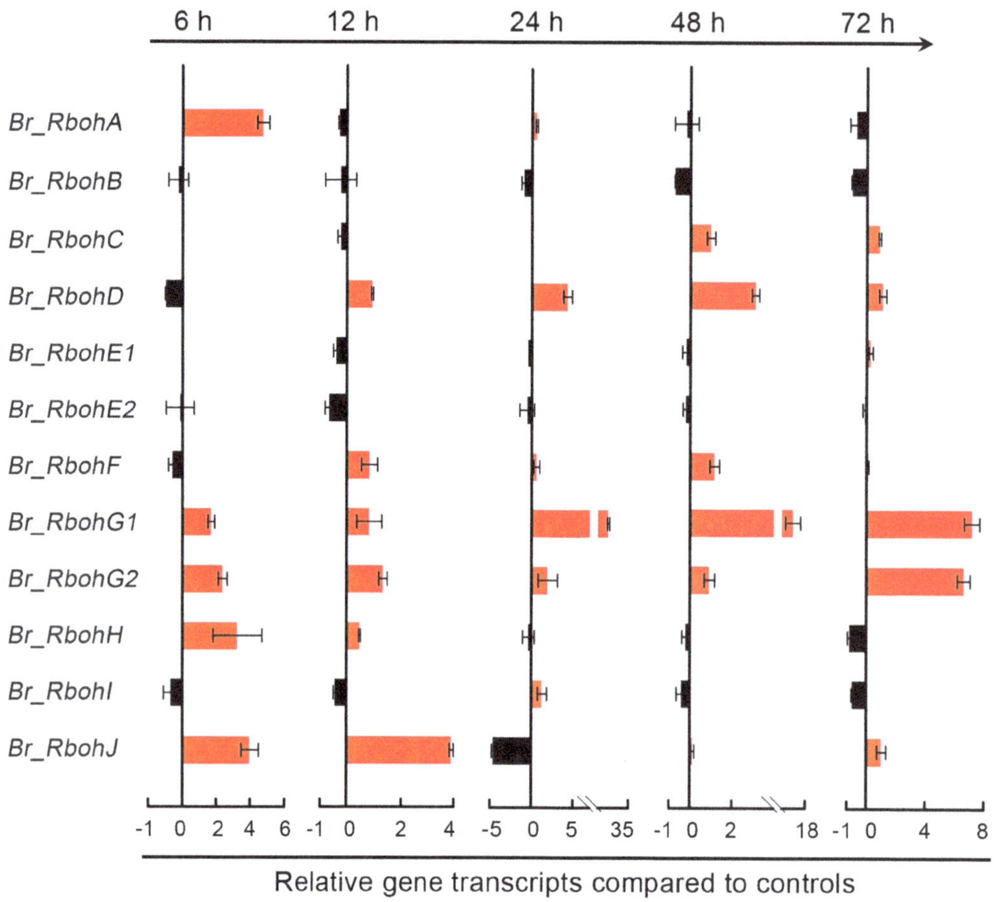

Figure 4. Effect of Se(IV) on the transcripts of *Br_RbohA-J*. The roots of seedlings were exposed to 0.06 mM of Se(IV) solution for 6, 12, 24, 48, and 72 h, respectively. The total RNA was extracted from roots for qRT-PCR analysis. *Actin* was used for cDNA normalization. The data were presented as the relative fold-change in transcript abundance of the target genes. *Red* column means up-regulation. *Black* column means down-regulation.

Therefore, 0.06 mM of Se(IV) was used for the further experiments. In a time-course experiment, the root growth was inhibited significantly after treatment with 0.06 mM of Se(IV) for 24 h. Se(IV) at 0.06 mM showed the continuous inhibitory effect on root growth after treatment for 24–72 h (Figure 1C, D).

Root tip is the important expansion zone for root elongation [50]. In order to test whether Se(IV) stress could induce ROS accumulation in root tips, we performed *in situ* detection of intracellular ROS generation by using specific fluorescent probe DCFH-DA. Compared to the control, the strong fluorescent density was observed in root tips in the presence of Se(IV) in a dose-dependent manner (Figure 2A, B). In a 24 h time-course experiment, treatment with 0.06 mM of Se(IV) resulted in the continuous accumulation of total ROS after 3 h of treatment (Figure 2C, D). The correlation analysis suggested that ROS generation was negatively correlated with root length under Se(IV) stress (Figure S1A).

Since $O_2^{\bullet -}$ is one of the most important ROS produced by NADPH oxidase, the generation of $O_2^{\bullet -}$ was detected histochemically using NBT. As expected, the generation of $O_2^{\bullet -}$ in roots increased considerably with the increase in Se(IV) concentration (Figure 2E). Because the over-generation of ROS is closely related to the oxidative injury to plant cells, the oxidation of membrane lipids and the loss of plasma membrane integrity were determined using histochemical staining with Shiff's regent and Evans blue, respectively [28]. The roots of *B. rapa* treated with Se(IV) were stained extensively while the control group had only light staining (Figure 2F, G). These results confirmed the oxidative stress induced by Se(IV) in the roots of *B. rapa*.

Inhibition of NADPH oxidase activity attenuated Se(IV)-induced oxidative injury and growth inhibition

Three NADPH oxidase inhibitors (DPI, PY, and IM) were used to determine the involvement of NADPH oxidase in Se(IV)-induced oxidative stress and growth inhibition in roots. Compared to Se(IV) treatment alone, simultaneous treatment with NADPH oxidase inhibitors remarkably reduced the over-generation of $O_2^{\bullet -}$ (Figure 3A) and the accumulation of total ROS (Figure 3B, C). In addition, blocking NADPH oxidase activity also alleviated Se-induced membrane lipids oxidation (Figure 3D), the loss of membrane integrity (Figure 3E), and growth inhibition (Figure 3F). These results suggested that Se(IV)-induced oxidative injury and growth inhibition resulted from NADPH oxidase-mediated over-generation of ROS.

Se(IV) stress induced the expression of *Br_Rbohs* in roots

In order to investigate the effect of Se(IV) stress on the expression of *Rboh* genes coding NADPH oxidase, 12 *Br_Rboh* genes were identified from the genome of *B. rapa* by using *Rboh*

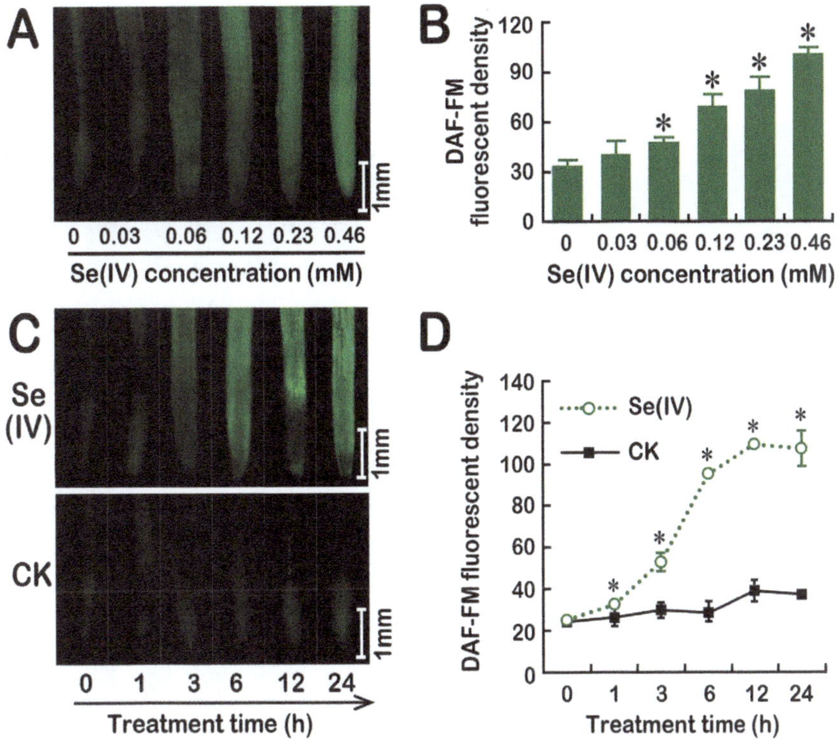

Figure 5. Effect of Se(IV) on the generation of endogenous NO in the roots of *B. rapa*. The roots of seedlings were exposed to 0, 0.03, 0.06, 0.12, 0.23, and 0.46 mM of Se(IV) solution for 48 h. Afterwards, the roots were loaded with DAF-FM DA for 15 min and immediately photographed (A). The relative DAF-FM fluorescent density in roots was estimated (B). (C–D) The image and density of DAF-FM fluorescence were obtained when the roots of seedlings were exposed to 0.06 mM of Se(IV) solution for 0, 1, 3, 6, 12, and 24 h, respectively.

homologues from Arabidopsis as baits for BLAST search. According to the sequence identity and phylogenetic relationship with other plant *Rbohs*, these 12 *Br_Rboh* genes were designated as *Br_RbohA* (Bra009266), *Br_RbohB* (Bra031658), *Br_RbohC* (Bra029194), *Br_RbohD* (Bra020724), *Br_RbohE1* (Bra025721), *Br_RbohE2* (Bra031070), *Br_RbohF* (Bra027764), *Br_RbohG1* (Bra019189), *Br_RbohG2* (Bra019191), *Br_RbohH* (Bra020270), *Br_RbohI* (Bra033151), and *Br_RbohJ* (Bra038274), respectively (Figure S2). The alignment of deduced amino acid sequences revealed that the obtained Br_Rbohs had many typically structural features of plant Rbohs (Figure S3). For instance, two Ca^{2+}-binding EF-hand motifs are located in the N-terminal region of each Br_Rboh. The C-terminal regions of Br_Rbohs contain conserved FAD and NADPH binding sites. Two pairs of histidine residues indicated by blue box are conserved haem binding sites in both human and plant Rboh [51,52].

Next, the expression pattern of *Br_Rbohs* in roots was monitored under 0.06 mM of Se(IV) stress in a time-course experiment using qRT-PCR (Figure S4). The analysis of the fold-change of the relative transcript abundance suggested that most *Br_Rbohs* could be up-regulated in roots under Se(IV) treatment (Figure 4). Three *Br_Rbohs* (*A*, *H*, and *J*) were up-regulated in the early stage upon Se(IV) exposure (6–12 h) (Figure 4). However, the transcription of *Br_RbohC* was stimulated in the late stage upon Se(IV) exposure (48–72 h) (Figure 4). The transcription of four *Br_Rbohs* (*D*, *F*, *G1*, and *G2*) showed up-regulation through almost the whole tested period of Se(IV) exposure (Figure 4). These results indicated that NADPH oxidase-mediated over-generation of ROS might resulted from the up-regulation of *Br_Rbohs* in *B. rapa* under Se(IV) stress.

Se(IV) stress stimulated the generation of endogenous NO in roots

The intracellular of NO was visually detected using NO specific fluorescent probe DAF-FM DA. Compared to the control, Se(IV) treatment stimulated the over-generation of endogenous NO in the root tips in a dose-dependent manner (Figure 5A, B). In a time-course experiment, treatment with 0.06 mM of Se(IV) resulted in the continuous accumulation of endogenous NO after treatment for 1–24 h (Figure 5C, D). The correlation analysis suggested that NO generation was negatively correlated with root length under Se(IV) stress (Figure S1B).

To further ascertain the key enzyme responsible for endogenous NO generation induced by Se(IV), the effects of L-NMMA (NOS inhibitor) and tungstate (NR inhibitor) on the level of endogenous NO in the roots of *B. rapa* under Se(IV) exposure were evaluated. Compared to the control, the content of NO increased considerably under Se(IV) treatment. However, both Se(IV)+L-NMMA and Se(IV)+tungstate significantly reduced the NO abundance as compared to the treatment of Se(IV) alone (Figure 6A, B). The addition of cPTIO, a NO scavenger, could decrease the over-generation of NO induced by Se(IV) as well. The enzymatic assay indicated that treatment with Se(IV) induced the increases in NR and NOS activity as compared to their control groups, respectively (Figure 6C, D). These results suggested that Se(IV) exposure could induce the rapid accumulation of endogenous NO through both NOS- and NR-dependent pathway.

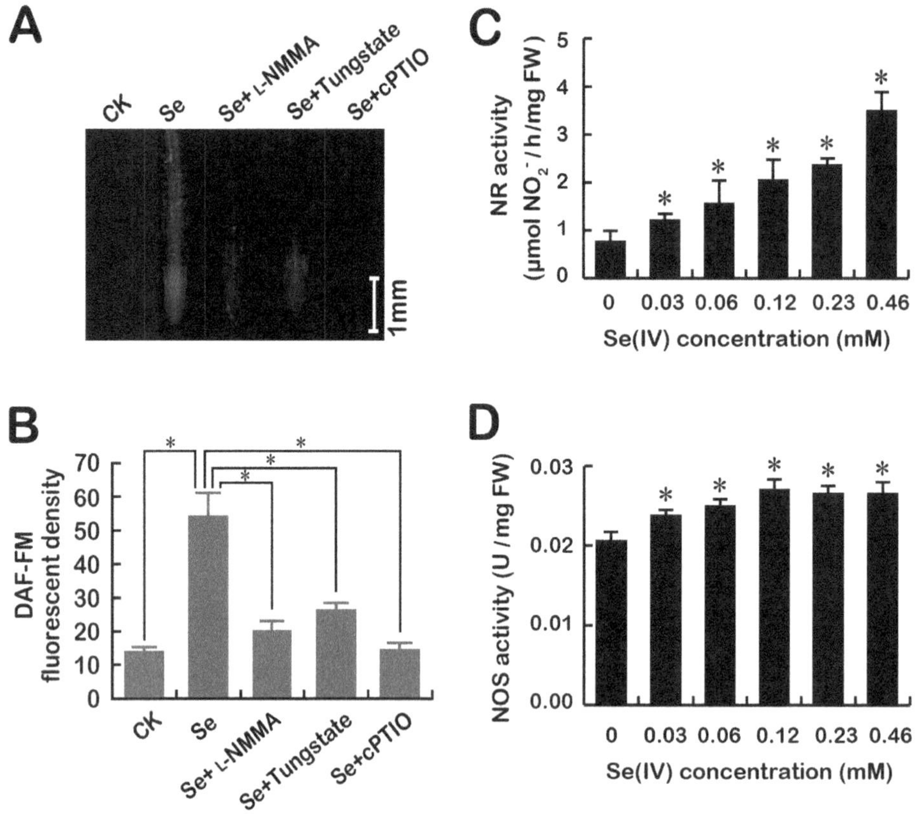

Figure 6. Identification of the sources of Se(IV)-induced NO generation in the roots of *B. rapa*. (A–B) The image and density of DAF-FM fluorescence were obtained when the roots of seedlings were exposed to Se(IV) (0.06 mM), Se(IV) (0.06 mM)+L-NMMA (200 μM), Se(IV) (0.06 mM)+ Tungstate (30 μM), and Se(IV) (0.06 mM)+cPTIO (0.1 mM) for 48 h. *Asterisk* in (B) indicates that mean values of three replicates are significantly different between different treatments ($P<0.05$). The roots of seedlings were exposed to 0, 0.03, 0.06, 0.12, 0.23, and 0.46 mM of Se(IV) solution for 48 h. After that, NR activity (C) and NOS activity (D) in roots were measured, respectively. *Asterisk* in (C) and (D) indicates that mean values of three replicates are significantly different between Se(IV) treatment and the control ($P<0.05$).

Blocking endogenous NO generation reduced Se(IV)-induced ROS accumulation and the expression of *Br_Rbohs* in roots

The correlation analysis suggested that NO generation was postively correlated with ROS accumulation in roots under Se(IV) stress (Figure S1C). In order to verify whether NO is involved in Se(IV)-induced ROS accumulation, the abundance of endogenous ROS in roots was determined when Se(IV)-induced NO generation was scavenged using cPTIO or blocked using L-NMMA and tungstate. The content of total ROS indicated DCF fluorescence increased remarkably upon exposure to 0.06 mM of Se(IV), while the addition of L-NMMA, tungstate, and cPTIO could decrease the accumulation of ROS in roots, respectively (Figure 7A, B). Both L-NMMA and tungstate could prohibited Se(IV)-induced the over-generation of $O_2^{\bullet -}$ by NBT staining indicated by as well (Figure 7C). The analysis of the transcripts of *Br_Rbohs* showed that L-NMMA and tungstate could inhibit Se(IV)-induced increase in the expression of *Br_RbohD, F, G1, G2,* and *I* (Figure 7D), which suggested that both NOS- and NR-dependent NO generation were responsible for Se(IV)-induced ROS accumulation by regulating the expression of several *Br_Rbohs* in roots.

Analysis of the nitric oxide-responsive cis-elements in the promoters of *Br_Rbohs*

According to the identification of NO-responsive *cis*-elements (NREs) from higher plants [53,54], several NREs (e.g. ACGT Box, MYCL, and W-BOX) were obtained from the promoter region of tomato *Br_RbohD, F, G1, G2,* and *I* (Table S2). This result suggested that NO induced the expression of *Br_Rbohs* by possibly regulating these NREs in their promoters.

Blocking endogenous NO generation attenuated Se(IV)-induced oxidative injury and growth inhibition in roots

To obtain more solid evidence for the involvement of NO in Se(IV)-induced phytotoxicity, we determined the effect of the decrease in NO generation on Se(IV)-induced oxidative injury. The lipid peroxidation and loss of membrane integrity induced by Se(IV) could be attenuated by the addition of L-NMMA, tungstate, and cPTIO, respectively (Figure 8A, B). Se(IV)-inhibited root growth could be significantly recovered by the addition of L-NMMA and cPTIO, respectively (Figure 8C).

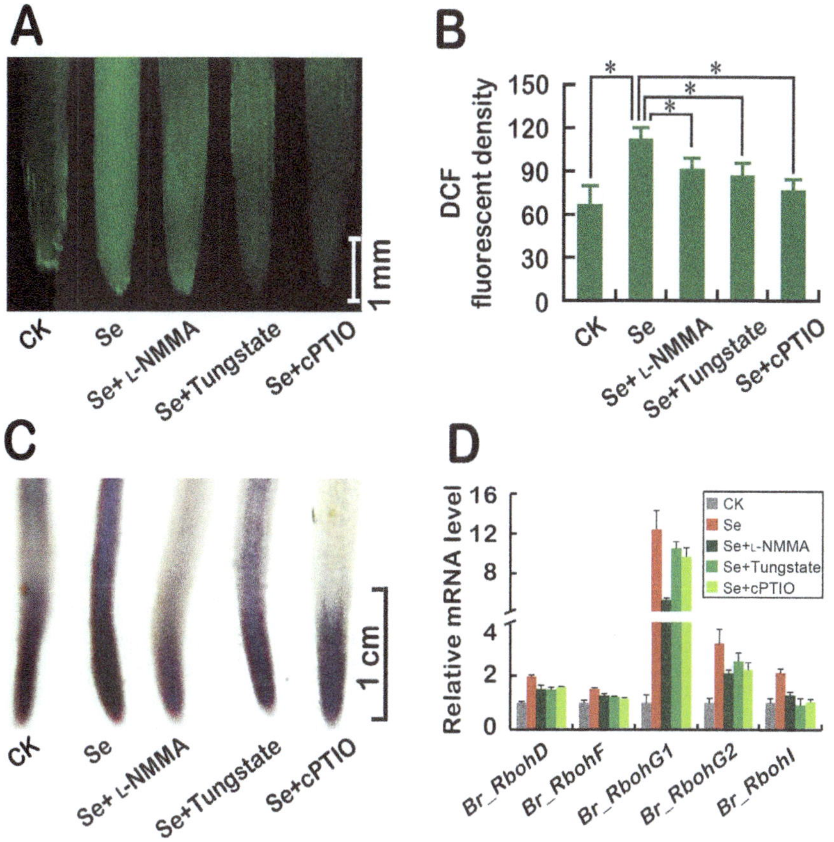

Figure 7. Effects of NO synthesis inhibitors (L-NMMA and tungstate) and NO scavenger (cPTIO) on the generation of endogenous ROS (A–B), $O_2^{\cdot -}$ (C), and the expression of several *Br_Rbohs* (D) in Se(IV)-treated roots of *B. rapa*. The roots of seedlings were exposed to Se(IV) (0.06 mM), Se(IV) (0.06 mM)+L-NMMA (200 μM), Se(IV) (0.06 mM)+Tungstate (30 μM), and Se(IV) (0.06 mM)+cPTIO (0.1 mM) for 72 h. The roots were loaded with DCFH-DA for 15 min and immediately photographed (A). The relative DCF fluorescent density in roots was estimated (B). *Asterisk* indicates that mean values of three replicates are significantly different between the different treatments ($P<0.05$). The roots were stained with NBT for the detection of $O_2^{\cdot -}$ (C). The total RNA was extracted from roots under the above treatments for 24 h for qRT-PCR analysis of the transcripts of *Br_RbohD*, *F*, *G1*, *G2*, and *I*. *Actin* was used for cDNA normalization (D).

Treatment with SNP aggravated Se(IV)-induced phytotoxicity

SNP is a frequently used NO donor. Treatment with SNP alone showed similar inhibitory effect with Se(IV) on root growth. Compared to the Se(IV) treatment alone, simultaneous treatment with Se(IV) and SNP resulted in more inhibitory effect on root growth (Figure 9).

Discussion

Previous studies have demonstrated that Se modifies various physiological process in plants [55,56]. The major adverse effect of Se on plants is the suppression of growth and induction of oxidative stress [16,17,57]. Recently, NO is shown to be a possible mediator of oxidative stress in plants [41]. Whether and how NO triggers heavy metal-induced oxidative stress in plants is little known. In the present study, four lines of evidence indicated that both NOS- and NR-dependent NO generation mediated Se(IV)-induced oxidative stress and growth stunt in the roots of *B. rapa* by regulating a group of *Br_Rbohs* coding for NADPH oxidase. First, Se(IV)-induced ROS accumulation, oxidative stress, and growth inhibition were NADPH oxidase-dependent. Second, the expression of most *Br_Rboh* homologues were up-regulated in *B. rapa* under Se(IV) stress. Third, Se(IV) stimulated both NOS- and NR-dependent generation of endogenous NO. Fourth, blocking NOS- and NR-dependent NO generation reduced Se(IV)-induced ROS accumulation, which further resulted in the alleviation of Se(IV)-induced oxidative injury and growth stunt.

The suppression of plant endogenous NO induced by heavy metals has been extensively reported [58], but the accumulation of NO may also be linked to heavy metal-induced toxicity in plants [34,35,59]. In a recent study, Al-induced growth inhibition of wheat root was dependent on the accumulation of endogenous NO [38]. In our present study, Se(IV) stress induced rapid accumulation of endogenous NO in both NR- and NOS-dependent pathway, which was closely associated with growth inhibition of roots. Besides of NR and NOS, polyamine is suggested to be a potential trigger of NO generation in plants [60]. Since heavy metals and other abiotic stresses (e.g. salt and osmosis) can induce the biosynthesis of polyamine [61], thus it is of interest to further investigate whether and how Se(IV) induces polyamine-based NO generation.

Se(IV) stress induced faster accumulation of NO than that of ROS. The removal of endogenous NO could significantly attenuated Se(IV)-induced ROS accumulation and oxidative injury, suggesting that the ROS accumulation and the following oxidative injury were dependent on the rapid accumulation of endogenous NO in the roots of *B. rapa* under Se(IV) stress. Thus, we established a positive link between NO accumulation and oxidative injury in plants under Se(IV) stress. However, Se(IV)

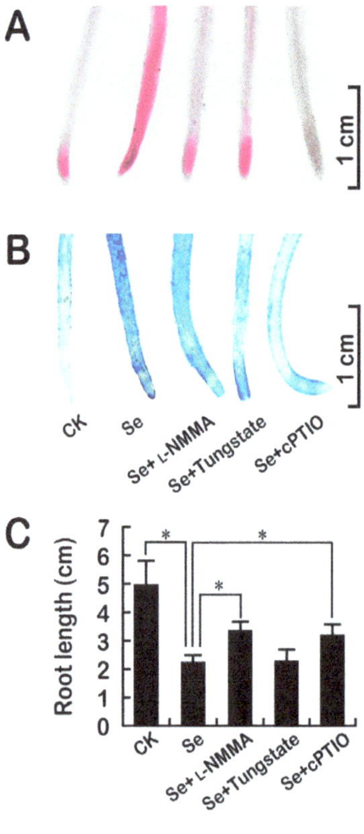

Figure 8. Effects of NO synthesis inhibitors (L-NMMA and tungstate) and NO scavenger (cPTIO) on lipid peroxidation (A), the loss of plasma membrane integrity (B), and root elongation under Se(IV) stress. The roots of seedlings were exposed to Se(IV) (0.06 mM), Se(IV) (0.06 mM)+L-NMMA (200 μM), Se(IV) (0.06 mM)+ Tungstate (30 μM), and Se(IV) (0.06 mM)+cPTIO (0.1 mM) for 72 h. Afterwards, the roots were stained with Schiff's reagent (A) and Evan blue (B), respectively. The root length was measured as well (C). *Asterisk* indicates that mean values of three replicates are significantly different between the different treatments ($P<0.05$).

stress can inhibit endogenous NO generation in the root of *Arabidopsis*, which further impacted plant auxin signalling [62]. Therefore, NO seems to play dual roles in regulating plant adaptions upon Se(IV) stress by modulating either hormone signalling or oxidative stress, which may be involved complicated regulatory mechanisms that needs to be illuminated further.

Plant NADPH oxidases have been considered as one of the important enzymatic complexes to deliberately generate ROS [63]. In the present study, pharmacological experiment suggested that Se(IV) induced oxidative injury in *B. rapa* through NADPH oxidase-dependent ROS accumulation. Se(IV)-induced root inhibition could be partially recovered by the addition of NADPH oxidase inhibitors, suggesting that NADPH oxidase-dependent ROS accumulation contributed to Se toxicity. *RbohD* has been closely linked to ROS generation in plants under copper (Cu), Cd, salt, and wound stress [64–67]. In *Zea mays*, abscisic acid (ABA)-triggered ROS generation has been linked to the *RbohA-D* [51]. *RbohF*-dependent ROS generation has been associated with pathogen recognition in multiple plant species [63]. In the present study, the induction of *Br_Rbohs* expression by NO under Se(IV) stress may resulted from the presence of NREs in the promoter regions of *Br_Rbohs*. The differential regulation of plant *Rbohs* by Se(IV) and other environmental stimuli may results from the diversity of upstream regulators of *Rbohs*.

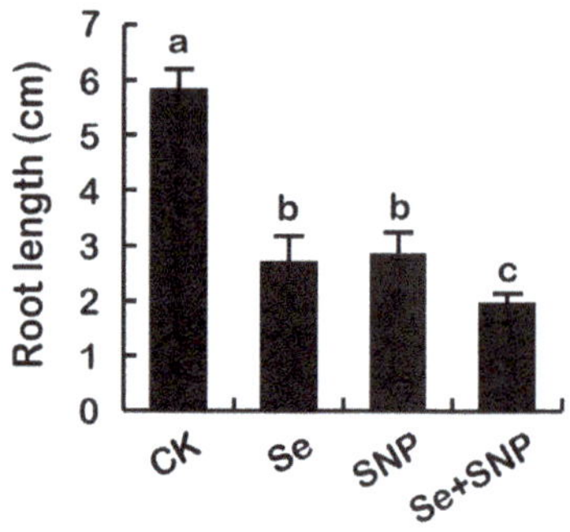

Figure 9. Effect of SNP on the root growth under Se(IV) stress. The roots of seedlings were exposed to Se(IV) (0.06 mM), SNP (250 mM), and Se(IV) (0.06 mM)+SNP (250 mM) for 72 h. After that, the length of roots were measured. The mean values of three replicates followed by the same letter were not significantly different at $P<0.05$.

In mammals, NO tends to lower the expression of *Nox1/2*, homologues of *Rboh* in human cells, leading to the inhibition of ROS production [68,69]. Our present study found the opposite results that NO could stimulate the expression of *Br_Rbohs*, leading to the over-generation of ROS in *B. rapa* under Se(IV) stress. Heavy metal-modulated phosphatidic acid (PA) and mitogen-activated protein kinase (MAPK) signalling cassettes have been extensively reported in plants [70,71]. PA and ABA are the well-characterized upstream regulators of *RbohD* and *RobhF* in *Arabidopsis* [72,73]. In pathogen defensive responses, NO induces ROS burst through PA signalling in plants [74]. In *Z. mays*, ABA-induced NO generation is indispensable for the activation of MAPK5 that positively regulates the expression of *Zmrbohs* for ROS production [51,75]. Thus, whether Se(IV)-triggered NO generation regulates the expression of *Br_Rbohs* through PA or MAPK signalling remains to be further illuminated.

Conclusions

The anthropogenic release of excessive Se gives rise to the increase in Se concentration in agricultural environment, which further shows adverse effect on crop growth. Here, we demonstrate that Se(IV)-induced oxidative injury resulted from NADPH oxidase-dependent ROS generation contributes to the growth inhibition of *B. rapa* roots. Endogenous NO seems to stimulate ROS generation by stimulating the expression of several *Br_Rbohs* coding for NADPH oxidase. These results not only propose the toxic role of NO in mediating heavy metal-induced oxidative stress, but also shed new light on the investigation of Se(IV)-induced phytotoxicity. Further identification of plant responses to Se stress will be helpful for crop breeding designed to improve plant tolerance to Se-contaminated environment.

Supporting Information

Figure S1 Correlation analysis among root growth, ROS accumulation, and NO accumulation in roots under Se(IV) stress at 0, 0.03, 0.06, 0.12, 0.23, and 0.46 mM. (A) Correlation between root length and ROS accumulation indicated by DCF fluorescence. (B) Correlation between root length and NO accumulation indicated by DAF-FM fluorescence. (C) Correlation between ROS accumulation indicated by DCF fluorescence and NO accumulation indicated by DAF-FM fluorescence.

Figure S2 The phylogenetic relationship of Br_RbohA-J and their related members of Rboh family. Species name: Aet, *Aegilops tauschii*; At, *Arabidopsis thaliana*, Bo, *Brassica oleracea*; Br, *Brassica rapa*; Ls, *Lepidium sativum*; Nt, *Nicotiana tabacum*; OsI, *Oryza sativa* Indica Group; OsJ, *Oryza sativa* Japonica Group; St, *Solanum tuberosum*; Tu, *Triticum urartu*.

Figure S3 Alignment of the predicted amino acid sequences of Br_RbohA-J. Multiple alignment of predicted Br_RbohA-J protein was made with Clustal Mega. *Dark shading* and *dull grey shading* reveal 100% and 75% sequence conservation, respectively. The *red line* indicates the conserved functional domains of NADPH oxidase, such as EF hand I and II, FAD-isoalloxazine site, NADPH-ribose site, and NADPH-adenine site. In two EF hands, dashes indicate variable amino acid residues. X, Y, Z, and –X, contain oxygen within their side chains. –Z is usually glutamic acid. *Blue box* indicates histidine residues involved in haem binding.

Figure S4 Effect of Se(IV) on the transcripts of *Br_RbohA-J*. The roots of seedlings were exposed to 0.06 mM of Se(IV) solution for 6, 12, 24, 48, and 72 h, respectively. The total RNA was extracted from roots for qRT-PCR analysis. *Actin* was used for cDNA normalization.

Table S1 Sequences of oligonucleotide primers for qRT-PCR. F: forward; R: reverse.

Table S2 Distribution of NO-responsive *cis*-elements (AGCT Box, MYCL, and W-BOX) in the promote region of *Br_RbohD*, *F*, *G1*, *G2*, and *I* in *B. rapa*. The plant motifs were predicted based on the publicly available *cis*-acting regulatory elements database PLACE (http://bioinformatics.psb.ugent.be/webtools/plantcare/html/) as a reference. Forward sequence is indicated as (+) while the complementary sequence is indicated as (−).

Author Contributions

Conceived and designed the experiments: JC. Performed the experiments: YC H-ZM L-BH. Contributed reagents/materials/analysis tools: YC Y-QL. Wrote the paper: JC YC L-FY.

References

1. Zhu YG, Pilon-Smits EA, Zhao FJ, Williams PN, Meharg AA (2009) Selenium in higher plants: understanding mechanisms for biofortification and phytoremediation. Trends Plant Sci 14: 436–442.
2. Silva M, Arruda M (2012) Identification of selenium in the leaf protein of sunflowers by a combination of 2D-PAGE and laser ablation ICP-MS. Microchimica Acta 176: 131–136.
3. Fang Y, Catron B, Zhang Y, Zhao L, Caruso JA, et al. (2010) Distribution and in vitro availability of selenium in selenium-containing storage protein from selenium-enriched rice utilizing optimized extraction. J Agric Food Chem 58: 9731–9738.
4. Sajedi NA, Ardakani MR, Madani H, Naderi A, Miransari M (2011) The effects of selenium and other micronutrients on the antioxidant activities and yield of corn (Zea mays L.) under drought stress. Physiol Mol Biol Plants 17: 215–222.
5. Hasanuzzaman M, Hossain MA, Fujita M (2011) Selenium-induced up-regulation of the antioxidant defense and methylglyoxal detoxification system reduces salinity-induced damage in rapeseed seedlings. Biol Trace Elem Res 143: 1704–1721.
6. Malik JA, Goel S, Kaur N, Sharma S, Singh I, et al. (2012) Selenium antagonises the toxic effects of arsenic on mungbean (*Phaseolus aureus* Roxb.) plants by restricting its uptake and enhancing the antioxidative and detoxification mechanisms. Environ Exp Bot 77: 242–248.
7. Cartes P, Jara AA, Pinilla L, Rosas A, Mora ML (2010) Selenium improves the antioxidant ability against aluminium-induced oxidative stress in ryegrass roots. Ann Appl Biol 156: 297–307.
8. Yao X, Chu J, He X, Liu B, Li J, et al. (2013) Effects of selenium on agronomical characters of winter wheat exposed to enhanced ultraviolet-B. Ecotoxicol Environ Saf 92: 320–326.
9. Djanaguiraman M, Prasad PV, Seppanen M (2010) Selenium protects sorghum leaves from oxidative damage under high temperature stress by enhancing antioxidant defense system. Plant Physiol Biochem 48: 999–1007.
10. Kumar M, Bijo AJ, Baghel RS, Reddy CR, Jha B (2012) Selenium and spermine alleviate cadmium induced toxicity in the red seaweed *Gracilaria dura* by regulating antioxidants and DNA methylation. Plant Physiol Biochem 51: 129–138.
11. Winkel LHE, Johnson CA, Lenz M, Grundl T, Leupin OX, et al. (2011) Environmental selenium research: from microscopic processes to global understanding. Environ Sci Technol 46: 571–579.
12. Umysova D, Vitova M, Douskova I, Bisova K, Hlavova M, et al. (2009) Bioaccumulation and toxicity of selenium compounds in the green alga *Scenedesmus quadricauda*. BMC Plant Biol 9: 58.
13. Molnárová M, Fargašová A (2009) Se(IV) phytotoxicity for monocotyledonae cereals (*Hordeum vulgare* L., *Triticum aestivum* L.) and dicotyledonae crops (*Sinapis alba* L., *Brassica napus* L.). J Hazard Mater 172: 854–861.
14. Hladun KR, Parker DR, Tran KD, Trumble JT (2013) Effects of selenium accumulation on phytotoxicity, herbivory, and pollination ecology in radish (*Raphanus sativus* L.). Environ Pollut 172: 70–75.
15. Schiavon M, Moro I, Pilon-Smits EA, Matozzo V, Malagoli M, et al. (2012) Accumulation of selenium in *Ulva* sp. and effects on morphology, ultrastructure and antioxidant enzymes and metabolites. Aquat Toxicol 122–123: 222–231.
16. Paciolla C, De Leonardis S, Dipierro S (2011) Effects of selenite and selenate on the antioxidant systems in *Senecio scandens* L. Plant Biosyst 145: 253–259.
17. Aggarwal M, Sharma S, Kaur N, Pathania D, Bhandhari K, et al. (2011) Exogenous proline application reduces phytotoxic effects of selenium by minimising oxidative stress and improves growth in bean (*Phaseolus vulgaris* L.) seedlings. Biol Trace Elem Res 140: 354–367.
18. Suzuki N, Miller G, Morales J, Shulaev V, Torres MA, et al. (2011) Respiratory burst oxidases: the engines of ROS signaling. Curr Opin Plant Biol 14: 691–699.
19. Torres MA, Dangl JL (2005) Functions of the respiratory burst oxidase in biotic interactions, abiotic stress and development. Curr Opin Plant Biol 8: 397–403.
20. Sharma SS, Dietz K-J (2009) The relationship between metal toxicity and cellular redox imbalance. Trends Plant Sci 14: 43–50.
21. Groppa MD, Ianuzzo MP, Rosales EP, Vázquez SC, Benavides MP (2012) Cadmium modulates NADPH oxidase activity and expression in sunflower leaves. Biol Plant 56: 167–171.
22. Źróbek-Sokolnik A, Asard H, Górska-Koplińska K, Górecki R (2009) Cadmium and zinc-mediated oxidative burst in tobacco BY-2 cell suspension cultures. Acta Physiol Plant 31: 43–49.
23. Hasanuzzaman M, Gill S, Fujita M (2013) Physiological role of nitric oxide in plants grown under adverse environmental conditions. In: Tuteja N, Singh Gill S, editors. Plant Acclimation to Environmental Stress: Springer New York. pp. 269–322.
24. Mur LAJ, Mandon J, Persijn S, Cristescu SM, Moshkov IE, et al. (2013) Nitric oxide in plants: an assessment of the current state of knowledge. AoB Plants 5: pls052.
25. Neill S, Bright J, Desikan R, Hancock J, Harrison J, et al. (2008) Nitric oxide evolution and perception. J Exp Bot 59: 25–35.
26. Gas E, Flores-Perez U, Sauret-Gueto S, Rodriguez-Concepcion M (2009) Hunting for plant nitric oxide synthase provides new evidence of a central role for plastids in nitric oxide metabolism. Plant Cell 21: 18–23.
27. Fröhlich A, Durner J (2011) The hunt for plant nitric oxide synthase (NOS): is one really needed? Plant Sci 181: 401–404.
28. Wang YS, Yang ZM (2005) Nitric oxide reduces aluminum toxicity by preventing oxidative stress in the roots of Cassia tora L. Plant Cell Physiol 46: 1915–1923.
29. Aftab T, Khan MMA, Naeem M, Idrees M, Moinuddin, et al. (2012) Exogenous nitric oxide donor protects *Artemisia annua* from oxidative stress generated by boron and aluminium toxicity. Ecotoxicol Environ Saf 80: 60–68.
30. Rodríguez-Serrano M, Romero-Puertas MC, Pazmiño DM, Testillano PS, Risueño MC, et al. (2009) Cellular response of pea plants to cadmium toxicity: cross talk between reactive oxygen species, nitric oxide, and calcium. Plant Physiol 150: 229–243.
31. Srivastava S, Dubey RS (2012) Nitric oxide alleviates manganese toxicity by preventing oxidative stress in excised rice leaves. Acta Physiol Plant 34: 819–825.
32. Verma K, Mehta SK, Shekhawat GS (2013) Nitric oxide (NO) counteracts cadmium induced cytotoxic processes mediated by reactive oxygen species (ROS) in *Brassica juncea*: cross-talk between ROS, NO and antioxidant responses. Biometals 26: 255–269.
33. Li L, Wang Y, Shen W (2012) Roles of hydrogen sulfide and nitric oxide in the alleviation of cadmium-induced oxidative damage in alfalfa seedling roots. Biometals 25: 617–631.

34. De Michele R, Vurro E, Rigo C, Costa A, Elviri L, et al. (2009) Nitric oxide is involved in cadmium-induced programmed cell death in Arabidopsis suspension cultures. Plant Physiol 150: 217–228.
35. Besson-Bard A, Gravot A, Richaud P, Auroy P, Duc C, et al. (2009) Nitric oxide contributes to cadmium toxicity in Arabidopsis by promoting cadmium accumulation in roots and by up-regulating genes related to iron uptake. Plant Physiol 149: 1302–1315.
36. Gallego SM, Pena LB, Barcia RA, Azpilicueta CE, Iannone MF, et al. (2012) Unravelling cadmium toxicity and tolerance in plants: Insight into regulatory mechanisms. Environ Exp Bot 83: 33–46.
37. Leterrier M, Airaki M, Palma JM, Chaki M, Barroso JB, et al. (2012) Arsenic triggers the nitric oxide (NO) and S-nitrosoglutathione (GSNO) metabolism in *Arabidopsis*. Environ Pollut 166: 136–143.
38. Chen M, Cui W, Zhu K, Xie Y, Zhang C, et al. (2014) Hydrogen-rich water alleviates aluminum-induced inhibition of root elongation in alfalfa via decreasing nitric oxide production. J Hazard Mater 267: 40–47.
39. Jiang M, Zhang J (2003) Cross-talk between calcium and reactive oxygen species originated from NADPH oxidase in abscisic acid-induced antioxidant defence in leaves of maize seedlings. Plant Cell Environ 26: 929–939.
40. Hu X, Wang W, Li C, Zhang J, Lin F, et al. (2008) Cross-talks between Ca^{2+}/CaM and H_2O_2 in abscisic acid-induced antioxidant defense in leaves of maize plants exposed to water stress. Plant Growth Regul 55: 183–198.
41. Chen J, Zhong YM, Zhang HQ, Shi ZQ (2012) Nitrate reductase-dependent nitric oxide production is involved in microcystin-LR-induced oxidative stress in *Brassica rapa*. Water, Air, & Soil Pollution 223: 4141–4152.
42. Sun C, Lu L, Liu L, Liu W, Yu Y, et al. (2013) Nitrate reductase-mediated early nitric oxide burst alleviates oxidative damage induced by aluminum through enhancement of antioxidant defenses in roots of wheat (*Triticum aestivum*). New Phytol: n/a-n/a.
43. Chen J, Zhang H-Q, Hu L-B, Shi Z-Q (2013) Microcystin-LR-induced phytotoxicity in rice crown root is associated with the cross-talk between auxin and nitric oxide. Chemosphere 93: 283–293.
44. Foreman J, Demidchik V, Bothwell JH, Mylona P, Miedema H, et al. (2003) Reactive oxygen species produced by NADPH oxidase regulate plant cell growth. Nature 422: 442–446.
45. Guo K, Xia K, Yang ZM (2008) Regulation of tomato lateral root development by carbon monoxide and involvement in auxin and nitric oxide. J Exp Bot 59: 3443–3452.
46. Frahry G, Schopfer P (2001) NADH-stimulated, cyanide-resistant superoxide production in maize coleoptiles analyzed with a tetrazolium-based assay. Planta 212: 175–183.
47. Yamamoto Y, Kobayashi Y, Matsumoto H (2001) Lipid peroxidation is an early symptom triggered by aluminum, but not the primary cause of elongation inhibition in pea roots. Plant Physiol 125: 199–208.
48. Xiong J, Lu H, Lu K, Duan Y, An L, et al. (2009) Cadmium decreases crown root number by decreasing endogenous nitric oxide, which is indispensable for crown root primordia initiation in rice seedlings. Planta 230: 599–610.
49. Li JL, Jiang CY, Li S, Xu SW (2013) Cadmium induced hepatotoxicity in chickens (*Gallus domesticus*) and ameliorative effect by selenium. Ecotoxicol Environ Saf 96: 103–109.
50. Dupuy L, Gregory PJ, Bengough AG (2010) Root growth models: towards a new generation of continuous approaches. J Exp Bot 61: 2131–2143.
51. Lin F, Ding H, Wang J, Zhang H, Zhang A, et al. (2009) Positive feedback regulation of maize NADPH oxidase by mitogen-activated protein kinase cascade in abscisic acid signalling. J Exp Bot 60: 3221–3238.
52. Wong HL, Pinontoan R, Hayashi K, Tabata R, Yaeno T, et al. (2007) Regulation of rice NADPH oxidase by binding of Rac GTPase to its N-terminal extension. Plant Cell 19: 4022–4034.
53. Palmieri MC, Sell S, Huang X, Scherf M, Werner T, et al. (2008) Nitric oxide-responsive genes and promoters in *Arabidopsis thaliana*: a bioinformatics approach. J Exp Bot 59: 177–186.
54. Gomez-Ros LV, Gabaldon C, Lopez Nunez-Flores MJ, Gutierrez J, Herrero J, et al. (2012) The promoter region of the *Zinnia elegans* basic peroxidase isoenzyme gene contains *cis*-elements responsive to nitric oxide and hydrogen peroxide. Planta 236: 327–342.
55. Watanabe M, Hubberten H-M, Hoefgen R (2012) Plant Response to Mineral Ion Availability: Transcriptome Responses to Sulfate, Selenium and Iron. In: De Kok LJ, Tabe L, Tausz M, Hawkesford MJ, Hoefgen R et al., editors. Sulfur Metabolism in Plants: Springer Netherlands. pp. 123–134.
56. Schiavon M, Pittarello M, Pilon-Smits EAH, Wirtz M, Hell R, et al. (2012) Selenate and molybdate alter sulfate transport and assimilation in *Brassica juncea* L. Czern.: Implications for phytoremediation. Environ Exp Bot 75: 41–51.
57. Seppänen M, Turakainen M, Hartikainen H (2003) Selenium effects on oxidative stress in potato. Plant Sci 165: 311–319.
58. Xiong J, Fu G, Tao L, Zhu C (2010) Roles of nitric oxide in alleviating heavy metal toxicity in plants. Arch Biochem Biophys 497: 13–20.
59. Zhou Y, Xu XY, Chen LQ, Yang JL, Zheng SJ (2012) Nitric oxide exacerbates Al-induced inhibition of root elongation in rice bean by affecting cell wall and plasma membrane properties. Phytochemistry 76: 46–51.
60. Tun NN, Santa-Catarina C, Begum T, Silveira V, Handro W, et al. (2006) Polyamines induce rapid biosynthesis of nitric oxide (NO) in *Arabidopsis thaliana* seedlings. Plant Cell Physiol 47: 346–354.
61. Sharma SS, Dietz K-J (2006) The significance of amino acids and amino acid-derived molecules in plant responses and adaptation to heavy metal stress. J Exp Bot 57: 711–726.
62. Lehotai N, Kolbert Z, Pető A, Feigl G, Ördög A, et al. (2012) Selenite-induced hormonal and signalling mechanisms during root growth of *Arabidopsis thaliana* L. J Exp Bot 63: 5677–5687.
63. Marino D, Dunand C, Puppo A, Pauly N (2012) A burst of plant NADPH oxidases. Trends Plant Sci 17: 9–15.
64. Remans T, Opdenakker K, Smeets K, Mathijsen D, Vangronsveld J, et al. (2010) Metal-specific and NADPH oxidase dependent changes in lipoxygenase and NADPH oxidase gene expression in *Arabidopsis thaliana* exposed to cadmium or excess copper. Funct Plant Biol 37: 532–544.
65. Jakubowicz M, Gałgańska H, Nowak W, Sadowski J (2010) Exogenously induced expression of ethylene biosynthesis, ethylene perception, phospholipase D, and Rboh-oxidase genes in broccoli seedlings. J Exp Bot 61: 3475–3491.
66. Xie YJ, Xu S, Han B, Wu MZ, Yuan XX, et al. (2011) Evidence of *Arabidopsis* salt acclimation induced by up-regulation of HY1 and the regulatory role of RbohD-derived reactive oxygen species synthesis. Plant J.
67. Miller G, Schlauch K, Tam R, Cortes D, Torres MA, et al. (2009) The plant NADPH oxidase RBOHD mediates rapid systemic signaling in response to diverse stimuli. Sci Signal 2: ra45.
68. Harrison CB, Drummond GR, Sobey CG, Selemidis S (2010) Evidence that nitric oxide inhibits vascular inflammation and superoxide production via a p47phox-dependent mechanism in mice. Clin Exp Pharmacol Physiol 37: 429–434.
69. Brandes RP, Weissmann N, Schröder K (2010) NADPH oxidases in cardiovascular disease. Free Radic Biol Med 49: 687–706.
70. Opdenakker K, Remans T, Vangronsveld J, Cuypers A (2012) Mitogen-Activated Protein (MAP) Kinases in plant metal stress: regulation and responses in comparison to other biotic and abiotic stresses. Int J Mol Sci 13: 7828–7853.
71. Smékalová V, Doskočilová A, Komis G, Šamaj J (2013) Crosstalk between secondary messengers, hormones and MAPK modules during abiotic stress signalling in plants. Biotechnol Adv 32: 2–11.
72. Zhang Y, Zhu H, Zhang Q, Li M, Yan M, et al. (2009) Phospholipase Dα1 and phosphatidic acid regulate NADPH oxidase activity and production of reactive oxygen species in aba-mediated stomatal closure in *Arabidopsis*. Plant Cell 21: 2357–2377.
73. Kwak JM, Mori IC, Pei ZM, Leonhardt N, Torres MA, et al. (2003) NADPH oxidase AtrbohD and AtrbohF genes function in ROS-dependent ABA signaling in *Arabidopsis*. EMBO J 22: 2623–2633.
74. Bellin D, Asai S, Delledonne M, Yoshioka H (2012) Nitric oxide as a mediator for defense responses. Mol Plant-Microbe Interact 26: 271–277.
75. Zhang A, Jiang M, Zhang J, Ding H, Xu S, et al. (2007) Nitric oxide induced by hydrogen peroxide mediates abscisic acid-induced activation of the mitogen-activated protein kinase cascade involved in antioxidant defense in maize leaves. New Phytol 175: 36–50.

5

Inspection of the Grapevine BURP Superfamily Highlights an Expansion of *RD22* Genes with Distinctive Expression Features in Berry Development and ABA-Mediated Stress Responses

José Tomás Matus[1,2], Felipe Aquea[2], Carmen Espinoza[2], Andrea Vega[3], Erika Cavallini[4], Silvia Dal Santo[4], Paola Cañón[2], Amparo Rodríguez-Hoces de la Guardia[2], Jennifer Serrano[2], Giovanni Battista Tornielli[4], Patricio Arce-Johnson[2]*

1 Center for Research in Agricultural Genomics CSIC-IRTA-UAB-UB, Bellaterra, Barcelona, Spain, **2** Departamento de Genética Molecular y Microbiología, Pontificia Universidad Católica de Chile, Santiago, Chile, **3** Departamento de Ciencias Vegetales, Facultad de Agronomía e Ingeniería Forestal, Pontificia Universidad Católica de Chile, Santiago, Chile, **4** Department of Biotechnology, University of Verona, Verona, Italy

Abstract

The *RESPONSIVE TO DEHYDRATION 22* (*RD22*) gene is a molecular link between abscisic acid (ABA) signalling and abiotic stress responses. Its expression has been used as a reliable ABA early response marker. In Arabidopsis, the single copy *RD22* gene possesses a BURP domain also located at the C-terminus of USP embryonic proteins and the beta subunit of polygalacturonases. In grapevine, a *RD22* gene has been identified but putative paralogs are also found in the grape genome, possibly forming a large *RD22* family in this species. In this work, we searched for annotations containing BURP domains in the *Vitis vinifera* genome. Nineteen proteins were defined by a comparative analysis between the two genome predictions and RNA-Seq data. These sequences were compared to other plant BURPs identified in previous genome surveys allowing us to reconceive group classifications based on phylogenetic relationships and protein motif occurrence. We observed a lineage-specific evolution of the RD22 family, with the biggest expansion in grapevine and poplar. In contrast, rice, sorghum and maize presented highly expanded monocot-specific groups. The Vitis RD22 group may have expanded from segmental duplications as most of its members are confined to a region in chromosome 4. The inspection of transcriptomic data revealed variable expression of BURP genes in vegetative and reproductive organs. Many genes were induced in specific tissues or by abiotic and biotic stresses. Three *RD22* genes were further studied showing that they responded oppositely to ABA and to stress conditions. Our results show that the inclusion of RNA-Seq data is essential while describing gene families and improving gene annotations. Robust phylogenetic analyses including all BURP members from other sequenced species helped us redefine previous relationships that were erroneously established. This work provides additional evidence for *RD22* genes serving as marker genes for different organs or stresses in grapevine.

Editor: Heping Cao, USDA-ARS, United States of America

Funding: This work was supported by the Chilean Wine Consortium 05CTE01-03, the Fruit Consortium, 07Genoma01 and Millennium Nucleus for Plant Functional Genomics (P06-009-F). FA was supported by a Postdoctoral Project "Programa Bicentenario de Ciencia y Tecnología/CONICYT-Banco Mundial" (PSD74-2006). AV, PC and JS were supported by a PhD fellowship from CONICYT, Chilean National Council of Science and Technology and VRAID, Pontificia Universidad Católica de Chile. This work also benefited from the networking activities within the European-funded COST ACTION FA1106 'An integrated systems approach to determine the developmental mechanisms controlling fleshy fruit quality in tomato and grapevine'. The funders had no role in study design, data collection and analysis, decision to publish, or preparation of the manuscript.

Competing Interests: The fact the authors received funding from professional associations "Chilean Wine Consortium 05CTE01-03" and "the Fruit Consortium" does not alter their adherence to PLOS ONE policies on sharing data and materials.

* Email: parce@bio.puc.cl

Introduction

A plant's adaptive response to overcome any abiotic or biotic stress requires the initiation of various biochemical and physiological measures, which will ultimately allow the organism to survive. Many advances in understanding plant adaptation to abiotic stress have arisen from the study of the phytohormone abscisic acid (ABA), which accumulates under cold, salinity and drought [1]. Changes in its concentration can lead to a number of adaptations including stomatal closure [2], growth inhibition and senescence or flowering induction, all of which can be regulated at a transcriptional level [3]. These events comprise changes in the expression of hundreds of genes that are influenced by the degree, extent and rate of each stress [4].

Genes induced by water-deficit stress belong to different functional categories, as reviewed by Bray [5], such as cell signalling, rescue and detoxification, cell maintenance, pathogenesis-related functions, solute and water relocation. Several genes which respond to dehydration (named *RD*) have been isolated in *Arabidopsis thaliana* [6]. The *RESPONSIVE TO DEHYDRATION 22* (*RD22*) protein is induced by an ABA-dependent signalling mechanism, requiring *de novo* protein synthesis [7]. The Arabidopsis MYC2 (also known as RD22-BP1) and MYB2 transcription factors bind cis-elements in the *RD22* promoter and cooperatively activate its transcription in response to drought, salinity and exposure to exogenous ABA [8].

The high and rapid inducible expression of *RD22* genes provides strong evidence for their involvement in stress tolerance. Over-expression of both AtMYC2 and AtMYB2, causes an ABA-hypersensitive phenotype, improves the osmotic-stress tolerance of transgenic plants and accelerates *RD22* expression [9]. Wang et al [10] showed that the soybean *Glycine max* GmRD22 protein could directly improve stress tolerance when overexpressed in rice and it was also able to interact with an apoplastic peroxidase linked to lignin biosynthesis and cell wall strengthening in response to salt stress. Wang et al [11] reported how the *Gossypium arboreum GaRDL1* gene was transactivated by the GLABRA1-like Ga-MYB2 and had a role in cotton fiber production. More recently, another cotton RD22 ortholog from *Gossypium hirsutum (GhRDL1)* was shown to interact with an α-expansin, promoting seed mass and fiber (seed trichome) length [12].

RD22 proteins possess a BURP domain, found at the C-terminus of several other plant proteins such as USP embryonic abundant and polygalacturonase proteins (BURP: BNM2, USP, RD22, PG1β). Members of the BURP superfamily share some primary structural features, subcellular localization patterns (e.g. cell wall matrix) and possible mechanistic similarities. BURP proteins possess different modules in addition to the BURP domain: a hypothetical transit peptide (N-terminal hydrophobic region), a short segment and a segment of repeated motifs that are unique to each family.

BURP genes have been recently genome-wide identified and related to abiotic stress tolerance in several plant species. In mangrove, four genes encoding BURP domain-containing proteins (*BgBDC1, 2, 3,* and *4*) were all induced by salt, ABA, and drought stress [13]. In rice, *OsBURP03*, *OsBURP05* and *OsBURP17* were induced by at least one abiotic stress treatment, with ABA-dependent and independent pathways involved [14]. Recently, *BURP* genes were identified in soybean [15], maize [16], sorghum [16] and poplar [17], showing that their expression is differentially responsive to ABA and ABA-related stress conditions.

Grapevine is an interesting model for studying drought and ABA signalling responses as the commercial production of this species is usually controlled by regulated-deficit irrigation regimes. In addition, fruit ripening in this species is associated with a short and rapid increase in ABA synthesis (reviewed by Kuhn et al [18]). A grape *RD22* gene was identified [19], which is constitutively expressed at low levels in all tissues. Nevertheless, its expression was induced by drought and salt stress, ABA and sugar. In addition to this gene, other reports have shown the possible existence of other *RD22* genes in grape. This evidence comes from microarray experiments in different grapevine organs, such as shoot tips [20], berries [21] and virus-infected leaves [22]. However, these studies focused on understanding global transcriptomic networks responding to stress, and individual genes were not isolated or characterized. Finally, no previous studies have assessed the dimension of the Vitis BURP domain superfamily. In this work, we characterized the grapevine BURP superfamily by conducting *in silico* phylogenetic and transcriptomic analyses. Some members from the RD22 family were isolated and their expression profiles were studied in different conditions for testing them as putative marker genes for organs or different stresses in grapevine.

Methods

Search for BURP-domain containing homologues in the grape genome

An approximately 230 amino acid consensus BURP domain sequence was obtained from the alignment of Arabidopsis and rice BURP-domain proteins. This consensus was used in a BLAT search to identify homologous gene models in the Genoscope Grape 8X Genome Browser [23] and CRIBI's 12X V1 prediction (http://genomes.cribi.unipd.it/). Since a variable number of gene models were obtained from each genome version, we compared these annotations with previously published RNA-Seq data [24,25] with the use of alignment and contig assembly tools in Vector NTI v9 (Invitrogen). Nineteen proteins were defined and deduced by manual editing based on the Genoscope and CRIBI annotations, RNA-Seq data and the comparisons with corresponding expressed sequence tags and deduced protein sequences from paralogous genes.

Phylogeny reconstruction and bootstrap analysis

Grape BURPs were aligned against the full predicted amino acid sequences of proteins belonging to Arabidopsis thaliana, Populus trichocarpa, Glycine max, Vicia faba, Gossypium hirsutum, Gossypium arboreum, Brassica napus, Zea mays, Sorghum bicolour and Oryza sativa. Alignments were performed using the MUSCLE algorithm-based AlignX module from MEGA5 software [26]. Phylogenetic trees were constructed using the Neighbour Joining Tree, Maximum Parsimony and Maximum Likelihood methods in MEGA5 and computed using the "p-distance" and "no difference" methods, with uniform rates among sites and partial deletion gap treatment. The trees obtained were graphed in MEGA5 and FigTree. Tree nodes were evaluated by bootstrap analysis for 100 replicates. For the construction of the complete BURP tree, the RD29 protein was used as outgroup.

Identification of conserved protein motifs

The online MEME (Multiple EM for Motif Elicitation) Suite was employed to analyze the protein sequences of BURP members from different plant species with an expected value lower than 2×10^{-30} (http://meme.nbcr.net/; [27]). Seventy-three sequences were screened, excluding the Polygalacturonase family since it is the most divergent outside the BURP domain.

Clustering analyses of transcriptomic data

As a first approach, the expression profiles of grapevine BURP genes were assessed in the global *V. vinifera* cv Corvina (clone 48) gene expression ATLAS (Nimblegen platform) of different organs at various developmental stages [28]. The expression data were analyzed using T-MeV v4.81 [29], The fluorescence intensity values of each transcript in all tissues/organs were calculated as log2 and normalized, based on the median center genes/rows adjustment in order to generate a clustered heat map such as the ones generated by Dal Santo et al [30].

Secondly, we searched for grape Affymetrix microarray public data in PLEXdb (Plant Expression Database) [31]. Probe sets corresponding to the putative *VvBURP* genes were identified by BLASTN, version 2.2.15 (e value $<1e^{-45}$, see Table S1 for probe IDs). For each microarray experiment, raw data were normalized for further analysis. The CEL files were normalized with RMA (Robust Multi-Array Average) [32] using the affy R package [33]. For genes with more than one probe set, the median of the expression values was considered. To calculate the fold change in each experiment, normalized expression values of each experimental condition were compared with their control. In addition, with the aim of identifying *VvBURP* genes showing similar expression profiles, average-linkage hierarchical clustering was performed using the Cluster 2.11 software as described previously [34].

Isolation of Vv*RD22* genes

The mRNA sequences (including partial 5′ and 3′UTR regions) of two Vv*RD22* genes were isolated. These were named Vv*RD22b* (*VvBURP18*) and Vv*RD22c* (*VvBURP06*). Their sequences were amplified from mature seed and green berry skin cDNAs, respectively, using the primers VvRD22b-5′utr (5′-TAGCTTTTGAGCTTGAGTCCTT-3′) and VvRD22b-3′utr (5′-GAATAACCCACATCTCCAGCC-3′) and VvRD22c-5′utr (5′-AGCAAGCAAAGGTTCCAGTT-3′) and VvRD22c-3′utr (5′-TTTCAGCATGCTTCAACAT-3′). PCR products were cloned into pTOPO-SD (Invitrogen). Six clones for each gene were sequenced using the universal M13 forward and reverse primers. Vv*RD22b* and Vv*RD22c* sequences were deposited in Genbank, with the accession numbers FJ869893 and FJ869894, respectively.

Grapevine developmental samples

Reproductive grapevine organs (*Vitis vinifera* L. cv. Cabernet Sauvignon) were collected from a commercial vineyard in the Maipo Valley (Chile). Inflorescence clusters from two developmental stages (eight and twelve weeks post bud break, WPBB) were included. A total of nine grape clusters were sampled from three plants every two weeks throughout fruit development, beginning two-three weeks after fruit set (four weeks before véraison) and ending at eight weeks after véraison. Berries were immediately peeled and deseeded. Seeds and skins were frozen in liquid nitrogen, and stored at -80 °C until required for RNA extraction.

ABA and salinity treatments in grapevine seedlings

Two month old Cabernet Sauvignon seedlings grown *in vitro* on Murashige-Skoog (MS) medium [35] were transferred to a hydroponic system using half-strength MS (1/2MS) medium supplemented with or without 100 µM ABA (Sigma). Seedlings were maintained in a culture chamber with a long-day cycle at 25 ± 1°C. After 12 h of treatment, leaves were collected and frozen in liquid nitrogen.

Cabernet Sauvignon nodal explants (young shoots with three nodes) were placed vertically on sterile MS media and propagated for a month in a growth chamber (23 ± 2°C; 16 h photoperiod), under three experimental conditions: MS with 3 mM (control), 100 mM or 200 mM NaCl. At the end of the experiment, samples were photographed, collected and frozen at -80°C until total RNA was extracted from complete plantlets. Three biological replicates for each experiment were performed.

Virus infection and *Botrytis cinerea* inoculation

Healthy and virus infected *V. vinifera* cv. Cabernet Sauvignon plants were selected from an experimental field (P. Universidad Católica de Chile). For this, viral screening was carried out in leaf samples of the medial segment of main shoots. Leaf and fruit at maturation stage were frozen in liquid nitrogen and then stored at −80°C until RNA extraction. Virus detection was performed by RT-PCR. Ten viruses (the most prevalent grapevine viruses worldwide) were assayed in each sample using appropriate primers as described by Vega et al [36]: Grapevine Virus A (GVA), Grapevine virus B (GVB), Grapevine Fanleaf Virus (GFLV), Grapevine Fleck Virus (GFkV), Tomato Ringspot Virus (ToRSV), and Grapevine Leaf-Roll-Associated Viruses (GLRaV) 1, 2, 3, 4 and 7. After this analysis, negative plants for all tested viruses were considered as healthy, while only GLRaV3 positive plants were considered as infected plants for experimental purposes. Healthy and virus-infected plants were kept separately in the field with similar growth conditions and handling practices.

Botrytis cinerea B05.10 spores were grown as previously described by Mengiste et al [37]. Detached grapevine Cabernet Sauvignon fully expanded leaves and mature grapes were infected with a 50 µL drop containing 5×10^6 spores/mL in water and control tissues were inoculated with 50 µL of water. Both inoculated and control tissues were placed over a wet filter paper in square petri dishes (leaves) or in 24-well plates (grapes) to maintain high humidity and kept in a Percival growth chamber at 21°C day and 18°C night temperatures. Samples were collected at 96 h post inoculation, frozen in liquid nitrogen and stored at −80°C until RNA extraction.

Nucleic acid extraction and quantitative comparison of gene expression

Total RNA was isolated from all organs and treated tissues, according to the procedure of Reid et al [38] using a CTAB-Spermidine extraction buffer. For cDNA synthesis, one µg of total RNA was reverse transcribed with random hexamer primers using Superscript II™ First-strand Synthesis™ (Invitrogen) according to the manufacturer's instructions. Relative transcript quantification of *VvRD22-a*, *b* and *c* genes was achieved by RT-qPCR, using the SensiMix SYBR kit (Bioline) and the Mx3000P detection system (Stratagene) as described in the manufacturer's manual. Considering the high sequence similarity between their coding regions, primers for quantitative PCR analysis were designed to amplify 3′UTR fragments. The primers used for quantitative PCR were: qVvRD22aF (5′-GCACATCATTCGGTGTATCG-3′), qVvRD22aR (5′-GCAATGGGGTTTGAAGTATTA-3′), qVvRD22bF (5′-TGCCCGACCCAAAACCACTGCTTC-3′), qVvRD22bR (5′-GAATAACCCACATCTCCAGCC-3′), qVvRD22cF (5′-GTATTTCAACCTTCAGCACA-3′) and qVvRD22cR (5′-TTTCAGCATGCTTCAACAT-3′).

PCR conditions and standard quantification curves were conducted according to Matus et al [39]. Gene expression levels were normalized differentially for each experiment, against control sample or a specific developmental stage, in order to obtain a ΔCt for each gene. Amplification of the *UBIQUITIN1* [38], *GLYCERALDEHYDE PHOSPHATE DEHYDROGENASE (GPDH)* [36] and *ACTIN* [40] genes was used for calibrating gene expression. Experiments were performed with three biological replicates and three technical replicates. Reaction specificities were further confirmed with melting gradient dissociation curves, electrophoresis gels and cloning and sequencing of each PCR product. All data were statistically analyzed with MINITAB v14 software (Minitab Inc., PA, USA). One-Way ANOVA and Tukey's media comparison analyses were conducted. Statistical differences between means were based on $p<0.05$ values.

Results and Discussion

Genome-wide identification of BURP genes in grapevine

The genome of the near-homozygous PN40024 genotype of *V. vinifera* cv Pinot Noir was screened for BURP gene sequences. BURP domains possess conserved features: two N-terminal phenylalanine residues, two cysteine residues and four repeated cysteine-histidine motifs: CH-X(10)–CH–X(25–27)–CH–X(25–

26)–CH, where X can be any amino acid [41]. A consensus BURP sequence was obtained from the alignment of previously-isolated genes from *Arabidopsis thaliana* and rice (*Oryza sativa*). This was later used as a BLASTP query against the 8X (Genoscope) and 12X V1 (CRIBI) genome assemblies, allowing the identification of over 20 BURP-like proteins. The number of annotations between both genome versions varied, requiring the use of further bioinformatic data for accurate gene discovery.

As observed in gene models belonging to other protein families (e.g. MYB genes [39]), the *ab initio* prediction algorithm used to define gene annotations may split a single gene into two gene models or include intron sequences as exonic regions, among other annotation errors. To overcome these issues, we searched for mRNA sequences within a comprehensive RNA-Seq dataset from *Vitis vinifera* cv. Corvina [24,25]. The careful inspection of these sequences supports the existence of 19 gene identities (Figure 1, Table S1). The chromosomal distribution of redefined grapevine BURP genes is presented in Figure 1A, showing that paralog genes are present in chromosomes 1, 3, 4, 11 and 17, with 13 of them in a cluster on chromosome 4. We compared the number of grapevine BURP genes to those found in other available angiosperm genomes. In soybean and rice, BURP families are composed of 16 and 17 members, respectively [14,42], while in sorghum, maize and poplar there are 11, 15 and 18 members respectively [16,17]. This data suggests that, even though the number of BURP genes between species is similar, the grape superfamily may have undergone a specific expansion of a particular group within chromosome 4. Family-specific expansions and diversification have been suggested previously, as in the case of genes related to wine characteristics [39] and grape expansins [30]. Additional molecular features of each BURP gene model and their predicted proteins are listed in Table S1, while DNA and protein alignments between the 8X, 12X V1 and RNA-Seq derived transcripts are shown in Figure S1. Nucleotide and amino acid polymorphisms may reflect cultivar differences between the reference genome and the RNA-Seq data (cv. Pinot Noir and cv. Corvina, respectively).

Previous comparisons between genome predicted sequences and available RNA-Seq data allowed the isolation of two sirtuin grape genes [43]. The use of such information has been important for building gene models *de novo* [44]. As an exemplifying case of the use of the RNA-Seq data for the re-interpretation of BURP gene model annotations, we show *VvBURP07* re-annotation in Figure 1B (also in Figure S1). The 12X V1 genome version denoted two independent gene models (VIT_04s0008g03990 and VIT_04s0008g04000), whose deduced amino acid sequences in fact represented incomplete BURP proteins. Two RNA-Seq derived transcripts (SV1 and SV2) allowed us to re-define and fuse these two gene models into one single BURP gene.

Phylogenetic analyses of the grapevine BURP proteins

To gain insight into the relationships between the grape BURP genes, we first constructed a phylogenetic tree including the 19 grapevine proteins identified in this work based on their deduced amino acid sequences together with the five Arabidopsis BURP proteins (Figure S2 and S3). BURP proteins have been grouped into different categories according to each genome-wide study that has been performed so far. In dicotyledonous species they are classified in: 1) The beta subunit of polygalacturonases (PGβ); 2) Brassica BNM2-like proteins; 3) RD22 proteins and 4) USP seed embryo abundant proteins [45] [46] [47]. The RD22, PGβ and BNM2 groups were found in Arabidopsis and grape. The grape RD22 family is clearly expanded, but up to this point it was not possible to distinguish if any of these models belonged to a USP lineage. We therefore constructed a phylogenetic tree containing the entire grapevine BURP superfamily and additional proteins from all previously surveyed BURP families that represented different monocot and eudicotclades (list of proteins in Table S2). We found that the previous classifications were maintained, as no Arabidopsis or grape models were present in the USP family (Figure 2). In fact, this clade may only form part of leguminous genomes. Arabidopsis AtUSPL1 (AtBNM2-L) belongs to the BNM2 family rather than the USP family although it is localized, as is VfUSP, in cellular compartments essential for seed protein synthesis and storage [47].

Polygalacturonases are responsible for cell wall pectin degradation in expanding tissues throughout development. They have being largely studied in tomato fruit ripening and flower development. They are formed by a catalytic subunit and a highly glycosylated β-subunit (PGβ). PGβs are known to regulate polygalacturonase enzymatic activity and thermostability [48]. Our phylogenetic analysis showed that the PGβ family was more conserved than other families, with fewer variations in gene number (Figure 2). With the exception of soybean (seven genes), between two and three PGβs were found in grape, poplar, Arabidopsis, rice, maize and sorghum. These grouped separately within the monocot and eudicot lineages.

A high number of *RD22* genes were found in soybean, grape and poplar, but only members from the last two species formed an additional RD22-like cluster, more distantly related with AtRD22 and GhRDL1, the α-expansin-interacting RD22 protein from cotton [12]. This massive duplication of *RD22* genes appears to be a woody-specific event within eudicots. Possibly as for many other gene families, tandem and segmental duplications in different plant lineages may lead to species-dependent expansion of BURPs. The BURP superfamily has clearly undergone a monocot-specific expansion, leading to groups VI, VII and VIII (Figure 2). However, none of these appeared closely related to the RD22 family. The rice-exclusive Group V, here represented as subgroups V and Vb, shows a closer relationship with the monocot RD22 group, though while inspecting their protein sequences for domain distributions, this group lacks some of the motifs found in the RD22 family.

To test and validate the phylogenetic relationships of Figure 2, we further analyzed the protein sequences of BURP members by using the online MEME Suite. Conserved motif identification with this tool allowed us to determine the number and consensus sequence of motifs constituting the BURP domain, as well as other motifs outside this region, some of which were specific to each group (Figure 3). As described by Ding et al [14], PGβs possess a series of exclusive motifs with few changes between species. For this reason, these sequences were not included in the analysis. The BURP domain generally consists of 8–9 motifs distributed mainly at the C-terminal of each protein. These motifs have been named differently in previous studies [14,16]. Here we present a standardized nomenclature and sequence description, as seen in Figure S4 (taking the AtRD22 protein as an example). The BURP domain is composed of motifs m8, m6, m11, m9, m2, m1, m10, m3 and m4. Eudicot RD22 and Monocot RD22 groups possess these nine motifs. In addition, they possess motif m7 at the N-terminal region. Motif m5 was also found outside of the BURP domain, showing variable (from one to nine) repetitions. Most of the proteins from the woody RD22 like clade presented the nine motifs from the BURP domain and m7 at the N-terminus. Only one m5 repetition was found in the case of VvBURP15. In the case of VvBURP13 and VvBURP14, more than ten repetitions of an m12 motif were found. Group V and Vb presented only six to eight motifs from the BURP domain, m7 and no repeated m5

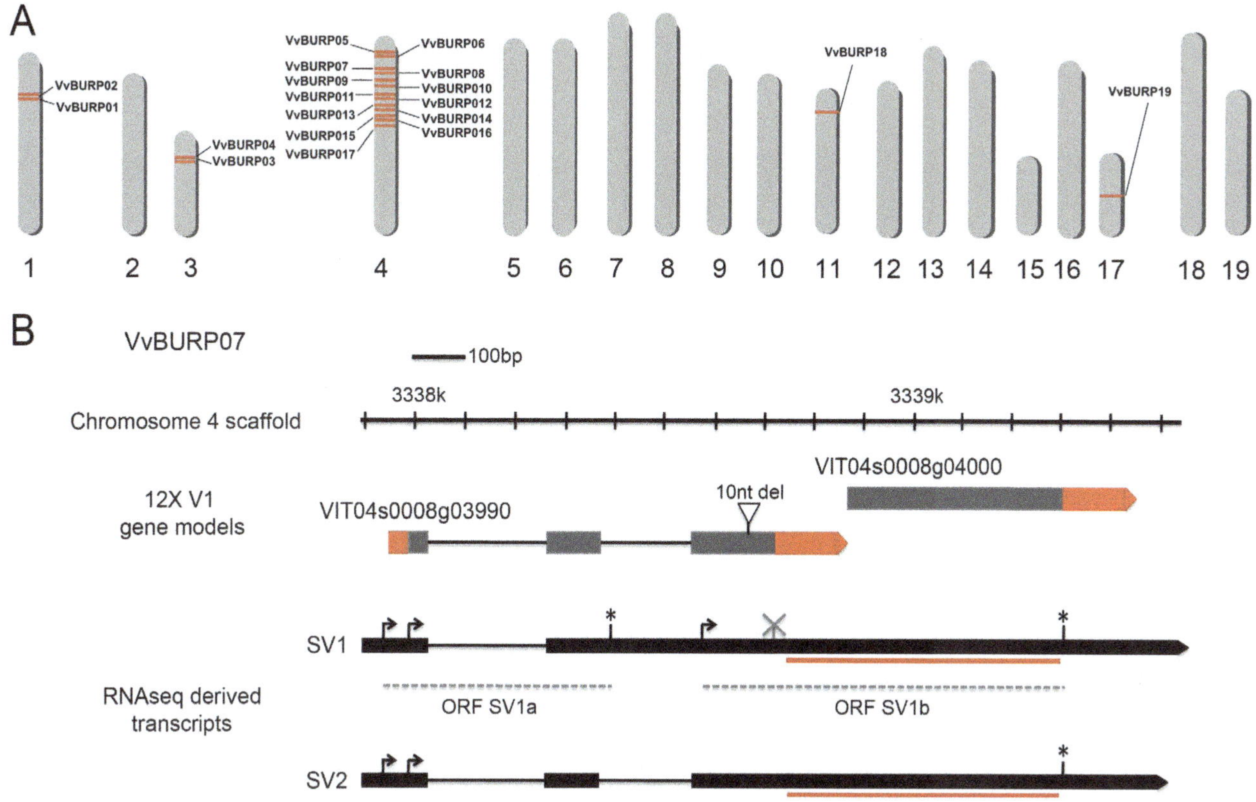

Figure 1. BURP domain-containing gene models in the grapevine genome. A) Gene positions in chromosomes 1, 3, 4, 11 and 17. An expanded cluster of closely related BURP genes (from the RD22 family) was found in chromosome 4. B) Assistance in gene model verification and re-annotation through the use of RNA-Seq data. The case for *VvBURP07* is shown as an example, where two incomplete BURP models from the 12X V1 prediction are no longer supported as independent units. The RNA-Seq data shows that a 10-nucleotide deletion was present in the 12X genome version, possibly corresponding to an error in the sequencing, as this deletion was not present in the 8X genome version. The two models actually represent one single annotation containing a complete BURP domain. SV: splicing variant. Red line under each SV represents the region coding for the BURP domain. Asterisks: stop codons in a corresponding ORF. Grey X: abolished stop codon in the SV ORF compared to the 12X annotation. For further inspection of this and other gene model re-annotations, see Table S1 and Figure S1.

motifs. OsBURP05 and OsBURP17 shared a long sequence between m7 and the BURP domain but no motif was found here.

Groups V to VIII were previously described [14,16,17] but in all of these cases, classification errors were assumed, which may have lead to misinterpretations of the extent of each one of these monocot clades. In general, few species were considered in the construction of these phylogenetic trees, which could have caused the inconsistencies observed in the members of each clade. The most common way to estimate the reliability of a phylogeny is through a bootstrap test, as it estimates the consistency of each node. However, bootstrap values are absent in several phylogenetic trees [16,17]. Additionally, we consider that poplar proteins that truly belong to the woody RD22 like cluster were erroneously assigned in the BURP V classification, as this group is monocot-specific [17]. Furthermore, two rice BURP genes (*OsBURP05* and *OsBURP17*) were mistakenly assigned as *VfUSP* like genes [16] and were later described as part of the RD22 group [14]; while we propose that both form part of group Vb as seen in Figure 2 and Figure 3. In our study, different combinations of models and algorithms were used to achieve the best phylogenetic relationships between BURP homologs. Another difference with previous studies was the use of the MUSCLE alignment algorithm instead of CLUSTALW, which applies a combination of both global and local alignments, re-optimizes the work as it progresses and performs better when sequence lengths are quite different even though they are from the same protein family [49]. Different trees were constructed and then compared by their bootstrap values. The best tree topologies were obtained from the Neighbor Joining Method under different substitution, rates and gap treatment models. The use of a large set of protein sequences from different species allowed us to construct a reliable tree, containing more accurate relationships. Finally, the use of protein domain search tools such as MEME allowed the validation of these associations.

Global expression features of the BURP domain superfamily

The *V. vinifera* cv. Corvina gene expression ATLAS [28] was screened to characterize the expression profiles of the BURP superfamily. This microarray dataset comprises both vegetative and reproductive tissues, as well as berries that have undergone post-harvest withering for up to three months. We retrieved the fluorescence intensity values of the 19 BURP transcripts, generating a bi-clustered heat map (Figure 4). The family's expression profiles were divided in four main clusters: 1) mainly expressed in floral organs, buds, tendrils and early stages of rachis development, with a significant reduction of expression in berry (pericarp, skin and flesh) development from véraison (V) onwards; 2) high expression in seedlings and roots (*VvBURP04*, *VvBURP013* and *VvBURP014*); 3) expressed in seedlings, roots, berry tissues at early stages close to fruit set (FS or PFS) and

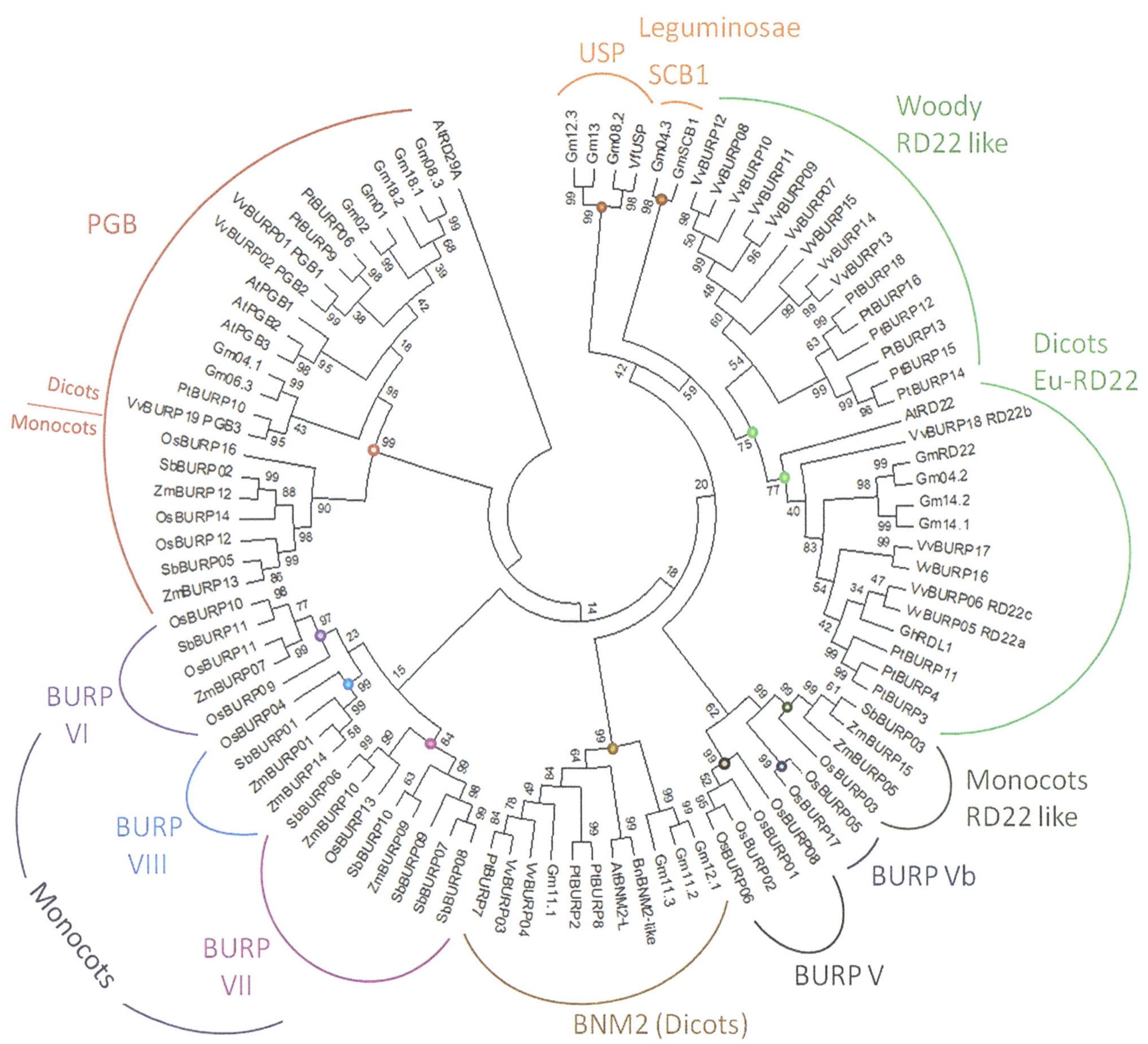

Figure 2. Phylogenetic tree of grapevine BURP proteins and their homologues in mono and dicotyledonous plant species. Highly resolutive NJ tree constructed under the "no differences" model, with uniform rates among sites and partial deletion gap treatment. At: *Arabidopsis thaliana*, Bn: *Brassica napus*, Gh: *Gossypium hirsutum*, Gm: *Glycine max*, Os: *Oryza sativa*, Pt: *Populus trichocarpa*, Sb: *Sorghum bicolour*, Vf: *Vicia faba*, Vv: *Vitis vinifera*, Zm: *Zea mays*. Protein IDs and descriptions for each gene are found in Table S1 and Table S2. Evolutionary distances are represented as amino acid substitutions per site. Clades with high bootstrap values (>75) that hold distinctive structural features are shown with coloured circles.

throughout rachis development (*VvBURP10*, *VvBURP12* and more distantly *VvBURP03*) and 4) floral tissues and some cases of berry (pericarp, skin) development. Certain BURP genes were characterized by unique expression profiles in certain tissues. Within cluster 4, *VvBURP15* shows high expression in stamen and pollen just before flowering. The genes with most distant expressions were *VvBURP07* and *VvBURP18 (VvRD22b)*, with the latter mainly expressed in mature seed stages. In general terms, and with the exception of *VvBURP07*, the BURP superfamily shared a common repression profile throughout flesh morphogenesis.

Despite the contradictory hypotheses regarding PGβ roles in polygalacturonase function [50,51], one certainty is that their expression is high in flower tissues and in ripening fruits. Grape *PGβ1* and *PGβ2* formed part of expression cluster 1. *PGβ3* (*VvBURP19*), closely related to cluster 1, had a similar expression profile in early flower development, tendrils, rachis and buds, but was not found in early stages of berry development. Instead it was expressed at pericarp ripening stages and in post harvest withering.

As suggested in all previous genome-wide studies, BURP genes may have a role in the response to abiotic stresses. However, their participation in biotic plant-pathogen interactions remains uncharacterized. As a second approach to study the expression profile of the BURP superfamily, we searched for public grape Affymetrix microarray data in PLEXdb [31]. One or two probe sets were found for some members of the family, either in the 16K or the *Vitis vinifera* GeneChip custom array (Table S1). As seen in Figure S5, different responses were found depending on the temporal extent of the stress and on the type of organ sampled. *VvBURP05* (*VvRD22a*) was induced at late stages after osmotic stress imposition in vegetative tissues (24 h after PEG or salt stress

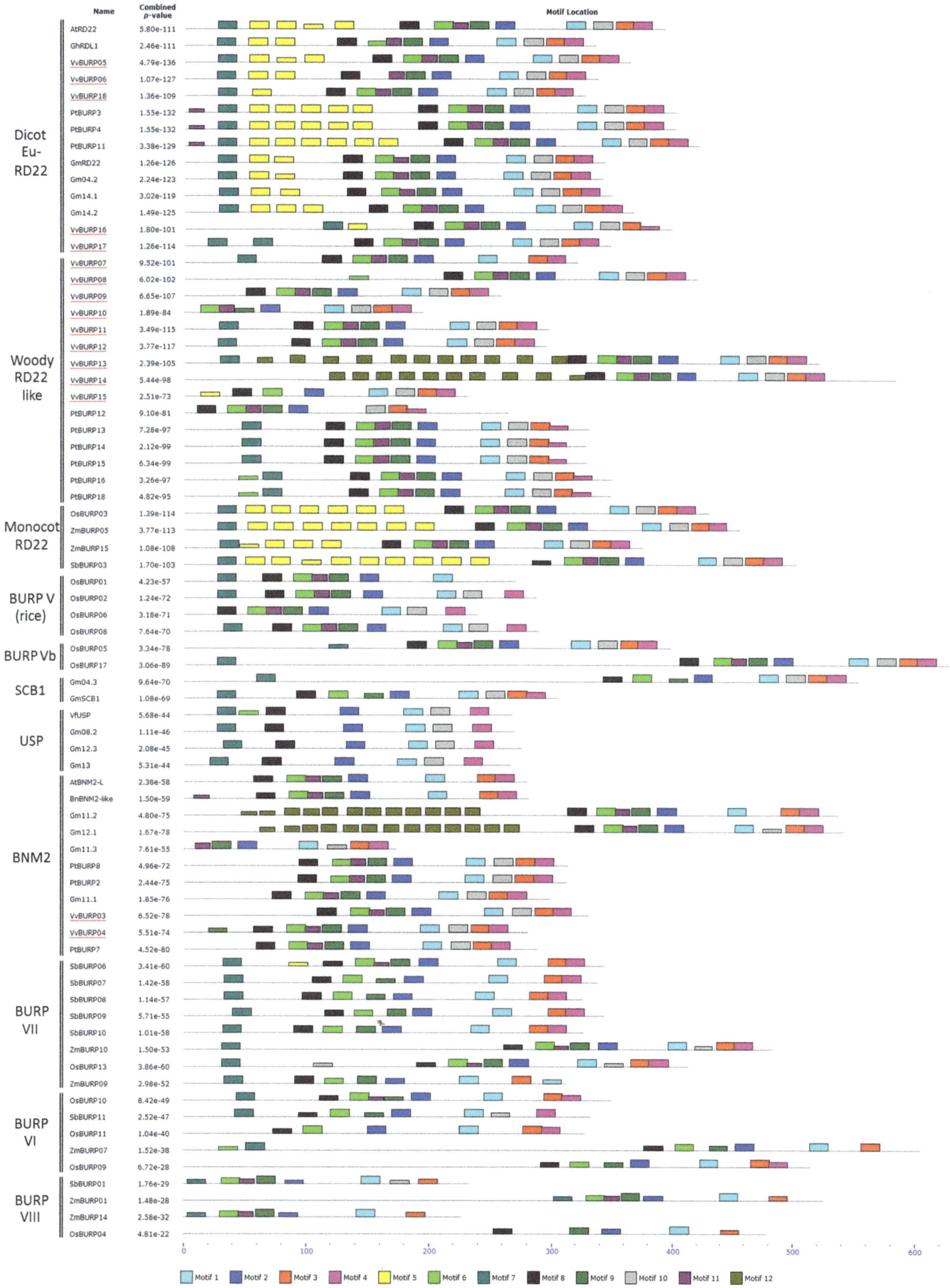

Figure 3. Motifs identified in the BURP proteins by MEME software. Polygalacturonase sequences were excluded from the analysis. Combined *p*-values are shown on the left of the figure.

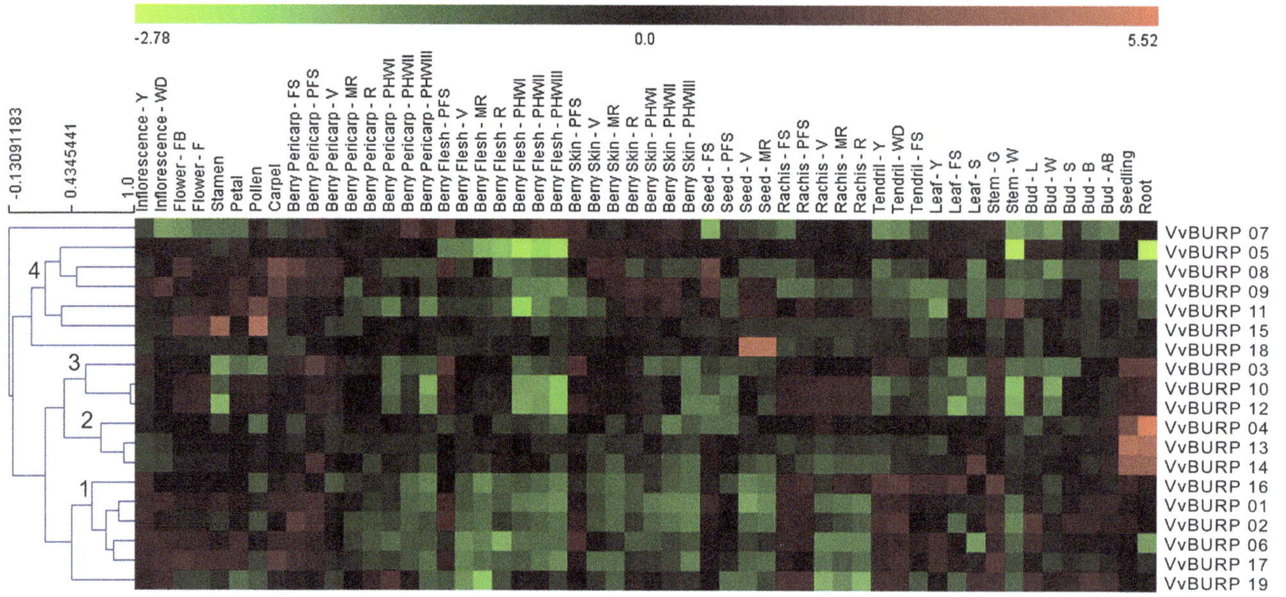

Figure 4. Expression features of the grapevine BURP superfamily. Log_2 fluorescence intensity values from the *V. vinifera* cv. Corvina ATLAS were normalized based on the mean expression value of each gene in all tissues/organs. Different organs/tissues are displayed vertically above each column. Gene names are displayed to the right of each row. The colour scheme used to represent expression level is red/green: black boxes indicate a low variation in expression, green boxes indicate a fold decrease and red boxes indicate a fold increase with respect to the mean value. Samples and genes were hierarchically clustered based on the average Pearson's distance. Abbreviations after organ names indicate the developmental stage. FS, fruit set; PFS, post fruit set; V, véraison; MR, mid-ripening; R, ripening; PHWI, post-harvest withering (1st month); PHWII, post-harvest withering (2nd month); PHWIII, post-harvest withering (3rd month), Bud - L, latent bud; Bud - W, winter bud; Bud - S, bud swell; Bud - B, bud burst; Bud - AB, bud after burst; Inflorescence - Y, young inflorescence with single flowers separated; Inflorescence - WD, well developed inflorescence; Flower - FB, flowering begins; Flower - F, flowering; Tendril - Y, young tendril; Tendril - WD, well developed tendril; Tendril - FS, mature tendril; Leaf - Y, young leaf; Leaf - FS, mature leaf; Leaf - S, senescing leaf; Stem - G, green stem; Stem - W, woody stem.

and 16 days after salt stress and water deficit, Figure S5A). *VvBURP12* was induced at different time points in response to osmotic stress. *VvBURP06, 09* and *17* were mainly induced in heat stress recovered tissues and *VvBURP18* was only induced at 24 h after PEG treatment. In berry tissues (Figure S5B), *VvBURP18* (*VvRD22b*) was highly induced by water deficit in cv. Cabernet-Sauvignon berries while *VvBURP06* (*VvRD22c*) was repressed in these samples, especially towards the last ripening stages. On the other hand, expression of *VvBURP05* (*VvRD22a*) was relatively stable in these samples, except for its induction in one late ripening stage of water stressed cv. Chardonnay berries. *VvBURP06* was induced by heat. As seen in Figure S5C (biotic stress in vegetative tissues), Bois Noir phytoplasma infection generally repressed BURP gene expression with the exception of *VvBURP05, VvBURP12* and *VvBURP18*. This observation is in agreement with the fact that infection increased physical barriers to limit phytoplasma spread, with the repression of genes responsible for cell wall degradation (e.g. *VvPGβ1*) and the induction of genes involved in cell wall reinforcement [52]. *VvBURP05* and *VvBURP12* were induced in powdery mildew-resistant grapevines while *VvBURP06* was induced in downy mildew-infected samples. Finally, virus infections repressed the expression of *VvBURP06* and *VvBURP09*. All these findings revealed that the BURP superfamily is differentially regulated by a variety of biotic stresses.

Analysis of three *VvRD22* genes with remarkably different expression profiles

As seen in the transcriptomic analysis, members of the grapevine RD22 family show several differences in their expression profiles, suggesting complementary or opposite roles in different organs. We further evaluated the expression of three Eudicot RD22 genes belonging to different expression clusters (Figure 4). These three closely-related genes (*VvBURP05, VvBURP06* and *VvBURP18*) were studied by means of Quantitative Real Time PCR (qPCR).

VvBURP18 and *VvBURP06* were isolated and named *VvRD22b* and *VvRD22c*, respectively. The grapevine RD22 gene *VvBURP05*, previously isolated by Hanana et al [19], is referenced here as *VvRD22a*. The three putative protein products possess between 50 and 70% similarity in their complete protein sequences, and over 90% similarity in their BURP domains. The conserved sequences and motifs found in and around the BURP domain are shown in Figure S6. These three genes share the same distance between the CH dipeptides within the BURP domain (X5–CH–X10–CH–X25–CH–X25–CH–X8–W). VvRD22a, VvRD22b and VvRD22c harbour 3, 1 and 2 motif m5 repeats, respectively, while the Arabidopsis homolog possesses four repeats. Although RD22 proteins share the consensus repeat sequence described by Hanana et al [19] (VGVGKGTGVNVHAGKGKPGGGTT), the most different features correspond to the number of these repetitions outside the BURP domain.

Our results revealed that these three grape *RD22* genes possess differential expression patterns in vegetative and reproductive organs (Figure 5A). None of these genes were expressed in roots, just like their homologues in Arabidopsis [6], soybean [42], and rice [14]. *VvRD22a* was the most expressed gene in all organs surveyed (seeds were not assessed). *VvRD22a* was highly expressed in leaves, late stages of inflorescence development and berry skins. Although *VvRD22b* was expressed at much lower levels than

VvRD22a, it showed a four-fold induction in berry skins compared to leaves or other organs. Interestingly, *VvRD22c* showed the most distinct pattern of the three genes, being repressed in ripened berry skins when compared to all other organs.

We further analyzed the expression profiles of each gene throughout skin and seed development (Figure 5B–C). The expression of *VvRD22a* and *VvRD22c* was higher in berry skins before the onset of ripening (−4 and −2 weeks after véraison, WAV). From véraison onwards, the expression of *VvRD22c* was practically undetectable, while the expression of *VvRD22a* was maintained during the early ripening stages (0–4 WAV), slightly increasing again at 6 and 8 WAV (Figure 5B). On the other hand, *VvRD22b* was highly expressed after véraison, although at much lower levels than the other two RD22 genes. *VvRD22a* and *VvRD22c* were expressed differently in berry skins compared to seeds (Figure 5C). Their expression declined from −4 WAV to 2 WAV and slightly increased at 6 and 4 WAV, respectively, but both were much less expressed than *VvRD22b*. At 6 WAV, *VvRD22b* transcript abundance was 30 times higher than at véraison. The opposite expression profiles of *VvRD22b* and *VvRD22c* confirm the Affymetrix-derived data found in PLEXdb (Figure S5).

VvRD22 gene expression is differentially regulated by ABA and abiotic stress

ABA governs several berry ripening processes in grape (reviewed by Kuhn et al [18]). The level of this hormone in berries decreases after anthesis and then increases significantly at véraison, as measured by its concentration or by the expression of genes related to its synthesis [53,54]. If ABA accumulates in berries at véraison, and if RD22 genes respond to ABA, as in diverse plant species [6], it is possible that the different expression patterns found for these *RD22* genes in berry tissues may be due to different concentrations of ABA in grape organs.

In order to evaluate whether ABA regulates the expression of grape *RD22* genes, seedlings cultured in hydroponic media were treated with this hormone and gene expression was analyzed 12 h after treatment. Figure 6A shows that after ABA treatment *VvRD22b* was induced by ABA, while *VvRD22c* was repressed; *VvRD22a* expression remained unaffected. These results confirm that both *VvRD22b* and *VvRD22c* genes are regulated by ABA. The expression of *VvRD22a* under our experimental condition seems not to be affected by ABA. However, Hanana et al [19] previously described that *VvRD22a* was indeed responsive to ABA, although those measurements were conducted in berry cell cultures, which may explain the difference in the ABA-responsiveness of this gene between these studies.

RD22 genes have been described as reliable ABA reporter genes in response to drought and salinity. As a first attempt to characterize these *VvRD22* genes in response to abiotic stress, we studied *VvRD22* expression in leaves from plantlets subjected to high salt in their culture medium. For this, Cabernet Sauvignon nodal segments were grown in different NaCl concentrations (3 mM, 100 mM and 200 mM) until roots and/or leaves were visible and fully expanded. After several weeks, leaves had emerged from all explants while roots had developed only in MS containing 3 mM NaCl. At the end of the experiment, plantlets from the 100 mM NaCl treatment had a small radicle and high leaf anthocyanin accumulation, as a clear signal of stress in the plant (Figure 6B). When explants were grown on 200 mM NaCl, they became necrotic, indicating a strong toxic effect of high salt stress. Under this condition, roots did not develop.

As expected, the three grape *VvRD22* genes responded to salt treatments (Figure 6B), but with different tendencies. The expression of *VvRD22a* was only induced at 200 mM, while *VvRD22b* was strongly induced in both salt concentrations. The expression of *VvRD22c* was inhibited by salt stress. Considering that salt stress responses are mediated by ABA as described in many plant species, the contrasting behaviour of *VvRD22a/b* and *VvRD22c* in response to salt stress directly correlates with the ABA responsiveness observed for these *RD22* genes. In addition, Hanana et al [19] reported that *VvRD22a* was rapidly induced under salt stress in a tolerant variety when compared to a sensitive one.

In order to understand global responses to drought, Deluc et al [55] performed transcriptomic analyses during cv. Cabernet Sauvignon berry development comparing normal irrigated and water deficit conditions in field vineyards. The general profiles of RD22 expression derived from those data (which can be viewed in the PLEX database) correlated with our results obtained by qPCR. Taken together, there is thus strong evidence that the opposite expression patterns of *VvRD22b* and *VvRD22c* during fruit development may be due to opposite ABA responsiveness.

Insights into the biotic stress responsiveness of grape *RD22* genes

Recent evidence suggests the existence of a significant overlap between signalling networks that control abiotic stress tolerance and disease resistance. On the basis of experiments with exogenous application of ABA, inhibition of ABA biosynthesis and/or the use of ABA-deficient mutants, it has been shown that enhanced ABA levels correlate with increased pathogen susceptibility, and that a reduction below normal levels increases resistance to many pathogens [56,57]. In order to evaluate the response of *VvRD22* genes to biotic stress, quantitative experiments were carried out in Grapevine Leaf Roll associated Virus 3 (GLRaV 3) infected and Botrytis inoculated grapevine leaves and fruits (Figure 7).

Under virus infection, the *UBIQUITIN* housekeeping gene does not behave homogenously between infected and non-infected samples [36]. For this reason, we used the *GPDH* gene as an internal control, as it does not show any significant variation between healthy and infected samples [22,36]. The expression of *VvRD22a* did not change when comparing leaf or ripened berries from healthy and virus-infected plants (Figure 7A-B). The expression of *VvRD22b* was strongly induced by virus presence in leaves while it was slightly induced in fruits. *VvRD22c* was inhibited in virus-infected leaves whereas no change was observed in infected berries. Affymetrix-derived analysis confirmed the expression of *VvRD22c* found in infected leaves but not the induction of *VvRD22b* (Figure S5). In compatible infections, such as those established between virus and grapevine, it has been reported that stress-related genes are affected in infected leaves [22]. *VvRD22b* and *VvRD22c* presented an opposite regulation in virus infected-leaves, similar to our experiment of salt stress, suggesting a regulation of these genes during general stress conditions.

In the case of Botrytis infection, we selected *ACTIN* as reference for RT-qPCR [40]. Grapevine response to Botrytis in terms of *RD22* gene expression varied depending on the organ studied (Figure 7C–D). The expression of *VvRD22a* was negatively affected during Botrytis infection in leaves whilst the expression of *VvRD22b* and *VvRD22c* did not change significantly. Unlike in leaves, Botrytis infection in berries significantly triggered the expression of *VvRD22a* and *VvRD22b*, and again *VvRD22c* expression was not modified by the pathogen (Figure 7D). Together, these results confirm the differential regulation of

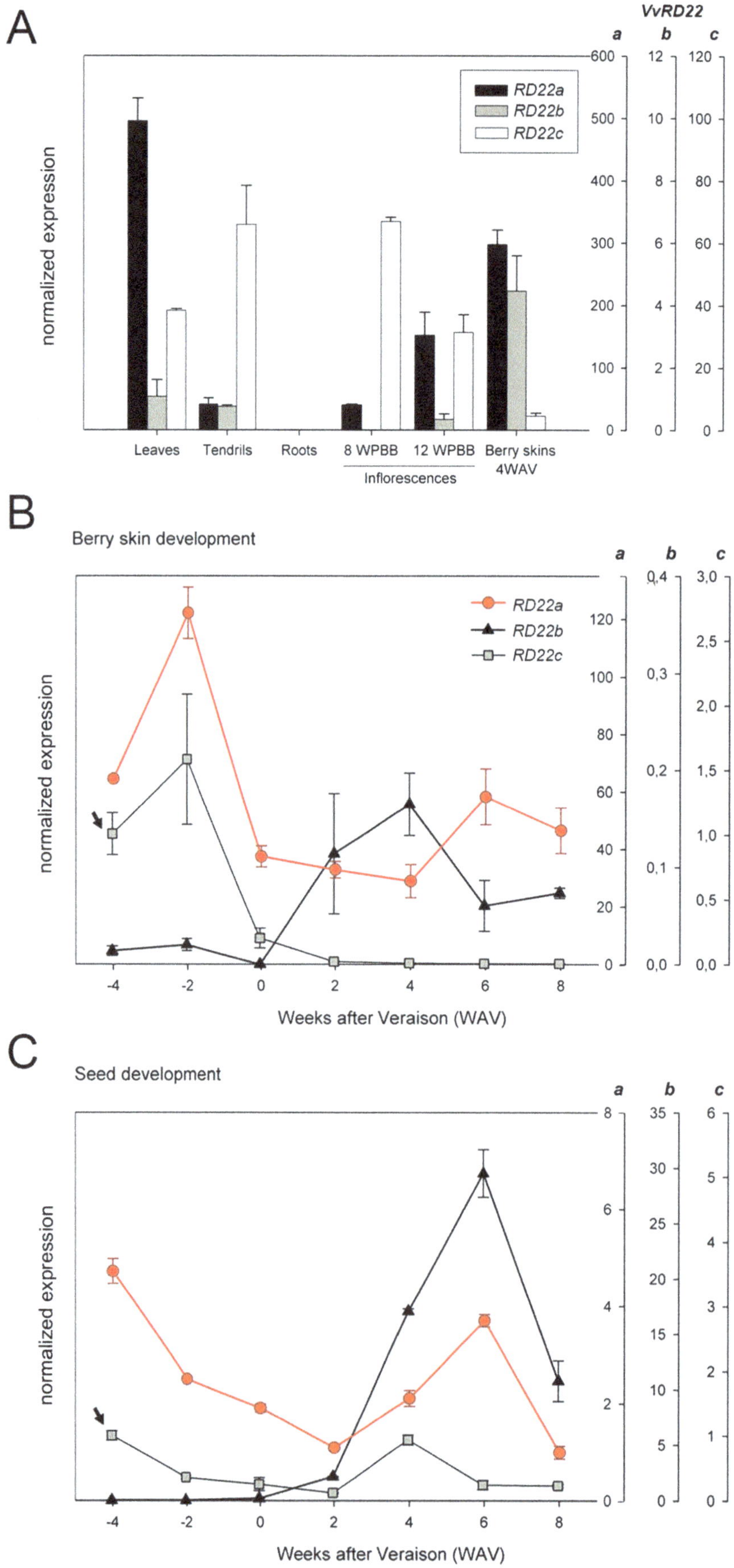
A
VvRD22
a b c
RD22a
RD22b
RD22c
normalized expression
Leaves
Tendrils
Roots
8 WPBB
12 WPBB
Inflorescences
Berry skins 4WAV
B
Berry skin development
Weeks after Veraison (WAV)
C
Seed development
Weeks after Veraison (WAV)

Figure 5. Expression profiles of RD22 genes in different grape organs and throughout berry development. A) Organ collection. All expression levels were normalized against *VvRD22b* expression in leaves. WPBB (weeks post-bud break) refers to inflorescence development. 4WAV: four weeks after véraison. Véraison: onset of ripening, when clusters are 30–50% coloured and the sugar concentration reaches 5° Brix (5% w/w soluble solids). Standard deviations (SD) are the result of three independent replicates. B–C) Expression of *VvRD22* genes in green and ripening stages of B) berry skin and C) seed development, beginning at -4WAV and ending at 8WAV. All expression levels were calibrated with the *VvUBIQUITIN1* housekeeping gene and normalized against *VvRD22c* expression at -4WAV. Stage 0WAV corresponds to véraison (onset of ripening). Standard deviations (SD) are the result of three independent replicates.

RD22 genes in a stimulus and tissue-specific manner, suggesting different/complementary functions in response to biotic stress. To our knowledge, this is the first report that associates *RD22* genes with biotic stresses.

The changes observed in viral and Botrytis infections could be a consequence of altered ABA levels following infection. ABA levels decreased in beans upon inoculation with rust [58]. In soybeans inoculated with Phytophthora, a decrease in ABA concentration occurred only during an incompatible interaction [59]. In contrast, viral infection in tobacco led to an increase in ABA levels [60]. These changes in hormone concentration were, however, modest compared to the dramatic changes in salicylate, jasmonate and/or ethylene production during pathogenesis, so these other hormones may also regulate RD22 expression. Recent

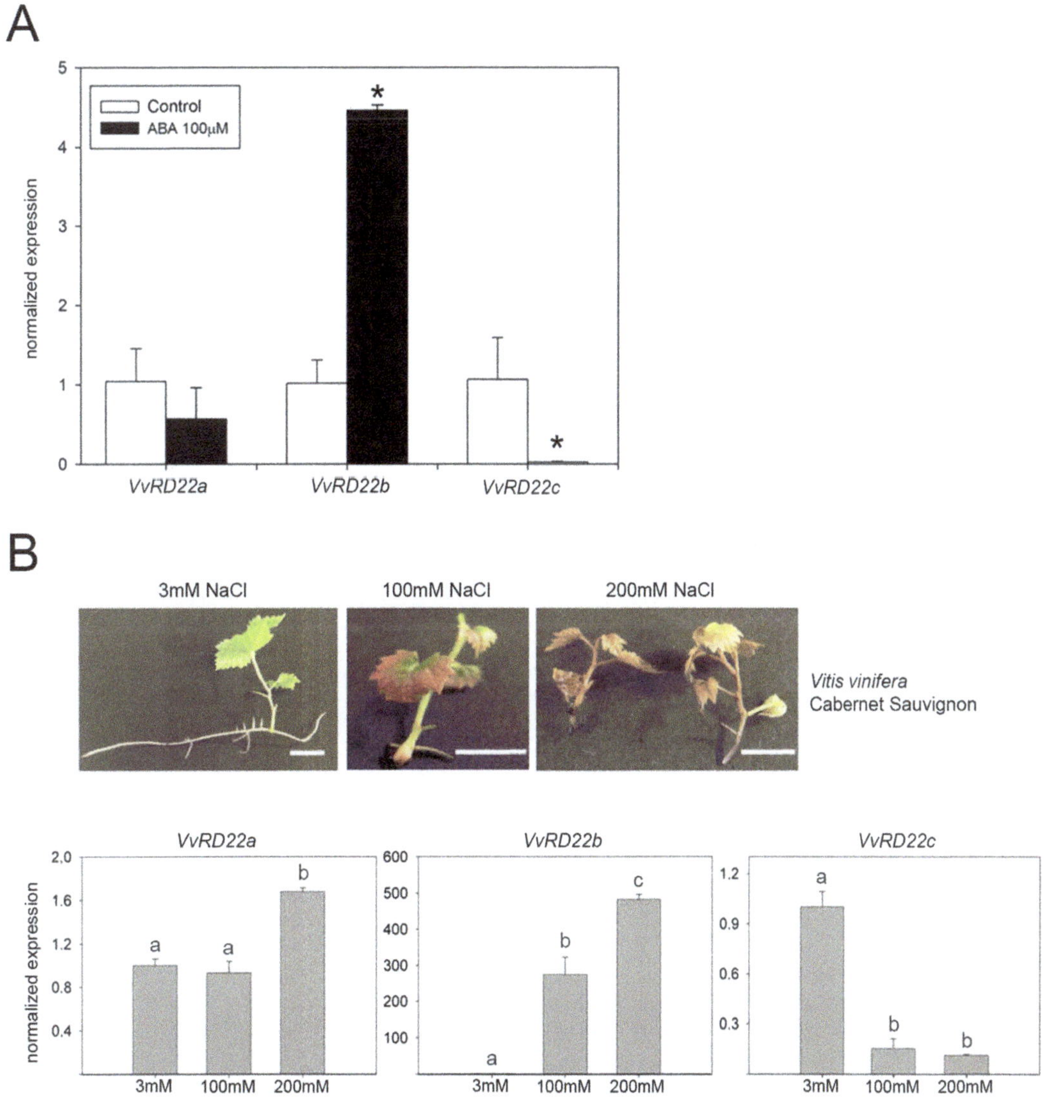

Figure 6. Expression of *VvRD22* genes in response to ABA and in grapevine plantlets exposed to salinity. A) Grapevine seedlings were treated with 100 μM ABA in a hydroponic solution. Gene expression was measured by RT-qPCR and data were normalized against the control (without ABA). B) Plantlets were grown for 30 days in standard MS or MS supplemented with 100 mM or 200 mM NaCl. Representative salinity stress phenotypes in Cabernet Sauvignon plantlets are shown in the upper part of the figure. Scale bar: 1 cm. qPCR expression levels were calibrated with the *VvGPDH* (for ABA samples) and *VvUBIQUITIN1* (for salt samples). Expression of each gene was normalized independently against the MS standard condition (3 mM NaCl). Standard deviations (SD) are the result of three independent replicates. Asterisks and lower case letters indicate significant differences between treatments as calculated by Tukey statistical analysis ($p<0.05$).

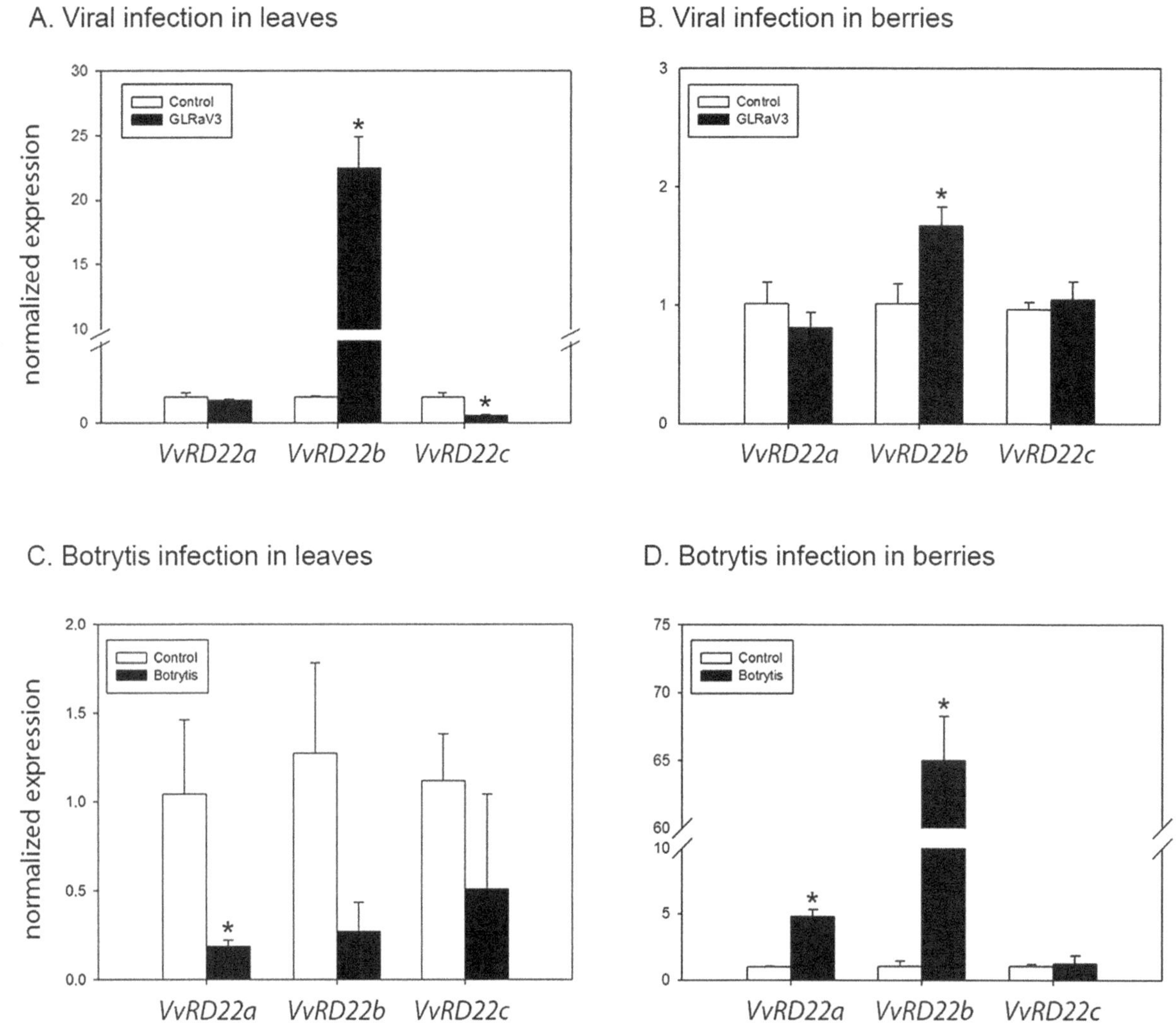

Figure 7. Expression of *VvRD22* genes in response to biotic stress. A) Vv*RD22* expression during viral infection in leaves. B) Vv*RD22* expression during viral infection in berries at maturity stage. Infected grapevine plants were positive for presence of GLRaV-3 virus only, while healthy plants were negative for all viruses tested. C) Expression analysis of Vv*RD22* genes in grapevine leaves inoculated with *Botrytis cinerea* mycelium. D) Expression analysis of Vv*RD22* genes in berries inoculated with *Botrytis cinerea* mycelium. Expression levels were calibrated with the *VvGPDH* and *VvACTIN* housekeeping genes for virus and Botrytis infected samples, respectively. Expression of each gene was normalized independently against its corresponding control. Standard deviations (SD) are the result of three independent replicates. Asterisks indicate significant differences between treatments as calculated by Tukey statistical analysis ($p<0.05$).

findings suggest that there is an impact of viral diseases in ABA concentrations and in the expression of its biosynthetic genes, but also on the effects of the hormone in the accumulation of viruses [61,62].

Final remarks on the ABA responsiveness of RD22 genes

ABA-induced transcriptional activation is mediated by the presence of different cis-acting sequences in the promoter regions of stress responsive genes. The promoter of *AtRD22* is activated by RD22-BP1 (AtMYC2) and AtMYB2 transcription factors [8]. In Arabidopsis, these proteins interact and bind to a 67 bp region (located between -207 and −149 bp from the transcription initiation site) responsible for the desiccation-induced transcription of *RD22* [7]. It is possible that different combinations of regulatory MYB and MYC elements in the RD22 promoters could explain the distinct patterns of transcription that are observed. However, it was not possible to establish a direct correlation between the presence of these elements and the different expression responses of the three grape *RD22* genes studied here (*data not shown*). Additional factors have been recently related to the *RD22* gene. In Arabidopsis, the overexpression of MYB15 conferred ABA hypersensitivity, improved abiotic stress tolerance and increased RD22 expression [63]. The MYC (bHLH) transcription factor AtAIG1 was able to bind the DNA E-box sequence and when abolished, RD22 expression was reduced [64]. Finding new elements or new factors in grape may help to understand additional forms of regulation for these *RD22* genes. Post-translational regulatory mechanisms may also impact total abundance and/or activity of RD22 proteins in response to stress. As an example, the soybean GmRD22 protein is processed and localized in the apoplast, and the presence of its BURP domain is required for this localization [10].

Carra et al [65] found several grape BURP genes from chromosome 4 as potential targets of four small interfering RNAs (siRNA), suggesting that *VvRD22*-derived siRNAs were components of a regulatory mechanism based on RNA silencing in grapevine. Among these gene models, *VvRD22c* and *VvBURP16* were predicted *trans* targets of the siRNA id12, which in addition matched the *VvBURP07* gene model. Since we show that *VvRD22c* suffers more dramatic differences compared to *VvRD22a* and *VvRD22b* in terms of expression and stress responsiveness, this possible siRNA-induced control may be crucial for *VvRD22c* expression.

Conclusions

This work reports the search for BURP genes in the *Vitis vinifera* genome. By using RNA-Seq data we re-defined those models originally found. Our study establishes new and improved phylogenetic classifications based on a big set of plant sequences and the search for protein motif occurrence. Our findings suggest that the expansion of *RD22* genes occurred with an increased rate within woody plant lineages. Although *RD22* genes have not been completely characterized, their expression is used as a direct indicator of an ABA-mediated response. Some evidence suggests that the specific induction of this gene in Arabidopsis (on a rhythmic basis) may be related to a process underlying memory functions of plants in response to ABA stimulus and light pulses in ABA-entrained plants [26]. From another perspective, *RD22* genes may be related to the maintenance of cell integrity, enhancing lignin polymerization and allowing plant cell to endure stressful conditions [10]. Each member of the grape *VvRD22* group presented a different expression pattern during organ development and in response to abiotic stress. Interestingly, we show they are induced still after long periods of stress and also in response to biotic stresses. Our expression analyses suggest that *VvRD22a* may respond to ABA while *VvRD22b* presumably responds to a higher degree. Nevertheless, these two genes show important differences in basal expression in non-stress conditions, where RD22a may have a predominant role in vegetative tissues. In contrast, *VvRD22c* was repressed by ABA, abiotic stresses and in berry development after the onset of ripening. These observations imply additional post-transcriptional regulation processes, such as the recently proposed siRNA mechanism, in addition to changes in protein processing and localization as described for other BURP superfamily members.

Supporting Information

Figure S1 DNA and protein alignments for each of the corresponding BURP gene models between the 8X and 12X genome versions and the RNA-Seq data.

Figure S2 Phylogenetic relationships of BURP homologues from Vitis and Arabidopsis. Colour bar represents bootstrap values for each node. Protein IDs and descriptions for each Arabidopsis gene are found in Table S2. Evolutionary distances are represented as amino acid substitutions per site.

Figure S3 Protein alignment between the grape and Arabidopsis BURP domains. Residues highlighted in yellow correspond to conserved aminoacids from all grape and Arabidopsis proteins. Coloured bars correspond to highly conserved segments, which were found by MEME software (Figure S4).

Figure S4 Schematic representation for AtRD22 motifs. A) A comparison of motif identification between this work and those conducted by Ding et al [14] and Gan et al [16]. B) Consensus sequences for each of the motifs found by MEME software. Adjusted *p*-values for each motif are shown in parenthesis. Motif 12 is only present in VvBURP13, VvBURP14, Gm 11.2 and Gm 12.1.

Figure S5 Heatmap clustering of *BURP* gene expressions for abiotic and biotic stress conditions in grapevine organs, obtained from the Affymetrix Plant Expression database (PLEXdb). A-B) short and long-term abiotic stress conditions in A) vegetative organs and B) berry tissues, C) biotic stress in vegetative tissues. h: hours, d: days, P: pulp, Sd: seed, Sk: skin. R means recovery (25 °C for 5 h) after exposure to 45 °C. Numbers ranging from 31 to 38 represent stages of the Modified Eichhorn-Lorenz system for pericarp samples taken at different developmental stages (35: véraison). CH: cv. Chardonnay, IM: cv. Incrocio Manzoni, B: genotype Rpv1(+)Rpv2(-), C: genotype Rpv1(-)Rpv2(+), D: genotype Rpv1(-)Rpv2(-), CA: cv. Carmenere, CS: cv. Cabernet-Sauvignon.

Figure S6 Protein alignment and motifs found in isolated grape RD22 proteins. Residues highlighted in yellow correspond to conserved aminoacids. Orange segments represent the repeated motif 5 identified by MEME. A discontinuous box shows an incomplete motif 6 in RD22c.

Table S1 Grapevine BURP genes identified in the PN40024 8X and 12X V1 predictions. A) List of BURP genes. The final protein sequence for each gene was defined from the DNA and amino acid alignments between the 8X and 12X gene models, together with the RNA-Seq derived sequences (Figure S1). Fernandez et al. [66] studied a partial cDNA from VvBURP03 and named it BURP1 (EST Accession BQ799859). B) Other grapevine models in the Chromosome 4 cluster which are not BURP genes or do not possess a complete BURP domain. C) Sequences from RNA-Seq derived splicing variants (SV).

Table S2 List of BURP domain genes from other genome-wide studied species used in our phylogenetic analyses (shaded in grey). Gene models corresponding to truncated splicing variants and models with less than five motifs in the BURP domain were not included in the phylogenetic trees (in white).

Acknowledgments

We thank Michael Handford (University of Chile) for critically reading the manuscript and Rodrigo Loyola for assistance in statistical analysis.

Author Contributions

Conceived and designed the experiments: JTM FA CE PAJ. Performed the experiments: JTM FA AV PC AR JS. Analyzed the data: JTM FA. Contributed reagents/materials/analysis tools: EC SDS GBT. Wrote the paper: JTM FA CE.

References

1. Leung J, Giraudat J (1998) Abscisic acid signal transduction. Annu Rev Plant Physiol Plant Mol Biol 49: 199–222.

2. Galbiati M, Matus JT, Francia P, Rusconi F, Canon P, et al. (2011) The grapevine guard cell-related VvMYB60 transcription factor is involved in the

regulation of stomatal activity and is differentially expressed in response to ABA and osmotic stress. BMC Plant Biol 11: 142.
3. Shinozaki K, Yamaguchi-Shinozaki K, Seki M (2003) Regulatory network of gene expression in the drought and cold stress responses. Curr Opin Plant Biol 6: 410–417.
4. Bray EA, Bailey-Serres J, Weretilnyk E (2000) Responses to abiotic stress. In: Buchanan BB GW, Jones RL, editor. Biochemistry and molecular biology of plants. Rockville: American Society of Plant Physiologists. pp.1158–1203.
5. Bray EA (2002) Classification of genes differentially expressed during water-deficit stress in Arabidopsis thaliana: an analysis using microarray and differential expression data. Ann Bot (Lond) 89: 803–811.
6. Yamaguchi-Shinozaki K, Shinozaki K (1993) The plant hormone abscisic acid mediates the drought-induced expression but not the seed-specific expression of rd22, a gene responsive to dehydration stress in Arabidopsis thaliana. Mol Gen Genet 238: 17–25.
7. Iwasaki T, Yamaguchi-Shinozaki K, Shinozaki K (1995) Identification of a cis-regulatory region of a gene in Arabidopsis thaliana whose induction by dehydration is mediated by abscisic acid and requires protein synthesis. Mol Gen Genet 247: 391–398.
8. Abe H, Yamaguchi-Shinozaki K, Urao T, Iwasaki T, Hosokawa D, et al. (1997) Role of arabidopsis MYC and MYB homologs in drought- and abscisic acid-regulated gene expression. Plant Cell 9: 1859–1868.
9. Abe H, Urao T, Ito T, Seki M, Shinozaki K, et al. (2003) Arabidopsis AtMYC2 (bHLH) and AtMYB2 (MYB) function as transcriptional activators in abscisic acid signaling. Plant Cell 15: 63–78.
10. Wang H, Zhou L, Fu Y, Cheung MY, Wong FL, et al. (2012) Expression of an apoplast-localized BURP-domain protein from soybean (GmRD22) enhances tolerance towards abiotic stress. Plant Cell Environ 35: 1932–1947.
11. Wang S, Wang JW, Yu N, Li CH, Luo B, et al. (2004) Control of plant trichome development by a cotton fiber MYB gene. Plant Cell 16: 2323–2334.
12. Xu B, Gou JY, Li FG, Shangguan XX, Zhao B, et al. (2012) A Cotton BURP Domain Protein Interacts With alpha-Expansin and Their Co-Expression Promotes Plant Growth and Fruit Production. Molecular Plant 6: 945–958.
13. Banzai T, Sumiya K, Hanagata N, Dubinsky Z, Karube I (2002) Molecular cloning and characterization of genes encoding BURP domain-containing protein in the mangrove, Bruguiera gymnorrhiza. Trees 16: 87–93.
14. Ding X, Hou X, Xie K, Xiong L (2009) Genome-wide identification of BURP domain-containing genes in rice reveals a gene family with diverse structures and responses to abiotic stresses. Planta 230: 149–163.
15. Xu H, Li Y, Yan Y, Wang K, Gao Y, et al. (2010) Genome-scale identification of soybean BURP domain-containing genes and their expression under stress treatments. BMC Plant Biol 10: 197.
16. Gan D, Jiang H, Zhang J, Zhao Y, Zhu S, et al. (2011) Genome-wide analysis of BURP domain-containing genes in maize and sorghum. Mol Biol Rep 38: 4553–4563.
17. Shao Y, Wei G, Wang L, Dong Q, Zhao Y, et al. (2011) Genome-wide analysis of BURP domain-containing genes in Populus trichocarpa. J Integr Plant Biol 53: 743–755.
18. Kuhn N, Guan L, Dai ZW, Wu BH, Lauvergeat V, et al. (2013) Berry ripening: recently heard through the grapevine. Journal of Experimental Botany.
19. Hanana M, Deluc L, Fouquet R, Daldoul S, Léon C, et al. (2008) Identification and characterization of "rd22" dehydration responsive gene in grapevine (Vitis vinifera L.). C R Biol 331: 569–578.
20. Cramer GR, Ergül A, Grimplet J, Tillett RL, Tattersall EA, et al. (2007) Water and salinity stress in grapevines: early and late changes in transcript and metabolite profiles. Funct Integr Genomics 7: 111–134.
21. Deluc LG, Grimplet J, Wheatley MD, Tillett RL, Quilici DR, et al. (2007) Transcriptomic and metabolite analyses of Cabernet Sauvignon grape berry development. BMC Genomics 8: 429.
22. Espinoza C, Vega A, Medina C, Schlauch K, Cramer G, et al. (2007) Gene expression associated with compatible viral diseases in grapevine cultivars. Funct Integr Genomics 7: 95–110.
23. Jaillon O, Aury JM, Noel B, Policriti A, Clepet C, et al. (2007) The grapevine genome sequence suggests ancestral hexaploidization in major angiosperm phyla. Nature 449: 463–467.
24. Zenoni S, Ferrarini A, Giacomelli E, Xumerle L, Fasoli M, et al. (2010) Characterization of Transcriptional Complexity during Berry Development in Vitis vinifera Using RNA-Seq. Plant Physiol 152: 1787–1795.
25. Venturini L, Ferrarini A, Zenoni S, Tornielli GB, Fasoli M, et al. (2013) De novo transcriptome characterization of Vitis vinifera cv. Corvina unveils varietal diversity. BMC Genomics 14: 41.
26. Tamura K, Dudley J, Nei M, Kumar S (2007) MEGA4: Molecular Evolutionary Genetics Analysis (MEGA) software version 4.0. Mol Biol Evol 24: 1596–1599.
27. Bailey TL, Elkan C (1994) Fitting a mixture model by expectation maximization to discover motifs in biopolymers. pp28–36.
28. Fasoli M, Dal Santo S, Zenoni S, Tornielli GB, Farina L, et al. (2012) The Grapevine Expression Atlas Reveals a Deep Transcriptome Shift Driving the Entire Plant into a Maturation Program. Plant Cell 24: 3489–3505.
29. Saeed AI, Bhagabati NK, Braisted JC, Liang W, Sharov V, et al. (2006) TM4 microarray software suite. Methods Enzymol 411: 134–193.
30. Dal Santo S, Vannozzi A, Tornielli GB, Fasoli M, Venturini L, et al. (2013) Genome-wide analysis of the expansin gene superfamily reveals grapevine-specific structural and functional characteristics. Plos One 8: e62206.
31. Dash S, Van Hemert J, Hong L, Wise RP, Dickerson JA (2012) PLEXdb: gene expression resources for plants and plant pathogens. Nucleic Acids Research 40: D1194–D1201.
32. Irizarry RA, Hobbs B, Collin F, Beazer-Barclay YD, Antonellis KJ, et al. (2003) Exploration, normalization, and summaries of high density oligonucleotide array probe level data. Biostatistics 4: 249–264.
33. Gautier L, Cope L, Bolstad B, Irizarry RA (2004) affy-analysis of Affymetrix GeneChip data at the probe level. Bioinformatics 20: 307–315.
34. Eisen MB, Spellman PT, Brown PO, Botstein D (1998) Cluster analysis and display of genome-wide expression patterns. Proceedings of the National Academy of Sciences of the United States of America 95: 14863–14868.
35. Murashige T, Skoog F (1962) A revised medium for rapid growth and bioassays with tobacco tissue cultures. Physiologia Plantarum 15: 473–497.
36. Vega A, Gutierrez RA, Pena-Neira A, Cramer GR, Arce-Johnson P (2011) Compatible GLRaV-3 viral infections affect berry ripening decreasing sugar accumulation and anthocyanin biosynthesis in Vitis vinifera. Plant Mol Biol 77: 261–274.
37. Mengiste T, Chen X, Salmeron J, Dietrich R (2003) The BOTRYTIS SUSCEPTIBLE1 gene encodes an R2R3MYB transcription factor protein that is required for biotic and abiotic stress responses in Arabidopsis. Plant Cell 15: 2551–2565.
38. Reid K, Olsson N, Schlosser J, Peng F, Lund S (2006) An optimized grapevine RNA isolation procedure and statistical determination of reference genes for real-time RT-PCR during berry development. BMC Plant Biol 6: 27.
39. Matus JT, Aquea F, Arce-Johnson P (2008) Analysis of the grape MYB R2R3 subfamily reveals expanded wine quality-related clades and conserved gene structure organization across Vitis and Arabidopsis genomes. BMC Plant Biol 8: 83.
40. Gamm M, Heloir MC, Kelloniemi J, Poinssot B, Wendehenne D, et al. (2011) Identification of reference genes suitable for qRT-PCR in grapevine and application for the study of the expression of genes involved in pterostilbene synthesis. Mol Genet Genomics 285: 273–285.
41. Hattori J, Boutilier KA, van Lookeren Campagne MM, Miki BL (1998) A conserved BURP domain defines a novel group of plant proteins with unusual primary structures. Mol Gen Genet 259: 424–428.
42. Granger C, Coryell V, Khanna A, Keim P, Vodkin L, et al. (2002) Identification, structure, and differential expression of members of a BURP domain containing protein family in soybean. Genome 45: 693–701.
43. Cucurachi M, Busconi M, Morreale G, Zanetti A, Bavaresco L, et al. (2012) Characterization and differential expression analysis of complete coding sequences of Vitis vinifera L. sirtuin genes. Plant Physiol Biochem 54: 123–132.
44. Denoeud F, Aury JM, Da Silva C, Noel B, Rogier O, et al. (2008) Annotating genomes with massive-scale RNA sequencing. Genome Biol 9: R175.
45. Boutilier KA, Gines MJ, DeMoor JM, Huang B, Baszczynski CL, et al. (1994) Expression of the BnmNAP subfamily of napin genes coincides with the induction of Brassica microspore embryogenesis. Plant Mol Biol 26: 1711–1723.
46. Bäumlein H, Boerjan W, Nagy I, Bassüner R, Van Montagu M, et al. (1991) A novel seed protein gene from Vicia faba is developmentally regulated in transgenic tobacco and Arabidopsis plants. Mol Gen Genet 225: 459–467.
47. Van Son L, Tiedemann J, Rutten T, Hillmer S, Hinz G, et al. (2009) The BURP domain protein AtUSPL1 of Arabidopsis thaliana is destined to the protein storage vacuoles and overexpression of the cognate gene distorts seed development. Plant Mol Biol 71: 319–329.
48. Peeters L, Fachin D, Smout C, van Loey A, Hendrickx ME (2004) Influence of beta-subunit on thermal and high-pressure process stability of tomato polygalacturonase. Biotechnology and Bioengineering 86: 543–549.
49. Edgar RC (2004) MUSCLE: a multiple sequence alignment method with reduced time and space complexity. BMC Bioinformatics 5: 113.
50. Orozco-Cardenas ML, Ryan CA (2003) Polygalacturonase beta-subunit antisense gene expression in tomato plants leads to a progressive enhanced wound response and necrosis in leaves and abscission of developing flowers. Plant Physiology 133: 693–701.
51. Liu H, Ma Y, Chen N, Guo S, Guo X, et al. (2014) Overexpression of stress-inducible OsBURP16, the beta subunit of polygalacturonase 1, decreases pectin content and cell adhesion and increases abiotic stress sensitivity in rice. Plant Cell Environ 37: 1144–1158.
52. Albertazzi G, Milc J, Caffagni A, Francia E, Roncaglia E, et al. (2009) Gene expression in grapevine cultivars in response to Bois Noir phytoplasma infection. Plant Science 176: 792–804.
53. Davies C, Boss PK, Robinson SP (1997) Treatment of grape berries, a nonclimacteric fruit with a synthetic auxin, retards ripening and alters the expression of developmentally regulated genes. Plant Physiol 115: 1155–1161.
54. Zhang M, Leng P, Zhang G, Li X (2009) Cloning and functional analysis of 9-cis-epoxycarotenoid dioxygenase (NCED) genes encoding a key enzyme during abscisic acid biosynthesis from peach and grape fruits. J Plant Physiol 166: 1241–1252.
55. Deluc LG, Quilici DR, Decendit A, Grimplet J, Wheatley MD, et al. (2009) Water deficit alters differentially metabolic pathways affecting important flavor and quality traits in grape berries of Cabernet Sauvignon and Chardonnay. BMC Genomics 10: 212.
56. Mauch-Mani B, Mauch F (2005) The role of abscisic acid in plant-pathogen interactions. Curr Opin Plant Biol 8: 409–414.
57. Ton J, Flors V, Mauch-Mani B (2009) The multifaceted role of ABA in disease resistance. Trends Plant Sci 14: 310–317.

58. Ryerson E, Li A, Young JP, Heath MC (1993) Changes in abscisic acid levels during the initial stages of host and non-host reactions to the rust fungus. Physiol Mol Plant Pathol 43: 265–273.
59. Cahill DM, Ward EWB (1989) Rapid localized changes in abscisic acid concentrations in soybean in interactions with Phytophthora megasperma f. sp. glycinea or after treatment with elicitors. Physiol Mol Plant Pathol 35: 4583–4593.
60. Whenham RJ, Fraser RSS, Brown LP, Payne JA (1986) Tobacco mosaic virus-induced increase in abscisic acid concentration in tobacco leaves: intracellular location in light and dark green areas, and relationship to symptom development. Planta 168: 592–598.
61. Miozzi L, Napoli C, Sardo L, Accotto GP (2014) Transcriptomics of the interaction between the monopartite phloem-limited geminivirus tomato yellow leaf curl Sardinia virus and Solanum lycopersicum highlights a role for plant hormones, autophagy and plant immune system fine tuning during infection. PLoS One 9: e89951.
62. Alazem M, Lin KY, Lin NS (2014) The abscisic acid pathway has multifaceted effects on the accumulation of Bamboo mosaic virus. Mol Plant Microbe Interact 27: 177–189.
63. Ding Z, Li S, An X, Liu X, Qin H, et al. (2009) Transgenic expression of MYB15 confers enhanced sensitivity to abscisic acid and improved drought tolerance in Arabidopsis thaliana. J Genet Genomics 36: 17–29.
64. Kim J, Kim HY (2006) Molecular characterization of a bHLH transcription factor involved in Arabidopsis abscisic acid-mediated response. Biochim Biophys Acta 1759: 191–194.
65. Carra A, Mica E, Gambino G, Pindo M, Moser C, et al. (2009) Cloning and characterization of small non-coding RNAs from grape. Plant J 59: 750–763.
66. Fernandez L, Torregrosa L, Terrier N, Sreekantan L, Grimplet J, et al. (2007) Identification of genes associated with flesh morphogenesis during grapevine fruit development. Plant Mol Biol 63: 307–323.

6

The Rose (*Rosa hybrida*) NAC Transcription Factor 3 Gene, *RhNAC3*, Involved in ABA Signaling Pathway Both in Rose and *Arabidopsis*

Guimei Jiang[1], Xinqiang Jiang[2], Peitao Lü[1], Jitao Liu[1], Junping Gao[1], Changqing Zhang[1]*

1 Department of Ornamental Horticulture, College of Agriculture and Biotechnology, China Agricultural University, Beijing, PR China, 2 College of Landscape Architecture and Forestry, Qingdao Agricultural University, Qingdao, PR China

Abstract

Plant transcription factors involved in stress responses are generally classified by their involvement in either the abscisic acid (ABA)-dependent or the ABA-independent regulatory pathways. A stress-associated NAC gene from rose (*Rosa hybrida*), *RhNAC3*, was previously found to increase dehydration tolerance in both rose and *Arabidopsis*. However, the regulatory mechanism involved in RhNAC3 action is still not fully understood. In this study, we isolated and analyzed the upstream regulatory sequence of *RhNAC3* and found many stress-related *cis*-elements to be present in the promoter, with five ABA-responsive element (ABRE) motifs being of particular interest. Characterization of *Arabidopsis thaliana* plants transformed with the putative *RhNAC3* promoter sequence fused to the β-glucuronidase (GUS) reporter gene revealed that *RhNAC3* is expressed at high basal levels in leaf guard cells and in vascular tissues. Moreover, the ABRE motifs in the *RhNAC3* promoter were observed to have a cumulative effect on the transcriptional activity of this gene both in the presence and absence of exogenous ABA. Overexpression of *RhNAC3* in *A. thaliana* resulted in ABA hypersensitivity during seed germination and promoted leaf closure after ABA or drought treatments. Additionally, the expression of 11 ABA-responsive genes was induced to a greater degree by dehydration in the transgenic plants overexpressing *RhNAC3* than control lines transformed with the vector alone. Further analysis revealed that all these genes contain NAC binding *cis*-elements in their promoter regions, and RhNAC3 was found to partially bind to these putative NAC recognition sites. We further found that of 219 *A. thaliana* genes previously shown by microarray analysis to be regulated by heterologous overexpression *RhNAC3*, 85 are responsive to ABA. In rose, the expression of genes downstream of the ABA-signaling pathways was also repressed in *RhNAC3*-silenced petals. Taken together, we propose that the rose RhNAC3 protein could mediate ABA signaling both in rose and in *A. thaliana*.

Editor: Ji-Hong Liu, Key Laboratory of Horticultural Plant Biology (MOE), China

Funding: This work was supported by the National Natural Science Foundation of China (Grant No. 31372096) and Beijing Nova Program (Grant No. 2009B51). The funders had no role in study design, data collection and analysis, decision to publish, or preparation of the manuscript.

Competing Interests: The authors have declared that no competing interests exist.

* Email: chqzhang@cau.edu.cn

These authors contributed equally to this work.

Introduction

Drought, or dehydration, is one of the major limiting factors for plant growth, development, and productivity and plants have evolved a range of physiological, biochemical and molecular responses to promote drought stress tolerance [1]. One such response to drought stress is the production of the plant hormone abscisic acid (ABA), which mediates numerous downstream responses, including stomatal closure, thereby restricting water loss. Using genomic and transcriptomic analyses, the products of drought-inducible genes, including those regulated by ABA, have been classified into two groups: structural proteins and regulatory proteins [2], such as transcription factors (TFs).

Drought or dehydration-induced TFs have been isolated from many plant species and demonstrated to be involved in drought tolerance, such as *DREB2A*, *DREB2C*, *AREB1*, and *WRKY57* from *A. thaliana* [3,4], *OsbZIP46* from *Oryza sativa* [5], *TaMYB30-B* from *Triticum aestivum* [6], *RhNAC2* from *Rosa hybrida* [7], and *ThbZIP1* from *Tamarix hispida* [8]. TFs involved in stress responses are typically classified as being involved either in ABA-dependent or the ABA-independent regulatory pathways [9] and structure and sequence analyses of the promoters of ABA-dependent TF genes have identified many stress-related *cis*-elements. These include the G-box (CACGTG, a MYC recognition site), the dehydration-responsive element/C-repeat (DRE/CRT) and the ABA-responsive element [10,11]. Among these, the conserved ABA-responsive *cis*-element, PyACGTG/TGC, also named ABA-responsive element (ABRE), is a signature sequence for genes involved in the ABA signaling pathway [12,13], and is important for promoter activity under osmotic stress conditions, such as those resulting from dehydration and high salinity [12,14]. The functions of ABA-dependent TFs have also been investigated through overexpression in many plant species [11], which has been found to result in hypersensitivity to ABA during seed germination [5,15], constitutive stomatal closure [16] and severely inhibited root growth [17]. Microarray analysis further revealed that constitutive overexpression of ABA-related TF genes, such as *ABO3* and *MYB96*, generally increases the expression of downstream ABA-responsive genes, and results in drought tolerance [18,19].

Another class of ABA-related TFs are NAC (*N*AM, *A*TAF1 and 2, and *C*UC2) proteins, plant-specific transcriptional regulators that contain conserved N-terminal NAC domains and divergent C-terminal regions [20]. NAC TFs play important roles in regulating numerous aspects of growth and development, including cell division, and senescence, as well as responses to environmental stress stimuli [21]. Many are involved in ABA mediated signaling during their response to abiotic stresses, such as *A. thaliana* ANAC019 and ANAC055, soybean GmNAC011 and GmNAC020 and rice OsNAC5 [22,23,24], and their overexpression can result in enhanced ABA sensitivity at both the germination and post-germination developmental stages [25].

Different members of the NAC family have also been shown to be responsive to dehydration in rose petals [7]. Of these, *RhNAC2* has been found to promote petal cell expansion, in association with the regulation of cell wall-related genes [7], while *RhNAC3* regulates osmotic stress-related genes when exposed to drought stress [26]. However, nothing has been reported to date regarding the mechanism by which RhNAC3 participates in the ABA regulatory pathway. In this study, we characterized the upstream regulatory sequence of *RhNAC3* and found that ABREs in the *RhNAC3* promoter are needed for gene activity in both the presence and absence of exogenous ABA. Transgenic *A. thaliana* overexpressing *RhNAC3* showed enhanced ABA sensitivity during seed germination and during stomatal closure, and ABA-responsive genes were also up-regulated under dehydration conditions in both rose and the *A. thaliana* overexpressing *RhNAC* lines. These data indicate that ectopically expressed RhNAC3 enhances ABA sensitivity in *A. thaliana* and is involved in an ABA-dependent signaling pathway, at least some components of which are likely conserved between rose and *A. thaliana*.

Materials and Methods

Regulatory *cis*-element analysis

The upstream regulatory sequence of *RhNAC3* was isolated using PCR-based genome walking method [27]. (We state clearly that no specific permissions were required for these locations/activities and confirm that the field studies did not involve endangered or protected species). And the primers used are listed in Table S1. All amplified fragments were sub-cloned into the pGEM T-Easy Vector (Promega, Madison, WI, USA) and transformed into *Escherichia coli DH5a* cells after sequencing. The position of the translation start site was designated "0". The *cis*-acting elements were analyzed and annotated using two software programs from the Plant *Cis*-acting Regulatory DNA Elements (PLACE) [28] (http://www.dna.affrc.go.jp/PLACE/) and Plant *Cis*-acting Regulatory Elements (PlantCARE) [29] (http://bioinformatics.psb.ugent.be/webtools/plantcare/html/) software programs. For NAC-binding site analysis of RhNAC3 upregulated genes in *A. thaliana*, a 1,000 bp regulatory sequence upstream of the genes was searched and analyzed by TAIR Loci Upstream Seq –1,000 bp of 'Sequence Bulk Download and Analysis' at www.arabidopsis.org.

Construction of plant expression vectors and *Arabidopsis* transformation

The 977 bp upstream regulatory sequence of *RhNAC3* was amplified with 5′ ACC*AAGCTT*CATTCTACTTGTCCAAATCTGAACCTC 3′ and 5′ GCTCTAGACCGTATCAGAGAGATGAAACAGGAA 3′ (Table S1) and the product digested with *Hind* III and *Xba*I, and inserted into the pBI121 binary vector. The resulting Pro_{RhNAC3}:GUS plasmid was introduced into the *Agrobacterium tumefaciens* strain *GV3101* and transformed into *A. thaliana* (Columbia) by the floral dip method [30]. Ten independent lines of kanamycin-resistant transgenic plants were obtained. The homozygous T3 generation seeds of the transgenic lines were used for subsequent experiments.

Histochemical staining and quantitative GUS activity assay

Histochemical staining for GUS activity was performed as described by Li *et al.* (2009) [31]. Plant samples exposed to different treatments were immersed in GUS staining buffer (0.5 mM 5-bromo-4-chloro-3-indoly β-D-GlcA, 0.5 M NaH_2PO_4, pH 7.0, 1 mM EDTA, 0.5 mM potassium ferricyanide and 0.5 mM potassium ferrocyanide). After staining at 37°C for 3–10 h, the samples were immersed in 95% (v/v) ethanol at 37°C to remove chlorophyll. For histochemical analysis of the Pro_{RhNAC3}:GUS transgenic *A. thaliana* plants in response to ABA treatment, 9-day-old seedlings were grown on MS medium supplemented with 100 µM ABA for 4 days, before being sampled for histochemical GUS staining. The GUS staining patterns were examined under a microscope (BX51; Olympus) and analyzed using Photoshop CS6 software (Adobe, McLean, VA). Quantitative assays of GUS activity were performed as described by Jefferson *et al.* (1987) [32]. All experiments were performed three times to give three independent biological replicates.

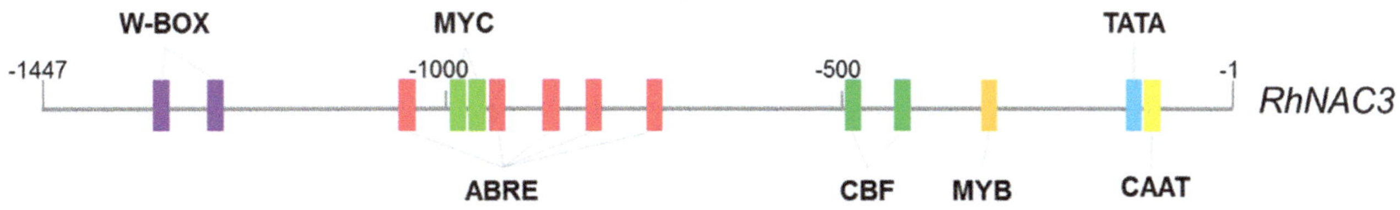

-484	acgtg	-516	acgtgc	-600	acgtgg	-914	acgtgc	-1134	acgtgtc

Figure 1. Schematic representation of the *RhNAC3* promoter. The major stress-related *cis*-acting elements in the 1447 bp promoter of *RhNAC3* are shown. The position and putative sequences of ABRE elements are listed.

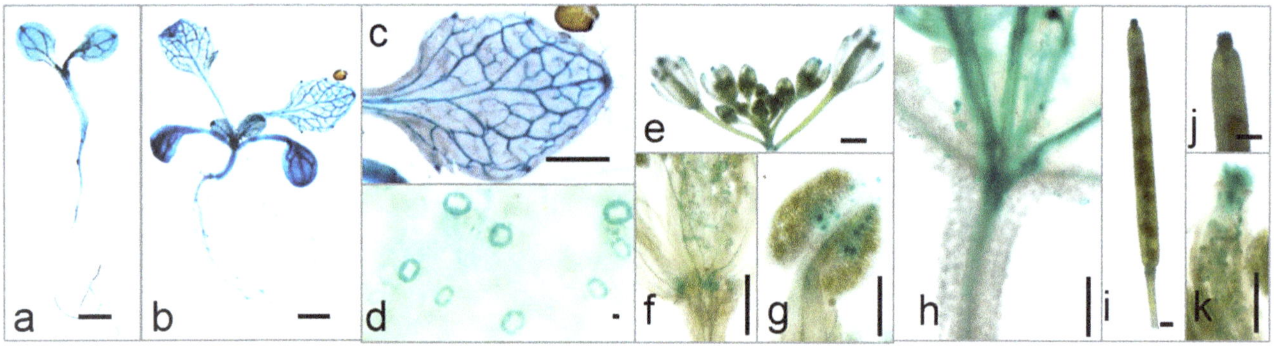

Figure 2. Histochemical analysis of *RhNAC3* expression in *A. thaliana*. The *GUS* gene driven by the *RhNAC3* promoter was expressed in 5-day-old *A. thaliana* seedlings (**a**), 14 day-old seedlings (**b**), young leaves (**c**), stomata of young leaves (**d**), flowers (**e**), petals (**f**), stigma (**g**), apical stems of inflorescences (**h**), mature siliques (**i** and **j**; the latter is magnified); and immature seeds (**k**), Scale bars = 1 mm.

Construction of the truncated *RhNAC3* promoter-GUS fusion and transient expression assays

Three truncated *RhNAC3* promoter fragments, N0 (−1447 to −160 bp), N1 (−707 to −160 bp) and N2 (−377 to −160 bp) were amplified from rose genomic DNA and cloned into a modified pUC19 plasmid containing the GUS reporter gene, as described by Dai *et al.* (2012) [7]. To mutate the ABRE cis-element, we replaced the ACGT of the ABRE core sequence with TTTA using overlap PCR methods [33]. Mutation fragments of N0 (mN0, five ABREs mutated) and N1 (mN1, three ABREs mutated) were amplified, and cloned into the modified pUC19 as described for N0 and N1. *A. thaliana* mesophyll protoplasts were transformed with the resulting vectors: N0, mN0, N1, mN1 and N2, and an empty (normal) vector control (NC). For the ABA treatment experiments, *Arabidopsis* mesophyll protoplasts harboring the different constructs (N0, N1, N2 and NC) were exposed to 10 µM ABA (Sigma, St. Louis, MO). GUS activity was measured in protoplast extracts after 24 h of incubation with ABA. Isolation of *A. thaliana* mesophyll protoplasts, transformation of protoplasts and GUS activity assays were carried out as previously described [32,34]. The primers are listed in Table S1, and the experiments were performed in triplicate.

Seed germination assay and root growth measurements

The *RhNAC3*-overexpressing plants (overexpressor OE#3, OE#6 and OE#12) had previously been generated [26], and wild type (WT) and vector (VC) plants were used as controls. Approximately 50 seeds were plated onto solid MS medium supplemented with either 0, 0.2, 0.4 or 0.8 µM ABA. After vernalization at 4°C for 3 days, the seeds were moved to a temperature controlled room at 23±1°C under long-day conditions (16 h light/8 h dark cycle), with a light intensity of 80–100 µmol/m^2/s and 40–60% relative humility. The rates of radicle emergence and cotyledon greening were measured after 7 days. All experiments were performed in triplicate. To measure seedling root growth, 5-day old seedlings of OE#3, OE#6 and OE#12 were transferred to plates of MS medium containing 5, 10, or 30 µM ABA respectively, and grown vertically. After growth for 10 d, primary root length and lateral root number was measured and analyzed using the Image J software (http://rsbweb.nih.gov/ij/). The WT and VC plants were used as controls.

Stomatal aperture measurements

For ABA-induced stomatal closure, mature leaves from light-grown 3-week-old control and *RhNAC3* transgenic plants were detached and incubated in stomatal opening solution (10 mM KCl, 100 µM $CaCl_2$ and 10 mM MES, pH 6.1) for 2 h at 22°C [16], before being transferred to fresh stomatal opening solution containing 0 µM or 10 µM ABA. Stomata on abaxial surfaces were photographed through a light microscope (BX51; Olympus), and the stomatal aperture (the ratio of width to length) was measured (n = 20). For drought-induced stomatal closure, 3-week-old seedlings of WT, VC and *RhNAC3*-overexpressing lines were grown for 10 d without water. Plants grown under normal conditions were used as control. Leaves in the same position on the plant were sampled, and the stomata on the leaf abaxial surfaces were immediately photographed. Stomatal aperture was measured (n = 20) and all experiments were repeated three times.

Quantitative reverse transcription PCR analysis

The detached leaves of 3-week-old vector (VC) and *RhNAC3* overexpressor (OE#3, 6 and 12) *A. thaliana* plants were dehydrated for 3 h at 23–25°C, 40–50% relative humidity, and 100 µmol m^{-2}s^{-1} light intensity, then sampled for quantitative reverse transcription polymerase chain reaction (qRT-PCR) analysis. Total RNAs were isolated from the leaf samples using the Trizol agent (Invitrogen, Carlsbad, CA). DNase-treated RNA (1 µg) was used for first-strand cDNA synthesis (Invitrogen, Carlsbad, CA) and the cDNA (2 µL) was used as the template in a 20 µL qRT-PCR using a qPCR Kit (Kapa Biosystems, Woburn, MA). The *A. thaliana Actin2* gene (GenBank accession no. NM_112764) was used as an internal control. The 11 selected genes and gene specific primers used for the qRT-PCR analysis are listed in Table S1. Each qRT-PCR evaluation was performed with three biological replicates.

For qRT-PCR analysis of downstream genes of *RhNAC3* action in rose, nine putative ABA signaling and downstream rose genes from the ABA-signaling pathways were selected from our rose transcriptome databases [7] (Table S2). The rose cDNAs from Tobacco Rattle Virus (TRV) and *RhNAC3*-silenced petals were obtained in our previous study [26]. *RhUbi1* (accession no. JK622648) was used as the internal control. The gene specific primers for qRT-PCR are listed in Table S1. Each qRT-PCR analysis was performed with three biological replicates.

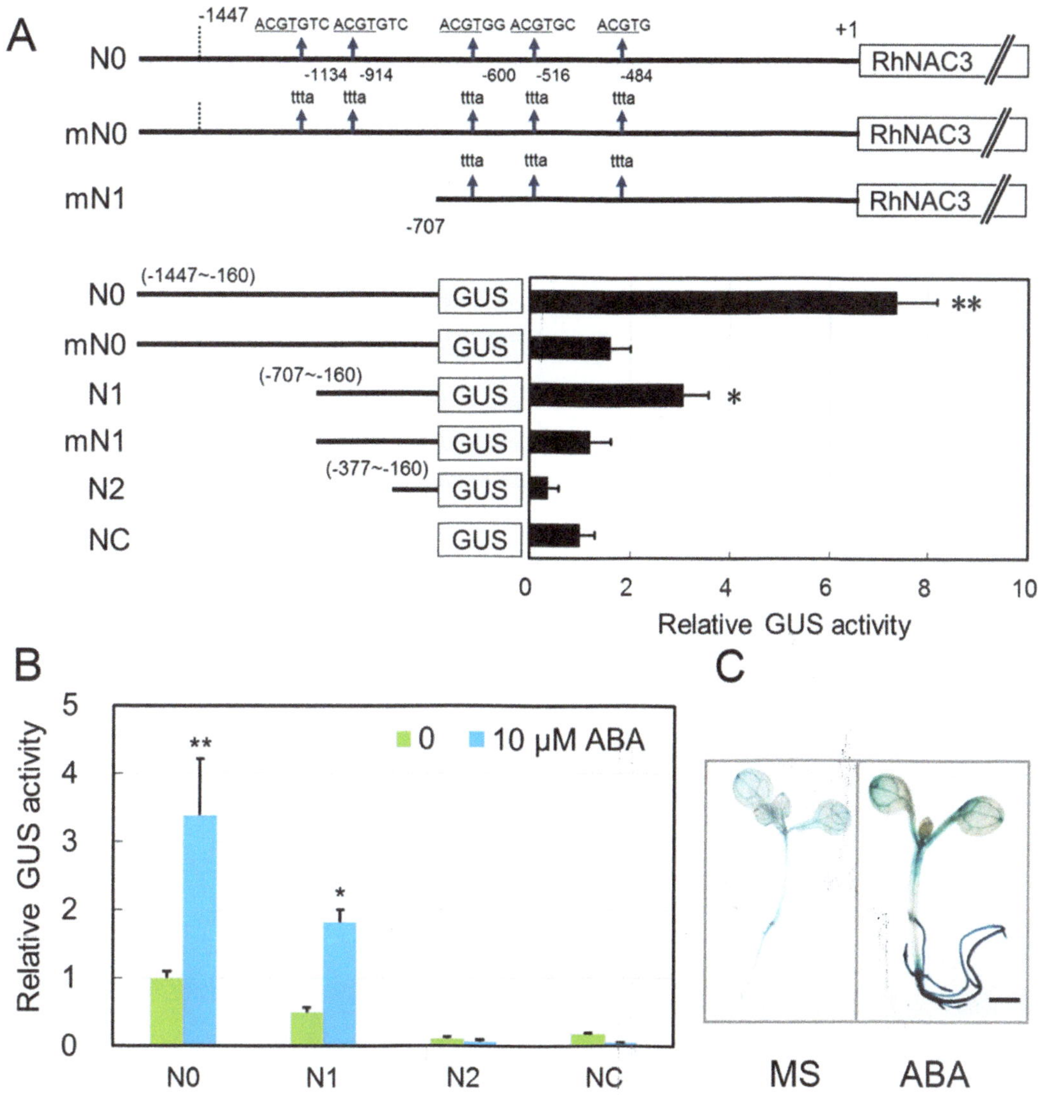

Figure 3. Deletion analysis and ABA dose dependent response of *RhNAC3* promoter activity. A, Assays of GUS activity in *A. thaliana* protoplasts containing RhNAC3 promoter deletion and ABRE mutation constructs. The numbers on top represent the positions of the ABRE *cis*-elements and mutations of ABREs in the *RhNAC3* promoter region. Relative GUS activity in transient expression experiments using five different constructs (N0, mN0, N1, mN1 and N2) and the vector control (NC) is shown at the bottom. GUS activity was determined after 24 h of incubation. Error bars represent standard error ($n=3$). **: $P<0.01$, *: $P<0.05$, *t* test. **B,** Effects of exogenous ABA on GUS activity in *A. thaliana* protoplasts containing *RhNAC3* promoter deletion constructs. The truncated *RhNAC3* promoter constructs (N0, N1 and N2) and vector control (NC) were transformed into *A. thaliana* protoplasts, which were then exposed to 0 and 10 μM exogenous ABA. GUS activity in protoplast extracts was measured after 24 h of incubation with ABA. Error bars represent standard error ($n=5$). **C,** Histochemical analysis of *RhNAC3* promoter::GUS expression in response to ABA. 9-day-old transgenic seedlings were grown on MS medium only or MS medium plus ABA (transferred to MS medium plus 100 μM ABA for 4 days) before being subjected to histochemical GUS staining. Scale bar = 1 mm.

Electrophoretic mobility-shift assay

The electrophoretic mobility-shift assay (EMSA) was performed according as previously described [26] with minor modifications. To construct the GST-RhNAC3 fusion protein, the N-terminal of RhNAC3 (RhNAC3^{N1-162}) was amplified by PCR (primers are listed in Table S1) and the PCR product was ligated into the pGEX-2T vector (Pharmacia LKB Biotechnology, Piscataway, NJ) via the *Bam*HI and *Sac*I sites and the recombinant vector was expressed in *Escherichia coli* BL21 cells. The fusion protein was induced by 0.2 mM isopropyl β-D-1-thiogalactopyranoside (IPTG), and the cells *E. coli* incubated at 28°C for a further 6 h. The recombinant protein was purified by GST-agarose affinity chromatography (GE Healthcare, http://www.gehealthcare.com/). Biotin-labeled DNA fragments used in the EMSA contain one or two putative NAC binding sequences [21]. The probes were incubated with the fusion protein at room temperature for 25 min in binding buffer (10× concentration: 100 mM Tris, 500 mM KCl, 10 mM dithiothreitol; pH 7.5). Each 20 μL binding reaction contained 0.2 pmol biotin probe and 2 μg fusion protein, and 1 μg Poly (dI•dC) was added to the reaction to minimize nonspecific interactions. The reaction products were analyzed using 5% native polyacrylamide gel electrophoresis and 0.5× Tris-borate/EDTA buffer. After electrophoresis, the DNA fragments on the gel were transferred to a nitrocellulose membrane using 0.5× Tris-borate/EDTA at 380 mA (~100 V) for 30 min at 4°C. After UV cross-linking, the membrane was

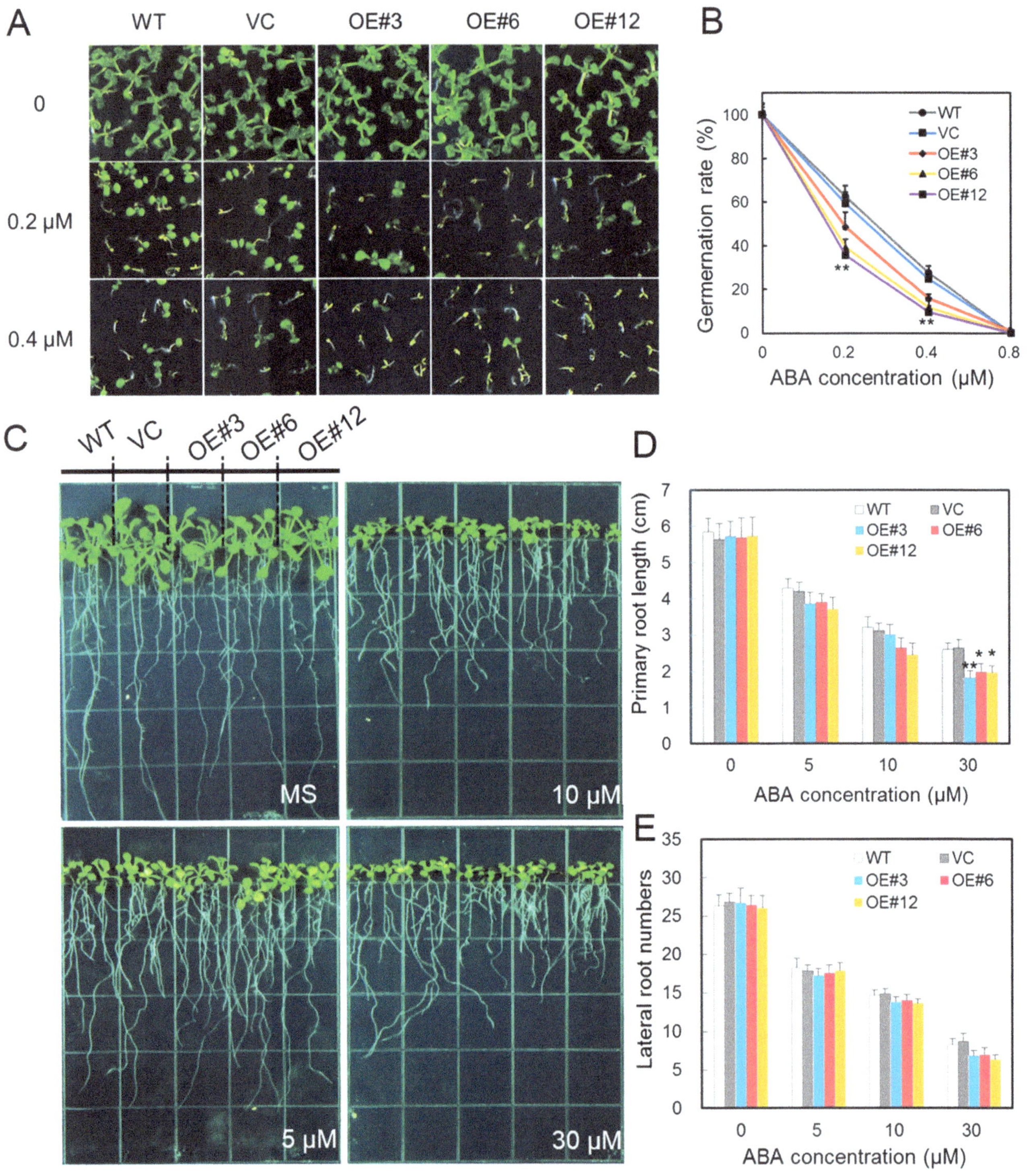

Figure 4. Effect of ABA concentration on seed germination and root growth in WT and *RhNAC3* overexpressing *A. thaliana* plants. A, Seed germination phenotypes. The homozygous T3 seeds of *RhNAC3*-overexpressing lines (OE#3, OE#6 and OE#12), wild type (WT) and vector plants were plated on MS supplemented with 0, 0.2 or 0.4 μM ABA. Images were obtained 7 days after planting. **B,** Seed germination rates. The germination rates were measured 7 days after planting. Error bars represent standard error ($n = 3$). **: $P<0.01$, *: $P<0.05$, t test. **C,** Root growth phenotypes. Five-day-old seedlings of WT, vector only and three *RhNAC3*-overexpressing *A. thaliana* lines (OE#3, #6 and #12) were transferred to MS plates supplemented with 0, 5, 10 and 30 μM ABA. Root phenotypes were visualized 10 days after planting. **D,** Primary root length analysis. **: $P<0.01$, *: $P<0.05$, t test. **E,** Lateral root number analysis. Both primary root length and lateral root number were measured after 10 days of growth. Three independent experiments were performed using 15 plants in each experiment in D and E. Error bars represent standard error ($n = 3$).

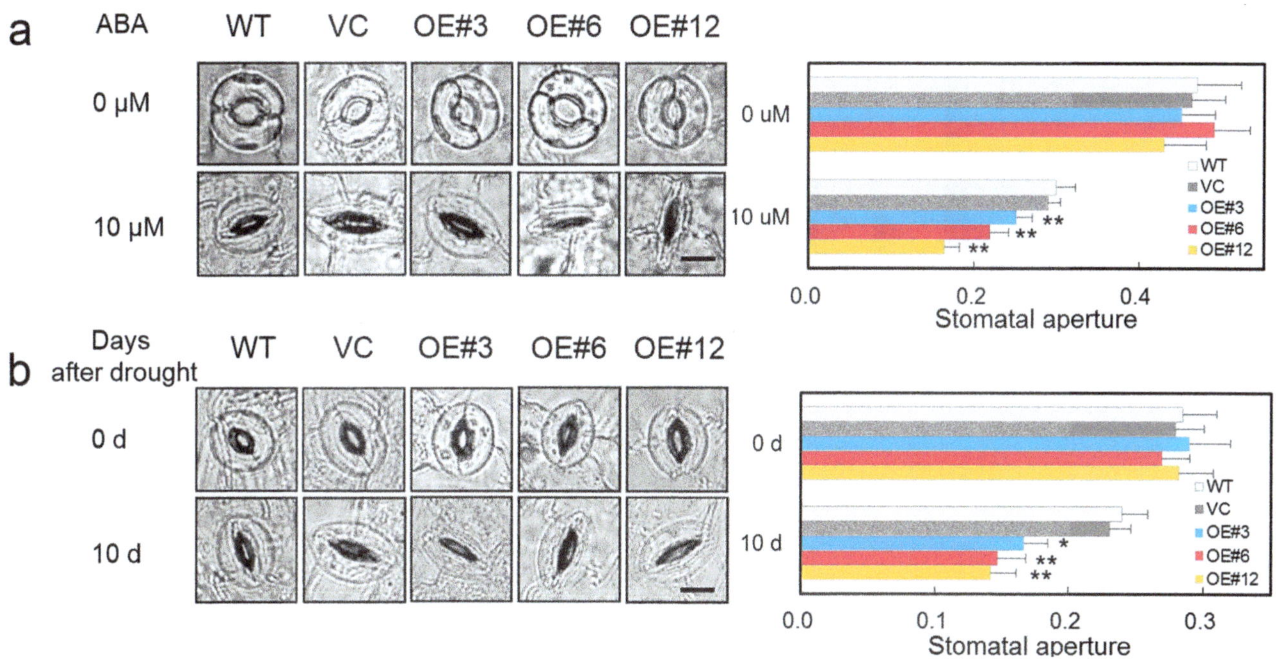

Figure 5. Stomatal aperture of the *RhNAC3* overexpressing *A. thaliana* plants in response to ABA and drought treatments. A, Stomatal aperture in response to ABA. Mature leaves from three-week old wild type (WT), vector control and independent *RhNAC3*-overexpressing plants (OE#3, OE#6 and OE#12) were treated with a stomatal opening solution for 2 h (0 μM) and incubated with 10 μM ABA for 2 h (10 μM). Stomata on the abaxial surfaces were imaged by light microscopy. Stomatal aperture (the ratio of width to length) was quantified using at least 20 guard cells from each sample. *Bar* 10 μm. **B,** Stomatal aperture of *RhNAC3* overexpressing lines in response to drought stress. Three-week old seedlings of WT, vector control and independent *RhNAC3*-overexpressing plants (OE#3, OE#6 and OE#12) were subjected to 10 days without water. Plants grown under normal well watered conditions were used as a control. The leaves were harvested and the stomata on the leaf abaxial surfaces were immediately photographed. Stomatal apertures were then quantified (n = 20). *Bar* 10 μm. **: $P<0.01$, *: $P<0.05$, *t* test.

transferred to conjugate/blocking buffer by mixing 16.75 μL stabilized streptavidin-horseradish peroxidase conjugate with 5 mL blocking buffer. After washing, biotin-labeled DNA was detected by chemiluminescence according to the manufacturer's protocol (Pierce, http://www.piercenet.com/).

Results

Structure and sequence analysis of the *RhNAC3* promoter

To elucidate the regulation of *RhNAC3* transcription, a 1,447 bp fragment corresponding to the sequence immediately upstream of its translational start site (TSS) (GenBank accession number: KJ000025) was isolated by PCR-based genome walking. Subsequent sequence analysis of this region revealed a number of putative *cis*-elements, including elements associated with ABA, cold, pathogen and wounding responses (Figure 1, Figure S1 and Table S2). The TATA box (TTATTT) was found −104 bp upstream of the TSS and a CAAT-box sequence (CAAT) was found 2 bp downstream of the TATA-box sequence. Five ABRE-related sequence motifs (ACGTG), located at positions −1140 to

Table 1. Analysis of putative NAC binding *cis*-elements in the promoter regions of genes downstream from *RhNAC3* action.

Gene name	NAC protein binding sites			
	CACG	CGTG	GTGC	CATGTG
RD29A	4	1	1	
RD29B	2	4	2	
RD20	7	4	2	2
RD26	5	6	3	
COR47	1	1		1
COR15A	4	1		
KIN2	6	3		
ABI1	2			
ABI3	1	2	1	
ABF4	3	1		
ABA3	4	4		

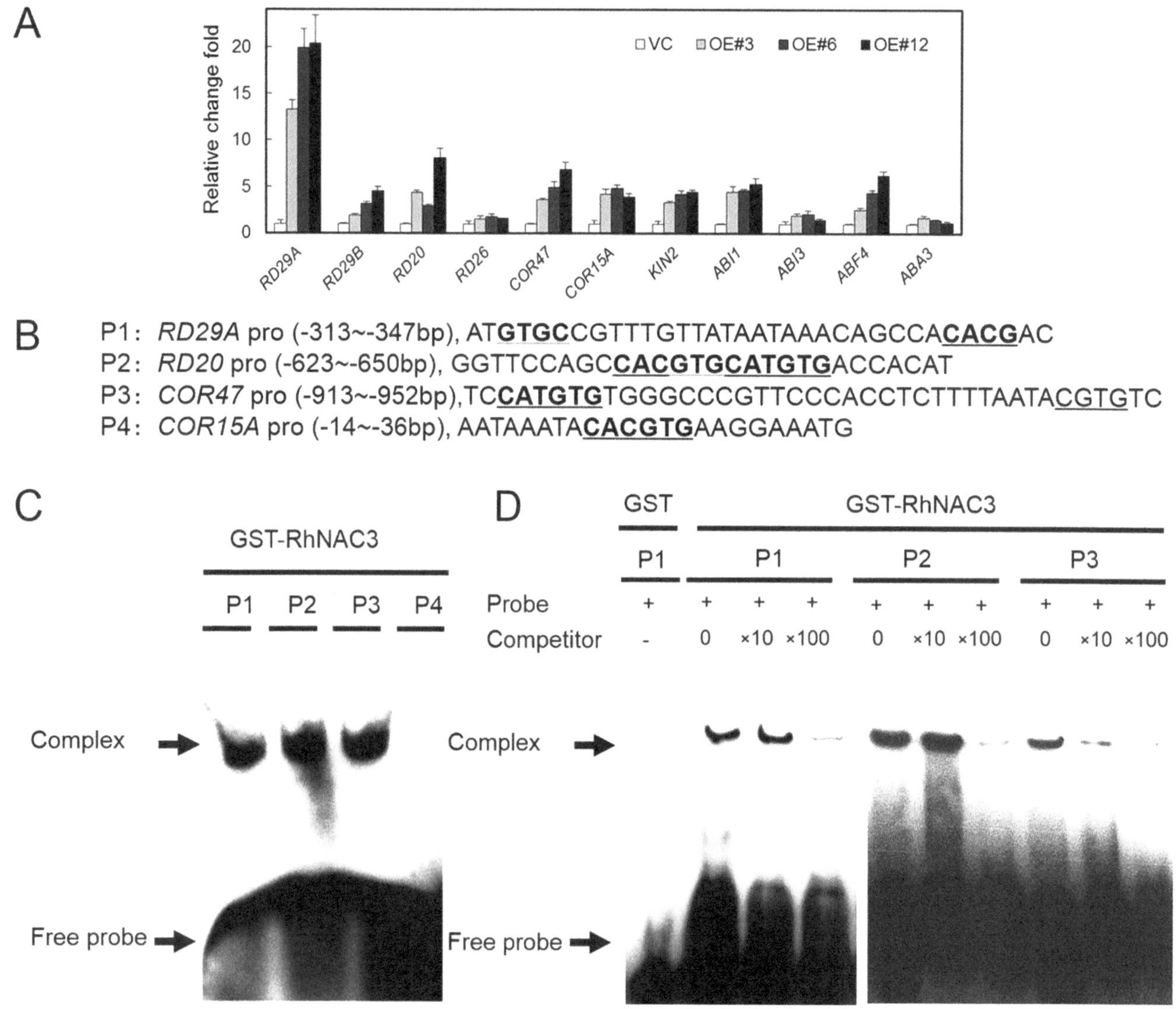

Figure 6. RhNAC3 binding to the regulatory sequences of ABA-related *A. thaliana* genes. A, ABA-related gene expression in *RhNAC3* overexpressing *A. thaliana* lines The aerial parts of light-grown, 3-week old vector control and three independent *RhNAC3* overexpressing *A. thaliana* lines (OE#3, OE#6 and OE#12) were dehydrated for 3 h and sampled (23–25°C, 40–50% relative humidity). The expression patterns of 11 ABA-responsive genes were analyzed by qPCR and the data represents the fold induction of each gene by dehydration relative to the control treatment. Mean values from three independent biological replicates were normalized to the levels of the internal control gene *Actin2*. **B,** Sequences and positions of putative RhNAC3 binding elements used in an electrophoretic mobility shift assay (EMSA). Probes were derived from the regulatory sequences of 4 selected ABA-responsive *A. thaliana* genes. Underlined letters indicate the core sequences of NAC protein targeted promoters. The sense strands of the oligonucleotide probes corresponding to the predicted RhNAC3 binding sites are shown. **C,** Interaction between GST-RhNAC3N$^{1-162}$ and biotin-labeled probes indicated in (B). **D,** DNA-binding specificity for RhNAC3 with interacting probes. The arrows indicate the positions of protein/DNA complexes and the free probes. Purified protein (2 μg) was incubated with 0.2 pmol of the biotin probe. GST incubated with the P1 probe was used as a control, and a 10 or 100 fold excess of the unlabeled P1, P2, or P3 probes was used for competitive binding.

−1136, −918 to −914, −605 to −601, −521 to −517 and −488 to −484, were identified, which we hypothesized might be involved in an ABA response. Other potential regulatory elements that were found include two CBF sequences (RYCGAC) at positions −493 to −488 and −404 to −399, and a Myb-type TF recognition sequence (GGATA) at position −291 to −287. There are also two Myc-type TF recognition sequences (CACATG) at positions −985 to −980 and −951 to −946, and two W-box sequences (TGACT) at positions −1300 to −1296 and −1231 to −1227, relative to the TSS. Collectively, the presence of these *cis*-elements suggests that *RhNAC3* may play a role in responses to a variety of stresses, such as cold, pathogen challenge or wounding, particularly via the ABA-dependent pathway.

Activity of the *RhNAC3* promoter in Pro$_{RhNAC3}$:*GUS* transgenic *A. thaliana* lines

We next examined the spatial expression pattern of the *RhNAC3* gene in 10 independent *A. thaliana* lines (Pro$_{RhNAC3}$:*GUS*) that had been transformed with a construct containing the *RhNAC3* promoter fused to the *GUS* reporter gene. The T3 generation homozygotes of six lines were selected to analyze by GUS staining. Histochemical GUS staining revealed that Pro$_{RhNAC3}$:*GUS* was expressed almost throughout the entire plant

Table 2. Up-regulated genes involved in the ABA response in *A. thaliana* lines overexpressing *RhNAC3*.

Affy ID[a]	Description[b]	AGI Code[c]	Fold change[d]	P-value[e]
	Signal transduction (21)			
267069_at	Calmodulin (CAM)-binding protein of 25 kDa	At2g41010	6.04	0.025
258947_at	Calcium-binding EF-hand family protein	At3g01830	2.58	0.045
255844_at	Protein kinase family protein / peptidoglycan-binding LysM domain-containing protein	At2g33580	2.48	0
255503_at	Concanavalin A-like lectin protein kinase family protein	At4g02420	2.39	0.045
251054_at	Lectin receptor kinase a4.3	At5g01540	2.39	0.02
261662_at	MAP kinase kinase 7	At1g18350	2.19	0.019
266371_at	Calcium-binding EF-hand family protein	At2g41410	2.14	0.006
257751_at	MAP kinase substrate 1	At3g18690	2.09	0.011
266037_at	Protein kinase superfamily protein	At2g05940	2	0.001
	Transcriptional regulation (32)			
261648_at	Salt tolerance zinc finger	At1g27730	10.1	0.011
257022_at	Zinc-finger protein 2	At3g19580	7.56	0.004
248448_at	Integrase-type DNA-binding superfamily protein	At5g51190	4.41	0.04
257053_at	Ethylene responsive element binding factor 4	At3g15210	3.69	0.022
266719_at	Circadian clock associated 1	At2g46830	3.44	0.049
246932_at	Integrase-type DNA-binding superfamily protein	At5g25190	3.04	0.044
252859_at	Integrase-type DNA-binding superfamily protein	At4g39780	2.76	0.032
266656_at	Zinc finger C-x8-C-x5-C-x3-H type family protein	At2g25900	2.41	0.018
258436_at	RING/U-box superfamily protein	At3g16720	2.4	0.001
245051_at	WRKY DNA-binding protein 15	At2g23320	2.34	0.003
259626_at	Basic region/leucine zipper motif 60	At1g42990	2.26	0.007
256426_at	RING/FYVE/PHD zinc finger superfamily protein	At1g33420	2.05	0.002
252009_at	A20/AN1-like zinc finger family protein	At3g52800	2.01	0.047
	Stress responsive (19)			
247708_at	Zinc finger (C3HC4-type RING finger) family protein	At5g59550	4.75	0.034
257763_s_at	Receptor like protein 38	At3g23110	3.9	0.001
262911_s_at	HSP20-like chaperones superfamily protein	At1g59860	2.45	0.03
262383_at	Toll-Interleukin-Resistance (TIR) domain-containing protein	At1g72940	2.34	0.043
259105_at	Rubber elongation factor protein (REF)	At3g05500	2.22	0.044
253046_at	Cytochrome P450, family 81, subfamily D, polypeptide 8	At4g37370	2.2	0
	Enzymes and metabolism (55)			
254975_at	2-oxoglutarate (2OG) and Fe(II)-dependent oxygenase superfamily protein	At4g10500	7.62	0.003
256933_at	Bifunctional inhibitor/lipid-transfer protein/seed storage 2S albumin superfamily protein	At3g22600	4.07	0.005
266993_at	Major facilitator superfamily protein	At2g39210	3.58	0.005
263852_at	Nudix hydrolase homolog 6	At2g04450	3.52	0.021
252908_at	Glycolipid transfer protein (GLTP) family protein	At4g39670	3.45	0
248330_at	NAD(P)-binding Rossmann-fold superfamily protein	At5g52810	3.33	0.003
248970_at	Solute:sodium symporters; urea transmembrane transporters	At5g45380	3.12	0.003
255630_at	C2 calcium/lipid-binding plant phosphoribosyltransferase family protein	At4g00700	2.94	0.048
253332_at	Peroxidase superfamily protein	At4g33420	2.86	0.028
249188_at	HXXXD-type acyl-transferase family protein	At5g42830	2.78	0.006
266761_at	NAD(P)-binding Rossmann-fold superfamily protein	At2g47130	2.73	0.001
252098_at	Eukaryotic aspartyl protease family protein	At3g51330	2.67	0.019
253806_at	RING membrane-anchor 2	At4g28270	2.61	0.037
249910_at	Arogenate dehydratase 2	At5g22630	2.58	0.013
245035_at	Acireductone dioxygenase 3	At2g26400	2.5	0.018
251422_at	Preprotein translocase Sec, Sec61-beta subunit protein	At3g60540	2.48	0.032
267337_at	HXXXD-type acyl-transferase family protein	At2g39980	2.38	0.02

Table 2. Cont.

Affy ID[a]	Description[b]	AGI Code[c]	Fold change[d]	P-value[e]
	Signal transduction (21)			
247604_at	COBRA-like protein 5 precursor	At5g60950	2.38	0.05
262237_at	Thioesterase superfamily protein	At1g48320	2.32	0
264843_at	2-oxoglutarate (2OG) and Fe(II)-dependent oxygenase superfamily protein	At1g03400	2.31	0.005
267300_at	UDP-Glycosyltransferase superfamily protein	At2g30140	2.24	0.006
253238_at	O-Glycosyl hydrolases family 17 protein	At4g34480	2.23	0.019
	Cell expansion related (4)			
247866_at	Xyloglucan endotransglucosylase/hydrolase 25	At5g57550	2.29	0.04
248263_at	Plant invertase/pectin methylesterase inhibitor superfamily	At5g53370	2.12	0.024
	Others (61)			
244966_at	Polyketide cyclase/dehydrase and lipid transport superfamily protein	At1g02470	5.58	0.041
256337_at	Serine-type endopeptidase inhibitors	At1g72060	4.84	0.008
257264_at	Receptor-like protein kinase-related family protein	At3g22060	3.77	0.015
258792_at	Glycine-rich protein	At3g04640	3.73	0.031
247193_at	MATE efflux family protein	At5g65380	3.2	0.049
250942_at	Legume lectin family protein	At5g03350	3.18	0.003
254832_at	Bifunctional inhibitor/lipid-transfer protein/seed storage 2S albumin superfamily protein	At4g12490	3.13	0.026
266097_at	SOUL heme-binding family protein	At2g37970	3.1	0.033
246289_at	VQ motif-containing protein	At3g56880	3.02	0.019
246495_at	Unknown protein	At5g16200	2.87	0.038
257690_at	SAUR-like auxin-responsive protein family	At3g12830	2.83	0.009
249769_at	Sigma factor E	At5g24120	2.75	0.027
263948_at	Late embryogenesis abundant (LEA) hydroxyproline-rich glycoprotein family	At2g35980	2.46	0.016
259502_at	Galactose oxidase/kelch repeat superfamily protein	At1g15670	2.32	0.002
248592_at	hydroxyproline-rich glycoprotein family protein	At5g49280	2.21	0.025
259410_at	Regulator of Vps4 activity in the MVB pathway protein	At1g13340	2.2	0.002
266247_at	Cysteine/Histidine-rich C1 domain family protein	At2g27660	2.15	0.019
252053_at	Syntaxin of plants 122	At3g52400	2.13	0.039
259507_at	P-loop containing nucleoside triphosphate hydrolases superfamily protein	At1g43910	2.11	0.03
264951_at	Target of Myb protein 1	At1g76970	2.11	0
262703_at	SAUR-like auxin-responsive protein family	At1g16510	2.1	0.025
258501_at	Glycine-rich protein	At3g06780	2.08	0.02
262571_at	Protein of unknown function (DUF1644)	At1g15430	2.08	0.024
251859_at	Proteophosphoglycan-related	At3g54680	2.03	0.02
	Unknown (27)			
253859_at	unknown protein	At4g27657	8.71	0.017
256891_at	unknown protein	At3g19030	3.98	0.031
260656_at	unknown protein	At1g19380	3.95	0.022
266017_at	unknown protein	At2g18690	3.53	0.008
265276_at	unknown protein	At2g28400	3.05	0.011
258188_at	unknown protein	At3g17800	2.57	0.021
258275_at	unknown protein	At3g15760	2.47	0.031
266259_at	unknown protein	At2g27830	2.19	0.009
252057_at	unknown protein	At3g52480	2.04	0.03

Genes derived from the *RhNAC3* up-regulated genes identified by the ATH1 microarray analysis in our previous study [26], classified to be responsive to ABA treatment according to the AtGenExpress global stress expression dataset [34].

[a] Affymetrix identification codes for the probes.

[b] Description as given by the Munich Information Center for Protein Sequences (MIPS) database.

[c] Represents a hyperlink to TAIR (www.arabidopsis.org) for more information.

[d] The ratio of three independent transgenic lines compared with the ratio of vector control plants. Genes expressed in RhNAC3 overexpressing transgenic plants with an up-regulation ratio higher than 2.0 are shown.

[e] Indicates one-way ANOVA of the differences in mean transcript expression levels between the transgenic and vector control plants at the 0.05 significance level.

A

Clone ID[a]	Accesion number	Description[b]
RU25535	JK622599	Responsive-to-dessication protein 29
RU07831	——	Responsive-to-dessication protein 28
RU01455	JK622958	Responsive-to-dessication protein 21
RU06450	JK620694	Cold regulated protein 47
RU04740	——	Kinesin motor protein (kin2)
RU22946	JK620897	ABI1 (ABA insensitive 1)
RU24499	JK620250	ABI2 (ABA insensitive 2)
RU03861	JK618281	ABF4 (ABRE binding factor 4)
RU26868	——	ABA3/ATABA3/LOS5/SIR3 (ABA deficient 3)

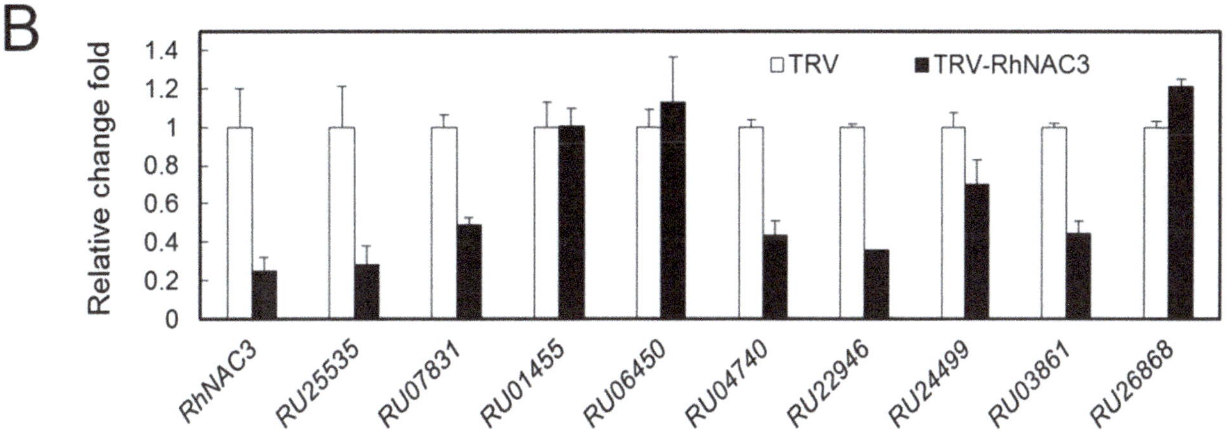

C

P1: *RU25535*-pro (-231~-201), 5'- GATTC**CACG**TCGCCTTTAAGC**CACGTG**GCGG -3'

P2: *RU04740*-pro (-444~-422), 5'- CCACGACGAC**CACGT**TCCCTTG -3'

P3: *RU03861*-pro (-543~-511), 5'- GAGCGGCCGCTTT**CGTG**GTTTCAATGGCTTAAA -3'

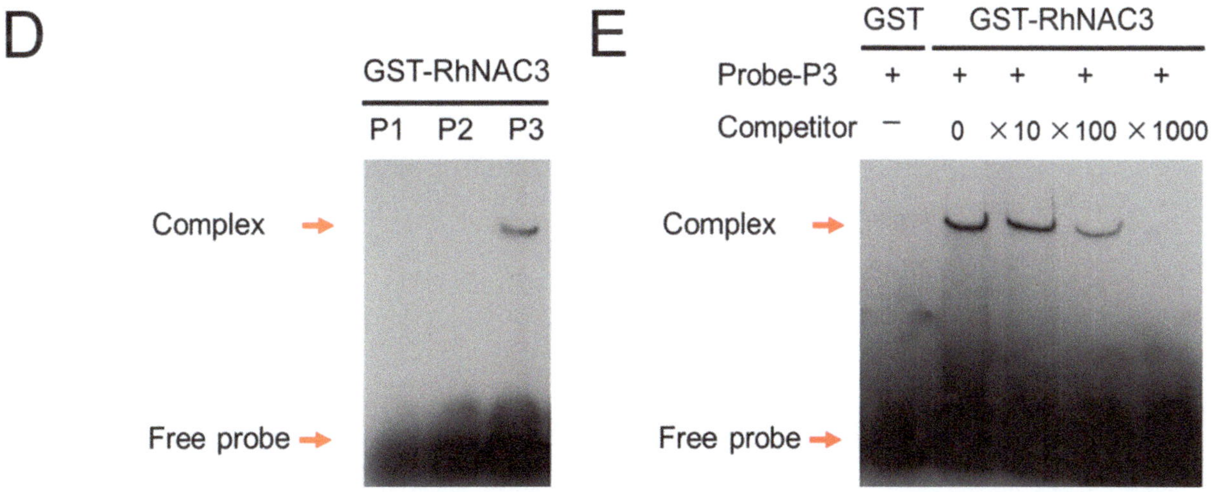

Figure 7. ABA-related gene expression in *RhNAC3*-silenced rose petals. A, The putative ABA signaling and downstream rose genes from the ABA-signaling pathway in rose. a, The clone ID from the rose transcriptome database [7]. b, Description of the *A. thaliana* homolog given by The Arabidopsis Information Resource (TAIR, http://www.arabidopsis.org). **B**, qRT-PCR analysis of *RhNAC3*-silenced rose petals. The rose cDNAs of TRV and *RhNAC3*-silenced (TRV-*RhNAC3*) petals were described in our previous report [26]. Data represent the fold change of each gene by TRV-*RhNAC3* relative to the TRV control. *RhUbi1* was used as the internal control. Error bars indicate SE ($n = 3$). **C**, Sequences and positions of putative RhNAC3 binding elements used for the EMSA. Probes were derived from the regulatory sequence of three selected ABA-related rose genes. Underlined letters indicate the core sequences of putative NAC protein-binding sites. The sense strands of oligonucleotide probes corresponding to the predicted RhNAC3 binding sites are shown. **D**, DNA-binding specificity for RhNAC3 with the probes indicated in **C**. The arrows indicate the positions of protein/

DNA complexes and the free probes, respectively. Purified protein (2 µg) was incubated with 0.2 pmol of biotin probe. **E**, DNA-binding specificity for RhNAC3 with *RU03861*. The *RU03861* (P3) probe incubated with GST was used as a control, and a 10, 100, and 1000 fold excess of the unlabeled P3 was used for competitive binding.

during the seedling stage of development (Figure 2a and b), and particularly strong staining was observed in the vascular system and leaf stomatal guard cells (Figure 2c and d). In addition, GUS staining was detected in the flower petals (Figure 2e and f), stigma (Figure 2g) and apical stem of the inflorescence (Figure 2h), while in mature siliques, staining was primarily localized to the stigma and immature seeds (Figure 2i, j and k). We conclude that the *RhNAC3* gene was expressed ubiquitously in a number of different plant tissues, with higher basal expression levels in leaf guard cells and areas of the vascular system.

ABREs are important for *RhNAC3* promoter activity

To assess the potential role of ABREs in the transcriptional activity of the *RhNAC3* promoter, we made three truncated promoter fragments containing five (N0), three (N1) or no (N2) ABREs, and two fragments, mN0 and mN1 with replacement of ACGT by TTTA in ABRE core sequence. These fragments were fused to the *GUS* reporter gene in the plant transient expression vector pUC19 (Figure 3A, top). The resulting constructs, as well as the vector control (NC), were transformed into *A. thaliana* protoplasts and relative GUS activity was measured. We observed that extracts from the protoplasts transformed with constructs containing a higher number of ABRE copies had higher GUS activity than control transformant extracts. Specifically, the N0 and N1 construct extracts had 7.2-fold and 3.1-fold greater GUS activity, respectively, than the NC extract, while constructs with ABRE mutations (mN0 and mN1) conferred only slight GUS activity and minimal activity was detected in the N2 construct extract (Figure 3A, bottom). We also investigated the transcriptional activity of three truncated *RhNAC3* promoter fragments in *A. thaliana* protoplasts exposed to exogenous ABA. Transcription of the *RhNAC3* promoter (N0 and N1) was found to be induced with a higher GUS activity for the N0 fragment (five ABREs) than for N1 (three ABREs), while no difference was seen for N2 (no ABRE) when ABA was added (Figure 3B). The potential role of the ABREs in the *RhNAC3* promoter in ABA induced transcription was also evaluated using the Pro_{RhNAC3}:*GUS* transgenic *A. thaliana* lines. Seedlings of thee lines had stronger GUS staining after treatment with ABA than those without ABA treatment (Figure 3C). Taken together these data indicate that ABREs in the *RhNAC3* promoter have a cumulative effect on the transcription activity of *RhNAC3* both in the presence and absence of ABA, and that ABA significantly induces *RhNAC3* transcription.

A. thaliana plants overexpressing *RhNAC3* show hypersensitivity to ABA during germination

Our previous study showed that three representative *RhNAC3*-overexpressing *A. thaliana* lines (OE#3, OE#6 and OE#12) had enhanced drought tolerance, with a higher water-retaining ability [26]. To understand the roles of *RhNAC3* in the ABA signaling pathway, we investigated the seed germination rates of *RhNAC3* overexpressing lines following ABA treatment. More than 98% of the seeds sown on control MS medium germinated well, while the germination rate of both the control and transgenic seeds decreased when grown for 7 days on MS medium plus ABA. In the presence of 0.2 µM ABA, the germination rates of WT and Vector (VC) seeds were 63% and 59%, while the rates for OE#3, OE#6 and OE#12 were 49%, 40% and 36%, respectively. A higher concentration of ABA (0.4 µM) resulted in lower germination rates for both the control and *RhNAC3*-overexpressing transgenic plants, and the latter showed a greater decrease (Figure 4A and 4B). We also compared the effects of ABA on the root architectures of the *RhNAC3*-overexpressing and control plants. Primary root growth of *RhNAC3*-overexpressors was inhibited more by 30 µM ABA treatment than that of the control plants, while no significant differences in the number of lateral roots was observed (Figure 4C–E). We conclude from these results that *RhNAC3* overexpression in *A. thaliana* results in ABA hypersensitivity at the seed germination stage.

RhNAC3 participated positively in ABA- and drought-induced stomatal closure

Since the leaves of *RhNAC3*-overexpressing *A. thaliana* plants have greater water-retaining capacity than those of WT plants [26], we examined ABA-dependent stomatal movement phenotypes. Expanded leaves of 3-week-old plants (12 h day/12 h night) were submerged in stomatal opening solution, treated with 10 µM ABA for 2 h, and then the stomatal apertures of the guard cells were measured in the focal planes of the outer edge in epidermal strips. None of the plants exhibited altered stomatal movement in the absence of ABA, and most of the guard cells examined were fully opened. No obvious difference in stomatal aperture (the ratio of width to length) were observed between the controls (WT and VC) and *RhNAC3* overexpressors (OE#3, OE#6 and OE#12). However, in the presence of 10 µM ABA, stomatal closure in the leaves of *RhNAC3* overexpressing transgenic plants was substantially enhanced compared with that of control plant leaves. The stomatal aperture ratios in the OE#3, OE#6, and OE#12 lines were approximately 0.16, 0.22 and 0.25, respectively, compared with 0.30 and 0.29 for the WT and VC plants (Figure 5A).

We also investigated stomatal movement in the leaves of *RhNAC3*-overexpressing *A. thaliana* plants grown under drought conditions. Water was withheld for 10 days from three-week-old plants that had previously been grown under normal conditions, after which the stomatal apertures of the expanded leaves were measured. Under normal growth conditions (0 days), both the controls (WT and VC) and overexpressor lines (OE#3, OE#6 and OE#12) showed no obvious difference, with stomatal apertures of 0.29, 0.28, 0.29, 0.27 and 0.28, respectively. However, after 10 days of exposure to drought conditions, the stomatal aperture ratios had decreased to 0.17, 0.15 and 0.12 in OE#3, OE#6 and OE#12 lines, respectively, but only 0.24 in WT and 0.23 in VC plants (Figure 5B). These results indicate that RhNAC3 is involved in stress responses in an ABA-dependent manner.

RhNAC3 activates ABA-responsive gene expression in rose and *Arabidopsis*

Given that the expression of RhNAC3 showed an association with both ABA sensitivity and drought tolerance, we investigated the expression profiles of drought-induced ABA-responsive genes. Eleven representative genes were selected, including ABA and stress-induced downstream marker genes (*RD29A*, *RD29B*, *RD20*, *RD26*, *COR47*, *COR15A*, and *KIN2*), the ABA-responsive protein phosphatase 2C gene (*ABI1*), the B3-domain transcription factor ABA-insensitive 3 (*ABI3*), the ABA-activated basic Leu zipper TF gene *ABF4*, and the ABA-biosynthesis gene

ABA3 (Table 1 and Figure 6). Under dehydrating conditions, *RD26*, *ABI3* and *ABA3* showed a slightly higher level of expression in *RhNAC3* overexpressing plants than in VC plants, but all the other tested genes showed more than a two-fold greater expression (Figure 6A). The promoters of these genes were then screened for putative NAC binding *cis*-elements and indeed they were identified in all the tested genes (Table 1). We then used an electrophoretic mobility shift assay (EMSA) to investigate whether the RhNAC3 protein directly binds to 4 of the selected genes: *RD29A*, *RD20*, *COR47* and *COR15A*. We observed that RhNAC3 bound to the putative NAC recognition sites of *RD29A*, *RD20* and *COR47*, but no binding signal was detected for *COR15A*, which may therefore be an indirect target of RhNAC3 following its overexpression in *A. thaliana* (Figure 6B–D). In our previous study, we used the ATH1 microarray to identify 219 *RhNAC3*-up-regulated genes [26] and in this current study, we further analyzed the response to ABA of these genes using the AtGenExpress global stress expression dataset [35]. In total, 85 of the 219 genes were found to be ABA responsive, including those encoding proteins involved in signal transduction (e.g. calmodulin-binding protein) and TFs (e.g. zinc-finger protein) (Table 2). In rose, we found 9 putative signaling and downstream genes of the ABA-signaling pathway from our rose transcriptome databases [7] (Figure 7A). qRT-PCR analysis revealed that the expression levels of 6 of these genes were substantially repressed in *RhNAC3*-silenced rose petals (with a fold change <0.8) (Figure 7B). Three ABA-responsive genes with putative NAC binding *cis*-elements (*RU25535*, *RU04740* and *RU03861*) were selected for RhNAC3 binding assays (Table S3). EMSA revealed that RhNAC3 could bind to the promoter region of *RU03861*, a rose homolog of *ABF4*, whereas no binding signals were detected for *RU25535* and *RU04740* (Figure 7C–E), which may therefore be indirect targets of RhNAC3 in rose. Collectively, these data suggest that RhNAC3 positively activates ABA-responsive gene expression and is involved in the ABA signaling pathway in rose and *A. thaliana*.

Discussion

RhNAC3 is involved in the ABA-dependent signaling pathway

Plants respond and adapt to drought stresses through a broad range of molecular and biochemical processes that result in cellular physiological changes [1]. Many drought-inducible genes with various functions, including a number of TFs that regulate stress-inducible gene expression, have been identified by molecular and genomic analyses of *A. thaliana*, rice and other plants, [9,18,23]. These TFs have been classified as being involved in one of two signal transduction pathways: ABA-independent or ABA-dependent [2,36]. ABA-dependent gene induction is controlled by at least five different classes of TFs at the transcriptional level: AREB (bZIPs), NACs, MYB/MYCs, AZF/STZs and DREB1D [2,37]. Many ABA-inducible genes contain a conserved ABRE motif in their promoter regions [38], which functions as a *cis*-element in ABA-regulated gene expression. In a previous study, we determined that *RhNAC3* expression is induced by exogenous ABA [26], implying RhNAC3 is involved in the ABA signaling pathway during stress responses. In this current study, we further analyzed the *RhNAC3* promoter, which was found to contain five ABRE motifs, as well as other stress-responsive elements (Figure 1, Figure S1 and Table S2). Promoter activity was detected ubiquitously in a number of tissues upon transformation into *A. thaliana*, and was highest in leaf guard cells and some areas of the vascular system (Figure 2). In rose, *RhNAC3* expression was also detected in sepals, petals, gynoecia, stamens and receptacles [26]. ABREs were found to be important for the activity of the *RhNAC3* promoter, and the multiple copies had a cumulative effect on transcriptional activity in both the presence and absence of exogenous ABA in *A. thaliana* protoplasts (Figure 3). It has been shown that multiple ABRE elements can collectively confer ABA responsiveness to a minimal promoter, whereas a single copy of ABRE is insufficient for the full ABA response of *AREB1* and *AREB2*, two basic leucine zipper TFs [14]. ABREs are also regarded as one of the major types of *cis*-acting elements in the promoter regions of stress-inducible genes during osmotic stress-responsive transcriptional regulation [12,39]. This is consistent with our previous findings that *RhNAC3* confers dehydration tolerance to rose petals, mainly through the regulation of osmotic adjustment-associated genes [26].

RhNAC3 overexpression in *A. thaliana* enhances ABA sensitivity

Many ABA-responsive TFs have been isolated and characterized from different plant species, including *A. thaliana* [16], rice [5] soybean [22], maize [40] and *Citrus reshni* [41]. Overexpression of these ABA-responsive TF genes has been reported to result in a range of phenotypic changes, including dwarfing [42,43], ABA hypersensitivity [44], lateral root formation [22] and stomatal closure [45]. In our study, *RhNAC3* overexpression in *A. thaliana* lead to ABA hypersensitivity during seedling germination and primary root growth (Figure 4), and promoted stomatal closure after exogenous ABA or drought treatments (Figure 5). We note that this differs from the effects of the soybean ABA-inducible gene *GmNAC20*, which promotes lateral root formation enhances salt and freezing tolerance when overexpressed in transgenic *Arabidopsis* [22].

RhNAC3 enhanced ABA-responsive gene expression in rose and *A. thaliana*

ABA-inducible TFs involved in ABA signaling pathways typically up-regulate ABA-responsive genes or stress-responsive genes [13,18] and such downstream genes have studied using qRT-PCR [16], cDNA microarrays [13] and other transcriptomic analyses [23]. Among these genes, AREB TFs play a primary role in the ABA-dependent signaling pathway [13,46], while other TFs (NAC, MYB/MYC and AZF/STZ TFs) can play additional direct or indirect regulatory roles. The diverse *cis*-elements in the promoter regions of these TFs suggest additional potential mechanisms of transcriptional regulation for ABA-signaling downstream genes as a consequence of abiotic stresses [47]. In the current study, 11 representative ABA-induced genes were investigated, all of which were found to be up-regulated in *RhNAC3* overexpressing *A. thaliana* plants (Figure 6A). Further analysis showed that NAC binding *cis*-elements were present in the upstream regulatory sequences of these genes (Table 1) and that RhNAC3 was able to bind to the putative NAC recognition sites of some of the tested genes (Figure 6B–D). We conclude that RhNAC3 may therefore directly or indirectly regulate their expression at the transcriptional level. These genes were selected based on their invovlement in the ABA-dependent signaling pathway and the fact that their overexpression in *A. thaliana* has been shown to result in an increased ABA sensitivity [25,48]. In a previous microarray study we found that the expression of 219 genes was up-regulated in *RhNAC3* overexpressing plants [26], of which 85 responded to ABA treatment in the current study (Table 2). These results suggest that these genes may contribute to ABA sensitivity in the *RhNAC3* overexpressing *A. thaliana* plants, an idea that is supported by previous experimental evidence. For example, *CYP81D8* expression has been suggested to be regulated by the ABA-dependent pathway under osmotic stress conditions [49], and loss-of-function mutations of *atrdufs* (At5g59550) resulted in hyposensitivity to ABA and reduced tolerance to drought stress [50]. Moreover, the expression of downstream genes in the ABA-signaling pathway was also repressed in *RhNAC3*-silenced rose petals (Figure 7). In our previous study, RhNAC3 was shown to bind to the promoter of *RU23063*, a rose homolog of *ABI2* [26], which encodes a protein phosphatase 2C involved in ABA signal transduction [51]. Here RhNAC3 was observed to bind to the promoter region of the ABA-responsive rose gene *ABF4*, which encodes the ABRE binding factor 4 (Figure 7C–E). Taken together, the data suggest that *RhNAC3* regulated genes that were responsive to osmotic stress, are also involved in the ABA-dependent signaling pathway in both rose and *A. thaliana*.

In conclusion, we found that: (1) ABRE elements in the *RhNAC3* promoter were necessary for, and had a cumulative effect on, its transcription activity in both the presence and absence of exogenous ABA; (2) *RhNAC3*-overexpressing *A. thaliana* lines showed ABA hypersensitivity during seed germination and constitutive leaf stomatal closure under ABA or drought treatment; and (3) RhNAC3 up-regulated the expression level of ABA-responsive genes, which were responsive to osmotic stress. These findings provide new evidence that RhNAC3 is a positive mediator of ABA signaling in the regulation of drought stress tolerance in rose and at least some components of the associated signaling pathways are conserved between rose and *A. thaliana*.

Supporting Information

Figure S1 The promoter sequence of the *RhNAC3* gene. A cumulative result of the 1447 bp promoter sequence showing the positions of important putative cis-acting elements deduced from PlantCARE and PLACE database. The regulatory elements identified by the programs are colorful boxed with appropriate annotations. TATA: TATA box, CAAT: CAAT box, ABRE: ABRE element, CBF: cold binding factor, MYB: MYB binding site, MYC: MYC binding site, WBOX: WRKY binding site.

Table S1 Primer sequences used in this study.

Table S2 *cis*-elements of the upstream regulatory region of *RhNAC3*.

Table S3 Upstream regulatory region of three ABA-related rose genes.

Acknowledgments

We thank PlantScribe (www.plantscribe.com) for editing this manuscript.

Author Contributions

Conceived and designed the experiments: CZ JG XJ. Performed the experiments: GJ XJ PL JL. Analyzed the data: CZ XJ. Contributed reagents/materials/analysis tools: CZ GJ. Contributed to the writing of the manuscript: CZ XJ.

References

1. Lawlor DW (2013) Genetic engineering to improve plant performance under drought: physiological evaluation of achievements, limitations, and possibilities. J Exp Bot 64: 83–108.
2. Shinozaki K, Yamaguchi-Shinozaki K (2007) Gene networks involved in drought stress response and tolerance. J Exp Bot 58: 221–227.
3. Lee S, Kang J, Park HJ, Kim MD, Bae MS, et al. (2010) DREB2C interacts with ABF2, a bZIP protein regulating abscisic acid-responsive gene expression, and its overexpression affects abscisic acid sensitivity. Plant Physiol 153: 716–727.
4. Sakuma Y, Maruyama K, Osakabe Y, Qin F, Seki M, et al. (2006) Functional analysis of an *Arabidopsis* transcription factor, DREB2A, involved in drought-responsive gene expression. Plant Cell 18: 1292–1309.
5. Tang N, Zhang H, Li X, Xiao J, Xiong L (2012) Constitutive activation of transcription factor OsbZIP46 improves drought tolerance in Rice. Plant Physiol 158: 1755–1768.
6. Zhang L, Zhao G, Xia C, Jia J, Liu X, et al. (2012) A wheat R2R3-MYB gene, TaMYB30-B, improves drought stress tolerance in transgenic *Arabidopsis*. J Exp Bot 63: 5873–5885.
7. Dai F, Zhang C, Jiang X, Kang M, Yin X, et al. (2012) *RhNAC2* and *RhEXPA4* are involved in the regulation of dehydration tolerance during the expansion of rose petals. Plant Physiol 160: 2064–2082.
8. Ji X, Liu G, Liu Y, Zheng L, Nie X, et al. (2013) The bZIP protein from *Tamarix hispida*, ThbZIP1, is ACGT elements binding factor that enhances abiotic stress signaling in transgenic *Arabidopsis*. BMC Plant Biol 13: 151.
9. Lata C, Yadav A, Prasad M (2012) Role of plant transcription factors in abiotic stress tolerance. abiotic stress response in plants-Physiological, Biochemical and Genetic Perspectives: 269–296.
10. Abe H, Urao T, Ito T, Seki M, Shinozaki K, et al. (2003) *Arabidopsis* AtMYC2 (bHLH) and AtMYB2 (MYB) function as transcriptional activators in abscisic acid signaling. Plant Cell 15: 63–78.
11. Cramer GR, Urano K, Delrot S, Pezzotti M, Shinozaki K (2011) Effects of abiotic stress on plants: a systems biology perspective. BMC Plant Biol 11: 163.
12. Kim JS, Mizoi J, Yoshida T, Fujita Y, Nakajima J, et al. (2011) An ABRE promoter sequence is involved in osmotic stress-responsive expression of the *DREB2A* gene, which encodes a transcription factor regulating drought-inducible genes in *Arabidopsis*. Plant Cell Physiol. 52: 2136–2146.
13. Yoshida T, Fujita Y, Sayama H, Kidokoro S, Maruyama K, et al. (2010) AREB1, AREB2, and ABF3 are master transcription factors that cooperatively regulate ABRE-dependent ABA signaling involved in drought stress tolerance and require ABA for full activation. Plant J 61: 672–685.
14. Uno Y, Furihata T, Abe H, Yoshida R, Shinozaki K, et al. (2000) *Arabidopsis* basic leucine zipper transcription factors involved in an abscisic acid-dependent signal transduction pathway under drought and high-salinity conditions. Proc Natl Acad Sci U S A 97: 11632–11637.
15. Cho SK, Ryu MY, Seo DH, Kang BG, Kim WT (2011) The *Arabidopsis* RING E3 ubiquitin ligase AtAIRP2 plays combinatory roles with AtAIRP1 in abscisic acid-mediated drought stress responses. Plant Physiol 157: 2240–2257.
16. Ryu MY, Cho SK, Kim WT (2010) The *Arabidopsis* C3H2C3-type RING E3 ubiquitin ligase AtAIRP1 is a positive regulator of an abscisic acid-dependent response to drought stress. Plant Physiol 154: 1983–1997.
17. Fujita Y, Fujita M, Satoh R, Maruyama K, Parvez MM, et al. (2005) AREB1 is a transcription activator of novel ABRE-dependent ABA signaling that enhances drought stress tolerance in *Arabidopsis*. Plant Cell 17: 3470–3488.
18. Ren X, Chen Z, Liu Y, Zhang H, Zhang M, et al. (2010) ABO3, a WRKY transcription factor, mediates plant responses to abscisic acid and drought tolerance in *Arabidopsis*. Plant J 63: 417–429.
19. Seo PJ, Xiang F, Qiao M, Park J, Lee YN, et al. (2009) The MYB96 transcription factor mediates abscisic acid signaling during drought stress response in *Arabidopsis*. Plant Physiol 151: 275–289.
20. Puranik S, Sahu PP, Srivastava PS, Prasad M (2012) NAC proteins: regulation and role in stress tolerance. Trends Plant Sci 17: 369–381.
21. Olsen AN, Ernst HA, Leggio LL, Skriver K (2005) NAC transcription factors: structurally distinct, functionally diverse. Trends in plant science 10: 79–87.
22. Hao Y, Wei W, Song Q, Chen H, Zhang Y, et al. (2011) Soybean NAC transcription factors promote abiotic stress tolerance and lateral root formation in transgenic plants. Plant J 68: 302–313.
23. Jeong JS, Kim YS, Redillas MC, Jang G, Jung H, et al. (2013) *OsNAC5* overexpression enlarges root diameter in rice plants leading to enhanced drought tolerance and increased grain yield in the field. Plant Biotechnol. J 11: 101–114.
24. Tran LSP, Nakashima K, Sakuma Y, Simpson SD, Fujita Y, et al. (2004) Isolation and functional analysis of *Arabidopsis* stress-inducible NAC transcription factors that bind to a drought-responsive *cis*-element in the *early responsive to dehydration stress 1* promoter. Plant Cell 16: 2481–2498.
25. Fujita M, Fujita Y, Maruyama K, Seki M, Hiratsu K, et al. (2004) A dehydration-induced NAC protein, RD26, is involved in a novel ABA-dependent stress-signaling pathway. Plant J 39: 863–876.
26. Jiang X, Zhang C, Lü P, Jiang G, Liu X, et al. (2014) RhNAC3, a stress-associated NAC transcription factor, has a role in dehydration tolerance through regulating osmotic stress-related genes in rose petals. Plant Biotechnol. J 12: 38–48.
27. Liu Y, Chen Y (2007) High-efficiency thermal asymmetric interlaced PCR for amplification of unknown flanking sequences. BioTechniques 43: 649–656.
28. Higo K, Ugawa Y, Iwamoto M, Korenaga T (1999) Plant cis-acting regulatory DNA elements (PLACE) database: 1999. Nucleic acids research, 27: 297–300.

29. Lescot M, Déhais P, Thijs G, Marchal K, Moreau Y, et al. (2002) PlantCARE, a database of plant cis-acting regulatory elements and a portal to tools for in silico analysis of promoter sequences. Nucleic acids research, 30: 325–327.
30. Clough SJ, Bent AF (1998) Floral dip: a simplified method for Agrobacterium-mediated transformation of *Arabidopsis thaliana*. Plant J 16: 735–743.
31. Li Y, Wu Z, Ma N, Gao J (2009) Regulation of the rose *Rh-PIP2;1* promoter by hormones and abiotic stresses in *Arabidopsis*. Plant Cell Rep 28: 185–196.
32. Jefferson RA, Kavanagh TA, Bevan MW (1987) GUS fusions: β-glucuronidase as a sensitive and versatile gene fusion marker in higher plants. EMBO J. 6: 3901–3907.
33. Urban A, Neukirchen S, Jaeger KE (1997) A rapid and efficient method for site-directed mutagenesis using one-step overlap extension PCR. Nuclei Acids Res 25: 2227–2228.
34. Sheen J (2001) Signal transduction in maize and *Arabidopsis* mesophyll protoplasts. Plant Physiol 127: 1466–1475.
35. Kilian J, Whitehead D, Horak J, Wanke D, Weinl S, et al (2007). The AtGenExpress global stress expression data set: protocols, evaluation and model data analysis of UV-B light, drought and cold stress responses. Plant J 50: 347–363.
36. Yamaguchi-Shinozaki K, Shinozaki K (2005) Organization of *cis*-acting regulatory elements in osmotic- and cold-stress-responsive promoters. Trends Plant Sci 10: 88–94.
37. Nakashima K, Yamaguchi-Shinozaki K (2013) ABA signaling in stress-response and seed development. Plant Cell Rep 32: 959–970.
38. Bonetta D, McCourt P (1998) Genetic analysis of ABA signal transduction pathways. Trends Plant Sci 3: 231–235.
39. Fujita Y, Fujita M, Shinozaki K, Yamaguchi-Shinozaki K (2011) ABA-mediated transcriptional regulation in response to osmotic stress in plants. J Plant Res 124: 509–525.
40. Yan F, Deng W, Wang X, Yang C, Li Z (2012) Maize (*Zea mays* L.) homologue of *ABA-insensitive* (*ABI*) *5* gene plays a negative regulatory role in abiotic stresses response. Plant Growth Regul 68: 383–393.
41. Xian L, Sun P, Hu S, Wu J, Liu JH (2014) Molecular cloning and characterization of *CrNCED1*, a gene encoding 9-cis-epoxycarotenoid dioxygenase in *Citrus reshni*, with functions in tolerance to multiple abiotic stresses. Planta 239: 61–77.
42. Vannini C, Locatelli F, Bracale M, Magnani E, Marsoni M, et al. (2004) Overexpression of the rice *Osmyb4* gene increases chilling and freezing tolerance of *Arabidopsis thaliana* plants. Plant J 37: 115–127.
43. Cominelli E, Sala T, Calvi D, Gusmaroli G, Tonelli C (2008) Over-expression of the *Arabidopsis AtMYB41* gene alters cell expansion and leaf surface permeability. Plant J 53: 53–64.
44. Lu G, Gao C, Zheng X, Han B (2009) Identification of OsbZIP72 as a positive regulator of ABA response and drought tolerance in rice. Planta 229: 605–615.
45. Jung C, Seo JS, Han SW, Koo YJ, Kim CH, et al. (2008) Overexpression of AtMYB44 enhances stomatal closure to confer abiotic stress tolerance in transgenic Arabidopsis. Plant Physiol 146: 623–635.
46. Fujita Y, Yoshida T, Yamaguchi-Shinozaki K (2013) Pivotal role of the AREB/ABF-SnRK2 pathway in ABRE-mediated transcription in response to osmotic stress in plants. Physiol Plant 147: 15–27.
47. Ithal N, Reddy AR (2004) Rice flavonoid pathway genes, *OsDfr* and *OsAns*, are induced by dehydration, high salt and ABA, and contain stress responsive promoter elements that interact with the transcription activator, OsC1-MYB. Plant Sci 166: 1505–1513.
48. Yamaguchi-Shinozaki K, Shinozaki K (1994) A novel *cis*-acting element in an *Arabidopsis* gene is involved in responsiveness to drought, low-temperature, or high-salt stress. Plant Cell 6: 251–264.
49. Narusaka M, Seki M, Umezawa T, Ishida J, Nakajima M, et al. (2004) Crosstalk in the responses to abiotic and biotic stresses in *Arabidopsis*: analysis of gene expression in cytochrome P450 gene superfamily by cDNA microarray. Plant Mol Biol 55: 327–342.
50. Kim SJ, Ryu MY, Kim WT (2012) Suppression of *Arabidopsis* RING-DUF1117 E3 ubiquitin ligases, AtRDUF1 and AtRDUF2, reduces tolerance to ABA-mediated drought stress. Biochem Biophys Res Commun 420: 141–147.
51. Rodriguez PL, Benning, G Grill E (1998) ABI2, a second protein phosphatase 2C involved in abscisic acid signal transduction in Arabidopsis, FEBS Lett 421: 185–190.

7

Removal of Fast Flowing Nitrogen from Marshes Restored in Sandy Soils

Eric L. Sparks[1,2]*, Just Cebrian[1,2], Sara M. Smith[1]

1 Dauphin Island Sea Lab, Dauphin Island, Alabama, United States of America, **2** Marine Sciences, University of South Alabama, Mobile, Alabama, United States of America

Abstract

Groundwater flow rates and nitrate removal capacity from an introduced solution were examined for five marsh restoration designs and unvegetated plots shortly after planting and 1 year post-planting. The restoration site was a sandy beach with a wave-dampening fence 10 m offshore. Simulated groundwater flow into the marsh was introduced at a rate to mimic intense rainfall events. Restoration designs varied in initial planting density and corresponded to 25%, 50%, 75% and 100% of the plot area planted. In general, groundwater flow was slower with increasing planting density and decreased from year 0 to year 1 across all treatments. Nevertheless, removal of nitrate from the introduced solution was similar and low for all restoration designs (3–7%) and similar to the unvegetated plots. We suggest that the low NO_3^- removal was due to sandy sediments allowing rapid flow of groundwater through the marsh rhizosphere, thereby decreasing the contact time of the NO_3^- with the marsh biota. Our findings demonstrate that knowledge of the groundwater flow regime for restoration projects is essential when nutrient filtration is a target goal of the project.

Editor: Fei-Hai Yu, Beijing Forestry University, China

Funding: Funding for this project provided by the Alabama Department of Conservation and Natural Resources, State Lands Division, Coastal Section, in part, by a grant from the National Oceanic and Atmospheric Administration, Office of Ocean and Coastal Resource Management, Award #10NOS4190206. The funders had no role in study design, data collection and analysis, decision to publish, or preparation of the manuscript.

Competing Interests: Just Cebrian is currently on the PLOS ONE editorial board. This does not alter the authors' adherence to PLOS ONE policies on sharing data and materials.

* Email: esparks4040@gmail.com

Introduction

Marsh restoration is a ubiquitous practice for mitigation of global marshland loss [1]. However, marsh restoration is expensive and labor intensive [2,3]. Compounded with the costly nature of marsh restoration, there is often inconsistency and discrepant outcomes among different techniques and designs [3]. Some studies have been conducted to evaluate cost-effectiveness of vegetative growth for restored marshes [2]; however, evaluations of the ecosystem services provided by different marsh restoration designs is scant, but should be evaluated to inform managers interested in maximizing the effectiveness of restoration projects [4–6].

Marshes provide important ecosystem services [7–10], and it has been suggested that nutrient filtration is the most economically valuable ecosystem service [4]. Processes such as denitrification and plant uptake can remove a large portion of nutrient inputs into marshes as groundwater percolates through the marsh rhizosphere [11,12].

Most nutrient filtration studies for marshes are conducted in mature natural marshes that are subjected to low to moderate flows of groundwater [12–14]. These studies have demonstrated that the presence of marsh plants increases nutrient removal through direct plant uptake as well as facilitating bacterial processes responsible for outgassing nitrogen (e.g., denitrification and anammox; [15]). However, marshes are subjected to varying groundwater flow rates from upland sources [16–18] and are dependent on factors such as rainfall intensity and soil permeability. In general, when areas are subjected to intense flow events (e.g., heavy rain), it is likely a smaller portion of the nutrients carried in these events can be removed than when the site is subjected to lower flow rates [19]. Along the northern Gulf of Mexico (nGOM) coast, there are frequent and intense rain events [20], thereby subjecting these marshes to a mixture of fast flow events, during and immediately after these rain events, and lower flow between events [16]. Assessment of nutrient removal by restored marshes under different scenarios of groundwater flow is important to improve the effectiveness of marsh restoration efforts targeting nutrient filtration as a primary goal.

In this study, we use black needlerush (*Juncus roemerinaus*) as our restored marsh plant. Black needlerush marshes are dominant on the nGOM coast [21] and have suffered significant loss over past decades primarily attributed to coastal development [22]. Due to the losses of marshes along the nGOM coast and prevalence of black needlerush, this marsh plant is the target for many restoration projects [2,23–25].

In this study, we compare groundwater flow rate and nitrate (NO_3^-) removal from an introduced groundwater solution in five black needlerush marsh restoration designs, varying in initial plant density, with unvegetated controls immediately after planting and one year after planting. Utilization of these marsh planting designs allows for comparisons of NO_3^- removal from fast flowing groundwater across designs that vary in the effort required to plant (i.e., time and cost). The groundwater plume introduced into the marsh mimics a pulse of groundwater derived from an intense rainfall event percolating through porous sediments. Expectations were groundwater flow rates, through the marsh rhizosphere, would decrease and NO_3^- removal would increase with increasing planting density over time. Results from this study can inform managers interested in maximizing restoration efficiency with the goal of reducing nutrient pollution into water bodies.

Materials and Methods

1. Site Construction

On June 11, 2010, we planted a black needlerush marsh on the outskirts of Camp Beckwith (30°23′16″ N, 87°50′31″ W) located on the eastern coast of Weeks Bay in Fairhope, Alabama, USA. The staff of this privately owned camp gave us permission to conduct this work on their property and they should be contacted for future permissions. The planting site was situated on a stretch of sandy beach with natural marsh nearby. This sandy beach was subjected to high wave energy from boat wakes. To reduce wave action at this site, a fence was constructed, prior to marsh restoration, ten meters offshore of the restoration site to reduce shoreline wave energy and erosion. The fence consisted of a wooden frame filled up with dead tree branches and trunks along with other natural debris. Black needlerush sods (approximately 20 cm long, 20 cm wide and 20 cm deep) were harvested from an adjacent marsh and planted at the restoration site. Individual sods had a black needlerush shoot density typical of nGOM salt marshes, with ranges from 1400–1800 shoots m^{-2} [2]. For experimental setup, we used a randomized block design with 3 blocks consisting of 6 plots each, yielding a total of 18 experimental plots (Fig. 1). Blocks were separated by 2 m and each plot had dimensions of 40 cm wide and 170 cm long (Fig. 1). Plots represented different restoration designs in terms of initial plant density (25%, 50%, 50%A, 75% and 100%) plus non-vegetated controls (0%) rendering 6 designs with 3 replicated plots each. Each plot contained 16 sod-sized units and the number of sods planted in the plot corresponded to the planting density treatment (e.g., 4 sods for the 25% planting density; Fig. 1). Plots planted in the 50%A design were arranged in an alternating "checkerboard" pattern (Fig. 1). To contain the introduced groundwater solution, each plot was enclosed on the top (i.e., upland) and the two lateral sides with vertical placement of rigid plastic sheeting. Thirty cm of the sheet height was buried below the sediment surface with 10 cm of sheet height above the sediment surface. Porewater collection wells, screened from 5 cm to 30 cm below the sediment surface, were placed at the bottom of each plot (experimental porewater wells) and 3 (natural porewater wells) on each lateral side of every block (Fig. 1). A diffuser plate was buried in the sand 10 cm from the upland planting edge to help disperse the introduced solution.

2. Experimental methodology and sampling

On June 21, 2010 and June 30, 2011 fluorescein tracing tests were conducted by releasing 15 L of a 20 mg L^{-1} fluorescein solution into each plot. Fluorescein was used because it is not actively removed through biological processes at high rates; therefore, the only factor that can change the fluorescein concentration is dilution [26]. These attributes of fluorescein allow it to be used to assess travel time and dilution rates of groundwater plumes through the plots [26]. The fluorescein solution was released during receding tides when the tide line was at the upland edge of the plots. It took approximately 15 min for the 15 L of solution to disperse through the diffuser plate. This quick pulse was intended to mimic groundwater inputs from an intense rain event over porous sediments and was equivalent to a 1.5 cm rainfall event drained off of a 1 hectare area through 100 linear meters of fringing marsh over a 12 hour period. Porewater samples were taken from each well in 15 min intervals after tracer release for 90 minutes and stored on ice in a cooler for transport back to Dauphin Island Sea Lab for analysis. All fluorescein samples were analyzed on a Turner Designs-700 fluorometer.

To determine how effective the marsh designs were at removing nutrient pollution from a fast flowing groundwater solution, four rounds of NO_3^- solution release and subsequent sampling were performed after the initial planting (June 27, July 8, 9, and 13, 2010) and 1 year post-planting (July 14, 21, 29 and August 2, 2011). The releases were done during receding tides when the tide line was at the upland edge of the plots (i.e., the same timing as the fluorescein tracing tests). For each round we released 15 L of a 200 μM NO_3^- solution into each plot. Water samples were taken from each input container and porewater well at the bottom of the plots. The sampling timeline for porewater samples was determined by the fluorescein experiments (i.e., sample at time fluorescein peak for each plot). To determine background nutrient levels, we also took samples from the natural porewater wells located outside of each block on every sampling day. Porewater samples were filtered in the field and transported to the lab on ice for analysis. NO_3^- concentrations were analyzed using cadmium reduction azo dye assays [27].

3. Calculations and statistical analyses

3.1. Flow rates. Flow rates were analyzed by recording the time when peak fluorescein concentration was observed in each plot. These peak times were analyzed using an ANOVA (treatment × year × block). If block and the interaction between treatment and year were found to be insignificant factors, data were pooled across block and reanalyzed with an ANOVA (treatment × year). For all statistical analyses, tests were conducted using Sigma Stat 3.5 and significance was considered at $p<0.05$ [28].

3.2. NO_3^- removal. As the introduced solution travels through the plots, it will be subjected to dilution through mixing with natural porewater (i.e., not subjected to the introduced solution) with lower background concentrations of NO_3^-. To determine the $[NO_3^-]$ in the portion of the porewater derived from the introduced solution, a dilution correction must be applied that accounts for the $[NO_3^-]$ found in natural porewater. Applying a dilution correction allows for the removal of NO_3^- from the introduced solution to be calculated as the introduced solution (*Input*) travels to the downland edge of the marsh. To calculate removal of NO_3^- from the introduced solution, we subtracted the dilution corrected $[NO_3^-]$ at the downland porewater well from the NO_3^- concentration in the input using the following equation: *Removal = Input - (Downland well – Natural × Dilution) ÷ (1-Dilution)*. The previous equation will be referred to as equation 1 throughout the manuscript. The term *Downland well* in equation 1 is the $[NO_3^-]$ measured at restored marsh porewater wells subjected to the simulated pollution plume. The *Natural* term is the $[NO_3^-]$ in the porewater wells outside of the restored marsh that was not subjected to the simulated pollution plume (Fig. 1). The *Dilution* term in equation 1 is the proportion of sample derived from natural porewater and was calculated as the proportional decrease in [fluorescein] from the input to the porewater collection well (i.e., peak % tracer contribution in Fig. 2). NO_3^- removal was converted to percent removal by dividing it by $[NO_3^-]$ of the input. Similar to the flow rate study, percent NO_3^- removal was first analyzed using an ANOVA (treatment × year × block). If block and the interaction between treatment and year were found to be insignificant factors,

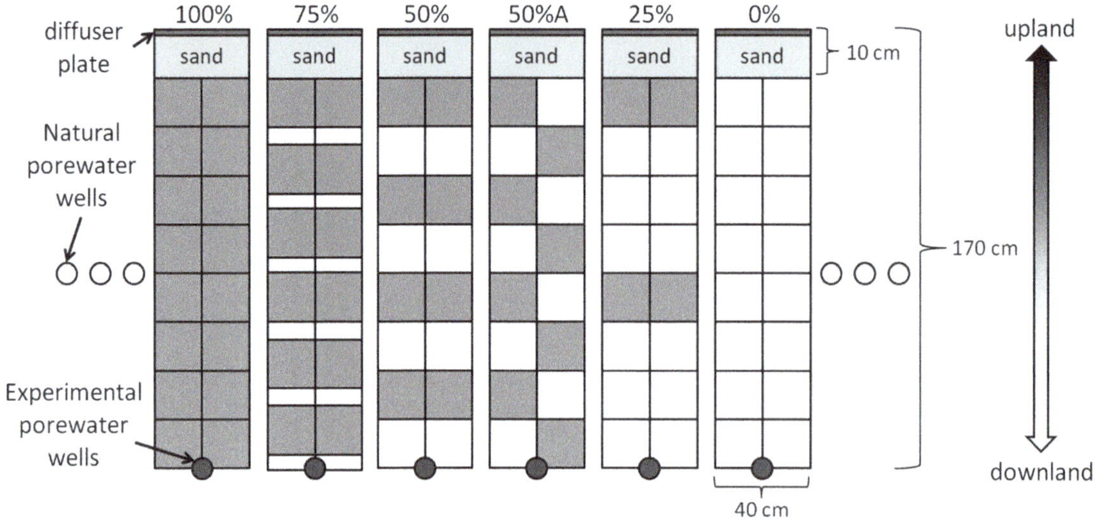

Figure 1. Schematic of 1 block of 6 marsh restoration designs (0%, 25%, 50%, 50%A, 75% and 100%). Shaded squares represent planted sods. There were a total of 3 blocks with each block consisted of a randomized arrangement of all 6 restoration designs. The groundwater solution was introduced at the diffuser plate and flowed down the plots toward the porewater collection well.

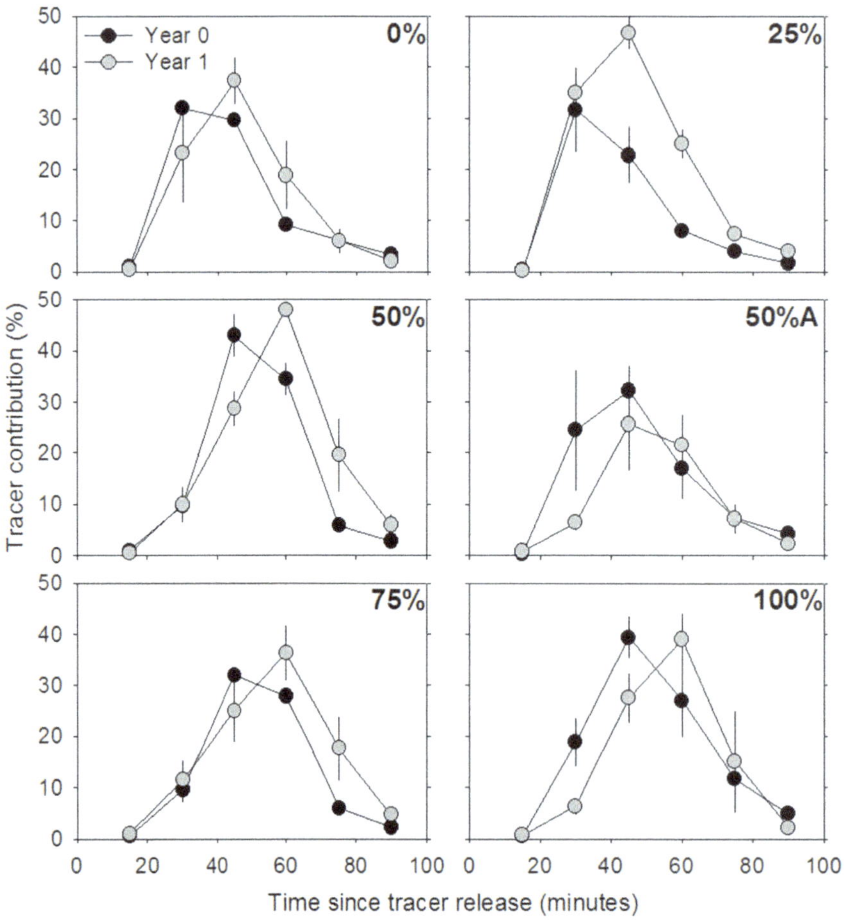

Figure 2. Fluorescein tracer porewater contribution (%) at the downland well through time. Black circles represent year 0 and grey circles represent year 1 samples. Percentages in the top right portion of each plot represents the planting density (0%, 25%, 50%, 50%A, 75% and 100%). Error bars indicate ±1 SE.

the data was pooled across block and reanalyzed with an ANOVA (treatment × year).

Results

For flow rate and NO_3^- removal, block was never a significant factor (flow rate - $p = 0.40$; NO_3^- removal - $p = 0.14$) and there were no significant interactions between treatment and year (flow rate - $p = 0.34$; NO_3^- removal - $p = 0.99$). Therefore, data were pooled across blocks and analyzed with an ANOVA for the effects of treatment and time (Table 1). Only the results of the ANOVA on the data pooled across blocks will be further discussed.

Most of the fluorescein solution traversed the plots in less than one hour across all treatments (Fig. 2). In general, the flow of the introduced solution decreased with increasing planting density (Table 1), as indicated by the later peaks in the dilution curves (Fig. 2). Furthermore, it took longer for the solution to cross the plots one year after planting than two weeks after planting (Table 1; Fig. 2). Longer retention time of the introduced solution within the plots at one year after planting than two weeks after planting implies that flow rates of the introduced solution, through the plots, decreased over time.

Concentrations of NO_3^- in porewater collections wells outside the plots ranged from 0.5 µM to 1.5 µM (i.e., natural or ambient porewater) and were low when compared to concentrations within the plots (i.e., subjected to the introduced solution). As expected, the input had $[NO_3^-]$ of 200 µM±10 µM, whereas wells at the downland edge of the plots had lower $[NO_3^-]$ ranging from 44 µM to 126 µM. When combining the observed changes in $[NO_3^-]$ with a dilution factor (equation 1), our calculations showed only a small percentage of NO_3^- was removed from the introduced solution (3–7%; Table 2). These small percentages of NO_3^- removal were similar across all treatments (Table 1) and sampling years (Table 1; Fig. 3). While NO_3^- processing increased slightly over time and with plant cover (Table 2) these differences were not statistically significant (Table 1).

Discussion

In this study, we found that the introduced groundwater traveled slowest through the most vegetated planting designs and slower one year after planting than immediately after planting. These results offer evidence that the presence of marsh plants increases the time required for groundwater to flow through the restoration area, likely through the presence of the marsh rhizosphere and accumulation of finer grained sediments [29]. As the planted plots mature and density increases, they will likely continue to decrease groundwater flow rates through binding sediments, expansion of the marsh rhizosphere [30] and reduction of wave energy that aids in the accumulation of finer grained sediments [31].

Increases in plant density and the accumulation of finer grained sediment are conducive to increased nutrient removal in marshes [15]. We did find some suggestive evidence that increases in vegetated area increased NO_3^- removal (Table 2), albeit not statistically significant (Table 1). Our range in planting density was large (0%–100%) and we measured NO_3^- removal in all of these planting densities twice over one year. If planting density was a primary driver of NO_3^- removal from this fast flowing solution, we would have captured it with this sampling design. Given no effect of planting density and overall findings that only a small portion of the input NO_3^- was removed across all of the planting designs through time (3–7%), it appears that the groundwater was traveling too quickly through the marsh rhizosphere for plants to have an impact on NO_3^- removal. In studies where groundwater traveled slower through the marsh rhizosphere, marsh plants had time to uptake and facilitate bacterial processes that fueled large removals of nitrogen [17]. Comparing these studies to our study suggests that flow rate can influence how effective marshes are at removing NO_3^- from groundwater. With these results, it is likely that nutrient processing in our restored marsh will remain similar to the unvegetated plots for several years to come [5]. An additional factor contributing to the negligible increases in NO_3^- removal over time is the typical slow growth of black needlerush [21]. This slow growth pattern was evident by visual observations of marginal increases in vegetated area (<5%) for the vegetated designs and no colonization of the 0% planting design at one year after planting. A timeframe when these marshes will decrease groundwater flow enough to allow for large proportions of input nutrients to be removed is unknown; however, previous studies have indicated restoring other ecosystem functions in restored marshes to natural levels takes many years [2,32,33].

A probable explanation to the high groundwater flow rates and low nutrient filtration is that the sediment at the restoration site was sandy, which is also the case for many restored marshes (e.g., Jamaica Bay marsh islands in New York, USA, Grand Bay National Estuarine Research Reserve boat launch marsh in Mississippi, USA, and the Labranche wetlands in Louisiana, USA). Groundwater flows more quickly through sand than sediments with higher mud and silt content typical of mature marshes [34]. Most other studies of nutrient filtration in salt marshes have been conducted mature natural marshes with fine sediments (mainly mud and silt). Finer sediments have smaller interstitial spaces and slow down groundwater flow in relation to the flow in sandy sediments [12–14,35]. Slower flow rates increase contact time between nutrients and reactive areas of the sediment and rhizosphere, thereby increasing opportunities for nutrient processing [19]. In addition, sandy sediments are primarily composed of inorganic material (e.g. quartz) and typically contain

Table 1. Results of ANOVA for flow rates and NO_3^- removal.

Test	Effect	Degrees of Freedom	F value	P value
Flow rates	Treatment	5	21.545	<0.001
	Year	1	76.818	<0.001
	Treatment × Year	5	1.200	0.339
NO_3^- removal	Treatment	5	1.195	0.317
	Year	1	0.197	0.658
	Treatment × Year	5	0.020	0.999

Table 2. Mean percent NO_3^- removal across 6 restoration designs directly after planting and 1 year after planting (±1 SE).

Year	Treatment 0%	25%	50%	50% A	75%	100%
0	3.22 (±0.87)	4.29 (±0.94)	5.44 (±1.81)	6.23 (±1.54)	5.95 (±1.22)	6.47 (±2.18)
1	3.73 (±1.19)	4.28 (±1.25)	5.99 (±1.27)	6.27 (±2.68)	6.89 (±1.12)	6.97 (±2.49)

little organic matter, which limits the biological processing of incoming nutrients [15].

While we did not find evidence for a strong role of these marsh planting designs as filters of runoff nutrient pollution, they provide other important services such as habitat and shoreline stabilization. We did not quantify these services, but we would expect these services to be small in magnitude for these young marshes and increase as the marshes age [6]. Similarly, we expect that these restored marshes will increase vegetated area and become effective nutrient filters through time.

Conclusions

In conclusion, our results suggest that groundwater flow rates and sediment type should be considered when planning marsh restoration efforts with the specific goal of runoff pollution removal. Despite finding slower groundwater flow rates with increasing vegetation, our introduced groundwater did still flow quickly through the restored plots (30–60 minutes). This fast flow is most likely attributable to the coarse texture of the sediment. We calculated that on average only 3–7% of the NO_3^- entering the restored marshes was processed by the marshes within one year since planting, and it appears that NO_3^- processing by these restored marshes will remain similar to the unvegetated plots for many years to come. However, the slower groundwater flow in the more vegetated plots suggests they will likely become effective nutrient filters prior to the less vegetated plots. Understanding the nature, control, and extent of nutrient processing by restored marshes requires additional research, such as more restoration designs across many different marsh environments, in order to help managers decide on best restoration practices given budget constraints.

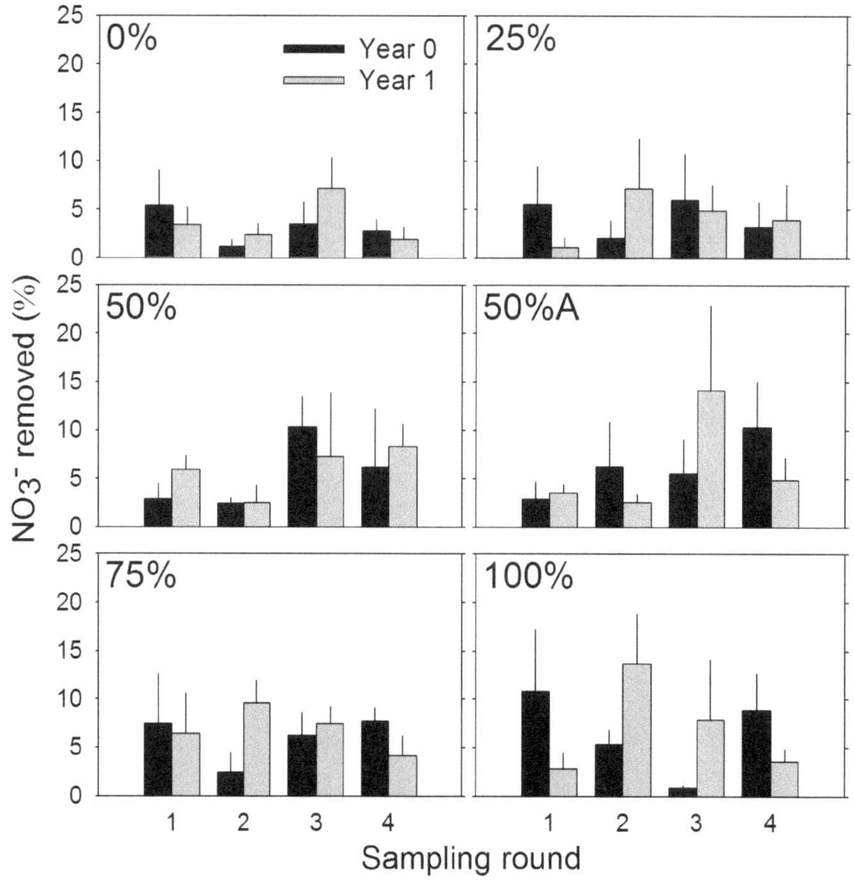

Figure 3. Percentage of NO_3^- removed from introduced solution directly after planting and 1 year post-planting for each sampling round. Black bars represent year 0 and grey bars represent year 1 samples. Percentages in the top left portion of each plot represent initial planting densities. Error bars indicate ±1 SE.

Acknowledgments

We would like to thank Jason Howard, Jelani Reynolds, Jennifer Hemphill and Amanda Pratt of the Ecosystems Lab at the Dauphin Island Sea Lab for their field assistance.

Author Contributions

Conceived and designed the experiments: ELS JC SMS. Performed the experiments: ELS SMS. Analyzed the data: ELS. Contributed reagents/ materials/analysis tools: JC. Contributed to the writing of the manuscript: ELS JC.

References

1. Bromberg Gedan K, Silliman BR, Bertness MD (2009) Centuries of Human-Driven Change in Salt Marsh Ecosystems. Annu Rev of Mar Sci 1: 117–141.
2. Sparks EL, Cebrian J, Biber PD, Sheehan KL, Tobias CR (2013) Cost-effectiveness of two small-scale salt marsh restoration designs. Ecol Eng 53(2013): 250–256.
3. Chapman MG, Underwood AJ (2000) The need for a practical scientific protocol to measure successful restoration. Wetlands (Australia) 19(1): 28–49.
4. Costanza R, d'Arge R, de Groot R, Farber S, Grasso M, et al. (1997) The value of the world's ecosystem services and natural capital. Nature 387: 253–260.
5. Ehrenfeld JG (2000) Defining the Limits of Restorations: The Need for Realistic Goals. Restor Ecol 8: 2–9.
6. Hilderbrand RH, Watts AC, Randle AM (2005) The myths of restoration ecology. Eco Soc 10: (online) URL: http://www.ecologyandsociety.org/vol10/iss1/art19/.
7. Beck M, Heck K Jr, Able K, Childers D, Eggleston D, et al. (2001) The identification, conservation, and management of estuaries and marine nurseries for fish and invertebrates. Biosci 51(8): 633–641.
8. Chmura GL, Anisfeld SC, Cahoon DR, Lynch JC (2003) Global carbon sequestration in tidal, saline wetland soils. Glob Biogeochem Cycles 17(44): 1–22.
9. Moeller I, Spencer T, French JR (1996) Wind wave attenuation over saltmarsh surfaces: Preliminary results from Norfolk, England. J Coast Res 12: 1009–1016.
10. Valiela I, Cole ML (2002) Comparative Evidence that Salt Marshes and Mangroves May Protect Seagrass meadows from Land-derived Nitrogen Loads. Ecosystems 5(1): 92–102.
11. Hammersley MR, Howes BL (2005) Coupled nitrification-denitrification measured in situ in a *Spartina alterniflora* marsh with a 15NH4 tracer. Mar Ecol Prog Ser 299: 123–135.
12. Tobias C, Macko S, Anderson I, Canuel E, Harvey J (2001) Tracking the fate of a high concentration nitrate plume through a fringing marsh: A combined groundwater tracer and in-situ isotope enrichment study. Limnol Oceanogr 46(8): 1977–1989.
13. Drake DC, Peterson BJ, Galvan KA, Deegan LA, Hopkinson C, et al. (2009) Salt marsh ecosystem biogeochemical responses to nutrient enrichment: a paired ^{15}N tracer study. Ecology 90(9): 2535–2546.
14. Tobias CR, Anderson IC, Canuel EA, Macko SA (2001) Nitrogen cycling through a fringing marsh-aquifer ecotone. Mar Ecol Prog Ser 210: 25–39.
15. Tobias CR, Neubauer SC (2009) Salt marsh biogeochemistry: An overview in Perillo G, Wolanski E, Cahoon D, Brinson M (eds). Coastal wetlands: An integrated ecosystem approach. Elsevier. p. 445–492.
16. Lehrter JC, Cebrian J (2009) Uncertainity propagation in an ecosystem nutrient budget. Ecol Appli 20: 508–524.
17. Tobias C, Harvey JW, Anderson IC (2001) Quantifying groundwater discharge through fringing wetlands to estuaries: Seasonal variability, methods comparison, and implication for wetland-estuary exchange. Limnol Oceanogr 46(3): 604–615.
18. Valiela I, Costa J, Foreman K, Teal JM, Howes B, et al. (1999) Transport of groundwater-borne nutrients from watersheds and their effects on coastal waters. Biodegradation 10(3): 177–197.
19. Barling RD, Moore ID (1994) Role of Buffer Strips in Management of Waterway Pollution: A Review Environ manag 18(4): 543–558.
20. Stout JP, Heck KL Jr, Valentine JF, Dunn SJ, Spitzer PM (1998) Preliminary Characterization of Habitat Loss: Mobile Bay National Estuary Program, MESC Contribution Number 301.
21. Eleuterius LN (1976) The distribution of *Juncus roemerianus* in the salt marshes of North America. Earth Environ Sci 17(4): 289–292.
22. Turner RE (1990) Landscape development and coastal wetland losses in the northern Gulf of Mexico Am Zool 30(1): 89–105.
23. LaSalle MW (1996) Assessing the functional level of a constructed intertidal marsh in Mississippi. Tech Rep WRP-RE-15, US Army Corps of Engineers, Waterways Experiment Station, Vicksburg, MS.
24. Lewis RR (1982) Creation & Restoration of Coastal Plant Communities. CRC Press, Boca Raton: 153–171.
25. Turner RE, Streever B (2002) Approaches to Coastal Wetland Restoration: Northern Gulf of Mexico. SPB Academic Publishing, Hague, The Netherlands.
26. Corbett DR, Dillon K, Burnett W (2000) Tracing groundwater flow on a barrier island in the north-east Gulf of Mexico. Estuar Coast Shelf Sci 51(2): 227–242.
27. Maynard DG, Kalra YP (1993) Nitrate and extractable ammonium nitrogen. In M.R. Carter, Ed. Soil Sampling and Methods of Analysis. Lewis Publishers, Boca Raton, FL, 25–38.
28. Quinn GP, Keough MJ (2002) Experimental design and data analysis for biologists. Experimental design and data analysis for biologists. Cambridge University Press, Cambridge, UK.
29. Vukovic M, Soro A (1992) Determination of Hydraulic Conductivity of Porous Media from Grain-Size Composition. Littleton, Colorado: Water Resources Publications.
30. Stumpf RP (1983) The process of sedimentation on the surface of a salt marsh. Estuar Coast Shelf Sci 17: 495–508.
31. Harrell J, Blatt H (1978) Polycrystallinity; effect on the durability of detrital quartz. J Sediment Res 48(1): 25–30.
32. Wilkins S, Keith DA, Adam P (2003) Measuring success: evaluating the restoration of a grassy eucalypt woodland on the Cumberland Plain, Sydney, Australia. Restor Ecol 11: 489–503.
33. Zedler JB, Callaway JC (1999) Tracking wetland restoration: do mitigation sites follow desired trajectories? Restor Ecol 7: 69–73.
34. Dingman SL (2008) Physical Hydrology. Waveland Press Inc, Long Grove, IL, USA.
35. Valiela I, Teal JM, Volkmann S, Shafer D, Carpenter EJ (1978) Nutrient and particulate fluxes in a salt marsh ecosystem: Tidal. Limnol Oceanogr 23(4): 798–812.

8

The Photoprotective Role of Spermidine in Tomato Seedlings under Salinity-Alkalinity Stress

Lipan Hu[1,2], Lixia Xiang[1,2], Li Zhang[1,2], Xiaoting Zhou[1,2], Zhirong Zou[1,2*], Xiaohui Hu[1,2*]

1 College of Horticulture, Northwest Agricultural & Forest University, Yangling, China, **2** Key Laboratory of Protected Horticultural Engineering in Northwest, Ministry of Agriculture, Shaanxi Yangling, China

Abstract

Polyamines are small, ubiquitous, nitrogenous compounds that scavenge reactive oxygen species and stabilize the structure and function of the photosynthetic apparatus in response to abiotic stresses. Molecular details underlying polyamine-mediated photoprotective mechanisms are not completely resolved. This study investigated the role of spermidine (Spd) in the structure and function of the photosynthetic apparatus. Tomato seedlings were subjected to salinity-alkalinity stress with and without foliar application of Spd, and photosynthetic and morphological parameters were analyzed. Leaf dry weight and net photosynthetic rate were reduced by salinity-alkalinity stress. Salinity-alkalinity stress reduced photochemical quenching parameters, including maximum photochemistry efficiency of photosystem II, quantum yield of linear electron flux, and coefficient of photochemical quenching (qP). Salinity-alkalinity stress elevated nonphotochemical quenching parameters, including the de-epoxidation state of the xanthophyll cycle and nonphotochemical quenching (NPQ). Microscopic analysis revealed that salinity-alkalinity stress disrupted the internal lamellar system of granal and stromal thylakoids. Exogenous Spd alleviated the stress-induced reduction of leaf dry weight, net photosynthetic rate, and qP parameters. The NPQ parameters increased by salinity-alkalinity stress were also alleviated by Spd. Seedlings treated with exogenous Spd had higher zeaxanthin (Z) contents than those without Spd under salinity-alkalinity stress. The chloroplast ultrastructure had a more ordered arrangement in seedlings treated with exogenous Spd than in those without Spd under salinity-alkalinity stress. These results indicate that exogenous Spd can alleviate the growth inhibition and thylakoid membrane photodamage caused by salinity-alkalinity stress. The Spd-induced accumulation of Z also may have an important role in stabilizing the photosynthetic apparatus.

Editor: Keqiang Wu, National Taiwan University, Taiwan

Funding: This work was supported by grants from the National Key Technology R&D Program in the 12th Five-year Plan of China (No. 2011BAD29B01), the Scientific Research Special Fund of Northwest Agriculture & Forestry University (QN2013018), and the China Agriculture Research System (No. CARS-25-D-02). The funders had no role in study design, data collection and analysis, decision to publish, or preparation of the manuscript.

Competing Interests: The authors have declared that no competing interests exist.

* Email: zouzhirong2005@ hotmail.com (ZZ); hxh1977@163.com (X-HH)

These authors contributed equally to this work.

Introduction

Salinity-alkalinity stress is becoming one of the most serious abiotic stresses affecting crop growth and yield [1], and causes considerable economic losses. Our previous studies clearly show that plants are more sensitive to salinity-alkalinity stress than to neutral salt stress [1,2]. Previous reports show that salinity-alkalinity stress strongly inhibits assimilation of photosynthetic carbon [3–5], which leads to an unfavorable dissipation of excess light energy. When photosystem II (PSII) is exposed to excess light, reactive oxygen species (ROS) such as singlet oxygen ($^1O_2$), hydroxyl ions (OH•), and H_2O_2 are formed by the interaction of molecular oxygen and triplet chlorophyll [6,7], which causes photoinhibition and photodamage [8]. ROS directly affects chloroplasts and inhibits the repair of photodamaged PSII [7]. Salinity stress reduces photosynthesis and electron transport activity, thereby exacerbating problems arising from photoinhibition and photodamage of the photosynthetic apparatus [9].

Polyamines (PAs) are organic, low molecular weight, nitrogen-containing compounds; they include spermidine (Spd), spermine (Spm), and their diamine obligate precursor putrescine (Put) [10]. PAs function as plant growth regulators or intracellular messengers that regulate plant growth, development, and responses to abiotic stresses [11]. They also function as antioxidant, scavengers of ROS and free radicals [12], inhibit lipid peroxidation reactions [13] due to a combination of their anion- and cation-binding properties [14], and respond to a wide range of abiotic stresses. PAs also stabilize the structure and function of the photosynthetic apparatus in response to unfavorable environment factors [15–17]. PAs protect the structure of chloroplasts by keeping the thylakoid membranes in an orderly arrangement and maintaining high photosynthetic efficiency [18]. Exogenous Spd may regulate the synthesis of xanthophyll components and promote the conversion of violaxanthin (V) to zeaxanthin (Z), which can protect LHCII reaction centers against salinity-induced oxidative damage and injury from excess light energy [19].

The xanthophyll cycle pigment Z is formed from V via the intermediate anthoxanthin (A) during abiotic stress. Z is thought to be involved in photoprotective thermal dissipation and nonphotochemical quenching (NPQ) [20,21]. The thermal dissipation process protects PSII against photodamage by decreasing excess redox potential of PSII and the electron transfer chain [22]. Z has an important antioxidant function in the lipid phase of the membrane and protects chloroplasts from photo-oxidative stress [23]. Z reduces the formation of ROS, or detoxifies already-formed ROS, to minimize photo-oxidative damage of the photosynthetic apparatus [21].

Previous studies showed that exogenous application of PAs partially alleviated the salinity-induced decline in photosynthetic efficiency [24]. However, the molecular mechanism of Spd-mediated photoprotection of the photosynthetic apparatus and the relationship between Spd and the xanthophyll cycle in response to salinity-alkalinity stress were not completely resolved. Therefore, we sought to determine whether exogenous Spd could protect tomato seedlings from photoinhibition and photodamage and interact with the endogenous xanthophyll cycle under salinity-alkalinity conditions.

Materials and Methods

Plant Materials and Growth Conditions

Tomato (*Solanum lycopersicum* cv. Jinpengchaoguan) seeds were germinated at 27°C in Petri plates lined with two layers of filter paper moistened with sterile distilled water. The germinated seeds were then sown in commix medium (Xintiandi Co., Yangling, Shanxi, China). Seedlings were grown in a controlled environmental greenhouse at Northwest A & F University, where the air temperature was maintained at 25±1°C/15±1°C (day/night). A 12-h light/12-h dark photoperiod was imposed with a daylight intensity of 800 $\mu mol \cdot m^{-2} s^{-1}$. Seedlings were transferred into 40-L containers of one-half-strength Hoagland solution (pH 6.5±0.1; electrical conductivity, 1.4−1.8 $ds \cdot m^{-1}$; dissolved oxygen, 6.0±0.2 $mg \cdot L^{-1}$) when the third true leaves were fully expanded.

Salinity-alkalinity Conditions and Spermidine Treatment

When seedlings had five true leaves, they were treated with 75 mM complex neutral and alkali salts [$NaCl:Na_2SO_4:NaHCO_3:Na_2CO_3 = 1:9:9:1$] and the foliage was sprayed with 0.25 mM Spd. The experiments included the following four treatments: 0 mM salinity-alkalinity plus 0 mM Spd (CK); 0 mM salinity-alkalinity plus 0.25 mM Spd (CS); 75 mM salinity-alkalinity plus 0 mM Spd (S); and 75 mM salinity-alkalinity plus 0.25 mM Spd (SS). Each treatment group included 48 plants and four replicates. Plant containers were arranged in a randomized block design. The solutions were renewed every 3 days. All experiments were performed at least three times with similar results.

Plant Growth Measurements

Four days after the salinity-alkalinity treatment, plants were washed with sterile distilled water and dissected into leaves, stems, and roots. The fresh weights of dissected tissues were determined. The leaves, stems, and roots were dried at 105°C for 15 min and then at 75°C for 72 h to constant mass, and their dry masses were recorded. The shoot weight was equal to the sum of the leaf weight and the stem weight. The root/shoot ratio (R/S) was calculated as (R/S) = root weight/shoot weight. Four plants were measured per treatment.

Gas-Exchange Measurements

Gas-exchange parameters were measured using a portable photosynthesis system (LI-6400, LI-COR Inc, USA). The photosynthetic rate was measured at 380±10 $\mu mol \cdot mol^{-1}$ CO_2, 25°C, 70% relative humidity, and 800 $\mu mol \cdot m^{-2} s^{-1}$ light intensity. Stomatal limitation (L_s) was calculated as $L_s = 1 - C_i/C_a$ (C_i and C_a represent the intercellular and ambient CO_2 concentration, respectively) according to Farquhar et al. [25].

Chlorophyll Fluorescence Parameters

Chlorophyll fluorescence was measured using a modulated fluorometer (PAM-2500; Walz, Effeltrich, Germany). The minimal (F_o) and maximal (F_m) Chl fluorescence emissions were determined after 30 min of dark adaptation, and F_o', F_s, and F_m' were measured after light adaption at 600 $\mu mol \cdot m^{-2} s^{-1}$. Fluorescence parameters were calculated as follows: maximum quantum yield of PSII, $(F_v/F_m) = (F_m - F_o)/F_m$; maximum photochemistry efficiency of PSII, $(F_v'/F_m') = (F_m' - F_o')/F_m'$; PSII operating efficiency, $\Phi PSII = (F_m' - F_s)/F_m'$; coefficient of photochemical quenching, $Qp = (F_m' - F_s)/(F_m' - F_o')$; and nonphotochemical quenching, $NPQ = (F_m - F_m')/F_m'$ [26].

Pigments Analysis

Pigment composition was determined according to the procedure of Chen et al. with some minor modifications [27]. Xanthophyll pigments were extracted with 100% acetone under weak light. The extract was centrifuged at 12,000×*g* for 10 min. The supernatant was passed through a 0.2-µm syringe filter into an amber HPLC vial for immediate analysis. Twenty-five microliters of the filtered supernatant was analyzed with a Shimadzu LC-20A HPLC system (Kyoto, Japan). The Waters

Table 1. Effects of exogenous spermidine on tomato seedling fresh weight under salinity-alkalinity stress.

Treatment	Shoot fresh weight (g)	Root fresh weight (g)	Stem fresh weight (g)	Leaf fresh weight (g)	R/S fresh weight
CK	40.726±2.066^{a}	6.743±0.276^{a}	17.968±1.263^{a}	22.759±0.946^{a}	0.166±0.007^{a}
CS	43.578±3.130^{a}	7.400±0.650^{a}	18.348±1.737^{a}	25.231±1.543^{ab}	0.170±0.007^{a}
S	22.482±0.711^{b}	3.718±0.338^{b}	5.649±0.555^{b}	16.834±1.232^{c}	0.167±0.020^{a}
SS	26.050±0.938^{b}	3.911±0.272^{b}	6.877±0.940^{b}	19.172±0.906^{bc}	0.150±0.008^{a}

Note: Data were measured after salinity-alkalinity treatment for 4 days. Each value represents mean ± standard error of four independent experiments ($n = 4$). Different letters indicate significant differences between treatments ($P < 0.05$).

Abbreviations: CK, 0 mM salinity-alkalinity plus 0 mM Spd; CS, 0 mM salinity-alkalinity plus 0.25 mM Spd; S, 75 mM salinity-alkalinity plus 0 mM Spd; SS, 75 mM salinity-alkalinity plus 0.25 mM Spd.

Table 2. Effects of exogenous spermidine on tomato seedling dry weight under salinity-alkalinity stress.

Treatment	Shoot dry weight (g)	Root dry weight (g)	Stem dry weight (g)	Leaf dry weight (g)	R/S dry weight
CK	3.049±0.312[a]	0.396±0.034[a]	0.978±0.139[a]	2.071±0.176[a]	0.131±0.009[b]
CS	3.018±0.250[a]	0.396±0.029[a]	0.979±0.086[a]	2.040±0.164[a]	0.132±0.004[b]
S	1.547±0.116[c]	0.289±0.037[b]	0.451±0.045[b]	1.097±0.084[c]	0.185±0.011[a]
SS	2.216±0.115[b]	0.264±0.023[b]	0.626±0.076[b]	1.591±0.046[b]	0.118±0.006[b]

Note: Data were measured after salinity-alkalinity treatment for 4 days. Each value represents mean ± standard error of four independent experiments ($n = 4$). Different letters indicate significant differences between treatments ($P < 0.05$).

Abbreviations: CK, 0 mM salinity-alkalinity plus 0 mM Spd; CS, 0 mM salinity-alkalinity plus 0.25 mM Spd; S, 75 mM salinity-alkalinity plus 0 mM Spd; SS, 75 mM salinity-alkalinity plus 0.25 mM Spd.

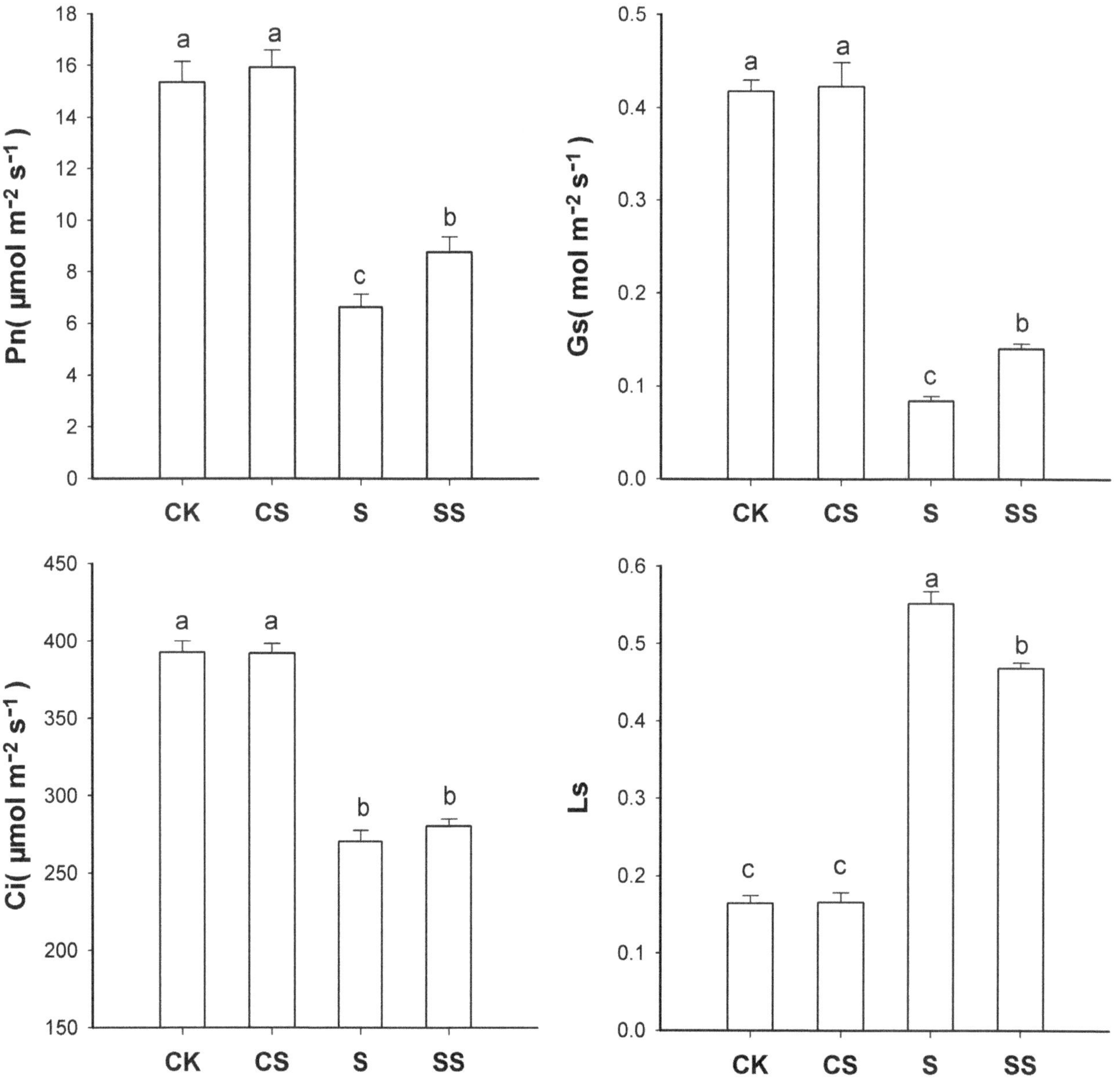

Figure 1. Effects of exogenous spermidine on gas-exchange parameters in tomato seedlings under salinity-alkalinity stress. Data were measured in the second expanded leaves (numbered basipetally) after salinity-alkalinity treatment for 4 days. Each histogram represents the mean ± standard error of four independent experiments ($n = 4$). Different letters indicate significant differences between treatments ($P < 0.05$).

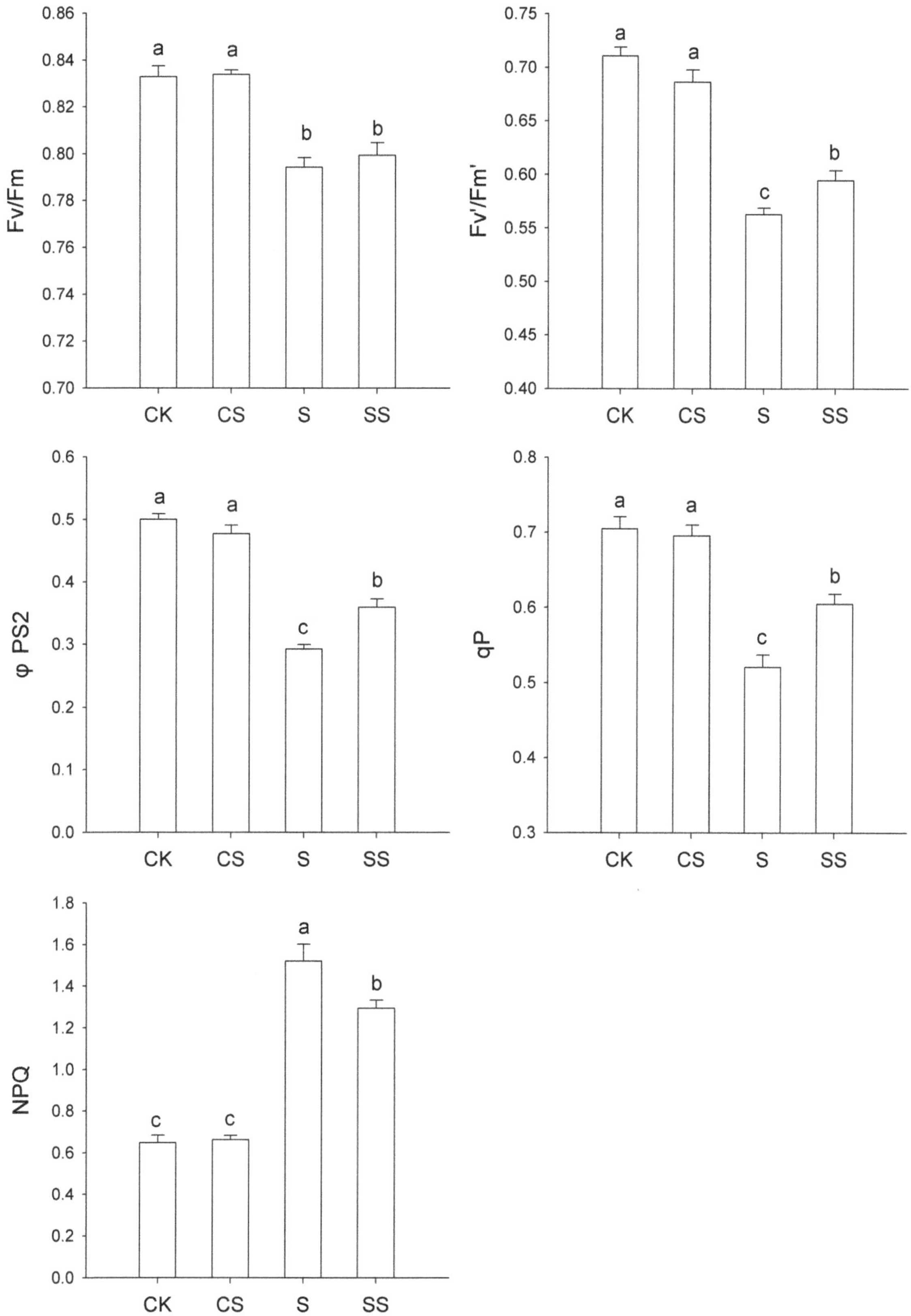

Figure 2. Effects of exogenous spermidine on chlorophyll fluorescence parameters in tomato seedlings under salinity-alkalinity stress. Data were measured in the second expanded leaves (numbered basipetally) after salinity-alkalinity treatment for 4 days. Each histogram represents the mean ± standard error of four independent experiments ($n = 4$). Different letters indicate significant differences between treatments ($P < 0.05$).

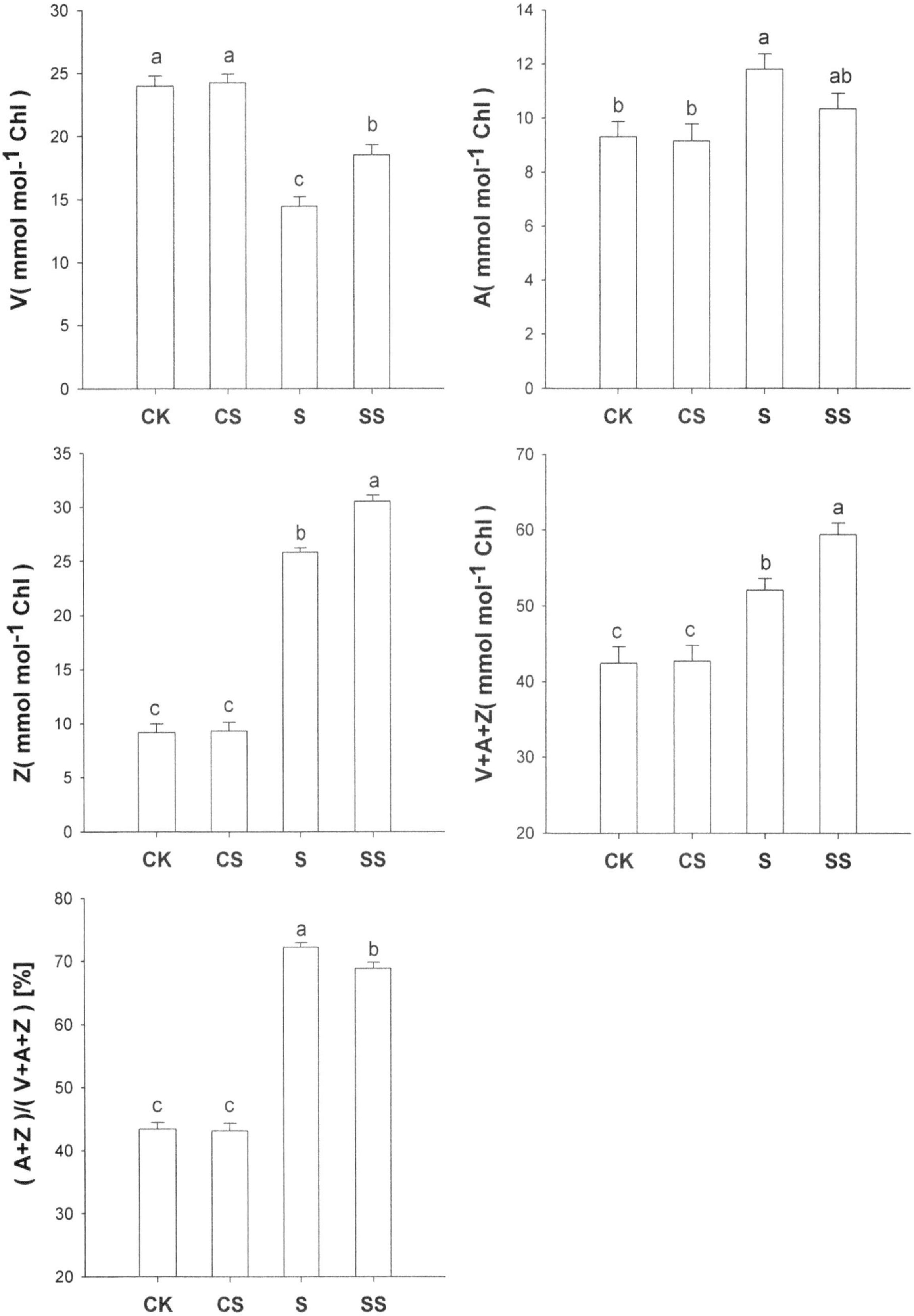

Figure 3. Effects of exogenous spermidine on xanthophyll cycle components in tomato seedlings under salinity-alkalinity stress. Data were measured in the second expanded leaves (numbered basipetally) after salinity-alkalinity treatment for 4 days. Each histogram represents the mean ± standard error of four independent experiments ($n = 3$). Different letters indicate significant differences between treatments ($P<0.05$).

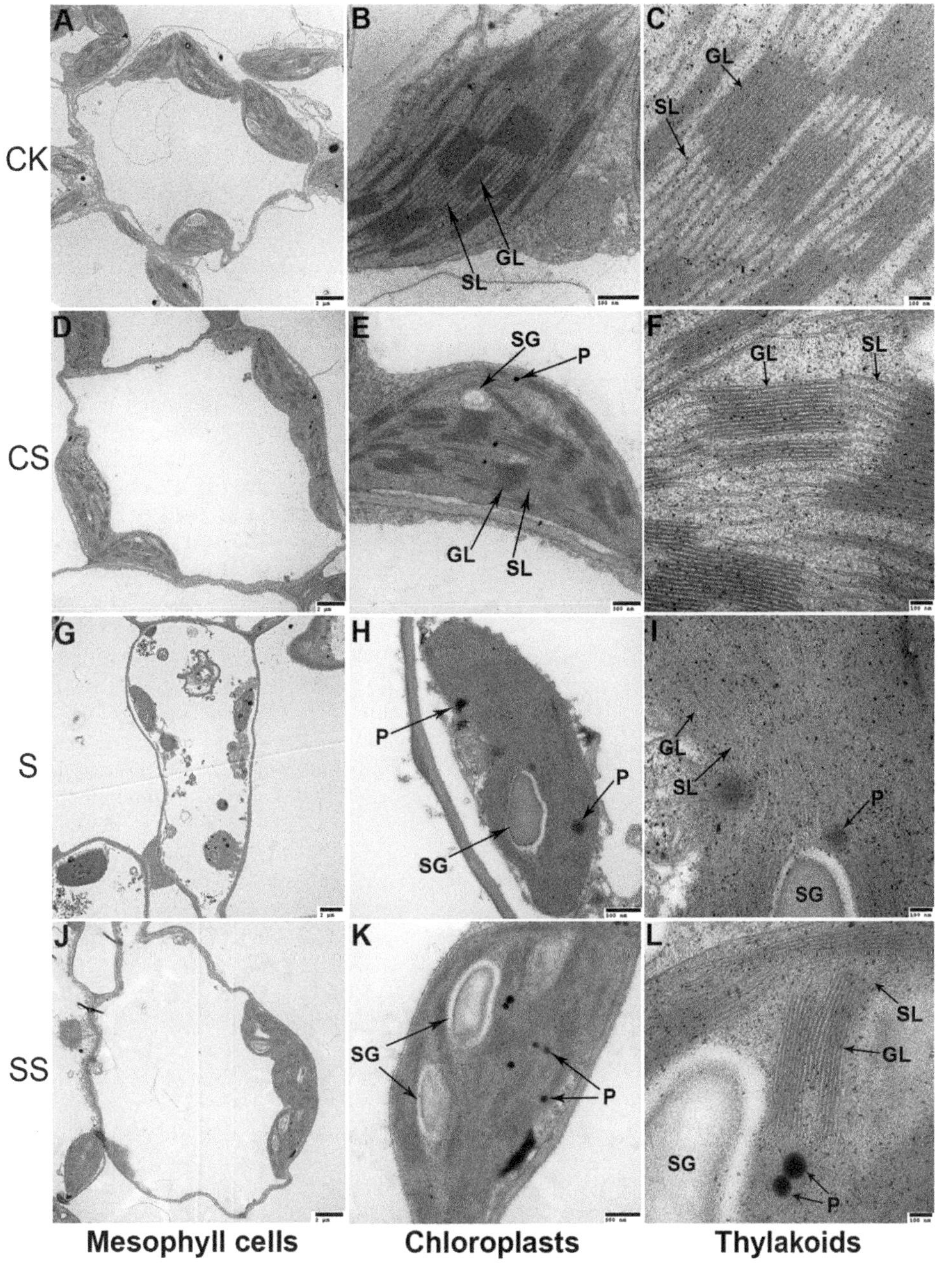

Figure 4. Effects of exogenous spermidine on chloroplast ultrastructure in tomato seedlings under salinity-alkalinity stress. Data were measured in the second expanded leaves (numbered basipetally) after salinity-alkalinity treatment for 4 days. SL, stroma lamellae; GL, grana lamellae; SG, starch grains; P, plastoglobuli. Scale bars for mesophyll cells, chloroplasts, and thylakoids are 2, 0.5, and 0.1 μm, respectively.

Spherisorb ODS1 column (5.0-μm particle size, 4.6×250 mm, Waters, USA) was used in the separation at 35°C. Solvent A [acetonitrile:methanol:50 mM Tris-HCl, 72:8:3 (v/v/v), pH 7.5] was used for the first 10 min, followed by a 2.5-min linear gradient to 100% solvent B [methanol:hexane, 5:1 (v/v)], with a run time of 30 min and a flow rate of 1.5 ml·min^{-1}. Pigments were detected by absorbance at 445 nm. The de-epoxidation state of the xanthophyll cycle was calculated as (Z+A)/(V+Z+A).

Transmission Electron Microscopy of Chloroplasts

The second fully expanded leaves from the top of the plants were randomly selected for microscopic examination. The leaf samples were cut into pieces of approximately 1 mm^2 and fixed overnight with 4% glutaraldehyde in 0.1 M phosphate-buffered saline (PBS, pH 7.4). The fixed samples were washed three times with the same solution for 10 min each. The samples were then post-fixed in 1% osmium tetroxide in cacodylate buffer for 2 h, and then washed three times in 0.1 M PBS (pH 7.4). The samples were dehydrated in a graded ethanol series (50%, 70%, 90%, and 100%), and then in absolute acetone for 15 min. The samples were embedded in Spurr's resin, and ultra-thin sections were cut and stained with uranium acetate and lead citrate, in series. The ultra-thin sections were mounted on copper grids and examined with a HITACHI HT7700 transmission electron microscope.

Statistical Analysis

All treatments and measurements were conducted at least in triplicate. All data were statistically analyzed with statistical software SAS (version 8.0, SAS Institute, Cary, NC) using Duncan's multiple range test at the *P* less than 0.05 level of significance.

Results

Plant Growth

The fresh and dry weights of tomato seedlings were significantly reduced by salinity-alkalinity stress compared with controls after 4 days of treatment (Tables 1 and 2). Reductions in dry leaf and shoot weights were alleviated by addition of exogenous Spd (Table 2). The S treatment had higher R/S than any other treatment (Table 2). There were no significant differences between S and SS treatments with respect to dry root weight and dry stem weight (Table 2). These results indicated that tomato seedling leaf growth was inhibited by salinity-alkalinity stress, and this inhibition was alleviated by exogenous Spd. However, exogenous Spd did not significantly affect the fresh weight measurements of tomato seedlings with or without salinity-alkalinity stress (Table 1).

Gas-Exchange Parameters

Salinity-alkalinity stress significantly reduced the net photosynthetic rate (P_n), stomatal conductance (G_s), and C_i, and increased L_s, compared with those of controls (Fig. 1). Exogenous Spd alleviated reductions in P_n and G_s, and increase in L_s. Exogenous Spd had no significant effect on C_i under salinity-alkalinity stress (SS treatment) compared with that of S treatment. No significant differences in P_n, G_s, C_i, or L_s were detected in CS and CK treatments.

Chlorophyll Fluorescence

F_v'/F_m', ΦPSII, and qP were substantially reduced under salinity-alkalinity conditions compared with controls (Fig. 2). The addition of exogenous Spd alleviated these effects (Fig. 2). The maximum quantum yield of PSII (F_v/F_m) was reduced in plants subjected to salinity-alkalinity stress. The addition of exogenous Spd did not protect the quantum yield of PSII. Plants subjected to salinity-alkalinity stress had significantly elevated NPQ, and this increase was alleviated by exogenous Spd. Exogenous Spd did not affect F_v/F_m, F_v'/F_m', ΦPSII, qP, or NPQ in the absence of salinity-alkalinity stress.

Xanthophyll Cycle Components

The Z content and the total V+A+Z pool size increased in response to salinity-alkalinity stress, and exogenous Spd induced an even higher increase in the contents of these metabolites. The V content markedly decreased in response to salinity-alkalinity stress, and this decrease was greater in leaves without exogenous Spd than in those with Spd. The (A+Z)/(V+Z+A) ratio increased in response to salinity-alkalinity stress in plants treated with exogenous Spd, but it was lower than in those without Spd. These results indicate that Spd increased the total V+A+Z pool size and the constitutive accumulation of Z in response to salinity-alkalinity stress.

Ultrastructural Changes in Chloroplasts

A typical elongated chloroplast was observed in controls (Fig. 3B), which had intact double membranes and a regular arrangement of granal and stromal thylakoids. Under salinity-alkalinity stress (Fig. 3G–I), chloroplasts were swollen and contained more plastoglobuli than those in controls (Fig. 3A–C). The chloroplast membrane structures became dilatant, undefined, and in some cases appeared disintegrated (Fig. 3H). The disoriented lamellar system of the granal and stromal thylakoids became swollen and unclear (Fig. 3I). The numbers of grana were considerably reduced. Exogenous Spd alleviated the salinity-alkalinity-mediated damage of the chloroplast photosynthetic apparatus, and more normal chloroplast ultrastructure was observed (Fig. 3J–L). No significant differences in chloroplast ultrastructure were observed in controls with or without the application of Spd (Fig. 3A–F).

Discussion

PAs are low molecular weight, ubiquitous, nitrogenous compounds that function in many cellular processes in numerous organisms [28]. Due to their polycationic nature and positive charge at physiological pH, PAs bind to macromolecules such as proteins and nucleic acids [14,29]. In this way, PAs are involved in regulating the physical and chemical properties of membranes, stabilizing nucleic acid structure and function, and modulating enzyme activities [30]. Chloroplasts are the primary sites of photosynthetic reactions in higher plants. Some types of stress reduce photochemical efficiency and electron transport activity, which might reflect changes in the structure of the photosynthetic apparatus [31]. Previous studies report that PAs stabilize the photosynthetic apparatus in response to stress [15].

In the present study, the deleterious effects of salinity-alkalinity stress on leaf and shoot dry weights in tomato seedlings were partly counteracted by the addition of exogenous Spd (Table 1). The improved dry matter accumulation in shoots and leaves might be associated with Spd-mediated improvement in P_n in tomato seedlings under salinity-alkalinity conditions (Fig. 1). A similar result was reported by Zhang et al. in salt-stressed cucumber plants sprayed with Put, which enhanced leaf area, plant height, and fresh and dry weight accumulation [24]. These authors demonstrated that Put likely enhanced photosynthetic production of salt-stressed plants, and this increased production appeared to effectively support the enhance growth. In tomato plants subjected to salinity-alkalinity stress and treated with exogenous Spd, our R/S results showed that most photosynthetic products were utilized to produce shoots rather than roots (Table 2).

Salinity-alkalinity stress reduced P_n and G_s, and increased L_s (Fig. 1). Previous studies evaluated the relationship between L_s and nonstomatal factors. If both C_i and G_s decreased, P_n was limited primarily by stomatal conductance. If G_s decreased but C_i did not change or increased, the P_n decrease could be ascribed to nonstomatal factors [19,24,32,33]. In the present study, salinity-alkalinity-induced reductions in P_n and G_s and increase in L_s were alleviated by exogenous Spd. However, exogenous Spd had no significant effect on C_i with or without stress (Fig. 1). These observed changes in G_s, C_i, and L_s imply that exogenous Spd alleviates both L_s and nonstomatal limitations under salinity-alkalinity conditions.

In general, F_v/F_m, F_v'/F_m', ΦPSII, and qP are parameters that reflect photochemical quenching, whereas NPQ reflects nonphotochemical quenching [26]. In this study, the values of F_v/F_m, F_v'/F_m', ΦPSII, and qP were reduced by salinity-alkalinity treatment, and these decreases were alleviated by exogenous Spd. A previous report suggested that salt-stress-induced reductions in F_v'/F_m' and ΦPSII might reflect damage to PSII electron transport [19]. These authors demonstrated that inhibition of electron transport may occur during transfer from the primary acceptor plastoquinone (QA) to the secondary acceptor plastoquinone (QB) at the acceptor side of PSII. Although exogenous Spd had no significant effect on F_v/F_m of tomato seedlings with or without salinity-alkalinity stress, it alleviated the stress-induced reductions in F_v'/F_m' and ΦPSII. This result indicates that exogenous Spd alleviates the stress-induced inhibition of photosynthetic electron transport.

Exogenous Spd alleviated the salinity-alkalinity-induced decrease in qP and increase in NPQ. This suggests that exogenous Spd may increase qP at the expense of NPQ under stress conditions, thereby mitigating the dissipation of excitation energy in the PSII antennae [34]. This is supported by the observation that exogenous Spd improved P_n and decreased the de-epoxidation status of the xanthophyll cycle in tomato seedlings under salinity-alkalinity stress (Fig. 2). The xanthophyll cycle serves as an essential photoprotective process in the dissipation of excess energy [35,36]. This thermal dissipation is an important function of NPQ to alleviate excess excitation energy on PSII reaction centers by converting excess excitation energy into heat [37]. Previous work showed that, to a certain extent, severe stress might result in higher levels of NPQ and (A+Z)/(V+Z+A) [38,39]. Consequently, under salinity-alkalinity conditions, leaves treated with Spd suffered less photoinhibition than those without Spd. Exogenous Spd may protect the photosynthetic apparatus against over-excitation, perhaps by preventing a loss of thylakoid membrane integrity [40]. The observed chloroplast ultrastructures support the proposal that exogenous Spd alleviates injury to the photosynthetic membrane caused by salinity-alkalinity stress (Fig. 4J–L). This result may be attributed to increases in the total V+A+Z pool size and the Spd-induced increases in Z content.

Z serves another important function as an antioxidant in the lipid phase of the thylakoid membrane to block photo-oxidative stress [21]. Photoinhibition occurs even under moderate light conditions if excess light energy cannot be appropriately dissipated [41]. $^1O_2$ is one of the most important ROS involved in photoinhibition and photo-oxidative damage [42]. Nonprotein-bound Z enhances photoprotection by blocking the formation of $^1O_2$ and $^1O_2$-related oxidation products [43,44], which reduces lipid peroxidation. ROS scavenging protects chloroplasts from direct injury and diminishes photon (electron)-related stress [7]. In our study, exogenous Spd increased the amount of Z under salinity-alkalinity stress, which stabilized thylakoid membrane structure and improved photosynthetic function.

Chloroplasts are crucial plant organelles because they are the site of photosynthesis. Chloroplast ultrastructure can be substantially disrupted by salinity stress [45]. In the present study, increased numbers of plastoglobuli were observed in mesophyll cells subjected to salinity-alkalinity stress, and some of these were larger than those in controls (Fig. 4G–I). Increases in plastoglobuli numbers may be caused by thylakoid membrane degradation [46], and individual plastoglobuli may be larger when chloroplasts are under oxidative stress [47]. Chloroplasts are a major source of ROS, including $^1O_2$, OH•, and H_2O_2 [7], which cause membrane lipid peroxidation and cellular abnormalities when plants are exposed to environmental stresses [48]. Tian et al. showed that exogenous Spd enhanced the activities of ascorbate peroxidase (APX), superoxide dismutase (SOD), peroxidase (POD), and catalase when cucumber seedlings were subjected to high temperatures [49]. Exogenous Spm alleviated salt-induced chloroplast membrane injury by increasing the levels of antioxidant metabolites (including dehydro-L-ascorbic acid, ascorbic acid, glutathione-S-S-glutathione, and glutathione-stimulating hormone) and enzyme activities (including SOD, POD, and APX) in chloroplasts [16]. PAs also bind membrane proteins, including polypeptides of the light-harvesting complex (LHC) [50]. Exogenous PAs with a high net positive charge can stabilize PSII proteins such as D1 and D2 under photoinhibition conditions [51]. PA binding to membrane proteins may provide stability for protein structure during stress, and consequently preserve photosynthetic activity. Our results indicate that Spd may serve a protective role in chloroplasts by enhancing ROS scavenging and stabilizing photosynthetic membrane proteins.

In conclusion, our results show that salinity-alkalinity stress induces photoinhibition and photodamage in tomato seedlings. Exogenous Spd alleviates the stress-induced inhibition of plant shoot growth and P_n by reducing ROS and stabilizing thylakoid membrane structure. Exogenous Spd indirectly improves total V+A+Z pool size, especially the amount of Z, which acts as an antioxidant in the lipid phase of the thylakoid membrane to block photo-oxidative stress. However, under salinity-alkalinity stress, Spd-induced increases in Z content did not increase NPQ compared with that in seedlings without Spd. This may be because Spd protects leaves from light-induced damage due to qP, which differs from heat dissipation.

Acknowledgments

The authors are grateful to Yanyan Zhao Ph.D. (NWSUAF) for advice regarding laboratory techniques, Fei Yan Ph.D. (NWSUAF) for insightful comments, and Hao Zhang (NWSUAF) for data analysis methods.

Author Contributions

Conceived and designed the experiments: X-HH ZZ. Performed the experiments: LH LZ XZ LX. Analyzed the data: LH LX. Contributed reagents/materials/analysis tools: LH LX. Contributed to the writing of the manuscript: LH XH.

References

1. Zhang Y, Hu XH, Shi Y, Zou ZR, Yan F, et al. (2013) Beneficial role of exogenous spermidine on nitrogen metabolism in tomato seedlings exposed to saline-alkaline stress. Journal of the American Society for Horticultural Science 138: 38–49.
2. Hu X, Zhang Y, Shi Y, Zhang Z, Zou Z, et al. (2012) Effect of exogenous spermidine on polyamine content and metabolism in tomato exposed to salinity-alkalinity mixed stress. Plant physiology and biochemistry: PPB/Societe francaise de physiologie vegetale 57: 200–209.
3. Yang J, Zheng W, Tian Y, Wu Y, Zhou D (2011) Effects of various mixed salt-alkaline stresses on growth, photosynthesis, and photosynthetic pigment concentrations of Medicago ruthenica seedlings. Photosynthetica 49: 275–284.
4. Li R, Shi F, Fukuda K, Yang Y (2010) Effects of salt and alkali stresses on germination, growth, photosynthesis and ion accumulation in alfalfa (Medicago sativa L.). Soil Science & Plant Nutrition 56: 725–733.
5. Deng C, Zhang G, Pan X (2011) Photosynthetic Responses in Reed (Phragmites australis (CAV.) TRIN. ex Steud.) Seedlings Induced by Different Salinity-Alkalinity and Nitrogen levels. Journal of Agricultural Science and Technology 13: 687–699.
6. Yadav DK, Pospíšil P (2012) Evidence on the formation of singlet oxygen in the donor side photoinhibition of photosystem II: EPR spin-trapping study. PloS one 7: e45883.
7. Asada K (2006) Production and scavenging of reactive oxygen species in chloroplasts and their functions. Plant physiology 141: 391–396.
8. Murata N, Takahashi S, Nishiyama Y, Allakhverdiev SI (2007) Photoinhibition of photosystem II under environmental stress. Biochimica et Biophysica Acta (BBA)-Bioenergetics 1767: 414–421.
9. Mittal S, Kumari N, Sharma V (2012) Differential response of salt stress on Brassica juncea: Photosynthetic performance, pigment, proline, D1 and antioxidant enzymes. Plant Physiology and Biochemistry 54: 17–26.
10. Bouchereau A, Aziz A, Larher F, Martintanguy J (1999) Polyamines and environmental challenges: recent development. Plant Science 140: 103–125.
11. Liu Y, Gu D, Wu W, Wen X, Liao Y (2013) The Relationship between Polyamines and Hormones in the Regulation of Wheat Grain Filling. PloS one 8: e78196.
12. Ha HC, Sirisoma NS, Kuppusamy P, Zweier JL, Woster PM, et al. (1998) The natural polyamine spermine functions directly as a free radical scavenger. Proceedings of the National Academy of Sciences 95: 11140–11145.

13. Bellé NAV, Dalmolin GD, Fonini G, Rubin MA, Rocha JBT (2004) Polyamines reduces lipid peroxidation induced by different pro-oxidant agents. Brain research 1008: 245–251.
14. Groppa MD, Benavides MP (2008) Polyamines and abiotic stress: recent advances. Amino acids 34: 35–45.
15. Demetriou G, Neonaki C, Navakoudis E, Kotzabasis K (2007) Salt stress impact on the molecular structure and function of the photosynthetic apparatus–the protective role of polyamines. Biochimica et Biophysica Acta (BBA)-Bioenergetics 1767: 272–280.
16. Shu S, Yuan LY, Guo SR, Sun J, Yuan YH (2013) Effects of exogenous spermine on chlorophyll fluorescence, antioxidant system and ultrastructure of chloroplasts in Cucumis sativus L. under salt stress. Plant physiology and biochemistry: PPB/Societe francaise de physiologie vegetale 63: 209–216.
17. Shu S, Guo SR, Sun J, Yuan LY (2012) Effects of salt stress on the structure and function of the photosynthetic apparatus in Cucumis sativus and its protection by exogenous putrescine. Physiologia plantarum 146: 285–296.
18. Li Y, Shi G, Wang H, Zhao J, Yuan Q (2009) Exogenous spermidine can mitigate the poison of cadmium in nymphoides peltatum. Chinese Bulletin of Botany 44: 571−577.
19. Sheng S (2012) Effects of exogenous spermidine on photosynthesis, xanthophyll cycle and endogenous polyamines in cucumber seedlings exposed to salinity. African Journal of Biotechnology 11: 6064–6074.
20. Demmigadams B, Adams III WW (2006) Photoprotection in an ecological context: the remarkable complexity of thermal energy dissipation. The New phytologist 172: 11–21.
21. Jahns P, Holzwarth AR (2012) The role of the xanthophyll cycle and of lutein in photoprotection of photosystem II. Biochimica et biophysica acta 1817: 182–193.
22. Demmigadams B, Adams III WW (1996) The role of xanthophyll cycle carotenoids in the protection of photosynthesis. Trends in Plant science 1: 21–26.
23. Zhang R, Kramer DM, Cruz JA, Struck KR, Sharkey TD (2011) The effects of moderately high temperature on zeaxanthin accumulation and decay. Photosynthesis research 108: 171–181.
24. Zhang RH, Li J, Guo SR, Tezuka T (2009) Effects of exogenous putrescine on gas-exchange characteristics and chlorophyll fluorescence of NaCl-stressed cucumber seedlings. Photosynthesis research 100: 155–162.
25. Farquhar GD, Sharkey TD (1982) Stomatal conductance and photosynthesis. Annual Review of Plant Physiology 33: 317–345.
26. Baker NR (2008) Chlorophyll fluorescence: a probe of photosynthesis in vivo. Annual Review of Plant Biology 59: 89–113.
27. Chen C, Zhang D, Li P, Ma F (2012) Partitioning of absorbed light energy differed between the sun-exposed side and the shaded side of apple fruits under high light conditions. Plant physiology and biochemistry: PPB/Societe francaise de physiologie vegetale 60: 12–17.
28. Hussain SS, Ali M, Ahmad M, Siddique KH (2011) Polyamines: natural and engineered abiotic and biotic stress tolerance in plants. Biotechnology advances 29: 300–311.
29. Childs AC, Mehta DJ, Gerner EW (2003) Polyamine-dependent gene expression. Cellular and molecular life sciences: CMLS 60: 1394–1406.
30. Galston AW, Sawhney RK (1990) Polyamines in plant physiology. Plant physiology 94: 406–410.
31. Parida A, Das A, Mittra B (2003) Effects of NaCl stress on the structure, pigment complex composition, and photosynthetic activity of mangrove Bruguiera parviflora chloroplasts. Photosynthetica 41: 191–200.
32. Liu ZX, Bie ZL, Huang Y, Zhen A, Lei B, et al. (2012) Grafting onto Cucurbita moschata rootstock alleviates salt stress in cucumber plants by delaying photoinhibition. Photosynthetica 50: 152–160.
33. Begcy K, Mariano ED, Gentile A, Lembke CG, Zingaretti SM, et al. (2012) A novel stress-induced sugarcane gene confers tolerance to drought, salt and oxidative stress in transgenic tobacco plants. PloS one 7: e44697.
34. Horton P, Ruban A, Walters R (1996) Regulation of light harvesting in green plants. Annual review of plant biology 47: 655–684.
35. Bilger W, Björkman O (1990) Role of the xanthophyll cycle in photoprotection elucidated by measurements of light-induced absorbance changes, fluorescence and photosynthesis in leaves of Hedera canariensis. Photosynthesis research 25: 173–185.
36. Goss R, Jakob T (2010) Regulation and function of xanthophyll cycle-dependent photoprotection in algae. Photosynthesis research 106: 103–122.
37. Nilkens M, Kress E, Lambrev P, Miloslavina Y, Muller M, et al. (2010) Identification of a slowly inducible zeaxanthin-dependent component of non-photochemical quenching of chlorophyll fluorescence generated under steady-state conditions in Arabidopsis. Biochimica et biophysica acta 1797: 466–475.
38. Yin Y, Li S, Liao W, Lu Q, Wen X, et al. (2010) Photosystem II photochemistry, photoinhibition, and the xanthophyll cycle in heat-stressed rice leaves. Journal of plant physiology 167: 959–966.
39. Guo YP, Guo DP, Zhou HF, Hu MJ, Shen YG (2006) Photoinhibition and xanthophyll cycle activity in bayberry (Myrica rubra) leaves induced by high irradiance. Photosynthetica 44: 439–446.
40. Ogweno JO, Hu WH, Song XS, Shi K, Mao WH, et al. (2010) Photoinhibition-induced reduction in photosynthesis is alleviated by abscisic acid, cytokinin and brassinosteroid in detached tomato leaves. Plant Growth Regulation 60: 175–182.
41. Demmigadams B, Adams III WW (1992) Photoprotection and other responses of plants to high light stress. Annual review of plant biology 43: 599–626.
42. Triantaphylidès C, Krischke M, Hoeberichts FA, Ksas B, Gresser G, et al. (2008) Singlet oxygen is the major reactive oxygen species involved in photooxidative damage to plants. Plant physiology 148: 960–968.
43. Wrona M, Różanowska M, Sarna T (2004) Zeaxanthin in combination with ascorbic acid or α-tocopherol protects ARPE-19 cells against photosensitized peroxidation of lipids. Free Radical Biology and Medicine 36: 1094–1101.
44. Havaux M, DallOsto L, Bassi R (2007) Zeaxanthin has enhanced antioxidant capacity with respect to all other xanthophylls in Arabidopsis leaves and functions independent of binding to PSII antennae. Plant physiology 145: 1506–1520.
45. Paramonova N, Shevyakova N, Kuznetsov VV (2004) Ultrastructure of chloroplasts and their storage inclusions in the primary leaves of Mesembryanthemum crystallinum affected by putrescine and NaCl. Russian Journal of Plant Physiology 51: 86–96.
46. Naeem MS, Warusawitharana H, Liu H, Liu D, Ahmad R, et al. (2012) 5-aminolevulinic acid alleviates the salinity-induced changes in Brassica napus as revealed by the ultrastructural study of chloroplast. Plant physiology and biochemistry: PPB/Societe francaise de physiologie vegetale 57: 84–92.
47. Austin JR, Frost E, Vidi PA, Kessler F, Staehelin LA (2006) Plastoglobules are lipoprotein subcompartments of the chloroplast that are permanently coupled to thylakoid membranes and contain biosynthetic enzymes. The Plant Cell Online 18: 1693–1703.
48. Gill SS, Tuteja N (2010) Reactive oxygen species and antioxidant machinery in abiotic stress tolerance in crop plants. Plant Physiology and Biochemistry 48: 909–930.
49. Tian J, Wang LP, Yang YJ, Sun J, Guo SR (2012) Exogenous spermidine alleviates the oxidative damage in cucumber seedlings subjected to high temperatures. Journal of the American Society for Horticultural Science 137: 11–19.
50. Hamdani S, Yaakoubi H, Carpentier R (2011) Polyamines interaction with thylakoid proteins during stress. Journal of Photochemistry and Photobiology B: Biology 104: 314–319.
51. Hamdani S, Gauthier A, Msilini N, Carpentier R (2011) Positive charges of polyamines protect PSII in isolated thylakoid membranes during photoinhibitory conditions. Plant and cell physiology 52: 866–873.

9

Influence of Soil Type, Cultivar and *Verticillium dahliae* on the Structure of the Root and Rhizosphere Soil Fungal Microbiome of Strawberry

Srivathsa Nallanchakravarthula*, Shahid Mahmood, Sadhna Alström, Roger D. Finlay

Uppsala BioCenter, Department of Forest Mycology and Plant Pathology, Swedish University of Agricultural Sciences, Uppsala, Sweden

Abstract

Sustainable management of crop productivity and health necessitates improved understanding of the ways in which rhizosphere microbial populations interact with each other, with plant roots and their abiotic environment. In this study we examined the effects of different soils and cultivars, and the presence of a soil-borne fungal pathogen, *Verticillium dahliae*, on the fungal microbiome of the rhizosphere soil and roots of strawberry plants, using high-throughput pyrosequencing. Fungal communities of the roots of two cultivars, Honeoye and Florence, were statistically distinct from those in the rhizosphere soil of the same plants, with little overlap. Roots of plants growing in two contrasting field soils had high relative abundance of *Leptodontidium* sp. C2 BESC 319 g whereas rhizosphere soil was characterised by high relative abundance of *Trichosporon dulcitum or Cryptococcus terreus*, depending upon the soil type. Differences between different cultivars were not as clear. Inoculation with the pathogen *V. dahliae* had a significant influence on community structure, generally decreasing the number of rhizosphere soil- and root-inhabiting fungi. *Leptodontidium* sp. C2 BESC 319 g was the dominant fungus responding positively to inoculation with *V. dahliae.* The results suggest that 1) plant roots select microorganisms from the wider rhizosphere pool, 2) that both rhizosphere soil and root inhabiting fungal communities are influenced by *V. dahliae* and 3) that soil type has a stronger influence on both of these communities than cultivar.

Editor: Kevin McCluskey, Kansas State University, United States of America

Funding: This work was funded by Grant No. 2007-1689 to RF from The Swedish Research Council for Environment, Agricultural Sciences and Spatial Planning. The funders had no role in study design, data collection and analysis, decision to publish, or preparation of the manuscript.

Competing Interests: The authors have declared that no competing interests exist.

* Email: srivathsa.nallan@gmail.com

Introduction

Plants roots influence the structure and diversity of soil microbial communities through their energy-rich exudates [1,2], creating diverse rhizosphere communities in close proximity to the roots [3]. The roots are colonized by a subset of these microorganisms, creating both a rhizoplane community on the surface and an endophytic compartment. Discrimination between pathogens and beneficial mutualists or commensals is controlled by an immune system [4], but the mechanisms underlying the selection of different core microbiomes [5] by different plant species are still poorly understood. Recent studies of bacteria associated with the roots or rhizosphere of *Arabidopsis thaliana* grown in geochemically distinct soils [6] have shown that they are both strongly influenced by soil type but that the root endophytic compartments from two soils contain overlapping, low complexity communities that are enriched in Actinobacteria and specific families belonging to the Proteobacteria. Differential relative abundance of *Actinocorallia* sp. has also been shown in different ecotypes of *A. thaliana* [7].

Plant host genetics plays a significant role in the selection of microbial communities associated with roots [8–10]. This selection has important implications for plant health and growth since rhizosphere microorganisms are known to have a variety of positive or negative effects. These include plant growth stimulation, antagonism against pathogens and effects on carbon sequestration and phytoremediation. Rhizosphere and root endophyte microorganisms are increasingly being used on a commercial scale for plant protection as biocontrol agents and biofertilizers [11–13]. Management of such microorganisms requires a thorough understanding of their interaction with other microorganisms sharing the same ecological niche. Some attempts have been made to study the effect of biocontrol agents on microbial communities [14–17], however studies focusing on the influence of soil-borne fungal plant pathogens on microbial community structures are less frequent [18–21]. Most of the inoculation studies with pathogens or biocontrol agents have been more focused on bacterial communities than on fungal communities.

The rhizosphere is a dynamic region for many biological processes and interactions [1,11,12,22–25] and empirically-based management strategies for sustainable cultivation need to be based on a more rigorous understanding of the functional basis of the mechanisms underlying the various biotic interactions taking place [26].

Microbial community structure in the rhizosphere and the root endophytic compartment is shaped by soil characteristics as well as by plant genotype [6,7,20,27–31]. Soils in organically managed strawberry fields have been shown to have greater functional activity and higher functional gene diversity than soils in conventionally managed fields [32]. The strawberry rhizosphere has been shown to harbour distinct microbial communities containing *Streptomyces*, *Rhizobium* and *Nocardia* [28,31] and in the presence of *Verticillium dahliae*, a soil-borne fungal pathogen, it is dominated by *Pseudomonas* spp. populations [27]. Studies have also shown that strawberry cultivars respond differently to the inoculation of arbuscular mycorrhizal fungi (AMF) [33,34]. Sporulation of *Phytophthora fragariae* associated with strawberry plants colonized by AMF has also been shown to be reduced in comparison to that in non-AMF colonized plants [35].

Strawberry (*Fragaria* ×*ananassa*) is grown for its berries in many countries and different cultivars have been developed for commercial production. Rhizosphere bacterial biomass and activity, fruit quality and yield have all been shown to vary under different management systems. Soil microorganisms have been shown to be indirectly involved in determining the fruit quality [32,36]. *Verticillium dahliae* is one of several plant pathogens that affect plant growth and yield in different plant species, including strawberry [37–41]. The pathogen forms conidia and microsclerotia that germinate in the presence of the root exudates, whereupon germ tubes enter the plant roots. The pathogen can persist as microsclerotia on dead plant tissues, or in the soil for more than 10 years [39–41]. In strawberry as few as two microsclerotia per gram of soil can cause 100% wilt [42] and this pathogen is difficult to combat with current management strategies. In a field trial, biocontrol using *Trichoderma harzianum* and *Trichoderma viride* against *V. dahliae* reduced wilting in strawberry by 60%. An indirect effect against *V. dahliae* through induced systemic resistance by *Paenibacillus alvei* K165 has also been observed in *Arabidopsis* sp. [43,44].

Studies have shown that the presence of fungal (*Phytophthora cinnamomi*) and bacterial pathogens, (*Liberibacter asiaticus* and *Erwinia carotovora* (*Pectobacterium carotovorum*)) increases the diversity of Bacteroidetes and Alphaproteobacteria in particular [18,20,21]. Such changes also facilitate the selection, by the plant, of taxa including those with antimicrobial properties [27,45].

Plant resistance is one of the effective strategies in plant protection and attempts have been made to study effects of resistant and susceptible cultivars on microbial communities [46–48]. The results of these studies demonstrate that plant genotype can have a significant impact on soil microbial community structure, and differences in the rhizosphere microbial community have been suggested to contribute to the differences in resistance to disease. Whether this impact is independent of the soil type is not evident from these studies.

Many earlier studies involving soil microbial community analyses have been conducted using either cultivation-dependent approaches or molecular methods such as T-RFLP, DGGE, cloning and Sanger sequencing which are now proven to underestimate the total microbial diversity [49,50]. More recent studies using high throughput techniques have revealed taxa that were not detected with previous methods [51].

In this study we investigated the impact of plant genotype on the fungal community structure in different soil types in the presence and absence of the soil-borne fungal pathogen, *V. dahliae*. We used 454 pyrosequencing to analyze the rhizosphere soil- and root fungal communities in different strawberry cultivars, grown in differently managed soils. We hypothesized that the structure of microbial communities in the rhizosphere soil and roots is determined by both the cultivar and soil type, and may also be re-structured in the presence of the pathogen.

Materials and Methods

Experimental soils and plant material

Four cultivars of strawberry, Florence, Honeoye, Senga Sengana, and Zephyr, and two of the three different soils used in this study were the same as those described in a previous study by Santos-González et al. [30]. The cultivars were selected on the basis of their breeding year, yield level and disease tolerance. Honeoye and Florence were bred after 1978, Zephyr and Senga Sengana before 1966. Honeoye is considered to be a high yielding cultivar and Florence and Senga Sengana are considered to be tolerant to *V. dahliae* [52–54]. In our figures the conventionally managed soil from Kristianstad is referred to as "conventional" and the organically managed soil from Hörby as "organic" [30]. The field in Hörby was managed organically with appropriate crop rotations since 1983 and pre-cropped with potatoes prior to sampling in 2009. The experimental field in Kristianstad is conventionally managed and was planted with strawberry in 2008. The two agricultural fields are located 38 km apart in Southern Sweden. These sites are on private land and permission to collect soil was obtained from the owners Sune Abrahamsson (Kristianstad) and Ingvar Åkesson (Hörby). The soil collection did not involve endangered or protected species. The Kristianstad site at (56° 06′ N, 14° 01′ E) and the Hörby site at (55° 50′ N, 13° 35′ E) have different soil properties. The physico-chemical traits of the two field soils are described by Santos-González et al. [30]. The soils had similar textural properties (8.5–11% clay, 29–35% silt, 51–58% sand) and bulk density (1.05–1.07 g cm^{-3}). However the pH, C, N and P levels of the Hörby soil were significantly ($P<$ 0.05) higher than those of the Kristianstad soil (6.36 vs. 5.97; 2.1% vs. 1.61%; 0.17% vs. 0.14%; 181 mg kg^{-1} vs. 113 mg kg^{-1}). The third "soil" was a peat-based growth substrate (Hasselfors Garden Special, Hasselfors Garden AB, Örebro, Sweden) containing 60% light peat, 25% black peat, 15% sand (0.5–4 mm) and 0.050 kg m^{-3} multisport/FTE36, 2.0 kg m^{-3} dolomite and 4.5 kg m^{-3} limestone, amended with 1.3 kg m^{-3} fertilizer (N14:P7:K15, pH-6.0), and is used for commercial greenhouse cultivation in Sweden.

Pathogen inoculum

The pathogen isolate used was *V. dahliae* 12086, originating from strawberry roots and available from the Food and Environment Agency, York, UK (see http://www.q-bank.eu/). It was purified and grown in potato dextrose broth (PDB per L, 12 g Accumedia, Michigan, USA) as stationary cultures for 20 days at 24°C. The mycelial mat thus obtained was macerated carefully in diluted potato dextrose broth (1:10 water v/v). A preparatory study was conducted with the suspension thus obtained to confirm the pathogenicity of the test pathogen. On the basis of the results from the preparatory study (data not shown), 30 ml mycelial suspension was inoculated in *V. dahliae*-treatments. The plant roots were injured by inserting a 0.9 cm diameter cork borer into the root volume in three locations to stimulate the pathogen infection. No dead-cell, or chitin controls were used but the control plants without the pathogen were treated in a similar manner but with suspension medium not containing fungal mycelium.

Experimental design

An outdoor experiment was conducted by growing the four cultivars in the three soils as specified in a previous study [30]. In brief, the soils were prepared before planting with one plantlet per pot (13×13 cm×25 cm deep), placed outdoors and irrigated as needed. The plantlets were taken from storage under refrigeration and were about 5 cm tall with three to four leaves and roots about 15 cm long.

The cumulative fresh berry weight per plant was recorded during the experimental period of 14 weeks. The rhizosphere soil was collected from three replicates 12 weeks after planting and 14 weeks after planting. To sample the rhizosphere soil, the plants were carefully removed from the pots and roots with tightly adhering soil were suspended in phosphate-buffered saline solution (PBS, 0.14 M NaCl, 0.0027 M KCl, 0.010 M phosphate buffer, pH 7.4, Medicago, Sweden) for 30 minutes. The rhizosphere soil thus collected was centrifuged at 8000 RPM for 10 minutes at 4°C (Biofuge PrimoR, Heraeus Sorvall), the roots were once again washed carefully under aerated tap water. The soil pellet and the roots were stored at −20°C for further analyses.

Analysis of fungal communities

DNA extraction and selection of samples for pyrosequencing. DNA extraction methods differ in their efficiency of extraction from different taxa and for the rhizosphere samples, we used two different nucleic acid extraction methods and the resultant DNA extracts were checked individually for quality and reproducibility on a DGGE (denaturing gradient gel electrophoresis) gel following PCR amplifications. The first method was based on 'CTAB' (hexadecyltrimethylammonium bromide) [55] that involved extraction of nucleic acids from 0.5 g of soil. The soil samples were weighed in 2 ml 'Lysing matrix B' tubes (Fisher Scientific, USA) containing 0.5 ml CTAB buffer (120 mM and pH 8.0) and 0.5 ml phenol: chloroform: isoamylalcohol (25:24:1 V/V). The cells were lysed at 5000 RPM for 30 seconds in a bead beater (Precellys 24; Bertin Technologies, France). Proteins and solvent residues were removed using chloroform and isoamyl alcohol (24:1) followed by further precipitation using polyethylene glycol (30%). The precipitated DNA was washed with ice-cold ethanol (70%), air dried and re-suspended in 30 µl of MilliQ water and stored at −20°C for further analysis.

The second nucleic acid extraction method was based on MOBIO 'RNA power soil' total RNA isolation and DNA elution accessory kits (MOBIO laboratories, California, USA) according to manufacturer's recommendations. The concentration and quality of extracted DNA were assessed using NanoDrop (ND-1000 Nanodrop technologies, USA). The banding profiles of these two methods differed so the DNA from these two extraction methods was pooled and used as a mixed template for generating PCR amplicons for pyrosequencing. Prior to pyrosequencing, the control samples were subjected to DGGE-based analysis following PCR amplification. The DGGE banding profiles showed higher numbers of bands at the first sampling, 12 weeks after planting, than at 14 weeks (data not shown). Pyrosequencing was therefore carried out on samples from 12 weeks.

For the analysis of root-associated fungi we chose to analyze two cultivars Honeoye, susceptible to *V. dahliae*, and Florence, tolerant to *V. dahliae*, growing in each of the two field soils. Nucleic acids from roots were extracted using DNeasy plant mini kit (Qiagen, Germany) according to the manufacturer's recommendations. Roots were carefully washed to remove the soil particles attached to the roots and soil removal was microscopically confirmed. They were gently washed three times with 0.1% Triton x100 for two minutes, thereafter washed with sterile milliQ water. All collected roots were subsequently freeze-dried for 48 h (CoolSafe, ScanLaf A/S, Denmark), homogenised by milling in 2 ml tubes containing 2.38 mm diameter stainless steel beads (MOBIO laboratories, California, USA) in a bead-beater for 30 s twice. 50 mg of the homogenised root material was used for nucleic acid extraction. The extracted DNA was quantified and used as a template for generating PCR amplicons for pyrosequencing.

Pyrosequencing

Pyrosequencing analysis of the fungal communities was carried out as described by Ihrmark et al. [56] using primers fITS7 (5′GTGARTCATCGAATCTTG-3′) and ITS4 (5′TCCTCC-GCTTATTGATATGC-3′) [57]. In brief, each 50 µl reaction mixture consisted of 200 µM of dNTPs, 2.75 mM $MgCl_2$, primers at 500 nM and 300 nM, 0.025 U polymerase (DreamTaq Green, Thermo Scientific, Waltham, MA) in buffer. The thermocycling conditions were: 5 min at 94°C; 35 cycles (30 s at 94°C; 30 s at 57°C; 30 s at 72°C); 7 min at 72°C. Three technical replicates were run for each sample. To ascertain PCR amplification efficiency and to rule out any possible contamination, all reactions were run with positive and negative controls. The PCR amplification products were analyzed on 1% (w/v) agarose gels prestained with Nancy red-520 (Sigma Aldrich, USA). The triplicate PCR products of each sample were pooled, purified using Agencourt AMPure, PCR purification kit (Beckman Coulter, Massachusetts, USA) and quantified using a Qubit Fluorometer (Invitrogen, USA). Equimolar concentrations of samples were pooled and the resultant mixture was freeze-dried (CoolSafe, ScanLaf A/S, Denmark) overnight. Pyrosequencing was carried out on 2×1/8th of a GS FLX Titanium Pico Titer Plate (LGC Genomics, Germany and MACROGEN, South Korea) according to the manufacturer's instructions (Roche, Branford, CT).

Bioinformatic analysis

The SCATA pipeline (Sequence Clustering and Analysis of Tagged Amplicons, http://www.scata.mykopat.slu.se) was used to analyze the sequence data from 72 rhizosphere samples and 24 root samples. The quality control and data filtering involved removal of singletons, low quality sequences, tags and primer missing sequences, followed by removal of sequences shorter than 200 base pairs (after primer and tag trimming). We used default parameters of the SCATA pipeline except that the proportion of primer match was set at 0.9. Sequences were assembled into clusters by single-linkage clustering with a 98.5% sequence similarity threshold. The most common genotype/sequence of the cluster was used to represent the operational taxonomic units (OTUs). All the representative sequences of the clusters were identified using the NCBI (National Center for Biotechnology Information) database and the nucleotide blast 'BLASTn' program. The most common sequence from each cluster was analyzed manually for blasting. With the exception of rarefaction analysis, only sequences with ≥95% identity were chosen for these analyses.

The pyrosequencing data have been deposited at ENA (European Nucleotide Archive) under accession number PRJEB6260.

Statistical analyses

Statistical analyses were conducted using a significance level of $p \geq 0.05$. The Chao-1 index (PAST (PAleontological STatistics), version 2.17b) was calculated to obtain the number of OTUs present in the four cultivars grown in three different soils with and without *V. dahliae*. The bar diagrams were generated using Microsoft Excel and the significance of the OTU data was estimated using ANOVA with MINITAB. The relative abundance of OTUs was used for Non-metric multi-dimensional scaling (NMDS) ordination performed with 'PAST' using Bray-Curtis dissimilarity indices. Numbers of OTUs were calculated for selected phyla and plotted in Microsoft Excel and their significance was calculated by using ANOVA. SIMPER (similarity percentage) analysis (PAST ver 2.17b) was performed using relative abundance data in order to identify the OTUs that were contributing to the differences among treatments. The rarefaction was estimated using Analytic Rarefaction (Ver 1.3, UGA Stratigraphy Lab, USA).

Results and Discussion

Bioinformatic analysis of root and rhizosphere material associated with the cultivars Honeoye and Florence revealed a total of 577 unique OTUs following quality control filtering and removal of singletons. In total 360 OTUs were associated with rhizosphere soil and 307 OTUs were associated with roots, with an overlap of 90 OTUs between the two compartments. Analysis of the rhizosphere soil associated with roots of all four cultivars revealed 128 OTUs common to both the field soils but only 16 fungal OTUs that were common to all three soils. In conventionally managed soil 63 OTUs were common to all four cultivars, with 28 common OTUs in organically managed soil and 18 common OTUs in peat-based growth substrate.

Cluster analysis of the community structure of the 50 most abundant fungal taxa was performed. These taxa constituted 90% of the reads in roots and 70% of the reads in rhizosphere soil. The analysis (Figure 1) revealed a clear separation between fungal communities from the rhizosphere soil and those from the roots. Within each of these compartments there was a clear separation according to soil type (ANOSIM $p = 0.0001$). The root associated community structure of the organically managed soil showed a statistically significant (ANOSIM $p = 0.015$) separation between *Verticillium* inoculated and uninoculated treatments. Rhizosphere soil communities in a peat-based substrate formed a distinct cluster with a clear separation between *V. dahliae* inoculated and equivalent, non-inoculated treatments. In a similar analysis of rhizosphere soil communities associated with four different strawberry cultivars, Honeoye, Florence, Senga-Sengana and Zephyr (Figure S1) clear effects of soil type were visible. Differences due to cultivar were less evident. These results are consistent with the finding of Santos-González *et al* [30] that soil type, but not cultivar, shapes the assemblages of arbuscular mycorrhizal fungi associated with strawberry roots in these soils.

Additional NMDS analysis of the fungal community structure in the rhizosphere soil of all four cultivars (Figure S2) showed a statistically significant effect of *Verticillium* inoculation in conventionally managed soil (ANOSIM $p = 0.005$) and peat based substrate (ANOSIM $p = 0.0004$). No such clear separation was evident in organically managed soil.

Rank abundance plots of the 30 most abundant fungal taxa in each of the rhizosphere and root compartments in the absence of *Verticillium* (Figure 2) revealed that the structure of the fungal microbiome in these two compartments is markedly different, with little overlap. The dominant OTUs from the roots were found at low frequencies in the rhizosphere, while OTUs that were dominant in the rhizosphere were absent or occurred at very low frequencies in the roots, in both field soils (Figure 2a–d). These results are in agreement with other recent studies of bacterial microbiomes of *Arabidopsis thaliana* and the fungal microbiome of *Pisum sativum* [6,7,58] that also show distinct communities in the rhizosphere and root endophyte compartments. Within each soil this pattern was the same for both Honeoye and Florence, however there were large differences between the rank abundance distributions of taxa between the two soils. The same lack of overlap between root and rhizosphere soil fungal communities was found in the presence of *Verticillium* (data not shown).

The top seven most abundant taxa in conventionally managed rhizosphere soil, irrespective of strawberry cultivar were *Cryptococcus terreus*, *Mortierella elongata*, *Cryptococcus terricola*, *Panaeolus foenisecii*, *Phoma eupyrena*, *Gibellulopsis nigrescens* and an unidentified *Leptosphaeria* species, together accounting for 33% and 37% of the total reads for the Honeoye and Florence treatments respectively. In the organically managed rhizosphere soil the top six fungal taxa accounted for 50% of the reads but *Trichosporon dulcitum* was the dominant taxon, accounting for 35% and 30% of the total reads in the Honeoye and Florence treatments respectively. *Trichosporon dulcitum* was also detected in the conventionally managed soil, but at much lower frequencies (3% and 2%). *Trichosporon dulcitum* is a soil-borne basidiomycetous yeast. Utilization of benzene compounds [59] and biodegradation of phenol [60] have been shown in *T. dulcitum*, but the ability of this yeast to interact with plants and fungal pathogens is still unclear. Other members of the Trichosporonales have been found in the rhizosphere and ectomycorrhizosphere in soil from a *Nothofagus pumilio* forest [61]. In the latter study *T. dulcitum* was found in the bulk soil but not in the rhizosphere and no other information appears to be available concerning the possible associations of this species with plant roots.

The most common fungal taxa found in roots included *Entrophospora* sp shylm 131, *Leptodontidium* sp. C2 BESC 319 g, *Ilyonectria macrodidyma*, *Fusarium* spp., *Phoma* sp BEA-2010, *Scytalidium lignicola*, *Acremonium strictum*, *Exophiala* sp. O 1 LILB 26 5 g, and *Paraconiothyrium sporulosum*. The composition of fungal taxa in roots of plants growing in the organically managed soil was characterised by high abundance of *Leptodontidium*, whereas both *Leptodontidium* and *Entrophospora* were dominant in roots of plants growing in the conventionally managed soil. Differences in the abundance of fungal taxa in relation to soil types and strawberry cultivar are discussed below in a separate SIMPER analysis.

The total numbers of fungal OTUs in the absence of *Verticillium* inoculation, were generally highest in the rhizosphere of plants grown in conventionally managed soil, intermediate in organically managed soil and lowest in peat-based growth substrate, indicating a strong effect of soil type on fungal diversity. The lowest fungal diversity in the peat based growth substrate is most likely due to its simple composition [62]. In the presence of *Verticillium* no clear difference between the field soils was evident but numbers of OTUs were lower in the peat-based growth substrate (ANOVA $p \leq 0.05$) (Figure 3a).

In conventionally managed soil, *Verticillium* inoculation resulted in a consistent reduction (20–45%) in the number of fungal OTUs in the rhizosphere that was statistically significant ($p \leq 0.05$) in Senga Sengana and Zephyr (Figure 3a). In contrast, in the organically managed soil, inoculation with *Verticillium* resulted in significant ($p \leq 0.05$) increases (114% and 125%) in the total numbers of ascomycete and basidiomycete OTUs respectively in the rhizosphere of Florence (Figure 3a). The reasons for this increase are not clear but may reflect differences in the quality and quantity of root exudates of Florence, which is considered to be a *Verticillium* tolerant cultivar [63]. Basidiomycota were less abundant in the rhizosphere but also showed a consistent reduction in response to *Verticillium* in plants grown in conventionally managed soil. This was statistically significant only in Honeoye, but evident as a trend in the other cultivars.

Ascomycota dominated, both in roots and in rhizosphere (Figure 3a & b). *Verticillium* inoculation had no statistically significant effects on the number of Ascomycotan OTUs in roots of Honeoye and Florence grown in conventional soil but these were significantly reduced by 21% in Honeoye in response to

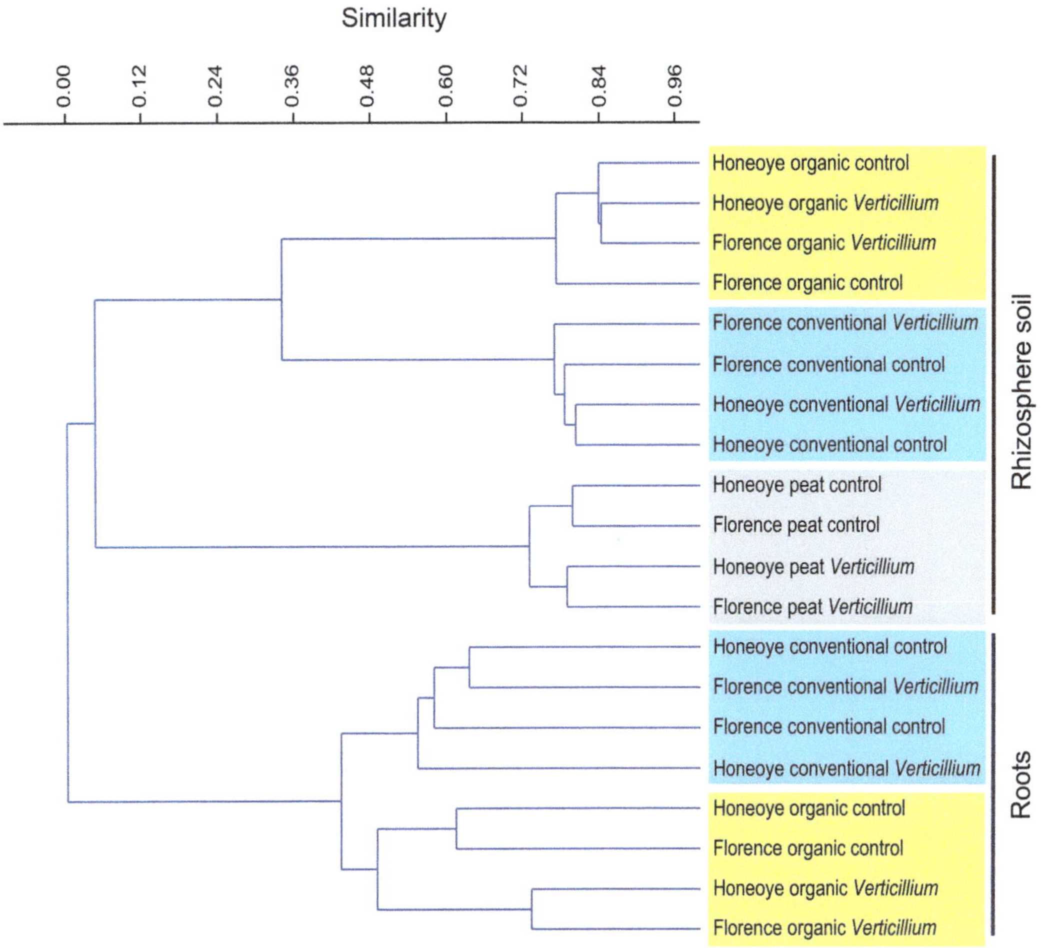

Figure 1. Cluster analysis of the 50 most abundant fungal operational taxonomic units (OTUs) in the rhizosphere and roots of two strawberry cultivars, Honeoye and Florence, grown in conventionally or organically managed soils or a peat based growth substrate, with and without *Verticillium dahliae*. These OTUs constituted 90% of the total reads in roots and 70% of the reads in rhizosphere soil. The clustering is based on paired group linkage using the Bray-Curtis similarity measure. Only OTUs with ≥95% identity are included.

Verticillium inoculation in organically managed soil. The mean number of Ascomycotan taxa in the roots of Florence was 38% lower in the presence of *Verticillium* but this reduction lacked statistical support.

Numbers of Basidiomycotan OTUs in roots were negligible, but 17 Glomeromycotan OTUs were found, including *Entrophospora* sp. shylm 131, *Rhizophagus intraradices* and *Rhizophagus clarus*. Numbers of Glomeromycotan OTUs in Honeoye roots in organically managed soil were significantly increased by 50% (Figure 3b) in response to *Verticillium* inoculation. Xu et al [58] have also shown that the relative abundance of different fungal phyla varies in root and rhizosphere samples of *Pisum sativum* depending upon whether the plant is healthy or infected with the pathogen *Aphanomyces euteiches*.

Rarefraction analysis of the numbers of OTUs associated with different numbers of sequences from root and rhizosphere samples are shown in Figure S3. The root treatment samples approached saturation, with asymptotes between 80 and 120 OTUs, although fewer root treatments were analysed in this study since only the roots of Honeoye and Florence plants were examined. The species accumulation curves for rhizosphere soil samples included three growth substrates and four cultivars and were not saturated, suggesting that not all taxa were detected. However there was a high degree of consistency between the most abundant taxa in the rhizosphere for both soils (Figure 2), suggesting that the dominant taxa in these communities were adequately described.

SIMPER (Similarity Percentage) analysis of OTUs from rhizosphere and root samples was performed to identify the species primarily contributing to the difference in community structure between the two compartments and the results are summarized in Table S1.

In Florence the effect of *Verticillium* inoculation on fungal community structure was stronger in roots (71% in organically managed soil, 60% in conventionally managed soil) than in the rhizosphere, (46% in organically managed soil, 52% in conventionally managed soil). In Honeoye the contribution to *Verticillium* induced differences was lower but the values for roots (53% in organic soil and 54% in conventionally managed soil) were still higher than in the rhizosphere (41% in organic soil and 48% in conventionally managed soil).

The numbers of reads matching *Verticillium* genera were 564 in roots compared to only 35 in the rhizosphere soil suggesting that *V. dahliae* was more abundant in the roots than the rhizosphere. This is not surprising since *V. dahliae* is a root pathogen but the low number of reads in rhizosphere soil may reflect primer bias against *V. dahliae* or low template concentrations at the time of sampling.

Figure 2. Ranked abundance of the 30 most abundant fungal operational taxonomic units (OTUs) present in the rhizosphere and roots of two strawberry cultivars, Honeoye and Florence, grown in conventionally (A and B) and organically (C and D) managed field soils. Only OTUs with ≥95% identity are included.

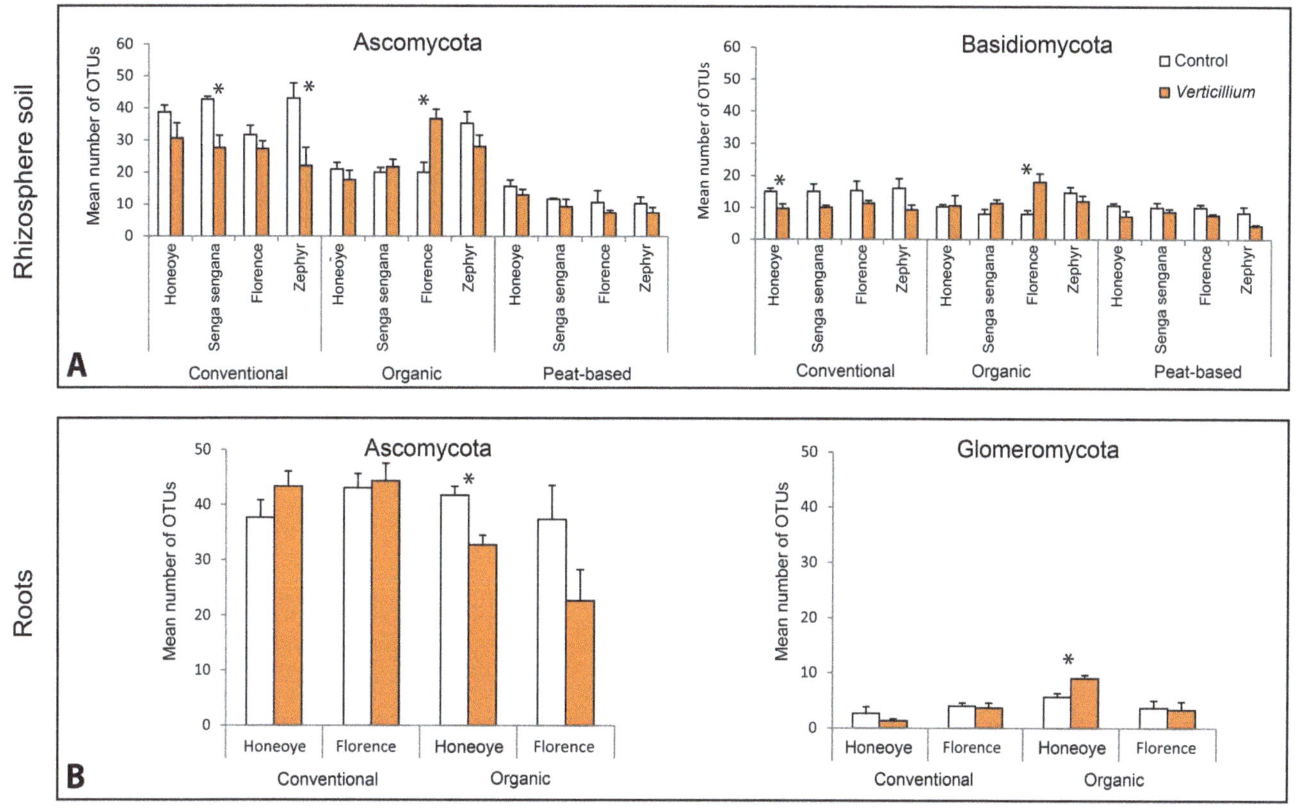

Figure 3. Distribution of fungal operational taxonomic units (OTUs) between different fungal phyla in the rhizosphere (A) and roots (B) of different strawberry cultivars grown in conventionally or organically managed soils or a peat-based growth substrate with and without *Verticillium dahliae*. Only OTUs with ≥95% identity are included. Asterisks indicate statistical significance between control and *Verticillium dahliae* treatments within each soil (* = $p \leq 0.05$, ANOVA). Vertical bars represent mean values and error bars show ±1 SE (n = 3).

Van der Mark et al [64] reported detection of significantly more pathogen DNA of *Aphanomyces euteiches* and *Phytophthora medicaginis* in a susceptible variety of alfalfa than in the resistant variety [64]. In our study, in the conventionally managed soil, significantly higher numbers of reads (345) matching *Verticillium* were detected in *Verticillium* inoculated treatments of Honeoye, the susceptible cultivar, than in Florence, a tolerant cultivar (17 reads). In the organically managed soil the numbers of reads were lower than 17 in both cultivars, suggesting that this soil was possibly suppressive in character, while the conventionally managed soil was more conducive to *Verticillium*.

The top ten OTUs contributing to the differences between control and *Verticillium* treatments across all treatments, identified using SIMPER analysis, are shown in Figures 4a–b. The dominant OTUs in roots occur at low frequencies in the rhizosphere soil while OTUs that dominate in the rhizosphere soil are either absent or occur at low frequencies in the roots. This discrepancy between root and rhizosphere soil communities is in agreement with recent observations in other plants [6,7,58]. *Cryptococcus* spp., *Gibellulopsis* sp., *Mortierella* spp. and *Phoma* sp. were abundant in rhizosphere in the conventionally managed soil while *Trichosporon dulcitum* and *Fusarium equiseti* were abundant in rhizosphere of organically managed soil. In contrast, *Leptodontidium* sp. dominated the roots in organically managed soil, while *Entrophospora* sp. dominated the roots of plants in conventionally managed soil (Figure 4a, b).

Verticillium dahliae was among the top ten species contributing to differences in fungal community structure between inoculated and uninoculated roots of Honeoye plants growing in conventionally managed soil and had a higher relative abundance in inoculated treatments than in uninoculated treatments (1.7%, 4.6%). The relative abundance of *Verticillium dahliae* was much lower in roots of Florence plants (0.56%, 0.27%). This result is consistent with the greater tolerance of Florence to *Verticillium* than Honeoye. The fungus was not present amongst the top ten taxa in organically managed soil, irrespective of strawberry cultivar or inoculation treatment.

Leptodontidium sp C2 BESC319g was one of the main contributors to differences in community structure in response to inoculation with *Verticillium*. In conventionally managed soil its relative abundance varied between 12 and 14% and was not affected by *Verticillium* inoculation. However in organically managed soil its relative abundance was much higher, increasing from 28.5% to 45.6% in Honeoye and from 17.3% to 49% in Florence, in the presence of *Verticillium* inoculation. This suggests that *Leptodontidium* sp C2 BESC319g may be one of the biological factors associated with the suppressive character of the organically managed soil. *Leptodontidium orchidicola* has been shown to decrease the negative effect of *V. dahliae* in tomato, but only at low levels of pathogen inoculation [65].

Root endophytes are known to produce secondary metabolites that may have anti-microbial effects on plant pathogens and biological control of *Verticillium dahliae* by root endophytes has been successfully demonstrated [66–70]. In the present study the absence of *V. dahliae* associated with increased abundance of *Leptodontidium* in roots of plants grown in organically managed soil, suggests that this fungus may be of functional significance in

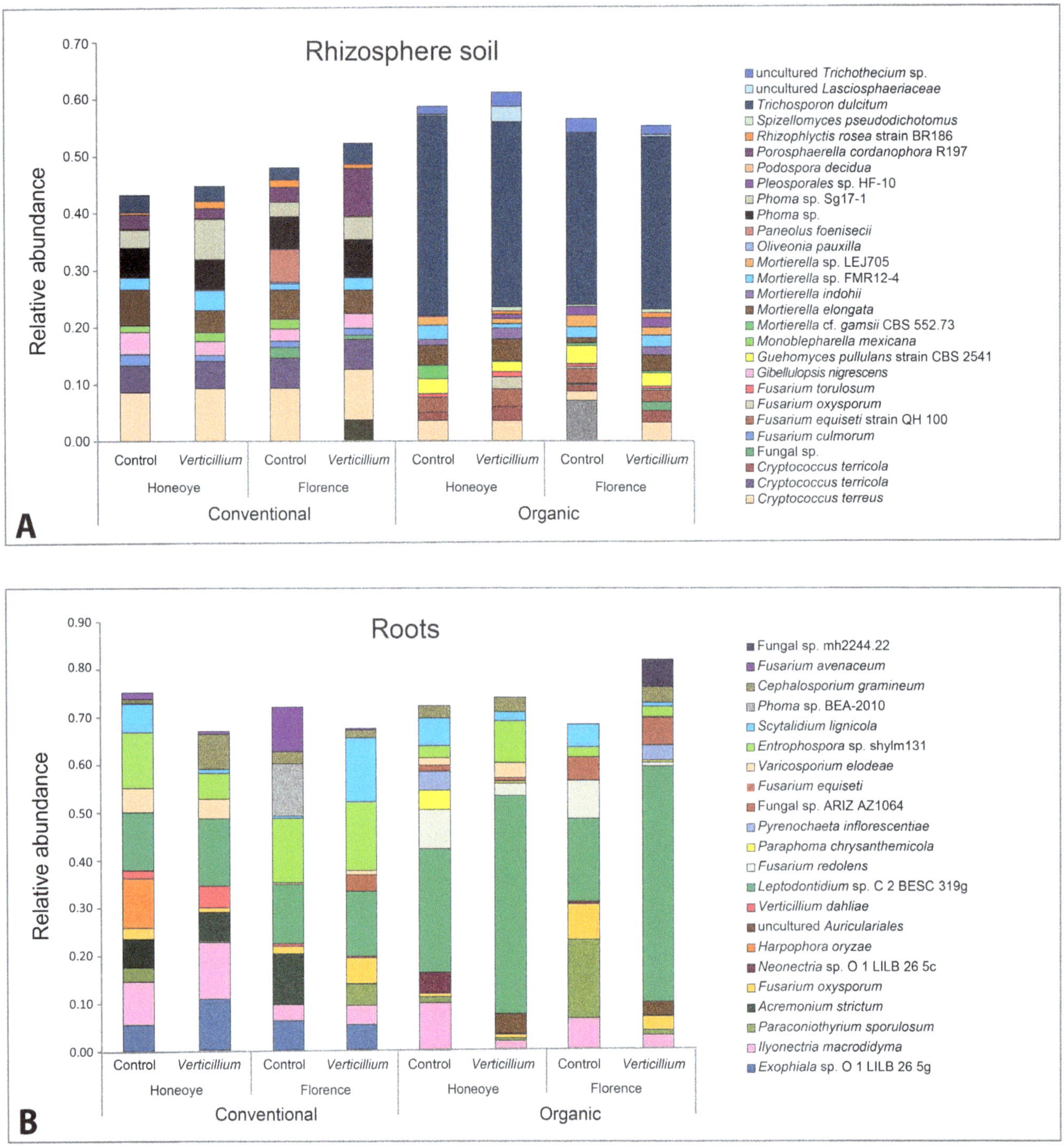

Figure 4. The top ten fungal operational taxonomic units (OTUs) (according to Similarity Percentage analysis) (SIMPER) contributing to the observed differences between control and *Verticillium dahliae* treatments in Honeoye and Florence grown in conventionally and organically managed soils. A) Rhizosphere soil (29–39%), B) Roots (61–74%).

protection against vascular pathogens. Further research on root endophytic communities should provide new biological control strategies for improving plant health [25].

OTUs of other broadly pathogenic taxa were also detected in our study. Examples of these include species belonging to the genera *Scytadilium, Fusarium, Pyrenochaeta, Phoma and Cephalosporium* but their pathogenicity to strawberry remains largely unknown. The relative abundance of these taxa was affected by both cultivar and soil type in the presence of *Verticillium* (Figure 4 a–b).

The effect of soil type, cultivar and pathogen inoculation on OTUs was also reflected in the plant performance measured as berry yield in this study (Figure S4). The strawberry yield differed significantly between the four cultivars in all three soils (conventional $p = 0.0014$, organic $p = 0.0087$ and peat $p = 0.0022$) in the absence of *V. dahliae*. Honeoye and Zephyr were selected because of their high yielding potential. In the present study these two cultivars produced the highest yields in the conventionally managed soil and peat based growth substrates but their mean yields decreased in response to inoculation with *V. dahliae*. The response of Florence to inoculation with the pathogen varied according to the soil type demonstrating a significant positive effect in conventionally managed soil and a significantly negative response in organically managed soil. Senga Sengana had the poorest yields in all three soils irrespective of the presence or absence of *V. dahliae* (Figure S4).

Concluding remarks

The results of this study show a marked difference between the fungal community structure in rhizosphere soil and that of fungi growing on or within the root tissues of strawberry plants. This suggests that plant roots select microorganisms from the wider rhizosphere soil pool, but further experiments with sampling at multiple time points are necessary to shed further light on this phenomenon. Strong differences in community structure were also observed between different soils, but differences between fungal communities colonizing roots of different cultivars were less clear. Inoculation with the pathogen *V. dahliae* had a significant influence on community structure and decreased the number of rhizosphere soil- and root-inhabiting fungi in several treatments. Lower numbers of *Verticillium* reads were detected in the roots of Florence, a tolerant cultivar, than in roots of Honeoye. Significant decreases in the number of ascomycete fungal OTUs occurred in roots of both Honeoye and Florence cultivars in response to *V. dahliae* in organically managed soil. The rhizosphere soil of these plants was characterised by high relative abundance of *Trichosporon dulcitum*, while the roots were dominated by *Leptodontidium* sp. C2 BESC 319 g. It is possible that the high abundance of these fungi may cause some suppression of pathogens such as *V. dahliae* but the role of these two fungi in plant protection is still unclear.

Supporting Information

Figure S1 Cluster analysis of the 50 most abundant fungal operational taxonomic units (OTUs) in the rhizosphere of four strawberry cultivars, Honeoye, Florence, Senga Sengana and Zephyr, grown in conventionally and organically managed soils or a peat-based growth substrate, with and without *Verticillium dahliae*. These OTUs constituted 90% of the total reads in roots and 70% of the reads in rhizosphere soil. The clustering was based on paired group linkage using the Bray-Curtis similarity measure. All OTUs with ≥95% identity are included.

Figure S2 Non-metric multidimensional scaling (NMDS) analysis of the effects of inoculation with *Verticillium dahliae* on community structure of fungi colonising the rhizosphere of four different strawberry cultivars, Honeoye, Florence, Senga Sengana and Zephyr, grown in three different soils.

Figure S3 Sample based rarefaction curves of all fungal operational taxonomic units (OTUs) in rhizosphere soil and roots of strawberry using Analytic Rarefaction (Ver 1.3, UGA Stratigraphy Lab, USA).

Figure S4 Fresh berry weight (g/pot) of four different strawberry cultivars, Honeoye, Florence, Senga Sengana and Zephyr, grown in conventionally and organically managed soils or a peat-based growth substrate, with and without *Verticillium dahliae*. Asterisks indicate statistically significant differences ($p<0.05$) between control and *Verticillium dahliae* inoculated treatments within each soil. Vertical bars represent mean values and error bars indicate ±1SE (n = 6).

Table S1 Similarity percentage (SIMPER) analysis of all fungal operational taxonomic units (OTUs) in rhizosphere and roots of four different strawberry cultivars, Honeoye, Florence, Senga Sengana and Zephyr, grown in conventionally and organically managed soils or a peat-based growth substrate, with and without *Verticillium dahliae*. All OTUs with ≥95% identity are included. (ND = not determined).

Acknowledgments

Prof. Christina Dixelius, Swedish University of Agricultural Sciences, Uppsala is acknowledged for providing the pathogen isolate.

Author Contributions

Conceived and designed the experiments: SN SM SA RDF. Performed the experiments: SN. Analyzed the data: SN SM SA RDF. Contributed reagents/materials/analysis tools: SN SM SA RDF. Contributed to the writing of the manuscript: SN SM SA RDF.

References

1. Morgan JAW, Bending GD, White PJ (2005) Biological costs and benefits to plant-microbe interactions in the rhizosphere. J Exp Bot 56: 1729–1739.
2. Drigo B, Pijl AS, Duyts H, Kielak AM, Gamper HA, et al. (2010) Shifting carbon flow from roots into associated microbial communities in response to elevated atmospheric CO_2. P Natl Acad Sci 107: 10938 –10942.
3. Philippot L, Raaijmakers JM, Lemanceau P, van der Putten WH (2013) Going back to the roots: the microbial ecology of the rhizosphere. Nat Rev Microbiol 11: 789–799.
4. Jones JDG, Dangl JL (2006). The plant immune system. Nature 444: 323–329.
5. Shade A, Handelsman J (2012) Beyond the Venn diagram: the hunt for a core microbiome. Env Microbiol 14: 4–12.
6. Lundberg DS, Lebeis SL, Paredes SH, Yourstone S, Gehring J, et al. (2012) Defining the core *Arabidopsis thaliana* root microbiome. Nature 488: 86–90.
7. Bulgarelli D, Rott M, Schlaeppi K, Ver Loren van Themaat E, Ahmadinejad N, et al. (2012) Revealing structure and assembly cues for *Arabidopsis* root-inhabiting bacterial microbiota. Nature 488: 91–95.
8. Wissuwa M, Mazzola M, Picard C (2008) Novel approaches in plant breeding for rhizosphere-related traits. Plant Soil 321: 409–430.
9. Smith KP, Goodman RM (1999) Host variation for interactions with beneficial plant associated microbes. Ann Rev Phytopathol 37: 473–491.
10. Andreote FD, Rocha UND, Araújo WL, Azevedo JL, Van Overbeek LS (2010) Effect of bacterial inoculation, plant genotype and developmental stage on root-associated and endophytic bacterial communities in potato (*Solanum tuberosum*). Antonie van Leeuwenhoek 97: 389–399.
11. Lodewyckx C, Vangronsveld J, Porteous F, Moore ERB, Taghavi S, et al. (2002) Endophytic bacteria and their potential applications. Crit Rev Plant Sci 21: 583–606.
12. Ryan RP, Germaine K, Franks A, Ryan DJ, Dowling DN (2008) Bacterial endophytes: recent developments and applications. FEMS Microbiol Lett 278: 1–9.
13. Berg G (2009) Plant-microbe interactions promoting plant growth and health: perspectives for controlled use of microorganisms in agriculture. App Microbiol Biotechnol 84: 11–18.
14. Edel-Hermann V, Brenot S, Gautheron N, Aimè S, Alabouvette C, et al. (2009) Ecological fitness of the biocontrol agent *Fusarium oxysporum* Fo47 in soil and its impact on the soil microbial communities. FEMS Microbiol Ecol 68: 37–45.
15. Gao G, Yin D, Chen S, Xia F, Yang J (2012) Effect of biocontrol agent *Pseudomonas fluorescens* 2P24 on soil fungal community in cucumber rhizosphere using T-RFLP and DGGE. PLoS ONE 7(2): e31806.
16. Zhao Y, Li W, Zhou Z, Wang L, Pan Y, et al. (2005) Dynamics of microbial community structure and cellulolytic activity in agricultural soil amended with two biofertilizers. Eur J Soil Bio 41: 21–29.

17. Hoitink H, Boehm M (1999) Biocontrol within the context of soil microbial communities: A substrate-dependent phenomenon. Ann Rev Phytopathol 37: 427–446.
18. Trivedi P, Duan Y, Wang N (2010) Huanglongbing, a systemic disease, restructures the bacterial community associated with citrus roots. Appl Environ Microb 76: 3427–3436.
19. Trivedi P, He Z, Van Nostrand JD, Albrigo G, Zhou J, et al. (2012) Huanglongbing alters the structure and functional diversity of microbial communities associated with citrus rhizosphere. ISME J 6: 363–383.
20. Reiter B, Pfeifer U, Schwab H, Sessitsch A (2002) Response of endophytic bacterial communities in potato plants to infection with *Erwinia carotovora* subsp. atroseptica. Appl Environ Microb 68: 2261–2268.
21. Yang C, Crowley DE, Menge JA (2001) 16S rDNA fingerprinting of rhizosphere bacterial communities associated with healthy and *Phytophthora* infected avocado roots. FEMS Microbiol Ecol 35: 129–136.
22. Kent AD, Triplett EW (2002) Microbial communities and their interactions in soil and rhizosphere ecosystems. Ann Rev Microbiol 56: 211–236.
23. Johansson JF, Paul LR, Finlay RD (2004) Microbial interactions in the mycorrhizosphere and their significance for sustainable agriculture. FEMS Microbiol Ecol 48: 1–13.
24. Hartmann A, Schmid M, Tuinen D van, Berg G (2008) Plant-driven selection of microbes. Plant Soil 321: 235–257.
25. Rodriguez RJ, White JF Jr, Arnold AE, Redman RS (2009) Fungal endophytes: diversity and functional roles. New Phytol 182: 314–330.
26. Mendes R, Kruijt M, de Bruijn I, Dekkers E, van der Voort M, et al. (2011) Deciphering the rhizosphere microbiome for disease-suppressive bacteria. Science 332: 1097–1100.
27. Berg G, Zachow C, Lottmann J, Gotz M, Costa R, et al. (2005) Impact of plant species and site on rhizosphere-associated fungi antagonistic to *Verticillium dahliae* Kleb. Appl Environ Microb 71: 4203–4213.
28. Costa R, Götz M, Mrotzek N, Lottmann J, Berg G, et al. (2006) Effects of site and plant species on rhizosphere community structure as revealed by molecular analysis of microbial guilds. FEMS Microbiol Ecol 56: 236–249.
29. Weller DM, Raaijmakers JM, Gardener BBM, Thomashow LS (2002) Microbial populations responsible for specific soil suppressiveness to plant pathogens. Ann Rev Phytopathol 40: 309–348.
30. Santos-González JC, Nallanchakravarthula S, Alström S, Finlay RD (2011) Soil, but not cultivar, shapes the structure of arbuscular mycorrhizal fungal assemblages associated with strawberry. Microb Ecol 62: 25–35.
31. Smalla K, Wieland G, Buchner A, Zock A, Parzy J, et al. (2001) Bulk and rhizosphere soil bacterial communities studied by denaturing gradient gel electrophoresis: plant-dependent enrichment and seasonal shifts revealed. Appl Environ Microb 67: 4742–4751.
32. Reeve JR, Schadt CW, Carpenter-Boggs L, Kang S, Zhou J, et al. (2010) Effects of soil type and farm management on soil ecological functional genes and microbial activities. ISME J 4: 1099–1107.
33. Fan L, Dalpé Y, Fang C, Dubé C, Khanizadeh S (2011) Influence of arbuscular mycorrhizae on biomass and root morphology of selected strawberry cultivars under salt stress. Botany 89: 397–403.
34. Norman JR, Atkinson D, Hooker JE (1996) Arbuscular mycorrhizal fungal-induced alteration to root architecture in strawberry and induced resistance to the root pathogen *Phytophthora fragariae*. Plant Soil 185: 191–198.
35. Norman JR, Hooker JE (2000) Sporulation of *Phytophthora fragariae* shows greater stimulation by exudates of non-mycorrhizal than by mycorrhizal strawberry roots. Mycol Res 104: 1069–1073.
36. Reganold JP, Andrews PK, Reeve JR, Carpenter-Boggs L, Schadt CW, et al. (2010) Fruit and soil quality of organic and conventional strawberry agroecosystems. PLoS ONE 5: e12346. doi:10.1371/journal.pone.0012346.
37. Ellis MA (2008) *Verticillium* wilt of Strawberry. Ohio State University Extension.
38. Fradin EF, Thomma BPHJ (2006) Physiology and molecular aspects of *Verticillium* wilt diseases caused by *V. dahliae* and *V. albo-atrum*. Mol Plant Pathol 7: 71–86.
39. Lord, Cheryl A S (1994) *Verticillium* wilt of Strawberry. Fact sheet. Durham, New Hampshire, USA: The University of New Hampshire Cooperative extension.
40. Heale J, Karapapa V (1999) The *Verticillium* threat to Canada's major oilseed crop: Canola. Can J Plant Pathol 21: 1–7.
41. Klosterman SJ, Atallah ZK, Vallad GE, Subbarao KV (2009) Diversity, pathogenicity, and management of *Verticillium* Species. Ann Review Phytopathol 47: 39–62.
42. Harris DC, Yang JR (1996) The relationship between the amount of *Verticillium dahliae* in soil and the incidence of strawberry wilt as a basis for disease risk prediction. Plant Pathol 45: 106–114.
43. Meszka B, Bielenin A (2009) Bioproducts in control of strawberry *Verticillium* wilt. Phytopathologia 52: 21–27.
44. Tjamos SE, Flemetakis E, Paplomatas EJ, Katinakis P (2005) Induction of resistance to *Verticillium dahliae* in *Arabidopsis thaliana* by the biocontrol agent K-165 and pathogenesis-related proteins gene expression. Molr Plant-Microb Interact 18: 555–561.
45. Rudrappa T, Czymmek KJ, Paré PW, Bais HP (2008) Root-secreted malic acid recruits beneficial soil bacteria. Plant Physiol 148: 1547–1556.
46. Azad HR, Davis JR, Schnathorst WC, Kado CI (1987) Influence of *Verticillium* wilt resistant and susceptible potato genotypes on populations of antagonistic rhizosphere and rhizoplane bacteria and free nitrogen fixers. Appl Microbiol Biotechnol 26: 99–104.
47. An M, Zhou X, Wu F, Ma Y, Yang P (2011) Rhizosphere soil microorganism populations and community structures of different watermelon cultivars with differing resistance to *Fusarium oxysporum* f. sp. niveum. Can J Microbiol 57: 355–365.
48. Yao H, Wu F (2010) Soil microbial community structure in cucumber rhizosphere of different resistance cultivars to *Fusarium* wilt. FEMS Microbiol Ecol 72: 456–463.
49. Rastogi G, Sani RK (2011) Molecular techniques to assess microbial community structure, function, and dynamics in the environment. In: Ahmad I, Ahmad F, Pichtel J, editors. Microbes and Microbial Technology. New York, NY: Springer New York. 29–57.
50. Torsvik V, Øvreås L (2002) Microbial diversity and function in soil: from genes to ecosystems. Curr Op Microbiol 5: 240–245.
51. Weinert N, Piceno Y, Ding, Meincke R, Heuer H, et al. (2011) PhyloChip hybridization uncovered an enormous bacterial diversity in the rhizosphere of different potato cultivars: many common and few cultivar-dependent taxa. FEMS Microbiol Ecol 75: 497–506.
52. Simpson D, Bell JA, Hammond K, Whitehouse AB (2002) The latest strawberry cultivars from horticultural research international, Tampere, Finland: Acta horticulturae 567: 165–168.
53. Olbricht K, Hanke M (2008) Strawberry breeding for disease resistance in Dresden. Small fruit: breeding and testing of cultivar. Hohenheim, Germany: Eco Fruit. 144–147.
54. Davik J, Daugaard H, Svensson B (2000) Strawberry production in the Nordic countries. Advances in Strawberry Research 19: 13–18.
55. Griffiths RI, Whiteley AS, O'Donnell AG, Bailey MJ (2000) Rapid method for co-extraction of DNA and RNA from natural environments for analysis of ribosomal DNA- and rRNA-based microbial community composition. Appl Environ Microb 66: 5488–5491.
56. Ihrmark K, Bödeker ITM, Cruz-Martinez K, Friberg H, Kubartova A, et al. (2012) New primers to amplify the fungal ITS2 region – evaluation by 454-sequencing of artificial and natural communities. FEMS Microbiol Ecol 82: 666–677.
57. White T, Bruns T, Lee S, Taylor J (1990) Amplification and direct sequencing of fungal ribosomal RNA genes for phylogenetics. PCR protocols: a guide to methods and applications. Innis MA, Gelfland DH, Sninsky JJ and White TJ. San Diego, CA, USA: Academic press. 315–322 p.
58. Xu L, Ravnskov S, Larsen J, Nicolaisen M (2012) Linking fungal communities in roots, rhizosphere, and soil to the health status of *Pisum sativum*. FEMS Microbiol Ecol 82: 736–745.
59. Middelhoven WJ, Koorevaar M, Schuur GW (1992) Degradation of benzene compounds by yeasts in acidic soils. Plant Soil 145: 37–43.
60. Margesin R, Fonteyne PA, Redl B (2005) Low-temperature biodegradation of high amounts of phenol by *Rhodococcus* spp. and basidiomycetous yeasts. Res Microbiol 156: 68–75.
61. Mestre MC, Rosa CA, Safar SVB, Libkind D, Fontenla SB (2011) Yeast communities associated with the bulk-soil, rhizosphere and ectomycorrhizosphere of a *Nothofagus pumilio* forest in northwestern Patagonia, Argentina. FEMS Microbiol Ecol 78: 531–541.
62. Girvan MS, Bullimore J, Pretty JN, Osborn AM, Ball AS (2003) Soil type is the primary determinant of the composition of the total and active bacterial communities in arable Soils. Appl Environ Microb 69: 1800–1809.
63. Wu Y, Fang W, Zhu S, Jin K, Ji D (2008) The effects of cotton root exudates on the growth and development of *Verticillium dahliae*. Frontiers of Agriculture in China 2: 435–440.
64. Vandemark GJ, Ariss JJ, Hughes TJ (2010) Real-time PCR Suggests that *Aphanomyces euteiches* is associated with reduced amounts of *Phytophthora medicaginis* in Alfalfa that is co-inoculated with both pathogens. J Phytopathol 158: 117–124.
65. Andrade-Linares DR, Grosch R, Restrepo S, Krumbein A, Franken P (2011) Effects of dark septate endophytes on tomato plant performance. Mycorrhiza 21: 413–422.
66. Raaijmakers JM, Mazzola M (2012) Diversity and natural functions of antibiotics produced by beneficial and plant pathogenic bacteria. Ann Rev Phytopathol 50: 403–424.
67. El-Deeb HM, Arab YA (2013) *Acremonium* as an endophytic bioagent against date palm *Fusarium* wilt. Arch Phytopathol Plant Prot: 1–8.
68. Fávaro LC, Sebastianes FL, Araújo WL (2012) *Epicoccum nigrum* P16, a sugarcane endophyte, produces antifungal compounds and induces root growth. PloS ONE 7(6): e36826. doi:10.1371/journal.pone.0036826.
69. Mousa WK, Raizada MN (2013) The diversity of anti-microbial secondary metabolites produced by fungal endophytes: an interdisciplinary perspective. Frontiers Microbiol 4: 65.
70. Aly AH, Debbab A, Proksch P (2011) Fungal endophytes: unique plant inhabitants with great promises. Appl Microbiol Biotechnol 90: 1829–1845.

10

Multiple NUCLEAR FACTOR Y Transcription Factors Respond to Abiotic Stress in *Brassica napus* L.

Li Xu[1], Zhongyuan Lin[1,2], Qing Tao[1,2], Mingxiang Liang[1,2]*, Gengmao Zhao[1,2], Xiangzhen Yin[1,2], Ruixin Fu[1,2]

1 College of Resources and Environmental Sciences, Nanjing Agricultural University, Nanjing, China, **2** Jiangsu Key Lab of Marine Biology, Nanjing, China

Abstract

Members of the plant NUCLEAR FACTOR Y (NF-Y) family are composed of the NF-YA, NF-YB, and NF-YC subunits. In *Brassica napus* (canola), each of these subunits forms a multimember subfamily. Plant NF-Ys were reported to be involved in several abiotic stresses. In this study, we demonstrated that multiple members of thirty three *BnNF-Ys* responded rapidly to salinity, drought, or ABA treatments. Transcripts of five *BnNF-YAs*, seven *BnNF-YBs*, and two *BnNF-YCs* were up-regulated by salinity stress, whereas the expression of thirteen *BnNF-YAs*, ten *BnNF-YBs*, and four *BnNF-YCs* were induced by drought stress. Under NaCl treatments, the expression of one *BnNF-YA10* and four *NF-YBs* (BnNF-YB3, BnNF-YB7, BnNF-YB10, and BnNF-YB14) were greatly increased. Under PEG treatments, the expression levels of four *NF-YAs* (BnNF-YA9, BnNF-YA10, BnNF-YA11, and BnNF-YA12) and five *NF-YBs* (BnNF-YB1, BnNF-YB8, BnNF-YB10, BnNF-YB13, and BnNF-YB14) were greatly induced. The expression profiles of 20 of the 27 salinity- or drought-induced *BnNF-Ys* were also affected by ABA treatment. The expression levels of six *NF-YAs* (BnNF-YA1, BnNF-YA7, BnNF-YA8, BnNF-YA9, BnNF-YA10, and BnNF-YA12*)* and seven *BnNF-YB* members (BnNF-YB2, BnNF-YB3, BnNF-YB7, BnNF-YB10, BnNF-YB11, BnNF-YB13, and BnNF-YB14) and two *NF-YC* members (BnNF-YC2 and BnNF-YC3) were greatly up-regulated by ABA treatments. Only a few *BnNF-Ys* were inhibited by the above three treatments. Several NF-Y subfamily members exhibited collinear expression patterns. The promoters of all stress-responsive *BnNF-Ys* harbored at least two types of stress-related *cis*-elements, such as ABRE, DRE, MYB, or MYC. The *cis*-element organization of *BnNF-Ys* was similar to that of *Arabidopsis thaliana,* and the promoter regions exhibited higher levels of nucleotide sequence identity with *Brassica rapa* than with *Brassica oleracea*. This work represents an entry point for investigating the roles of canola NF-Y proteins during abiotic stress responses and provides insight into the genetic evolution of *Brassica* NF-Ys.

Editor: Keqiang Wu, National Taiwan University, Taiwan

Funding: This research was supported by grants from the Natural Science Foundation of Jiangsu province (BK2011635), the Doctoral Program of Higher Education of China (20120097120015), Fundamental Research Funds for the Central Universities (KYZ201206), the Scientific Research Foundation of the State Human Resource Ministry, and the Priority Academic Program Development of Jiangsu Higher Education Institutions (RAPD program (809001)). The funders had no role in study design, data collection and analysis, decision to publish, or preparation of the manuscript.

Competing Interests: The authors have declared that no competing interests exist.

* Email: liangmx@njau.edu.cn

Introduction

Abiotic stress, such as salinity, drought or dramatic temperature change, adversely affects plant growth and productivity. Plants, being sessile, have evolved a range of strategies to adapt to detrimental environmental conditions. Plants respond to unfavorable conditions through two pathways, i.e., the abscisic acid (ABA)-dependent and ABA-independent pathway [1–3]. Transcription factors (TFs) play vital roles in mediating stress tolerance in both of these pathways.

NUCLEAR FACTOR Y (NF-Y) TFs are ubiquitous in higher eukaryotes and belong to the CCAAT-binding factor (CBF) family, also known as the Heme Activator Protein (HAP) family in yeast. The NF-Y heterocomplex consists of at least three subunits, namely NF-YA (also named CBF-B or HAP2), NF-YB (CBF-A or HAP3), and NF-YC (CBF-C or HAP5) [4]. In planta, each NF-Y subunit is encoded by a multi-member family, and there are at least 10 annotated members in each NF-Y subfamily in *Arabidopsis* [5–7]. Rice harbors at least 10 *NF-YA* genes, 12 *NF-YB* genes, and 8 *NF-YC* genes [8]. In a previous study, we identified 14 *NF-YA* genes, 14 *NF-YB* genes, and 5 *NF-YC* genes in *Brassica napus* L. [9]. NF-Y family members are extensively involved in plant development and stress responses, as summarized in a recent review [10].

Increasing evidence suggests that NF-Y subunits are important regulators of abiotic stress responses [11–14]. Overexpression of *NF-YA1* in *Arabidopsis* was shown to result in hypersensitivity to salt stress during early postgerminative growth [15]. Real-time PCR analysis revealed that *Glycine max* (soybean) *GmNF-YA3* was induced by ABA and other abiotic stresses, and that overexpression of *GmNF-YA3* in *Arabidopsis* enhanced drought resistance [16]. A microarray analysis identified *Arabidopsis NF-YB2* as one of the most strongly induced genes under multiple treatments, including cold, mannitol (dehydration stress), and NaCl (salinity) [17]. Further analysis confirmed that AtNF-YB2 functions in early flowering under osmotic stress [12]. AtNF-YB1 was found to improve plant performance under drought conditions and its ortholog in maize, ZmNF-YB2, was shown to resist drought stress in the field [11]. Several drought-related *NF-YB* genes were identified in *Triticum aestivum* (wheat), *Populus euphratica* (poplar), and *Hordeum vulgare* (barley) [18–20]. In the NF-YC subfamily, the transcript level of *AtNF-YC2* was highly induced by light, oxidative, heat, cold, and drought stress [21]. Overexpression of *PwHAP5* (homolog of *Arabidopsis NF-YC2*) from the conifer (*Picea wilsonii*) partially rescued the increased sensitivity of *nf-yc2* to salt, drought, and ABA treatments [22].

In addition, several members of the NF-Y family were shown to be regulated by the microRNA169 (miR169) family, suggesting that a complex regulatory cascade is activated under stress conditions [13,14,23]. Li et al. (2008) demonstrated that *AtNF-YA5* is a target of miRNA169a or miRNA169c [13]. At least six other miR169 targets were predicted to exist in the AtNF-YA family [24]. Of those six NF-YAs, NF-YA2, 3, 7, and 10 were found to be involved in multiple abiotic stresses, especially those involving low nutrient availability [25]. In rice, an *NF-YA* gene (*Os03g29760*) known to be a target of the miR169 family was found to be induced by high salinity [14]. The above results suggest that plant NF-Ys play a crucial role in the plant's response to abiotic stress. However, the roles of BnNF-Ys under stress have not been examined in detail.

Canola (*B. napus*) constitutes the third largest vegetable oil source in the world. However, salinity and drought stress have a major limiting impact on its production and yield stability. Drought accounts for at least a 30% yield loss in canola every year [26]. In this study, we examined the expression patterns of *BnNF-Ys* under drought, high salinity, and ABA conditions. Our results suggest that 27 of 33 putative *BnNF-Ys* were induced by salt or drought stresses. Furthermore, 20 *BnNF-Ys* may be involved in the ABA-related pathway. We cloned the promoters of 28 *BnNF-Ys* from canola, which mostly included regions ~1000 bp upstream of the coding sequences, and identified their stress-related *cis*-elements. Taken together, our results suggest that most *BnNF-Y* members are regulated by abiotic stresses in an abscisic acid (ABA)-dependent or ABA-independent manner and serve as an entry point to investigate the roles of these genes in the plant's response to environmental stresses.

Materials and Methods

Plant materials and growth conditions

Seeds of oilseed rape (*Brassica napus* 'Nanyanyou 1', kindly provided by Professor Zhaopu Liu at Nanjing Agricultural University) were kept in darkness for 2 days and germinated in Hoagland nutrient solution. The plants were grown for 3 weeks in the greenhouse with a light intensity of 392~415 μmol m^{-2} s^{-1} during a daily cycle consisting of 16 h of light at 25°C and 8 h of darkness at 18°C. For dehydration and salt stress treatments, plants were grown hydroponically in solution containing 15% (w/v) PEG6000 and 150 mM NaCl plus ½ Hoagland nutrient solutions for 1 or 3 h, respectively. For ABA treatments, 100 μM ABA was added to the ½ Hoagland nutrient solution and the plants were treated for 1 or 3 h. Seedlings growing on ½ Hoagland nutrient solution for various periods of time were used as the control. The harvested samples were immediately frozen in liquid nitrogen and stored at −80°C until used for gene expression analysis. All experiments were performed in biological triplicate.

Genomic DNA extraction, total RNA isolation, and primary cDNA synthesis

Genomic DNA was extracted from oilseed rape using a method from Rogers and Bendich's CTAB-based protocol [27]. Total RNA of each sample was extracted using an E.Z.N.A. Plant RNA Kit (Omega Biotek, Cat#R6827, USA). The quality and quantity of extracted DNA and RNA from all samples were confirmed by both agarose gel electrophoresis and spectrophotometry (Thermo Scientific, NanoDropTM 1000, USA). Total RNA samples were pretreated with an RNase-Free DNase Set (Omega Biotek, Cat#E1091, USA) to eliminate residual genomic DNA. Primary cDNA was synthesized using a PrimeScript RT-PCR Kit (TaKaRa Code: D6110A, Japan), according to the manufacturer's instructions.

Analysis of Gene Expression by Quantitative real-time PCR (qRT-PCR)

Quantitative real-time PCR was performed on an Applied Biosystems 7500 real-time PCR system using the SYBR Premix ExTaqTM Kit (TaKaRa Code: DRR041A, Japan). Sequence data, primers, and the PCR procedure used in this study were provided in our previous study [9]. All sequences of qRT-PCR amplification products were verified by sequencing. 18S rRNA was used as an internal control for expression analysis. Data were processed using the $2^{-\Delta\Delta CT}$ method [28,29].

The transcript levels of each BnNF-Y gene in each tissue were first normalized to those of the housekeeping gene 18S and then compared to the transcript levels of each time-point control to obtain the relative gene expression level (i.e., the levels of expression in the same untreated tissue at the corresponding time point). Each data point represents the mean ±SE of three independent experiments. For the figures S1-S5, the transcript levels of each *BnNF-Y* genes were first normalized to those of the housekeeping gene 18S and then compared to the levels in the 0-h leaf control.

Statistical Analysis and Correlation Analysis

Significant differences between three treated samples and three untreated controls (same tissue and time-point only) are indicated by a single ($P<0.05$) or double ($P<0.01$) asterisk, according to Dunnett's method of one-way ANOVA in SPSS 18.0 (SPSS Corp., Chicago, IL, USA). Figures were created using Sigma Plot 10.0 (Systat Software, Inc. Germany). Correlation analysis of expression patterns was determined using relative gene expression levels from Fig. S1, Fig. S2, and Fig. S3. The regression coefficient was calculated in Microsoft Excel.

Measurements of Relative Water Content (RWC) from Drought-Stressed Plants

RWC measurements under PEG treatments were conducted as previously described [20].

Isolation of *NF-Y* promoters and Identification of their regulatory elements

To identify the promoter regions of *NF-Ys* from canola, each *BnNF-Y* coding region was used as a query in a BLASTN search against the *Brassica* database (http://brassicadb.org/brad/index.php), which includes *B. rapa* and *B. oleracea* sequences, and the highest hits were identified. Based on *BnNF-Y* coding sequences and the promoter sequences of the retrieved orthologous *Brassica NF-Ys*, most promoter regions of *BnNF-Y* sequences could be amplified with the primers listed in Table S1. Others promoter regions were amplified using a Genome Walking Kit (TaKaRa

Code: D316, Table S1). In addition, the promoters of *Arabidopsis NF-Y* sequences were downloaded from the Plant Promoter Database (http://133.66.216.33/ppdb/cgi-bin/index.cgi). Sequences ~1000 bp upstream of the ATG start codon of each gene were retrieved from the genome, and known elements were identified using Plant *Cis*-element Regulatory DNA Elements (PLACE) (http://www.dna.affrc.go.jp/PLACE/) and PlantCARE (http://bioinformatics.psb.ugent.be/webtools/plantcare/html/).

Phylogenetic tree analysis

The phylogenetic tree based on the promoter regions of plant NF-Ys was constructed using the neighbor-joining method implemented in MEGA software (version 4.1) with the following parameters: Jones–Taylor–Thornton (JTT) model and 1,000 bootstrap replicates.

Results

Expression Patterns of *BnNF-Y* Genes in the Leaves and Roots of Canola in Response to Salinity Stress

We previously identified BnNF-Y TF families in canola (*B. napus* L.) [9]. Since numerous reports have demonstrated that plant NF-Ys function in abiotic stress responses [13,15], we examined whether BnNF-Ys had similar roles. To investigate whether *BnNF-Y* genes were responsive to salt stress, we monitored the transcript levels of all 33 *BnNF-Ys* in both the leaves and roots up to 3 h of exposure to high salinity. First, we quantified the transcript levels of each *BnNF-Y* gene in different tissues and expressed these levels relative to those in 0-h untreated leaves. In this study, a gene was considered to be induced or down-regulated if its expression levels were both significantly different from those of an untreated control and >2-fold higher or lower than the control values, respectively. As shown in Fig. S1A, S2A and S3A, in the absence of any stress, the expression levels of most *BnNF-YA* subfamily members were not significantly altered during a 3-h observation period, except for *BnNF-YA4/5*, *BnNF-YA11*, *BnNF-YA12,* and *BnNF-YA13*. The transcription levels of *BnNF-YA4/5* (*BnNF-YA4* and *BnNF-YA5* are alternative splice variants, with *BnNF-YA4* being 11 amino acids shorter than *BnNF-YA5*) [9], and *BnNF-YA11* were up-regulated in the 3-h root samples, while levels of *BnNF-YA12* and *BnNF-YA13* were down-regulated in 1-h leaf samples. This implies that these *BnNF-YA*s are subject to circadian clock regulation. To minimize the effect of circadian-mediated changes, we recalculated the expression levels by dividing the expression level of each gene to the corresponding gene expression level in the untreated control at each time point and comparing values in the same tissues. As shown in Fig. 1A, the transcript levels of *BnNF-YA1*, *BnNF-YA2*, *BnNF-YA8* and *BnNF-YA10* in leaves were higher than those of the control after 1 h of NaCl treatment and remained elevated up to 3 h of treatment, whereas those of, *BnNF-YA7* and *BnNF-YA9* in leaves were higher than those of the control up to 3 h of treatment. In roots, two *BnNF-YA* subfamily members (*BnNF-YA9* and *BnNF-YA10*) were induced at 1 h or 3 h. Most salt-responsive *BnNF-YA*s were only moderately up-regulated (2~4 fold) upon exposure to salt stress, whereas *BnNF-YA3* and *BnNF-YA4/5* were slightly down-regulated in roots upon NaCl treatment.

We found that the expression of only two members of the *BnNF-YB* subfamily (i.e., *BnNF-YB10*, and *BnNF-YB13*) increased during a 3-h period in the absence of any treatments (Fig. S1B, S2B, and S3B). In leaves, the transcript levels of six *BnNF-YB*s (i.e., *BnNF-YB3, BnNF-YB7, BnNF-YB8, BnNF-YB9, BnNF-YB10,* and *BnNF-YB14*) were elevated after 1 h or 3 h of NaCl stress (Fig. 1B). In roots, NaCl treatments increased the transcript levels of *BnNF-YB3*, *BnNF-YB13*, and *BnNF-YB14*. Our results show that the levels of *BnNF-YB1*, *BnNF-YB5*, *BnNF-YB6*, *BnNF-YB9*, *BnNF-YB11,* and *BnNF-YB12* transcripts were mostly not affected by salinity. Only the expression level of *BnNF-YB4* in the roots was decreased at 3 h.

None of the five *BnNF-YC* subunits was regulated by the biological clock, since the expression levels of these genes remained constant in the absence of any treatments (Fig. S1C, S2C, and S3C). Only *BnNF-YC5* was induced in the leaf samples both at the 1- and 3-h time-points under salinity (Fig. 1C). *BnNF-YC2* transcripts were more abundant in roots subjected to NaCl treatments than in untreated roots.

Expression Patterns of *BnNF-Y* Genes in Leaves and Roots of Canola in Response to drought Stress

As shown in Fig. 2, all *BnNF-YA* members were responsive to PEG6000 treatments, which mimic drought stress, in leaf tissues. A 15% PEG treatment generally caused a 72% relative water content loss within 24 h (Table S2). The expression levels of *BnNF-YA10*, *BnNF-YA11* and *BnNF-Y12* in leaves were dramatically up-regulated to more than four-fold the levels observed in the untreated control (Fig. 2A). *BnNF-YA8*, *BnNF-YA9*, *BnNF-YA10* and *BnNF-YA13* were induced in roots under drought stress. Four *BnNF-YA*s (*BnNF-YA8*, *-YA9*, *-YA10,* and *-YA13*) responded to drought stress both in leaves and roots, particularly up to 3 h of treatment. The expression levels of five *BnNF-YA*s (*BnNF-YA1*, *BnNF-YA2*, *-YA3*, *-YA4/5,* and *-YA6*) decreased in leaves at 1 h and increased at 3 h.

Most members of the *BnNF-YB* subfamily also responded to PEG treatments, except for *BnNF-YB12,* while the expression level of *BnNF-YB14* increased 80-fold in leaves after 1 h of treatment (Fig. 2B). As in the *BnNF-YA* subfamily, the transcripts of *BnNF-YB* were induced more strongly in leaves than in roots, since only 3 of 14 *BnNF-YB* members (*BnNF-YB7, BnNF-YB10,* and *BnNF-YB14*) were induced in roots subjected to drought stress. Four BnNF-YBs (*BnNF-YB4*, *-B9*, *-B10* and *-B11*) exhibited decreased expression in the roots after 1 h of treatment.

The transcript levels of four *BnNF-YC* members (*BnNF-YC1, BnNF-YC2, BnNF-YC4,* and *BnNF-YC5*) were elevated in both the leaf and root samples up to 3 h of drought treatment, while those of three *BnNF-YC*s (*BnNF-YC1*, *BnNF-YC2,* and *-YC3*) were initially inhibited in leaves at the 1-h time point (Fig. 2C).

Expression Patterns of *BnNF-Y* Genes of Canola in Leaves and Roots in Response to ABA Stress

Plants respond to abiotic stresses via both the ABA-dependent and ABA-independent pathways. Some NF-Ys were reported to be involved in ABA-dependent pathways, whereas others were not [11,13,15]. To determine whether canola NF-Ys function via the ABA pathway, we further examined the expression profiles of *BnNF-Ys* that were found to be responsive to salinity or drought stress (i.e., all *BnNF-YA* members and most *BnNF-YB*s and *BnNF-YC*s; Figs 1 and 2) under ABA treatments. All *BnNF-YA*s, except for *BnNF-YA4/5* and *BnNF-YA13,* were responsive to 3-h of ABA treatment (Fig. 3A). The transcript levels of *BnNF-YA1*, *-YA2*, *-YA3*, *-YA6*, *-YA7*, *-YA8*, *-YA9*, *-YA10* and *BnNF-YA11* were elevated at 1 h. Also, most NaCl- or PEG- responsive *BnNF-YB* members were induced by ABA treatments, with the exception of *BnNF-YB1* and *BnNF-YB9* (Fig. 3B). The transcript level of *BnNF-YB2* was inhibited in leaves after 1 h of ABA treatment. Interestingly, several *BnNF-YB*s (*BnNF-YB2, -YB3, -YB7, -YB10, -YB11, -YB13,* and *-YB14*) were greatly up-regulated by ABA treatment. Amongst the *BnNF-YC* members, only the transcripts

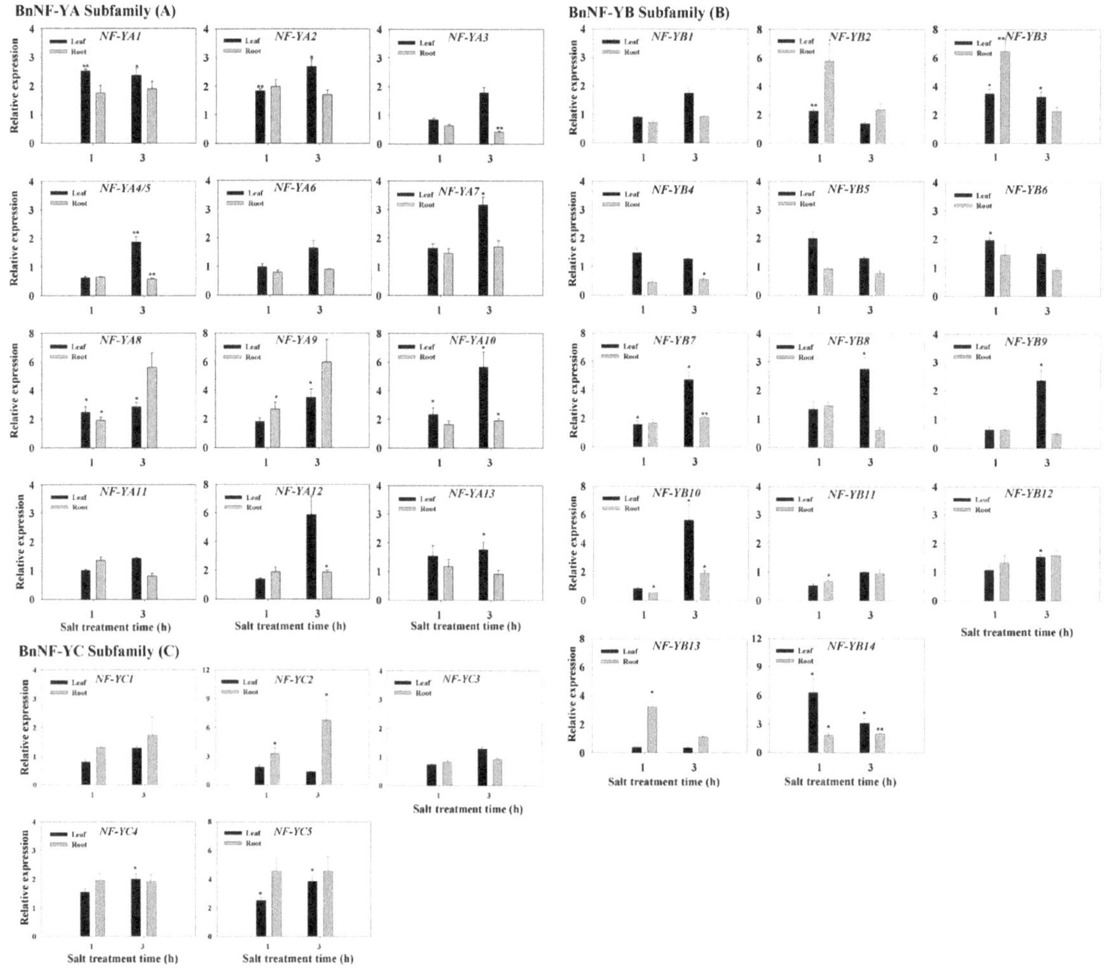

Figure 1. Expression pattern of the *BnNF-Y* genes in the leaves and roots of plants subjected to salinity stress. The expression of *BnNF-YA* (A), *BnNF-YB* (B), and *BnNF-YC* (C) genes in the leaves and roots of plants exposed to 150 mM NaCl. Three-week-old canola seedlings were treated with 150 mM NaCl for 1 and 3 h. Total RNA was extracted from leaves (L) and roots (R) for quantitative PCR (qRT-PCR) analysis. Transcript levels of each BnNF-Y were first normalized to those of the housekeeping gene 18S and then compared to each time-point control (i.e., the level of untreated sample at the corresponding time point). Each data point represents the mean ±SE of three independent experiments. Significant differences between treated samples and untreated controls (same tissue only) are indicated by a single ($P<0.05$) or double ($P<0.01$) asterisk, according to Dunnett's method of one-way ANOVA in SPSS. Expression levels of untreated samples (C, 1-h or 3-h leaf and root samples) were arbitrarily set to 1.0.

of *BnNF-YC2* and *BnNF-YC5* were increased in response to ABA treatment, and the upregulation was greater in the roots than in the leaves (Fig. 3C). The transcript levels of *BnNF-YC1* remained almost constant from 1 h to 3 h of ABA treatment in all tissues examined.

To compare the gene expression levels of different *BnNF-Ys* in canola, the expression level of each *NF-Y* gene in the 0-h leaf or root samples (arbitrarily chosen as baseline in the real-time PCR analysis) was also determined by semi-quantitative RT-PCR analysis (35 cycles; Fig. S4). The *BnNF-Ys* showed diverse expression profiles. Interestingly, most *BnNF-Ys* were expressed at similar levels in 0-h leaves and 0-h roots.

Correlation between the gene expression patterns of *BnNF-Y* genes

Since NF-Y TFs are known to form functional complexes in plants [30,31], one would expect that NF-Ys with correlated expression profiles would be involved in the same regulatory processes under abiotic stress conditions. Extensive correlations were found to exist in the expression profiles of *BnNF-Ys* not only within the same subfamily, but also between the three subfamilies. The expression profiles of *BnNF-YA6* and *BnNF-YA11,* of *BnNF-YA8* and *BnNF-YB14,* and of *BnNF-YC4* and *BnNF-YB12* were highly correlated under NaCl treatment ($R = 0.93$, 0.948 and 0.961, respectively. Fig. 4A). Under drought stress, the expression of *BnNF-YA10* and *-YB3, -YC4,* or *-YC5,* of *-YA12* and *-YB3,* and of *-YC4* and *-YC5* were highly correlated ($R = 0.915$, 0.927, 0.913, 0.983, and 0981, respectively. Fig. 4B). Under ABA treatments, *BnNF-YB3* and *BnNF-YC5* showed very similar expression profiles ($R = 0.99$. Fig. 4C).

The Promoters of *NF-Y* Genes Contain Diverse Stress-responsive Elements

Since most BnNF-Y members were found to be regulated by abiotic stress, we sought to identify the promoter sequences of their corresponding genes and to identify stress-relative *cis*-elements. We cloned the promoters of 12 *BnNF-YAs*, 13 *BnNF-YBs*, and 3 *BnNF-YCs* that were responsive to one or more of the three treatments used above (i.e., NaCl, PEG6000, and ABA). Using a BLASTN search and genome walking, we successfully acquired all

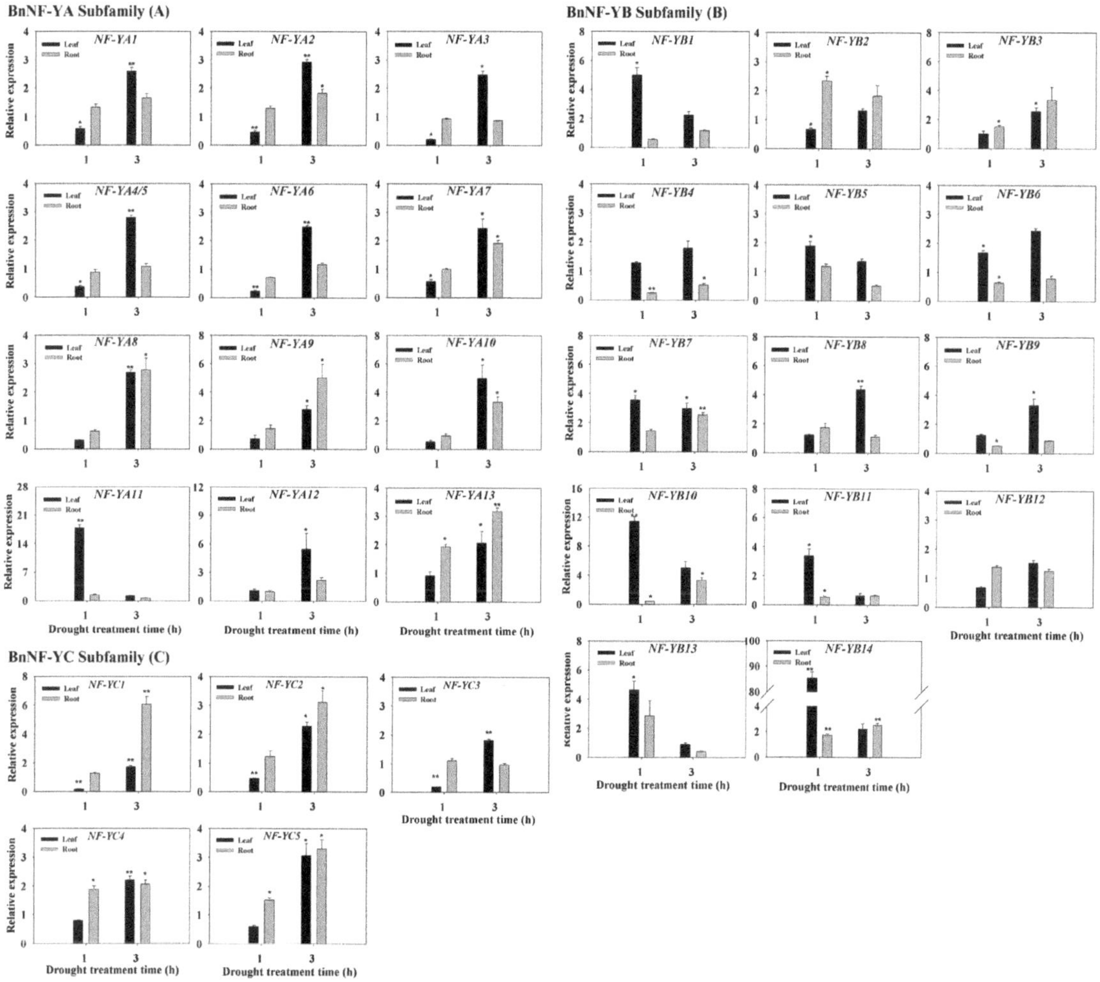

Figure 2. Expression pattern of the *BnNF-Y* genes in plants subjected to drought stress. The expression of *BnNF-YA* (A), *BnNF-YB* (B), and *BnNF-YC* (C) genes in the leaves and roots of plants exposed to 15% (w/v) PEG6000. Three-week-old canola seedlings were treated with 15% (w/v) PEG-6000 for 1 and 3 h. Total RNA was extracted from leaves (L) and roots (R) for quantitative PCR (qRT-PCR) analysis. Transcript levels of each BnNF-Y were first normalized to those of the housekeeping gene 18S and then compared to each time-point control. Each data point represents the mean ±SE of three independent experiments. Significant differences between treated samples and untreated controls (same tissue only) are indicated by single ($P<0.05$) or double ($P<0.01$) asterisks, according to Dunnett's method of one-way ANOVA in SPSS. Expression levels in untreated samples (CK, 1-h or 3-h leaf and root samples) were arbitrarily set to 1.0.

the promoter sequences, most of which consisted of around 1000 bp upstream of the ATG start codon. However, we were unable to acquire the promoter sequences of two *BnNF-YA* members (*BnNF-YA4/5* and *BnNF-YA6*) and two *BnNF-YB* members (*BnNF-YB1* and *BnNF-YB8*) longer than 1000 bp. We then identified *cis*-elements in these promoters using PLACE and PlantCARE software (Fig. 5), focusing on stress-related *cis*-elements, such as ABRE, CBF/DRE, MYB, and MYC. The promoters of each of the above-mentioned *BnNF-Ys* included at least two types of *cis*-elements, with the exception of the truncated *BnNF-YA4/5* promoter sequences. All of the promoters, except for the promoter of *BnNF-YA4/5*, harbored MYB and MYC elements. More than half of the *BnNF-YA* and *BnNF-YC* promoters harbored at least one ABRE element, while only a few BnNF-Ys had DRE elements. We also constructed *Arabidopsis cis*-elements in the homologous *NF-Y* family and compared the arrangement of *cis*-elements in canola and *Arabidopsis*. Most canola *NF-Y* members had analogous *cis*-elements in *Arabidopsis*, but the number and arrangement of *cis*-elements differed (Fig. S5).

Canola promoter sequences showed greater similarities with those of *B. rapa* than with those of *Arabidopsis thaliana*

We previously showed that canola NF-Ys were more similar to those of the dicot *Arabidopsis* than to those of the monocot *Oryza sativa* (rice) based on putative amino acid sequences. In this study, we also used the promoter sequences to further explore the phylogenetic relationship amongst members of this family. All canola promoters grouped with promoters of their own homologs and formed clades with the corresponding subfamily members in the other species (Fig. 6). The promoter sequence of each canola *NF-Y* usually had the highest level of identity with *B. rapa,* and the lowest with *Arabidopsis,* with the exceptions of *BnNF-YA6*, *BnNF-YB2* and *BnNF-YB8*. The promoter regions of these three *BnNF-Ys* were more similar to those of *B. oleracea* than to those of *B. rapa*.

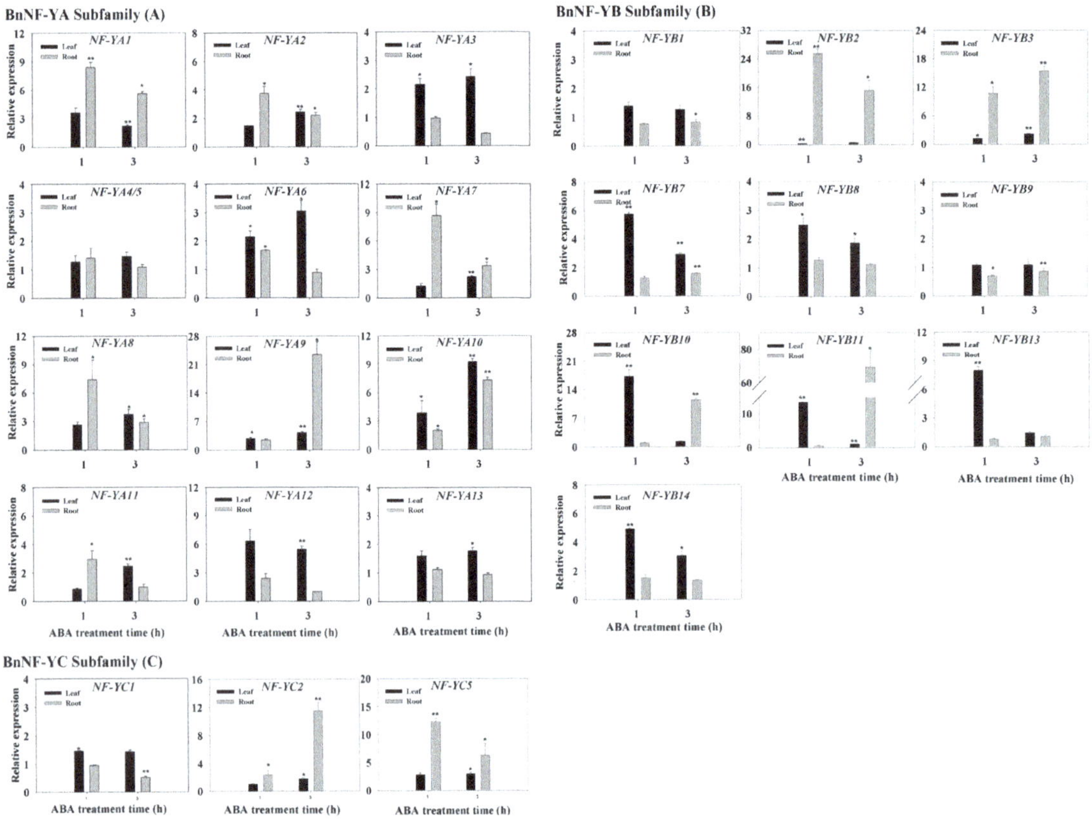

Figure 3. Expression pattern of the *BnNF-Y* genes in the leaves and roots of plants exposed to 100 μM ABA. Transcript levels of *BnNF-Y* (A), *BnNF-YB* (B), and *BnNF-YC* (C) genes after exposure to 100 μM ABA. Three-week-old canola plantlets were treated with 100 μM ABA for 1 h and 3 h. Total RNA was extracted from leaves and roots for quantitative PCR (qRT-PCR) analysis. Transcript levels of each BnNF-Y were first normalized to those of the housekeeping gene 18S and then compared to levels at each time point in the control. Each data point represents the mean ±SE of three independent experiments. Significant differences between treated samples and untreated controls (same tissue only) are indicated by a single ($P<0.05$) or double ($P<0.01$) asterisk, according to Dunnett's method of one-way ANOVA in SPSS. Expression levels in untreated samples (CK, 1-h or 3-h leaf and root samples) were arbitrarily set to 1.0.

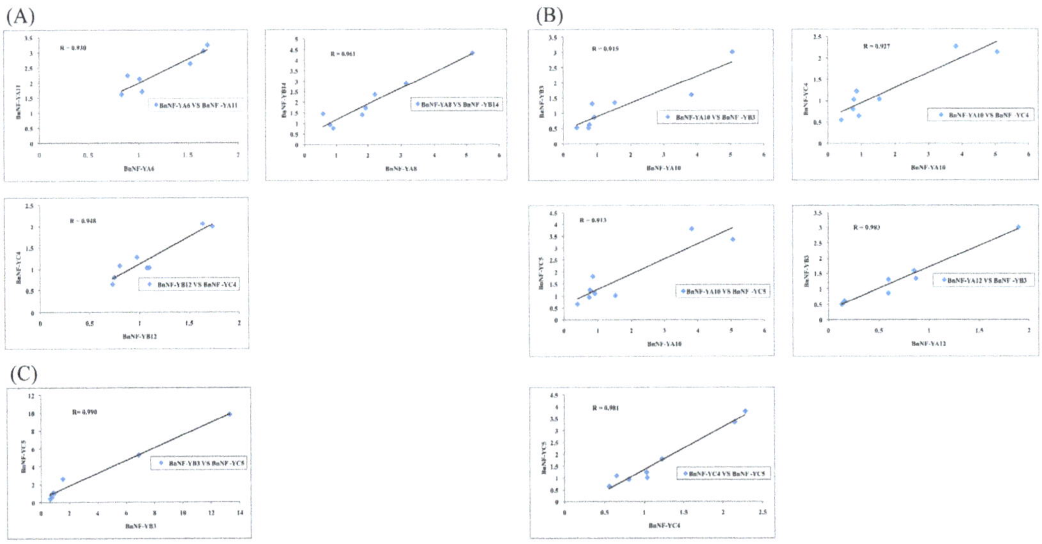

Figure 4. Correlation in expression levels of the three BnNF-Y subfamily members. Correlated gene expression under salinity (A); drought (B); and ABA (C) treatments. Relative expression levels including 8 data points (relative NF-Y gene expression levels with and without treatments at 1 h or 3 h compared to the same 0-h samples) are used for each gene according to Fig. S1, S2, and S3. These data were fitted using linear regression analysis.

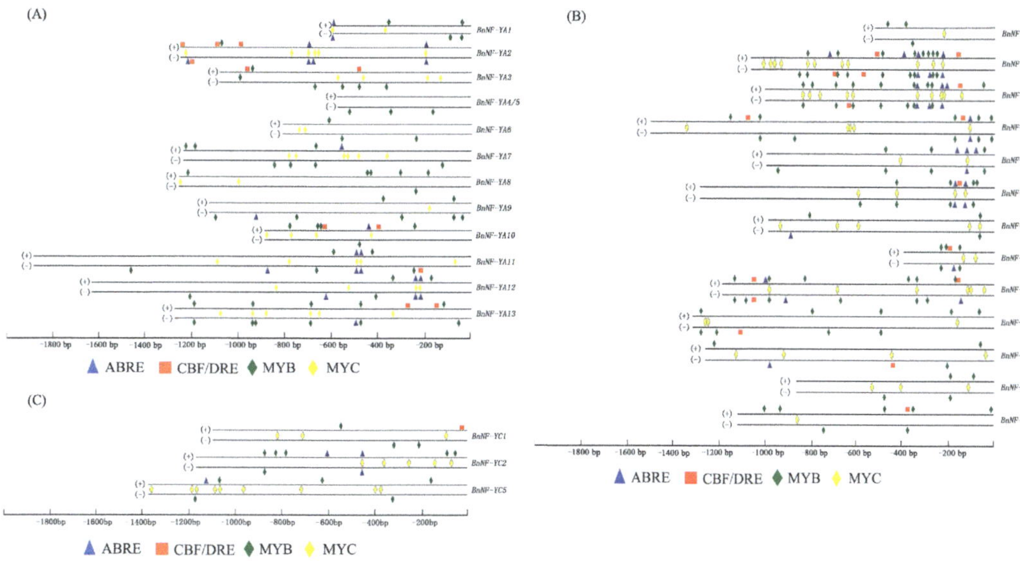

Figure 5. Diagrammatic representation of stress-related elements in the *BnNF*-Y promoters. Elements in the *BnNF*-YA (A), *BnNF-YB* (B), and *BnNF-YC* (C) promoters. Symbols above the top line indicate elements that are in the forward orientation, and those below the bottom line indicate elements in the reverse orientation. MYC(CANNTG) sequences are between the double lines, to indicate their palindromic nature. Numbers are the distance in nucleotides between the sequences and ATG start codon.

Discussion

The fact that multiple members of the NF-Y family exist in plant genomes implies that these proteins have redundant or differentiated functions. In this study, we extensively explored the response of *BnNF-Y*s to abiotic stresses through qRT-PCR analysis. The expression profiles of the *BnNF-Y* TF family provide insight into their roles during the abiotic stress response.

It is not surprising that most identified *BnNF-YA*s are involved in salinity, drought, or ABA treatments, as most previously reported stress-responsive plant NF-Y members belong to the NF-YA subfamily [15]. Quantitative RT-PCR analysis indicated that transcripts of *Arabidopsis NF-YA1* were increased by salinity, PEG, and ABA treatments. It was previously found that overexpression of *Arabidopsis NF-YA1* rendered the transgenic plants more sensitive than the control to a 200 mM NaCl treatment, whereas the corresponding *Arabidopsis* RNAi plants grew better than the control under these conditions [15]. *BnNF-YA1* and *BnNF-YA2*, which showed the highest level of similarity with *Arabidopsis NF-YA1* [9], were induced by salinity, PEG, and ABA treatments. Overexpression of *Arabidopsis NF-YA2*, *NF-YA7*, and *NF-YA10* caused a dwarf phenotype and enhanced the plant's resistance to multiple abiotic stresses [25]. We previously showed that BnNF-YA3, BnNF-YA13, and BnNF-YA14 formed a clade with AtNF-YA2 and AtNF-YA10 [9]. Expression of *Arabidopsis NF-YA5* was up-regulated up to 14 days of drought treatment or within 24 h of ABA treatment [13]. The ectopic expression of *NF-YA5* in *Arabidopsis* increased the plant's drought tolerance [13]. BnNF-YA8 or BnNF-YA9, which clustered with *Arabidopsis* NF-YA5, exhibited similar expression patterns up to 3 h of PEG treatment. Using an inducible estradiol system, the overexpression of *NF-YA1, 5, 6*, or *9* was shown to result in increased ABA sensitivity during seed germination [32]. One NF-YA homolog, *GmNF-YA3,* from soybean was also shown to be involved in the abiotic stress response. Transcripts of *GmNF-YA3* were induced after as little as 0.5 h of NaCl, ABA, or cold treatment, and GmNF-YA3 showed a high level of similarity with *Arabidopsis* NF-YA3, 5, 6, and 8. Of the reported members of the plant NF-YA subfamily, the transcript levels of several members (including *NF-YA1, 5, 6,* and *9*) were affected by ABA, implying that these genes regulate the plant's response to abiotic stress via an ABA-dependent pathway. This may also be the case for *BnNF-YA1*, *BnNF-YA8*, and *BnNF-YA9,* which exhibit similar expression patterns under salinity, drought, or ABA stress.

AtNF-YB1 and its maize homologue ZmNF-YB2 were shown to enhance drought resistance, and maize plants overexpressing *ZmNF-YB2* performed well in the field [11]. Barley *HvNF-YB5,* similar to *Arabidopsis* NF-YB1, showed around a 70-fold expression increase under salinity treatment [20]. *Triticum aestivum* (wheat) *TaNF-YB2*, homologous to *Arabidopsis* NF-YB1, was also upregulated in response to drought stress in the leaves [18]. *BnNF-YB1*, which is homologous to *Arabidopsis NF-YB1*, was induced only by drought in our study. *Arabidopsis NF-YB2* was first found to be up-regulated by NaCl, mannitol, or cold (4°C) treatment [17]. Interestingly, *NF-YB2* was identified as the top third gene to respond to osmotic stress. The expression of poplar *NF-YB7* increased during a 15-d PEG or ABA treatment [33]. Transgenic *Arabidopsis* plants overexpressing poplar *NF-YB7* exhibited an increased germination rate, root length, and photosynthetic rate under drought stress. Poplar NF-YB7 showed a high level of amino acid sequence identity with *Arabidopsis* NF-YB3, while *Arabidopsis nf-yb3* mutant plants were more sensitive than the wild type to environmental stimuli. Canola BnNF-YB2 and BnNF-YB3, similar to *Arabidopsis* NF-YB2, were found to be responsive to NaCl and ABA treatments in this study. The expression of several poplar *NF-YBs*, such as *NF-YB6*, *NF-YB11*, and *NF-YB13,* was altered in the leaves of plants treated with PEG-6000 [19], while the expression of *BnNF-YB7* and *BnNF-YB14* (high identity with Poplar NF-YB11 and NF-YB13 in terms of the conserved core domain) was also up-regulated in the leaves under PEG or NaCl treatment (Fig. 1 and Fig. 2).

Previous studies revealed that a few NF-YC members were involved in the stress response [21,22]. One NF-YC member, *NF-YC2,* from both *Arabidopsis* and *Nicotiana tabacum* (tobacco), was found to be activated by photooxidative stress [21]. This gene

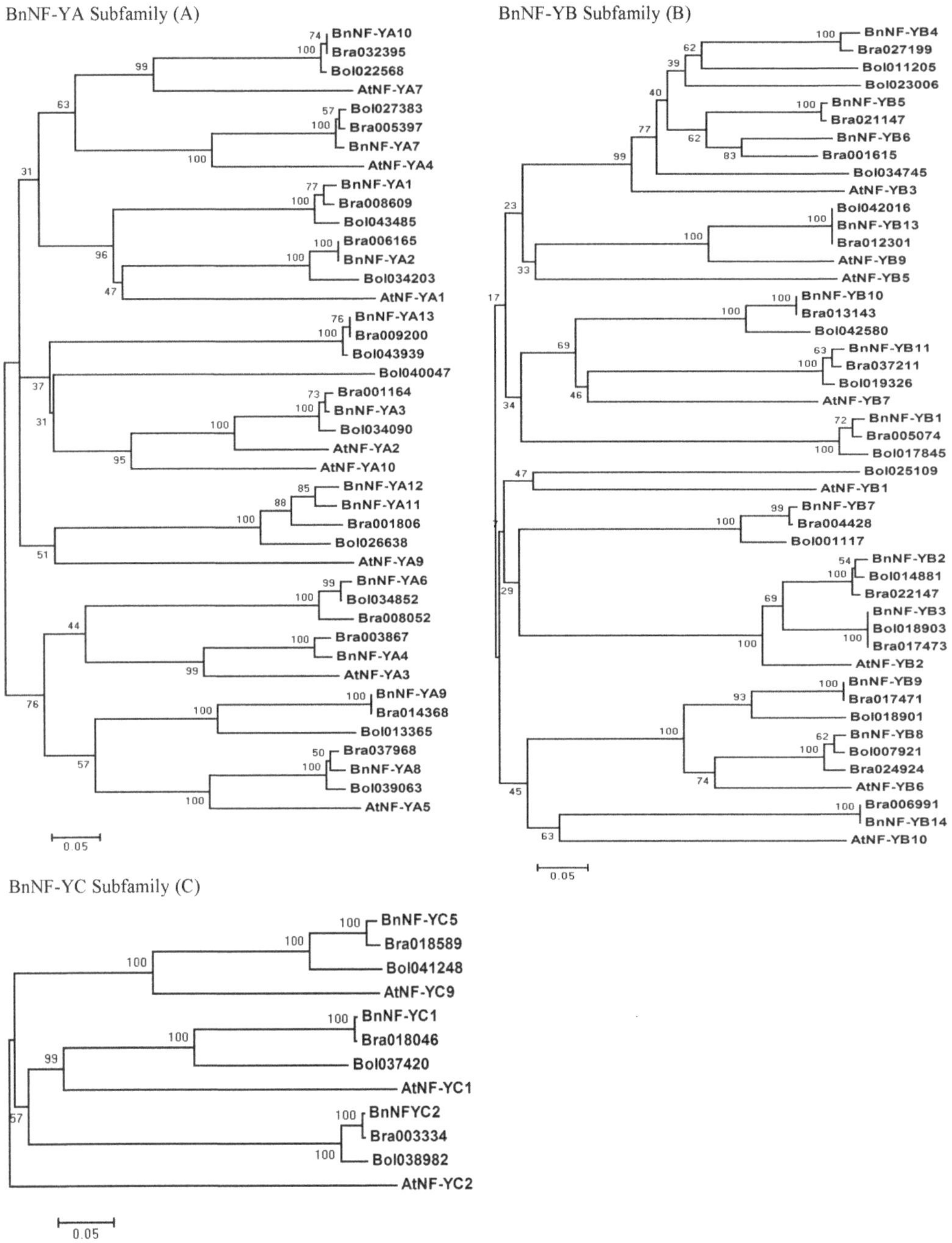

Figure 6. Phylogenetic trees of *Brassica* and *Arabidopsis* NF-Y families based on the nucleotide sequences in the promoter regions. The phylogenetic trees were constructed by the neighbor-joining method implemented by MEGA software, version 4.1. The numbers at each branch point represent the bootstrap scores (1,000 replicates). A branch with a bootstrap score of below 50 was usually considered unreliable. a Phylogenetic tree of *Brassica* and *Arabidopsis* NF-YA promoters. Due to low levels of nucleotide similarity, *BnNF-YA4* homologues in *B. oleracea* were not included here. b Phylogenetic tree of *Brassica* and *Arabidopsis* NF-YB promoters. Due to low levels of nucleotide similarity, *BnNF-YB14* homologues in *B. oleracea* were not included here. c Phylogenetic tree of *Brassica* and *Arabidopsis* NF-YC promoters. Bol and Bra represent *B. oleracea* and *B. rapa*, respectively. The numbers after the species abbreviation correspond to the individual gene name. Most promoters were at least 1000 bp in length, except for those of *BnNF-YA1* (595 bp), *BnNF-YA4/5* (577), *BnNF-YA6* (806), *BnNF-YA10* (892), *BnNF-YB1* (519), *BnNF-YB7* (988), *BnNF-YB8* (391), and *BnNF-YB13* (867) and their homologues in *B. oleracea*, *B. rapa*, and *Arabidopsis*.

responded to multiple stresses, including heat, cold, and drought. In the same study, several *Arabidopsis NF-YC* members were shown to be induced by abiotic stress. The transcript levels of *Arabidopsis NF-YC3* and *NF-YC4* were increased under drought stress, while *NF-YC4* was also induced by cold. Furthermore, *Arabidopsis* NF-YC2 formed a complex with NF-YB3, bZIP28, and NF-YA4 and was shown to play a role during endoplasmic reticulum stress [30]. *PwHAP5* (*Picea wilsonii*) was up-regulated by NaCl, dehydration, and ABA treatment [22]. Ectopic expression of *PwHAP5*, which is homologous to *Arabidopsis*

NF-YC2, improved salinity tolerance in plants. In our previous study, we showed that BnNF-YC2 and BnNF-YC5 grouped with NF-YC2 and NF-YC3 [9] and, like these homologs in *Arabidopsis*, responded to salinity and drought stress.

Since plants have so many NF-Ys, functional redundancy seems inevitable. *Arabidopsis* NF-YB2 and NF-YB3 function additively in the long-day flowering pathway [34]. LEAFY COTYLEDON 1 (LEC1) and LEAFY COTYLEDON1-LIKE (L1L) were shown to be involved in embryo development [35,36]. Many of the BnNF-Ys characterized in our study were responsive to salinity, drought, or ABA treatment. Some of these members in the same subfamily were clustered in the same phylogenetic clade based on their promoter sequences, such as BnNF-YA1 and BnNF-YA2 (Fig. 6). Further evidence is needed to confirm the roles of these proteins during the abiotic stress response. It would also be interesting to explore whether these proteins from the three subfamilies combine to form trimeric complexes, as reported in yeast and animal systems [37]. It is not known whether the NF-Y complex always has fixed components or whether the components differ under different conditions. Even though extensive correlations in the expression patterns of *TaNF-Y* subfamily members were identified [18], no three members from the three different BnNF-Y subfamilies exhibited the same expression pattern under all three treatments in our study. This phenomenon implies that the canola BnNF-Y complex does not always consist of the same monomers under different conditions.

The plant's response to abiotic stress involves the transcriptional regulation of genes via their *cis*-regulatory elements. ABRE, MYB, and MYC elements are known to be involved in the ABA-dependent stress pathway, while the DRE element plays a role in the ABA-independent pathway [38–40]. Two ABRE elements were identified in the promoter region of *Arabidopsis* drought-responsive *NF-YA5,* which functioned in an ABA-dependent manner [13]. Interestingly, *GmNF-YA3*, a homolog of *NF-YA5*, harbored an additional DRE *cis*-acting element in its promoter, which suggests that soybean NF-YA3 may be involved in both pathways [16]. A previous study of the global expression patterns of rice plants subjected to various abiotic stresses identified more ABRE and DRE elements in the promoter regions of genes responsive to both drought and salinity than in those specifically responsive to drought or salinity stress [41]. In our study, the promoter regions of *BnNF-YA11* and *BnNF-YA12*, which were strongly induced by salinity and drought stress, each harbored 4 ABRE elements. Canola *NF-YB2* and *NF-YB3* each possessed at least 5 ABRE elements and were strongly up-regulated by ABA treatments. *BnNF-YC2,* which contains 2 ABRE elements in its promoter region, was strongly induced by ABA stress. In contrast, the promoter regions of canola *NF-YB11* and *NF-YB14,* which were also strongly induced by ABA or drought treatment, had few DRE or ABRE elements, but several MYB or MYC elements, suggesting that MYB or MYC play roles in the abiotic stress response.

According to the well-known triangle theory [42,43], canola, an allopolyploid, originated from the hybridization of *B. rapa* (the A genome) and *B. oleaacea* (the C genome), while all *Brassica* species basically arose from common *Arabidopsis* ancestors. Based on our NF-Y stress-related *cis*-element analysis and promoter sequence alignments, the upstream regulatory regions of *NF-Y* sequences of canola were found to be similar to those of *Arabidopsis* (Fig. 5 and 6). Through extensive comparisons based on nucleotide sequences, homoeologous segments conserved in canola and *Arabidopsis* were found to exhibit perfect collinearity [44]. Our study revealed that the promoters of canola *NF-Ys* were more similar to those of *B. rapa* than to *B. oleracea*. Canola (*B. napus*) was proposed to have multiple origins, and natural canola species were found to be more closely related to *B. rapa* species than to *B. oleracea* species, according to an Restriction fragment length polymorphisms (RFLP) analysis of nuclear, chloroplastic, and mitochondrial DNA [45]. A recent study found that two canola self-incompatibility genes (S-locus glycoproteins, SLGs) and an S-locus receptor kinase (SRK) had higher levels of amino acid sequence identity with their *B. rapa* homologues than with those from *B. oleracea* [46], supporting the notion that a subset of canola sequences are more closely related to *B. rapa* than to *B. oleracea*.

In conclusion, this study represents an extensive evaluation of BnNF-Y family members under salinity, drought, or ABA stress. The results presented here offer a useful foundation for further studies of BnNF-Y proteins under abiotic stress conditions. Several *BnNF-Y* members in each subfamily showed similar expression patterns, indicating that these genes may have redundant functions. Members of different families were found to have similar expression patterns, suggesting that BnNF-Ys form a heterocomplex. Our BnNF-Y promoter analysis shows that multiple BnNF-Y members contain abiotic stress-responsive elements and provides clues as to the evolution of BnNF-Ys in *Brassica* species.

Supporting Information

Figure S1 Expression pattern of *BnNF-Y* genes in the leaves and roots of plants subjected to salinity stress for 1 or 3 h. The expression of *BnNF-YA* (A), *BnNF-YB* (B), and *BnNF-YC* (C) in the leaves and roots of plants exposed to 150 mM NaCl for the indicated periods of time. The transcript levels of each *BnNF-Y* gene were first normalized to those of the housekeeping gene 18S and then compared to the control (0-h level in the leaf). The expression levels of untreated samples (C, 0-h leaf samples) were arbitrarily set to 1.0. L, leaves; R, roots. CK, no treatment; Salt, NaCl treatment. Significant differences between different samples and 0-h samples (same tissue only) are indicated by a single ($P<0.05$) or double ($P<0.01$) asterisk, according to Dunnett's method of one-way ANOVA in SPSS.

Figure S2 Expression pattern of *BnNF-Y* genes exposed to osmotic stress. The expression of *BnNF-YA* (A), *BnNF-YB* (B), and *BnNF-YC* (C) genes in the leaves of plants exposed to treatment with 15% (w/v) PEG-6000 for the indicated periods. The transcript levels of each *BnNF-Y* gene were first normalized to those of the housekeeping gene 18S and then compared to the levels in the 0-h leaf control. Expression levels in untreated samples (C, 0-h leaf samples) were arbitrarily set to 1.0. L, leaves; R, roots. CK, no treatment; Drought, PEG6000 treatment. Significant differences between different samples and 0-h samples (same tissue only) are indicated by a single ($P<0.05$) or double ($P<0.01$) asterisk, according to Dunnett's method of one-way ANOVA in SPSS.

Figure S3 Expression pattern of the *BnNF-Y* genes after exposure to 100 μM ABA. The expression of *BnNF-YA* (A), *BnNF-YB* (B), and *BnNF-YC* (C) genes in the leaves and roots of plants exposed to 100 μM ABA. Transcript levels of each *BnNF-Y* were first normalized to those of the housekeeping gene 18S and then compared to levels at 0 h in the control (untreated) leaves. Expression levels in untreated samples (CK, 0-h leave samples) were arbitrarily set to 1.0. L, leaves; R, roots. CK, no treatment; ABA, ABA treatment. Significant differences between different samples and 0-h samples (same tissue only) are indicated by a single ($P<0.05$) or double ($P<0.01$) asterisk, according to Dunnett's method of one-way ANOVA in SPSS.

Figure S4 Semi-quantitative RT-PCR analysis of BnNF-Y expression in control leaves and roots. RT-PCR analysis was performed on untreated leaf (L) and root (R) tissue samples. The 18S housekeeping gene and *BnNF-Ys* were amplified for 28 cycles and 35 cycles, respectively.

Figure S5 Diagrammatic representation of NF-Y promoter regions. *Arabidopsis* NF-YA (A), NF-YB (B), and NF-YC (C) promoter regions. Symbols above the top line indicate elements in the forward orientation, and those below the bottom line are in the reverse orientation. MYC(CANNTG) sequences are between the double lines, to indicate their palindromic nature. Numbers are the distance in nucleotides between the sequences and ATG start codon.

Table S1 Primers for BnNF-Y promoters. F indicates the PCR forward primer and R indicates the PCR reverse primer. Particularly, SP1, SP2, SP3 are three specific primers for Genome Walking.

Table S2 Relative water content (RWC) of drought-stressed plants. Plants were treated with 10%, 15% and 20% (w/v) PEG6000 solution for 24 h.

Acknowledgments

We thank Prof. Zhaopu Liu from Nanjing Agricultural University for kindly supplying the canola ('Nanyanyou 1') seeds. We also thank Kathleen Farquharson for valuable comments on the manuscript revision. We also express our sincere thanks to the anonymous reviewers.

Author Contributions

Conceived and designed the experiments: ML LX GZ. Performed the experiments: ZL QT LX. Analyzed the data: LX ZL ML. Contributed reagents/materials/analysis tools: RF XY. Contributed to the writing of the manuscript: LX ML ZL.

References

1. Chen WJ, Zhu T (2004) Networks of transcription factors with roles in environmental stress response. Trends Plant Sci 9: 591–596.
2. Kasuga M, Liu Q, Miura S, Yamaguchi-Shinozaki K, Shinozaki K (1999) Improving plant drought, salt, and freezing tolerance by gene transfer of a single stress-inducible transcription factor. Nat Biotechnol 17: 287–291.
3. Shinozaki K, Yamaguchi-Shinozaki K (1997) Gene expression and signal transduction in water-stress response. Plant Physiol 115: 327–334.
4. Kim IS, Sinha S, de Crombrugghe B, Maity SN (1996) Determination of functional domains in the C subunit of the CCAAT-binding factor (CBF) necessary for formation of a CBF-DNA complex: CBF-B interacts simultaneously with both the CBF-A and CBF-C subunits to form a heterotrimeric CBF molecule. Mol Cell Biol 16: 4003–4013.
5. Gusmaroli G, Tonelli C, Mantovani R (2001) Regulation of the CCAAT-binding NF-Y subunits in *Arabidopsis thaliana*. Gene 264: 173–185.
6. Gusmaroli G, Tonelli C, Mantovani R (2002) Regulation of novel members of the *Arabidopsis thaliana* CCAAT-binding nuclear factor Y subunits. Gene 283: 41–48.
7. Siefers N, Dang KK, Kumimoto RW, Bynum WE, Tayrose G, et al. (2009) Tissue-specific expression patterns of *Arabidopsis* NF-Y transcription factors suggest potential for extensive combinatorial complexity. Plant Physiol 149: 625–641.
8. Thirumurugan T, Ito Y, Kubo T, Serizawa A, Kurata N (2008) Identification, characterization and interaction of *HAP* family genes in rice. Mol Gen Genet 279: 279–289.
9. Liang M, Yin X, Lin Z, Zheng Q, Liu G, et al. (2014) Identification and characterization of NF-Y transcription factor families in Canola (Brassica napus L.). Planta 239: 107–126.
10. Petroni K, Kumimoto RW, Gnesutta N, Calvenzani V, Fornari M, et al. (2012) The promiscuous life of plant NUCLEAR FACTOR Y transcription factors. Plant Cell 24: 4777–4792.
11. Nelson DE, Repetti PP, Adams TR, Creelman RA, Wu J, et al. (2007) Plant nuclear factor Y (NF-Y) B subunits confer drought tolerance and lead to improved corn yields on water-limited acres. Proc Natl Acad Sci U S A 104: 16450–16455.
12. Chen NZ, Zhang XQ, Wei PC, Chen QJ, Ren F, et al. (2007) AtHAP3b plays a crucial role in the regulation of flowering time in *Arabidopsis* during osmotic stress. J Biochem Mol Biol 40: 1083–1089.
13. Li WX, Oono Y, Zhu J, He XJ, Wu JM, et al. (2008) The *Arabidopsis* NFYA5 transcription factor is regulated transcriptionally and posttranscriptionally to promote drought resistance. Plant Cell 20: 2238–2251.
14. Zhao B, Ge L, Liang R, Li W, Ruan K, et al. (2009) Members of miR-169 family are induced by high salinity and transiently inhibit the NF-YA transcription factor. BMC Mol Biol 10: 29.
15. Li YJ, Fang Y, Fu YR, Huang JG, Wu CA, et al. (2013) NFYA1 Is Involved in Regulation of Postgermination Growth Arrest Under Salt Stress in Arabidopsis. Plos One 8: e61289.
16. Ni Z, Hu Z, Jiang Q, Zhang H (2013) *GmNFYA3*, a target gene of miR169, is a positive regulator of plant tolerance to drought stress. Plant Mol Biol: 1–17.
17. Kreps JA, Wu YJ, Chang HS, Zhu T, Wang X, et al. (2002) Transcriptome changes for *Arabidopsis* in response to salt, osmotic, and cold stress. Plant Physiol 130: 2129–2141.
18. Stephenson TJ, McIntyre CL, Collet C, Xue GP (2007) Genome-wide identification and expression analysis of the NF-Y family of transcription factors in Triticum aestivum. Plant Mol Biol 65: 77–92.
19. Yan DH, Xia X, Yin W (2013) NF-YB Family Genes Identified in a Poplar Genome-wide Analysis and Expressed in Populus euphratica Are Responsive to Drought Stress. Plant Mol Biol Report: 1–8.
20. Liang M, Hole D, Wu J, Blake T, Wu Y (2012) Expression and functional analysis of NUCLEAR FACTOR-Y, subunit B genes in barley. Planta: 1–13.
21. Hackenberg D, Keetman U, Grimm B (2012) Homologous NF-YC2 subunit from Arabidopsis and tobacco is activated by photooxidative stress and induces flowering. Int J Mol Sci 13: 3458–3477.
22. Li L, Yu Y, Wei J, Huang G, Zhang D, et al. (2013) Homologous HAP5 subunit from Picea wilsonii improved tolerance to salt and decreased sensitivity to ABA in transformed Arabidopsis. Planta: 1–12.
23. Combier JP, Frugier F, de Billy F, Boualem A, El-Yahyaoui F, et al. (2006) *MtHAP2-1* is a key transcriptional regulator of symbiotic nodule development regulated by microRNA169 in *Medicago truncatula*. Genes Dev 20: 3084–3088.
24. Jones-Rhoades MW, Bartel DP (2004) Computational identification of plant microRNAs and their targets, including a stress-induced miRNA. Mol Cell 14: 787–799.
25. Leyva-González MA, Ibarra-Laclette E, Cruz-Ramírez A, Herrera-Estrella L (2012) Functional and Transcriptome Analysis Reveals an Acclimatization Strategy for Abiotic Stress Tolerance Mediated by Arabidopsis NF-YA Family Members. PloS one 7: e48138.
26. Farooq M, Wahid A, Kobayashi N, Fujita D, Basra SMA (2009) Plant drought stress: effects, mechanisms and management. Agron Sustain Dev 29: 185–212.
27. Rogers SO, Bendich AJ (1985) Extraction of DNA from milligram amounts of fresh, herbarium and mummified plant tissues. Plant Mol Biol 5: 69–76.
28. Livak KJ, Schmittgen TD (2001) Analysis of Relative Gene Expression Data Using Real-Time Quantitative PCR and the $2^{-\Delta\Delta CT}$ Method. Methods 25: 402–408.
29. Schmittgen TD, Livak KJ (2008) Analyzing real-time PCR data by the comparative CT method. Nat Protoc 3: 1101–1108.
30. Liu JX, Howell SH (2010) bZIP28 and NF-Y Transcription Factors Are Activated by ER Stress and Assemble into a Transcriptional Complex to Regulate Stress Response Genes in Arabidopsis. Plant Cell 22: 782–796.
31. Wenkel S, Turck F, Singer K, Gissot L, Le Gourrierec J, et al. (2006) CONSTANS and the CCAAT box binding complex share a functionally important domain and interact to regulate flowering of *Arabidopsis*. Plant Cell 18: 2971–2984.
32. Mu J, Tan H, Hong S, Liang Y, Zuo J (2013) *Arabidopsis* transcription factor genes *NF-YA1*, *5*, *6*, and *9* play redundant roles in male gametogenesis, embryogenesis, and seed development. Mol Plant 6: 188–201.
33. Han X, Tang S, An Y, Zheng DC, Xia XL, et al. (2013) Overexpression of the poplar NF-YB7 transcription factor confers drought tolerance and improves water-use efficiency in Arabidopsis. J Exp Bot: ert262.
34. Kumimoto RW, Adam L, Hymus GJ, Repetti PP, Reuber TL, et al. (2008) The nuclear factor Y subunits NF-YB2 and NF-YB3 play additive roles in the promotion of flowering by inductive long-day photoperiods in *Arabidopsis*. Planta 228: 709–723.
35. Lotan T, Ohto M, Yee KM, West MAL, Lo R, et al. (1998) *Arabidopsis* LEAFY COTYLEDON1 is sufficient to induce embryo development in vegetative cells. Cell 93: 1195–1205.

36. Kwong RW, Bui AQ, Lee H, Kwong LW, Fischer RL, et al. (2003) LEAFY COTYLEDON1-LIKE defines a class of regulators essential for embryo development. Plant Cell 15: 5–18.
37. Dolfini D, Gatta R, Mantovani R (2012) NF-Y and the transcriptional activation of CCAAT promoters. Crit Rev Biochem Mol Biol 47: 29–49.
38. Yamaguchi-Shinozaki K, Shinozaki K (1994) A novel *cis*-acting element in an Arabidopsis gene is involved in responsiveness to drought, low-temperature, or high-salt stress. Plant Cell 6: 251–264.
39. Abe H, Yamaguchi-Shinozaki K, Urao T, Iwasaki T, Hosokawa D, et al. (1997) Role of *Arabidopsis* MYC and MYB homologs in drought- and abscisic acid-regulated gene expression. Plant Cell 9: 1859–1868.
40. Yamaguchi-Shinozaki K, Shinozaki K (2005) Organization of *cis*-acting regulatory elements in osmotic- and cold-stress-responsive promoters. Trends Plant Sci 10: 88–94.
41. Zhou J, Wang X, Jiao Y, Qin Y, Liu X, et al. (2007) Global genome expression analysis of rice in response to drought and high-salinity stresses in shoot, flag leaf, and panicle. Plant Mol Biol 63: 591–608.
42. Jules J (2009) Plant Breeding Reviews Wiley.
43. Nagaharu U (1935) Genome analysis in Brassica with special reference to the experimental formation of B. napus and peculiar mode of fertilization. Japan J Bot 7: 389–452.
44. Cheung F, Trick M, Drou N, Lim YP, Park JY, et al. (2009) Comparative Analysis between Homoeologous Genome Segments of Brassica napus and Its Progenitor Species Reveals Extensive Sequence-Level Divergence. Plant Cell 21: 1912–1928.
45. Song K, Osborn TC (1992) Polyphyletic Origins of Brassica Napus: New Evidence Based on Organelle and Nuclear Rflp Analyses. Genome 35: 992–1001.
46. Zhang XG, Yin DM, Ma CZ, Fu TD (2011) Phylogenetic Analysis of S-Locus Genes Reveals the Complicated Evolution Relationship of S Haplotypes in Brassica. Plant Mol Biol Report 29: 481–488.

11

Simultaneous Over-Expression of *PaSOD* and *RaAPX* in Transgenic *Arabidopsis thaliana* Confers Cold Stress Tolerance through Increase in Vascular Lignifications

Amrina Shafi[1,2], Vivek Dogra[1], Tejpal Gill[1¤], Paramvir Singh Ahuja[1,2], Yelam Sreenivasulu[1,2*]

1 Division of Biotechnology, CSIR-Institute of Himalayan Bioresource Technology, Palampur, Himachal Pradesh, India, **2** Academy of Scientific and Innovative Research (AcSIR), CSIR- Institute of Himalayan Bioresource Technology, Palampur, Himachal Pradesh, India

Abstract

Antioxidant enzymes play a significant role in eliminating toxic levels of reactive oxygen species (ROS), generated during stress from living cells. In the present study, two different antioxidant enzymes namely copper-zinc superoxide dismutase derived from *Potentilla astrisanguinea* (*PaSOD*) and ascorbate peroxidase (*RaAPX*) from *Rheum austral* both of which are high altitude cold niche area plants of Himalaya were cloned and simultaneously over-expressed in *Arabidopsis thaliana* to alleviate cold stress. It was found that the transgenic plants over-expressing both the genes were more tolerant to cold stress than either of the single gene expressing transgenic plants during growth and development. In both single (*PaSOD*, *RaAPX*) and double (*PaSOD* + *RaAPX*) transgenic plants higher levels of total antioxidant enzyme activities, chlorophyll content, total soluble sugars, proline content and lower levels of ROS, ion leakage were recorded when compared to the WT during cold stress (4°C), besides increase in yield. In the present study, Confocal and SEM analysis in conjunction with qPCR data on the expression pattern of lignin biosynthetic pathway genes revealed that the cold stress tolerance of the transgenic plants might be because of the peroxide induced up-regulation of lignin by antioxidant genes mediated triggering.

Editor: Girdhar K. Pandey, University of Delhi South Campus, India

Funding: This work was supported by the Council of Scientific and Industrial Research, New Delhi, India. A.S., T.G. and V.D. acknowledge Fellowship by the CSIR, New Delhi, India. The funders had no role in study design, data collection and analysis, decision to publish, or preparation of the manuscript.

Competing Interests: The authors have declared that no competing interests exist.

* Email: sree_yelam@yahoo.com

¤ Current address: National Institute of Arthritis and Musculoskeletal and Skin Diseases, National Institute of Health, Bldg 10 CRC, 1–5256, Bethesda, MD, United States of America

Introduction

Abiotic stresses such as cold, drought, metal and salt stress are the major factors adversely affecting development and productivity in plants. Plants growing in different climatic niches have their own protective mechanisms to cope with the adverse conditions. Metabolic activities in various cellular compartments lead to the production of reactive oxygen species (ROS) such as superoxide anion (O_2^{-}), hydrogen peroxide (H_2O_2), hydroxyl radical (OH·), and singlet oxygen (O_2) [1]. These ROS are highly toxic, and are taken care of by the antioxidant system in the living cells. During abiotic and biotic stress conditions, the production of ROS increases to lethal levels, however, there is an up regulation of both enzymatic and non-enzymatic antioxidants, which modify the ROS to non toxic forms. Oxidative stress response involving antioxidant enzymes includes superoxide dismutase (SOD), ascorbate peroxidase (APX), catalase (CAT), glutathione reductase (GR) and peroxidase (POD) [2]. SOD (EC1.15.1.1) is the first enzyme in the plant's antioxidative defense mechanism that converts superoxide anion radicals ($O_2^{\cdot-}$) to hydrogen peroxide (H_2O_2) and water, thereby imparting protection against the harmful effects of highly reactive superoxide radical ($O_2^{\cdot-}$). APX is an integral component of the glutathione-ascorbate cycle [3] which detoxifies hydrogen peroxide in the presence of ascorbate to produce dehydroascorbate and water [4]. Genetic manipulation of antioxidant enzymes is one of the effective measures to impart stress tolerance in plants. Activities of superoxide dismutase and catalases were increased in plants subjected to cold stress [5]. It is commonly known that lignin evolved together with the plants adaptation to a terrestrial life to provide them with the structural support needed for an erect growth habit. Secondarily thickened cell walls were predominantly deposited by the lignin polymers, and making them rigid. In addition to developmentally programmed deposition of lignin, its biosynthesis can also be induced upon various biotic and abiotic stress conditions. The core enzymes of the phenylpropanoid pathway namely PAL, C4H and 4CL, converts the phenylalanine to *p*-Coumaroyl CoA. Thereafter, that the pathway flows to three directions, each of which is responsible for the production of individual components of monomers of lignin polymer i.e. *p*-hydroxyphenyl (H), guaiacyl (G) and syringyl (S) units. The generalized lignin biosynthetic pathway is shown in Fig. S1 and important enzymes which regulate the lignin pathway are also highlighted.

High altitude plants habitats are very specific because of their extreme niche areas that have naturally selected the best fit traits for adaptation in extreme environments. Their genes/proteins are being used as molecular tools for engineering crop and other plants for better stress tolerance and adaptability against the present scenarios of climate change. In *Potentilla atrosanguinea* - a Himalayan high altitude (4200 m msl) alpine plant, induction of a new copper-zinc superoxide dismutase (SOD) isozyme was recorded with the cold stress [6]. This gene encodes an isozyme of copper-zinc superoxide dismutase (EU532614.1) (*PaSOD*) and shared a homology of 80% (Expect = 5e-83) with the cytosolic *AtSOD* (NM_100757.3) [6]. Earlier studies confirmed that the protein encoded by this gene was functional from sub-zero temperature to >50°C and also the protein retained activity after autoclaving (heating at 121°C, at a pressure of 1.1 kg per square cm for 20 min) [7,8]. Increased SOD activity has been positively correlated against different stress tolerance such as low/high temperature drought, salinity, high light intensity, and ozone [9,10]. Over-expression of *Potentilla atrosanguinea* copper zinc superoxide dismutase (*PaSOD*) has improved copper and salt stress tolerance in *Arabidopsis thaliana* [11,12]. Apparently, over-expression of the same SOD in potato, enhanced photosynthetic performance under drought stress [13]. In another study, over accumulation of lignin in vascular bundles was found to be the molecular mechanism which underlies the improved stress tolerance induced by the over-expression of *PaSOD* in *Arabidopsis thaliana* [14]. *Rheum australe* - another high altitude alpine plant, up regulation of ascorbate peroxidase was recorded against cold stress (unpublished data). This APX gene from Rheum (*RaAPX*)(DQ078123.2) was one of the cold tolerant genes, which shared a homology of 75% (Expect = 4e-162) with the peroxisomal *APX* gene of *Arabidopsis thaliana* (NM_119666.3). Hence it was thought that high altitude plants can be a good source of potential antioxident genes which can be successfully utilized for the reprogramming of stress tolerance in plants without disturbing native physiology, after functionally validated in the model plant i.e. *Arabidopsis*. In view of the above discussion, the present study illustrates the efficiency of over expressing two important antioxidant genes from high altitude plants namely superoxide dismutase from *Potentilla atrosanguinea* (*PaSOD*) and ascorbate peroxidase from *Rheum australe* (*RaAPX*) in *Arabidopsis thaliana*. It was established that simultaneous over expression of *PaSOD* and *RaAPX* was more effective in alleviating cold stress over independent over expression of either *PaSOD* or *RaAPX*.

Materials and Methods

Plant growth and cold stress treatment

Arabidopsis (ecotype *coloumbia*) plants were grown on soil mixture of vermiculite: peat moss: perlite (1:1:1) in the greenhouse under a 16 h light and 8 h dark cycle at 20±1°C. For stress treatment, 21d old seedlings of wild type, and hygromycin selected transgenic seedlings were transferred to cold room (4°C). Samples were collected 0 h, 6 h, 12 h, 24 h, 48 h (2d), 96 h (4d) and 192 h (8d) during cold treatment for the analyses of transcript levels, enzyme activity, proline and soluble sugars accumulation, electrolyte leakage, ROS, chlorophyll and biomass accumulation estimation etc,.

Plasmid construction and plant transformation

Full length cDNAs of Copper-Zinc Superoxide Dismutase (*PaSOD*) and Ascorbate Peroxidase (now onwards *RaAPX*) from high altitude plants *Potentilla atrosanguinea* (which grows at day-time air temperatures of 3–10°C in Lahaul and Spiti districts of Himachal Pradesh: altitude 4517 m; 32° 24′ 20″N; 077° 38′ 40″ E) and *Rheum australe* (Rohtang Pass in Himachal Pradesh: altitude 4000 m; 32° 22′ 19″ N; 077° 14′ 46″ E), respectively, from Western Himalaya, were cloned in *Arabidopsis thaliana* as described earlier by Gill et al. [14]. Briefly, coding nucleotide sequences of these genes were amplified using the gene specific primers with incorporated *Nco*1 and *Bgl*II restriction sites at 3′ end. PCR products were cloned into a cloning vector pGEMT easy (Promega) and then sub-cloned into binary plant vector pCAMBIA1302 under the Cauliflower mosaic 35 S promoter. The prepared plasmid construct was mobilized into *Arabidopsis* plants via *Agrobacterium* mediated vacuum infiltration method [15]. Seeds were collected and screened in Murashige and Skoog [16] medium supplemented with 20 µg ml^{-1} hygromycin. Homozygous (T3 generation) transgenic *PaSOD* and *RaAPX* plants were crossed to obtain dual transgenic plants. Integration and expression of these transgenes in the dual transgenics was assessed by PCR using gene specific primers with genomic DNA as a template and semi-quantitative PCR from the total RNA isolated from leaf samples.

Gene-specific semi-quantitative and real time PCR

Total RNA was isolated from 0 hrs and 2d cold treated transgenic and the wild type *Arabidopsis* plants using Total RNA extraction kit (Real Genomics). One microgram of total RNA was used for oligo (dT) primed first-strand cDNA synthesis in 20 µl reaction using of Superscript III Reverse transcriptase (Invitrogen). Transcripts of *PaSOD* and *RaAPX* were quantified with PCR using gene specific primers. Constitutively expressed *GAP-C* (Glyceraldehyde-3-phosphate dehydrogenase C subunit) was amplified simultaneously in 27 cycles to ensure equal amounts of cDNA used.

For real time expression analysis of lignin pathway genes during cold stress, primers were designed using Primer 3 software [17]. Details of the primers used in this study are given in Table S1. Gene expression was performed on a Stratagene Mx3000P system (Agilent Technologies, Germany) using 2× Brilliant III SYBR@ Green qPCR Master Mix (Agilent Technologies, Germany). All qPCRs were run in triplicates with a no-template control to check for contamination. PCR was conducted under the following conditions: 10 min at 95°C (enzyme activation), 40 cycles each of 30 s at 95°C, 30 s at 55°C and 72°C for 30 s and a final melting curve analysis was performed (55° to 95°C) to verify the specificity of amplicons. The raw threshold cycle (Ct) values were normalized against a housekeeping gene encoding Actin [18] to calculate both the difference in expression between control and respective treatment samples and tissue specific gene abundance using the Relative Expression Software Tool (REST; [19]). Expression values were transformed (log2) to generate expression profiles. This experiment was repeated three times for the reproducibility and for statistical significance calculation.

SOD and APX enzyme activity assay

Total enzyme activity of SOD and APX was estimated at different time points during cold stress i.e. 0–192 h. Total SOD activity was estimated as described earlier [14] while APX activity was determined according to Nakano and Asada [20]. Briefly, leaf samples (100 mg) were homogenized in a pre-cooled mortar in homogenizing buffer containing 2 mM EDTA, 1 mM DTT, 1 mM PMSF, 0.5% (v/v) Triton-X100 and 10% (w/v) PVPP in 50 mM phosphate buffer pH 7.8. For APX activity homogenizing buffer contained ascorbate in addition and the buffer pH was 7.0. The homogenate was transferred to 1.5 ml Eppendorf and centrifuged at 13,000 rpm for 20 min at 4°C. The supernatant was used to estimate total SOD and APX activities. The total SOD activity was measured by adding 5 µl enzyme extract to a reaction mixture (200 µl) containing 1.5 µm Riboflavin, 50 µm NBT, 10 mM DL-Methionine and 0.025% (v/v) Triton-X100 in

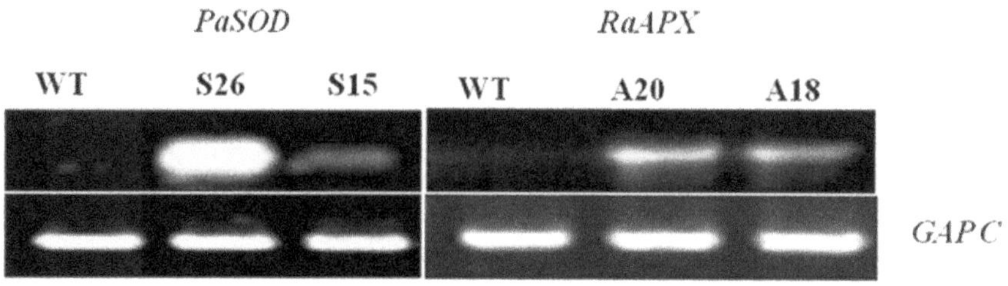

Figure 1. Confirmation of *PaSOD* and *RaAPX* gene expression in transgenic *Arabidopsis* plants at 20°C. The same plants were used for crossing to obtain the double transgenic plants. Expression of genes was (S26 and S15; A20 and A18) confirmed by estimating their transcripts from the leaf tissue by semi-quantitative RT-PCR. Constitutively expressed *GAP-C* was used as loading control.

50 mM phosphate buffer. One unit of enzyme activity was defined as the amount of enzyme required for 50% inhibition of NBT reduction at 25°C. Total protein content was estimated according to the dye binding method of Bradford [21] using BSA as standard. These experiments were repeated three times from three biological replicates.

In-situ ROS staining

In situ ROS staining was done in accordance with Beyer and Fridovich [22], on the basis of the principle of NBT (nitroblue tetrazolium) reduction to blue formazan by $O_2^{-\cdot}$ The intracellular concentration of ROS ($O_2^{-\cdot}$) was directly proportional to the development of intensity of blue color in the leaves. Briefly, leaf tissue was detached from the wild type and transgenic plants and vacuum infiltrated with 10 mM sodium azide (NaN_3) in 10 mM potassium phosphate buffer for 1 min. The infiltrated leaf tissue was incubated in 0.1% NBT (nitroblue tetrazolium) in 10 mM potassium phosphate buffer; pH 7.8 for 30 min. The stained leaf tissue was boiled in acetic acid:glycerol:ethanol (1:1:3) solution to remove other pigments and the stain content was visually documented under Carl-Zeiss Stereo DiscoveryV12 with Axiovision software. This experiment was repeated three times from three biological replicates.

Estimation of electrolyte leakage

Cold-induced cell membrane injuries in leaf tissues were quantified by estimating the electrolyte leakage from the leaf samples. Electrolyte leakage was measured using an electrical conductivity meter as described by Lutts et al. [23]. Briefly, leaves were excised and washed with de-ionized water. One gram of fresh leaves were cut into small pieces (about 1–2 cm) and then immersed in 20 mL of deionized water, incubated at 25°C. After 24 h, electrical conductivity (EC1) of the bathing solution was recorded. These samples were then autoclaved at 120°C for 20 min to completely kill the tissues. Samples were then cooled to 25°C and the final electrical conductivity (EC2) of all released electrolytes was measured. The electrolyte leakage (EL) was expressed following the formula $EL = EC1/EC2 \times 100$. This experiment was repeated three times from three biological replicates.

Estimation of total soluble sugars, proline and chlorophyll content

Total soluble sugar (TSS) was determined by anthrone method of Irigoyen et al. [24]. Free proline content was estimated using the acid ninhydrin method described by Bates et al. [25]. Fresh shoot and root samples (0.5 g) harvested at different time intervals were ground using mortar-pestle with 10 ml of 3% (w/v) sulfosalicylic acid aqueous solutions and the homogenate was filtered through Whatman no. 1 filter paper and taken for further analysis. Two ml

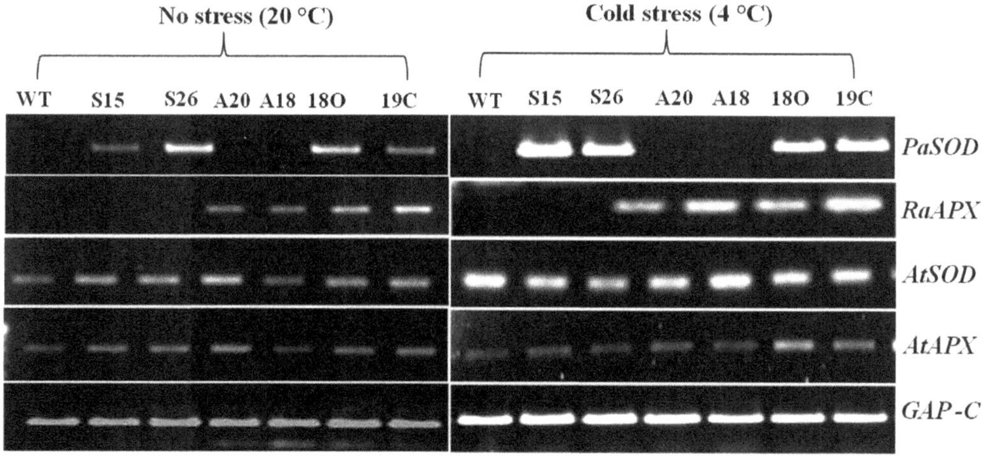

Figure 2. Assessment of native (*AtSOD, AtAPX*) and transgene (*PaSOD* and *RaAPX*) expression by estimating transcript levels under normal (20°C) and cold stress (4°C) conditions. Total RNA was isolated from the leaf samples of WT and transgenic plants, transcripts were amplified by using gene specific primers by semi-quantitative RT-PCR under normal and cold stress conditions. Constitutively expressed *GAP-C* was used as loading control.

Figure 3. Transgenic *Arabidopsis* plants showing tolerance to cold stress. Five weeks old seedlings were transferred to cold (4°C) and allowed to grow for 192 h and different stress tolerance parameters were estimated from the samples of these plants.

of filtered extract, combined with 2 ml acid ninhydrin (1.25 g ninhydrin warmed, in a mixture of 30 ml glacial acetic acid and 20 ml of 6 M phosphoric acid, until dissolved). The reaction mixture was incubated in a boiling water bath (100°C) for 1 h and the reaction was then terminated in an ice bath. Then 4 ml of toluene was added to the reaction mixture and the organic phase (chromophore containing toluene) was warmed to room temperature and its optical density was measured at 520 nm using toluene as blank in a UV-visible spectrophotometer. The amount of proline was determined from a standard curve.

Chlorophyll content was estimated by the method of Arnon [26]. The chlorophyll from the fresh leaves was extracted in 80% acetone and the absorbance was read spectrophotometer (Synergy HT Multi-Mode Microplate Reader, Bio-Tek, USA) at 663 and 645 nm. These experiments were repeated three times from three biological replicates.

Confocal and Scanning Electron Microscopy (SEM)

Confocal microscopy analysis was done as described earlier Gill et al. [14]. Stem of WT and transgenic lines were harvested and fixed in formalin, glacial acetic acid and 50% ethyl alcohol (FAA) (1:1:18) at room temperature. Samples were subsequently dehydrated in a tertiary butyl-alcohol series [27]; embedded in paraffin (melting point 58–60°C) and 8–10 mm sections were cut using a Finsee microtome. Sections were stained with 1% safranin in water and with 4% fast green in clove oil for 4 h and for 30 s, respectively. These were mounted in Canada balsam and examined using Confocal Laser Scanning Microscope (Zeiss LSM510 Meta Gmbh, Germany) equipped with a Zeiss Axiovert 100 M inverted microscope. Lignin auto fluorescence was collected by excitation/emission wave- lengths 488/505 nm.

For SEM analysis, stem cross-sections were fixed in two steps: first with a mixture of 2% paraformaldehyde and 2.5% glutaraldehyde in 0.1 mol/l cacodylate buffer, pH 7.4 for 1 h and then with 1% OsO_4 in 0.1 mol/l cacodylate buffer, pH 7.4 for 30 m. After critical point drying, the samples were sputter-coated with gold, and the coated samples were viewed with a Hitachi S-3400N field emission SEM using an accelerating voltage of 30 kV.

Statistical Analysis

All obtained data were analyzed statistically by Advance Linear/Non-linear models using General Linear Model (STATISTICA release 7, statsoft Wipro, Bangalore, Karnataka, India). Analysis of variance (ANOVA) followed by Duncan's multiple range tests was used for the analysis of the results. The means of each treatment (as mentioned in respective experiments) and their interactions were compared and found statistically significant ($P < 0.05$). The number of samples taken and their replicates are mentioned in respective experiments.

Results

Transgenic *Arabidopsis* over-expressing *PaSOD* in combination with *RaAPX* enhanced cold tolerance

Arabidopsis transgenic plants over expressing *PaSOD* under the control of CaMV35 S promoter were generated as reported earlier [12]. Similarly transgenic *Arabidopsis* plants over expressing ascorbate peroxidase gene from *Rheum* (*RaAPX*) under the control of CaMV35 S promoter were also developed. Presence of these transgenes was confirmed by PCR by using genomic DNA from these plants as template and their expression by the presence of its transcripts from leaf tissues by semi-quantitative RT-PCR (Fig. 1). On the basis of 3:1 segregation and with high enzyme activities, homozygous transgenic lines S26 (for *PaSOD*) and A20 (for *RaAPX*) were selected for further analysis as well as for crossings for raising the transgenics with both these genes. Transgenic *Arabidopsis* plants having both *PaSOD* and *RaAPX* genes were obtained by crossings of these individual gene

Table 1. Growth and biomass production under normal (20°C) and cold stress (4°C) conditions in WT and transgenic plants. ±mean SD values.

Plant type	No of leaves		Rosette diameter (cm)		Plant height (cm)		No of pods		Pod length (cm)		Total seeds	
	20°C	4°C	20°C	4°C	20°C	4°C	20°C	4°C	20°C	4°C	20°C	4°C
WT	18.6 ± 1.2^{c}	19.4 ± 1.2^{c}	3.9 ± 0.4^{d}	3.3 ± 0.3^{d}	26.5 ± 1.07^{ef}	20.8 ± 0.4^{g}	43.6 ± 1.1^{de}	31 ± 1.0^{g}	1.14 ± 0.1^{fghi}	1.08 ± 0.1^{fghi}	884 ± 14.7^{d}	620 ± 20.0^{g}
***PaSOD* (S15)**	25.6 ± 1.6^{ab}	25.4 ± 0.6^{ab}	5.5 ± 0.2^{bc}	6.6 ± 0.2^{a}	27.3 ± 0.94^{cde}	31.8 ± 0.9^{ab}	$39.2\pm0.3f$	56.4 ± 0.5^{c}	1.5 ± 0.1^{bcd}	1.62 ± 0.1^{abc}	784 ± 7.4^{f}	1128 ± 10.2^{c}
***PaSOD* (S26)**	24.2 ± 2.0^{b}	24 ± 1.1^{b}	5.3 ± 0.4^{c}	6.72 ± 0.1^{a}	26.5 ± 1.28^{ef}	31.4 ± 1.1^{ab}	37.4 ± 0.9^{f}	58.2 ± 1.0^{c}	1.38 ± 0.1^{cde}	1.68 ± 0.1^{abc}	748 ± 18.5^{f}	1164 ± 21.3^{c}
***RaAPX* (A20)**	17.4 ± 0.6^{c}	$27.8\pm.7^{ab}$	5.4 ± 0.2^{c}	6.9 ± 0.2^{a}	23.1 ± 1.12^{fg}	30.2 ± 1.3^{bc}	43.2 ± 1.0^{de}	68.4 ± 0.9^{ab}	1.22 ± 0.1^{fgh}	1.82 ± 0.1^{ab}	864 ± 20.4^{de}	1368 ± 8.5^{ab}
***RaAPX* (A18)**	18.2 ± 0.3^{c}	29 ± 1.0^{a}	5.5 ± 0.2^{bc}	6.68 ± 0.2^{a}	23.7 ± 0.99^{efg}	30 ± 1.4^{bcd}	41.6 ± 0.6^{e}	67.2 ± 0.3^{b}	1.3 ± 0.0^{efg}	1.94 ± 0.1^{a}	832 ± 12^{e}	1344 ± 7.4^{b}
***PaSOD*+ *RaAPX* (18O)**	20 ± 1.3^{c}	28.4 ± 0.5^{a}	5 ± 0.4^{c}	6.9 ± 0.5^{a}	25.8 ± 0.73^{ef}	34 ± 1.7^{a}	42.8 ± 0.6^{de}	70.6 ± 1.0^{a}	0.94 ± 0.0^{hi}	1.78 ± 0.1^{ab}	856 ± 13.2^{de}	1412 ± 21.5^{a}
***PaSOD*+ *RaAPX* (19C)**	20.4 ± 1.2^{c}	27.2 ± 0.9^{ab}	5.2 ± 0.3^{c}	6.5 ± 0.3^{ab}	26.6 ± 0.68^{ef}	33.2 ± 1.3^{ab}	43.6 ± 0.6^{de}	69.4 ± 0.4^{ab}	0.94 ± 0.1^{hi}	1.96 ± 0.04^{a}	872 ± 12^{de}	1388 ± 8.0^{ab}

Each value is a mean of three separate biological replicates, Mean ± Standard Deviation. Different parameters were analyzed using ANOVA to detect significant difference between means. Means were compared using Duncan's Multiple Range Test (DMRT) at $P<0.05$. Different combinations of alphabets abcdefghi denote the significantly different groups.

transgenic lines. Presence of transcripts for both genes in transgenic plants was confirmed by semi-quantitative RT-PCR (Fig. 2) and considered as transgenic line with both the genes. Two individual transgenic lines of each *PaSOD* (S15 and S26) and *RaAPX* (A18 and A20) were analysed and data was presented. Out of a number of transgenic lines having both the genes, the line 18O and 19C was selected for further analysis and data was presented.

Thus generated single and double transgenics *Arabidopsis* plants were found tolerant when exposed to cold (4°C) stress (Fig. 3). The morphological and vegetative growth of transgenic plants was more or less similar to the wild plants under normal growth conditions i.e. 20±1°C and 16 h/8 h light regime (Fig. S2) whereas, under cold stress the transgenic plants showed better growth (Fig. 3). Effects of cold stress were clearly observed in the growth of wild type plants, where leaves started yellowing which shows the symptoms of chlorosis with stunted growth but the transgenic plants showed better growth. *PaSOD* single transgenic plants (S26 and S15) and double transgenic plants (18O and 19C) were better than the *RaAPX* transgenic plants (A20 and A18). The same was confirmed by the accumulation of transcripts levels also (Fig. 2). *PaSOD* (S26, S15) and *RaAPX* (A20, A18) transgenic lines exhibited expression of *PaSOD* and *RaAPX*, respectively, whereas 18O and 19C exhibited expression of both genes under normal conditions. A low level of transcripts was observed for plants without stress at 20°C. Further, significant over accumulation of transcripts of transgenes were observed in single as well as in double transgenic plants under cold stress (Fig. 2).

PaSOD and *RaAPX* induced cold stress tolerance alters *Arabidopsis* root growth and biomass

Root growth and development pattern was also studied during cold stress. After 192 h of cold stress in the single PaSOD transgenics (S26, S15) and double transgenic lines (18O, 19C) showed increased root growth biomass (Fig. 4). But in A20 and A18 the APX transgenic Arabidopsis plants showed decreased root growth when compared to WT. In order to determine if this difference existed in young seedlings at the in vitro stage, root growth on normal medium without any stress was also measured. A marginal increase in root length was observed in between WT and transgenic plants under normal growth conditions (Fig. S3). Initially, WT and transgenic plants exhibited similar root growth under cold stress, but WT root growth rate was lower than that of transgenic, which had longer roots after a prolonged culture period (10 d). Transgenic plants efficiently tolerated the stress and exhibited better root growth and biomass production in cold stress conditions (Table 1). Transgenic *Arabidopsis* plants over expressing *RaAPX* perform more or less similar in certain aspects of vegetative growth like rosette diameter, number of leaves; reproductive growth in terms of number of pods, length and number of seeds per silique, with the transgenic over expressing both *PaSOD* and *RaAPX* genes under cold stress conditions (Table 1). But overall, the transgenic plants over expressing both the genes performed better under cold stress.

Reduced accumulation of ROS contents under cold stress

Accumulation of ROS may cause damage to many bio molecules of the cells. As the transgenic lines were more tolerant to cold stress than WT, an attempt was made to compare accumulation of ROS in their leaf samples at the end of the cold stress by staining with NBT. Histo-chemical staining of leaves from WT and transgenic plant with NBT revealed that it could also be stained ROS without cold stress also (Fig. 5). The figure showed that the transgenic lines had slightly lower ROS accumulation

Figure 4. Root growth of WT and transgenic plants under cold stress. A). Photograph showing root growth of transgenic plants after cold stress (192h). B). Graphical representation of root length of transgenic plants after cold stress (192 h). Values are the representation of Mean±SE of three biological replicates. Transgenic plants developed better root system and thus exhibited better stress tolerance.

relative to WT without stress. Cold resulted in significantly higher levels of ROS accumulation in WT leaves whereas the transgenic lines S26, S15, A20, A18, 18O and 19C low levels accumulation as evidenced by the lower intensity of the blue colour (Fig. 5). As expected, the double transgenic line 18O and 19C showed notable less blue colour than *PaSOD* (S26, S15) and *RaAPX* (A20, A18) transgenic plants. This implies that less ROS was accumulated in these transgenic lines.

Accumulation of *PaSOD* and *RaAPX* in transgenic *Arabidopsis* improved the cold stress tolerance

As the main objective of the present study was to improve the scavenging of ROS by the over expression of supplemental

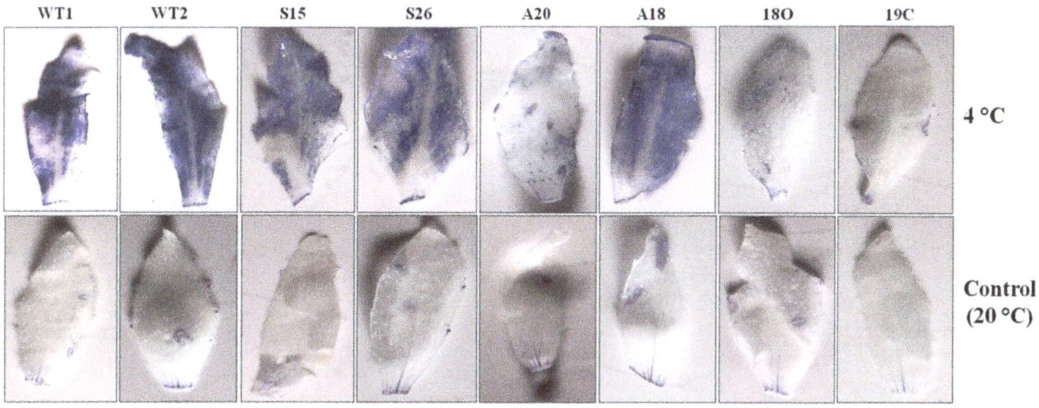

Figure 5. Histochemical staining assay of ROS accumulation with nitro blue tetrazolium (NBT) in the wild type and transgenics under cold stress. Leaves from potted plants that have been subjected to cold stress for 96 h were used for the accumulation of ROS.

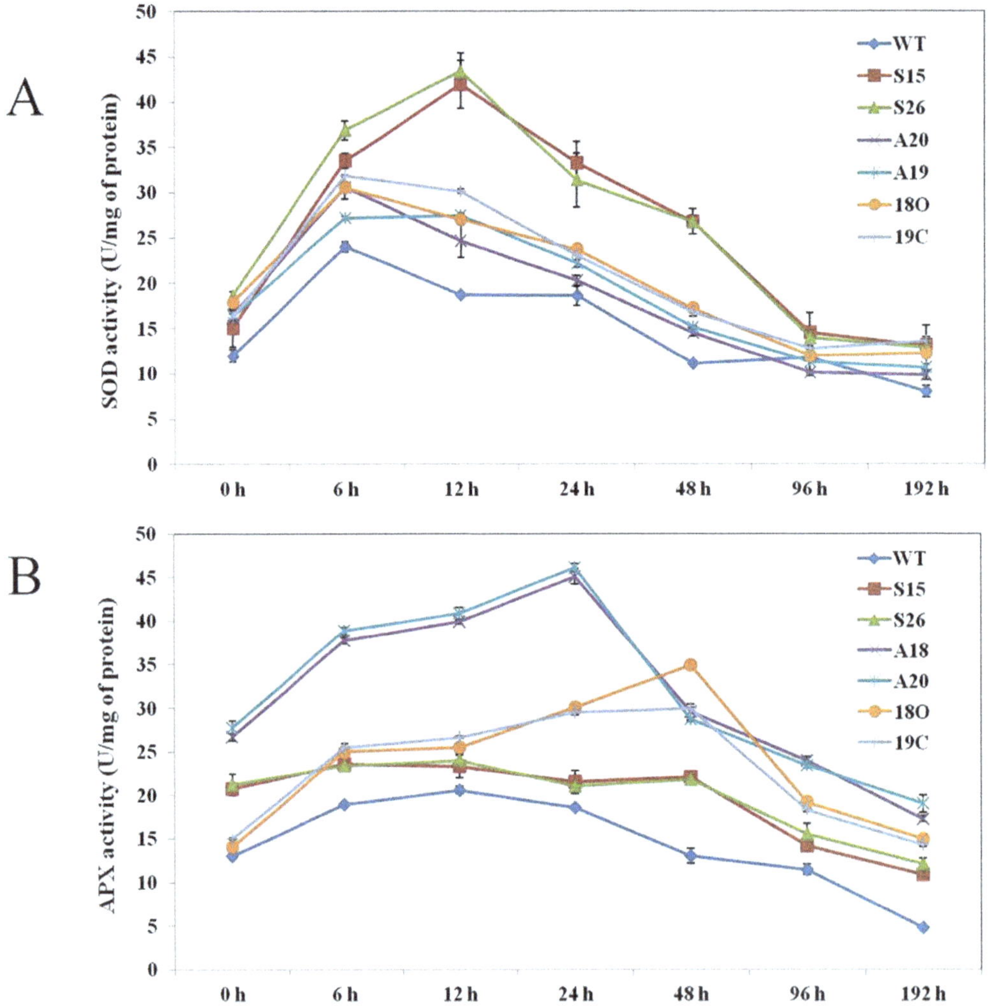

Figure 6. Estimation of total enzyme activities during cold stress both in WT and transgenics, A) superoxide dismutase and B) ascorbate peroxidase activity. Error bars represents ± SE of mean of three biological replicates.

antioxidant enzymes to the native ones, we checked the expression of the transgenes in the transgenics during cold stress. Superoxide dismutase and ascorbate peroxidase enzyme activities were estimated from the leaf samples of WT and transgenic plants at different time points during cold stress i.e. from 0–192 h under cold stress. As expected total SOD and APX activities were significantly higher in transgenic plants when compared to WT plants even under normal growth conditions, whereas 1.6, 1.2 and 1.3 folds increase in SOD levels was observed in SOD (S26, S15), APX (A20, A18) and double transgenic SOD and APX (18O, 19C) plants, respectively throughout 192 h of cold stress (Fig. 6). In the case of ascorbate peroxidase, 1.7-fold increase was recorded in transgenic plants A20 and A18 plants over WT plants under normal growth conditions and an increase of 2 fold was recorded throughout the cold stress. In double transgenic lines (18O and 19C) 1.3 folds higher activity was recorded. Abrupt elevation of these enzyme activities were recorded during the initial 12 h of cold in WT and all the transgenic lines and consequently a gradual decrease was recorded between 12 and 96 h, after which the minimal levels were maintained to sustain the growth and development of plants. Improved growth in terms of increase in biomass during cold stress in transgenic plants might be because of presence of the additional copy of antioxidant gene(s), implies increase in the activity of corresponding enzymes. Superoxide dismutase and ascorbate peroxidase enzyme activities were also estimated from the leaf samples of WT and transgenic plants at different time points without cold stress for comparisons. Slightly increased levels of SOD and APX activities were recorded in all the transgenic plants when compared to WT (Fig. S4).

Over expression of antioxidant enzymes protects cellular membranes from freezing injury

Stress accumulated ROS causes lipid peroxidation wherein the lipids in the cell membranes are damaged. Cold-induced cell membrane injuries in leaf tissues of transgenic lines were quantified by estimating the electrolyte leakage during *PaSOD* and *RaAPX* up regulation induced cold stress tolerance (Fig. 7). Chilling induced ion leakage was more in WT plants, when compared to transgenic plants. Whereas no difference was observed under normal conditions i.e. 20°C (Fig. S5). The relative electrolyte conductivity (REC) of transgenic plants was significantly lower than that of wild-type plants during cold stress. The results demonstrated that membrane damage was more in WT plants than that in the transgenic plants against cold stress. The ion leakage increased up to 48 h exposure to cold in *PaSOD*,

RaAPX and double transgenic lines, but after that there was a decline. After prolonged exposure to cold the ion leakage is more prominent in WT (Fig. 7), whereas transgenics were less affected by the cold. It seems that the over expressing antioxidant enzymes might be protecting the membrane damage.

Cold stress induces increase in aminoacid, soluble sugars and chlorophyll contents in transgenics

An increase in proline and simple sugars against stress is commonly known phenomena in *Arabidopsis* and other plants [28,29,30]. Under normal conditions, more or less similar amounts of proline content was recorded in all the plants, but higher levels of proline were found under stress in the transgenic lines when compared to corresponding WT plants (Fig. 8). Similar trend was observed in the case of soluble sugars under cold stress (Fig. 8). Overall, these findings suggested that the content of proline and soluble sugars, which are known osmo-protectants, accumulated more in transgenic plants under cold stress, thus providing better stress tolerance and better growth.

Chlorophyll content is susceptible to cold stress and usually continuously declines with cold stress [31]. In this study, chlorophyll content in the wild-type plants decreased at 4°C with the prolonged periods of cold stress. However, the transgenic plants have higher chlorophyll content than the WT plants under stress (Fig. 8). Thus cold tolerance of transgenic plants over expressing *PaSOD*, *RaAPX* and double transgenic plants showed increased chlorophyll accumulation.

Cold stress tolerance by simultaneous over-expression of *PaSOD* and *RaAPX* is because of the accumulation of lignin in the vascular bundles

Anatomical investigation of vascular changes during cold stress by using Confocal and scanning electron microscopy revealed that there was a dramatic alteration in lignin accumulation in transgenic plants as compared to the WT (Fig. 9A). Inter-fascicular lignification was evident under cold stress in single and double transgenics, whereas inflorescence stem of WT showed lignin deposition in the vascular bundles only (Fig. 9). Lignification of the inter-fascicular arcs was conspicuous along the entire stem and showed vascular bundles connected by lignified arcs in all transgenic lines, but more conspicuous in S26, S15 and 18O, 19C under stress condition only (Fig. 9). In addition, xylem elements of vascular bundles of the transgenic lines were larger than that of the WT plants.

The thickness of the walls of vascular bundles of the transgenic and WT were more or less the same (3μ) under normal conditions, whereas under cold stress the transgenic showed thicker walls. The thickness was more in double transgenics (18O and 19C) followed by *PaSOD* transgenics (S26 and S15) and *RaAPX* transgenics (A20 and A18) i.e. 12, 8 and 4 μ, respectively (Table 2).

PaSOD and *RaAPX* induced cold tolerance modulated lignin biosynthesis pathway in transgenic *Arabidopsis*

Higher vascular lignin thickenings were observed in transgenics, it was assumed that increased H_2O_2 levels might be responsible for these accumulations. Lignin is a polymer where different subunits of lignin (4-hydroxy cinnamyl aldehyde, coniferyl alcohol and sinapoyl alcohol) are produced via phenylpropanoid pathway and polymerized by peroxidase (*AtPRX9GE*) in to lignin. To understand the possible alterations in lignin biosynthesis pathway with the simultaneous over expression of *PaSOD and RaAPX* in transgenics, the expression profiles of different genes of this pathway namely *COMT*, *CAD*, *4CL*, *TRYT*, *PXR*, *C4 H*, *CcOMAT*, *C3 H*, *HCT*, *LAC*, *UGT*, *F5 H* and phenylalanine ammonia-lyase 1 (*PAL1*) were assessed by real time PCR. Generated expression profile data confirmed that in transgenic lines an up-regulation of the expression of all these genes tested was recorded under cold stress as compared to the WT. The heat map shows an overview of expression profiles of the phenylpropanoid pathway genes, and their interactions are shown in Fig. 10. Almost all the genes tested showed a basal level expression in

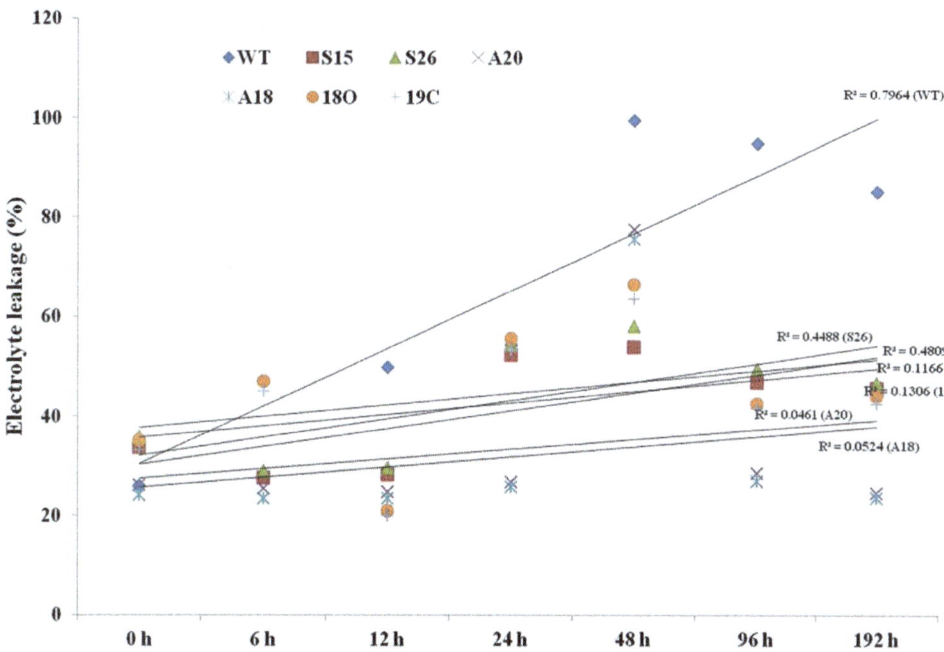

Figure 7. Change in relative electrolyte conductivity (electrolyte leakage) during cold stress in WT and transgenic plants. Error bars represents ± SE of mean of three biological replicates.

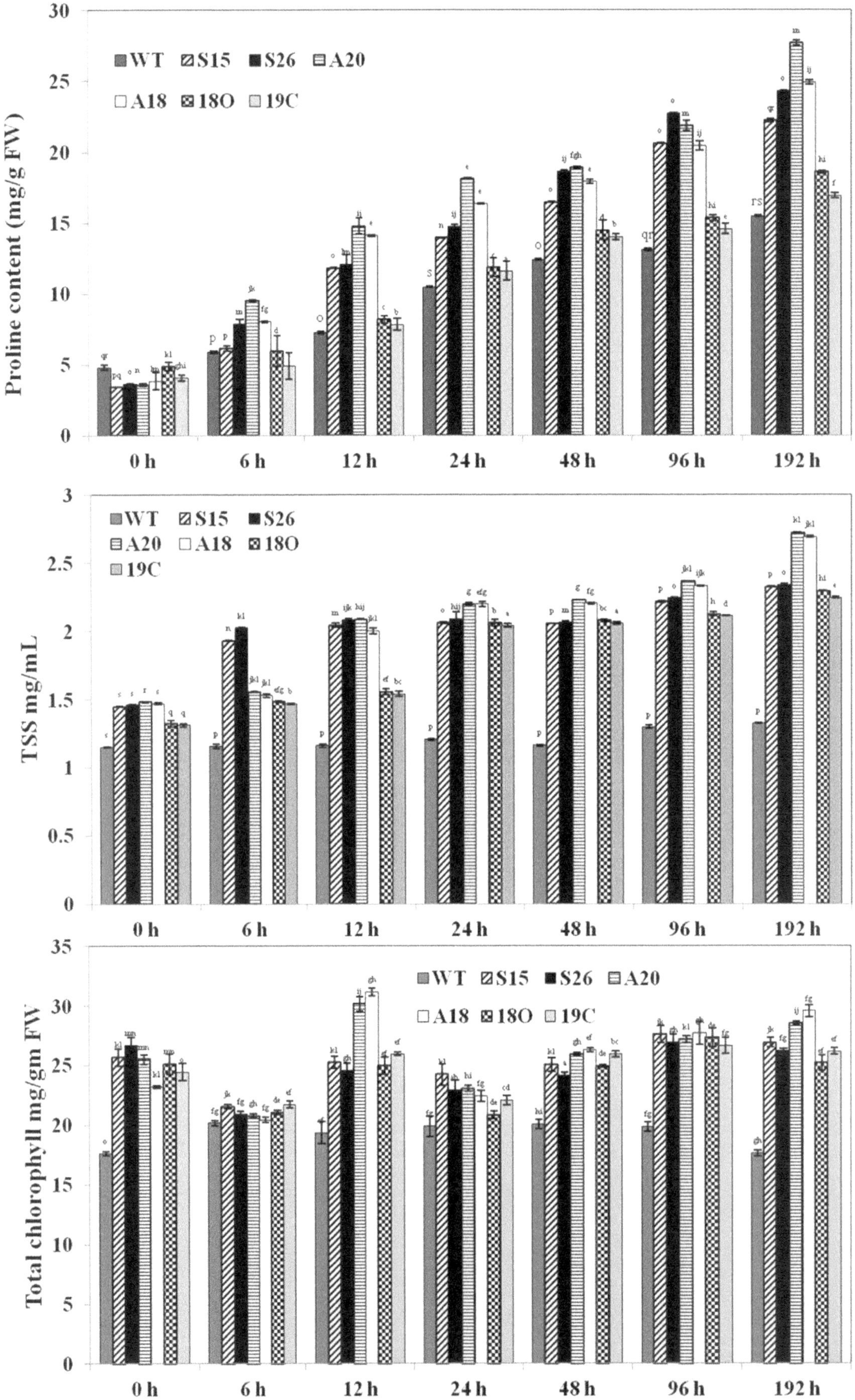
Proline content (mg/g FW)
WT S15 S26 A20
A18 18O 19C
TSS mg/mL
WT S15 S26
A20 A18 18O
19C
Total chlorophyll mg/gm FW
WT S15 S26 A20
A18 18O 19C
0 h
6 h
12 h
24 h
48 h
96 h
192 h

Figure 8. Estimation of different stress tolerance parameters in transgenics during cold stress (at 4°C). (A) Estimation of Proline content (B) Estimation of soluble sugars and (C) estimation of chlorophyll content. In all the cases, data is the mean of three biological replicates. Duncan's multiple range tests was used to establish the statistical significance at $P<0.05$.

PaSOD (S15 and S26) and *RaAPX* (A18 and A20) transgenics, whereas, comparatively higher expression was recorded except *CAD, COMT, C3 H, PXR1* in double transgenics (18O and 19C) under normal conditions (20°C). An up regulation was observed in the expression of the core phenylpropanoid pathway genes, including *PAL*, cinnamate 4-hydroxylase (C4 H), and 4-coumarate CoA ligase (4CL) under cold stress. After that, the lignin pathway divides into three specific branches for the formation of H, G and S lignin monomers. Most of the identified genes involved in these three pathways were up-regulated during cold stress. *F5 H, LAC4, C3 H, PAL1, PXR1, CoMT1* were the commonly up regulated both in *PaSOD* and double transgenic plants. In *RaAPX* over-expressed lines i.e. A18 and A20 a basal level expression was recorded in *4CL2, CAD2, COMT1* even in cold stress. Overall, the results indicated that the activation of the phenylpropanoid pathway led to the synthesis of lignin in response to cold stress by the simultaneous over expression of *PaSOD* and *RaAPX*.

Discussion

As compared to other living organisms, plants need to protect themselves from different environmental, abiotic and biotic stresses. These stresses cause oxidative damage, ion toxicity and disruption of cellular homeostasis through the production of reactive oxygen species (ROS). ROS induces various complex biochemical, molecular, cellular, and physiological changes in plants including damage of DNA, proteins and membrane lipids [14,32]. The responses to abiotic stresses can generally be separated into two categories: those that regulate signal transduction and gene expression in response to the stress, and those that protect against environmental stress by redirection of metabolism. Resistance to cold and salt stress requires a number of changes in gene expression [33,34,35]. Plants have developed a wide range of enzymatic and non-enzymatic mechanisms to scavenge ROS. Among the enzymatic methods, superoxide dismutase (SOD) is one of the crucial enzymes in the plant's defense mechanism that converts superoxide anion radicals ($O_2^{-\cdot}$) to hydrogen peroxide (H_2O_2), there by imparting protection against the harmful effects of highly reactive $O_2^{-\cdot}$ [14,36]. This in turn with the help of other antioxidant enzymes such as APX and catalase converts them in to

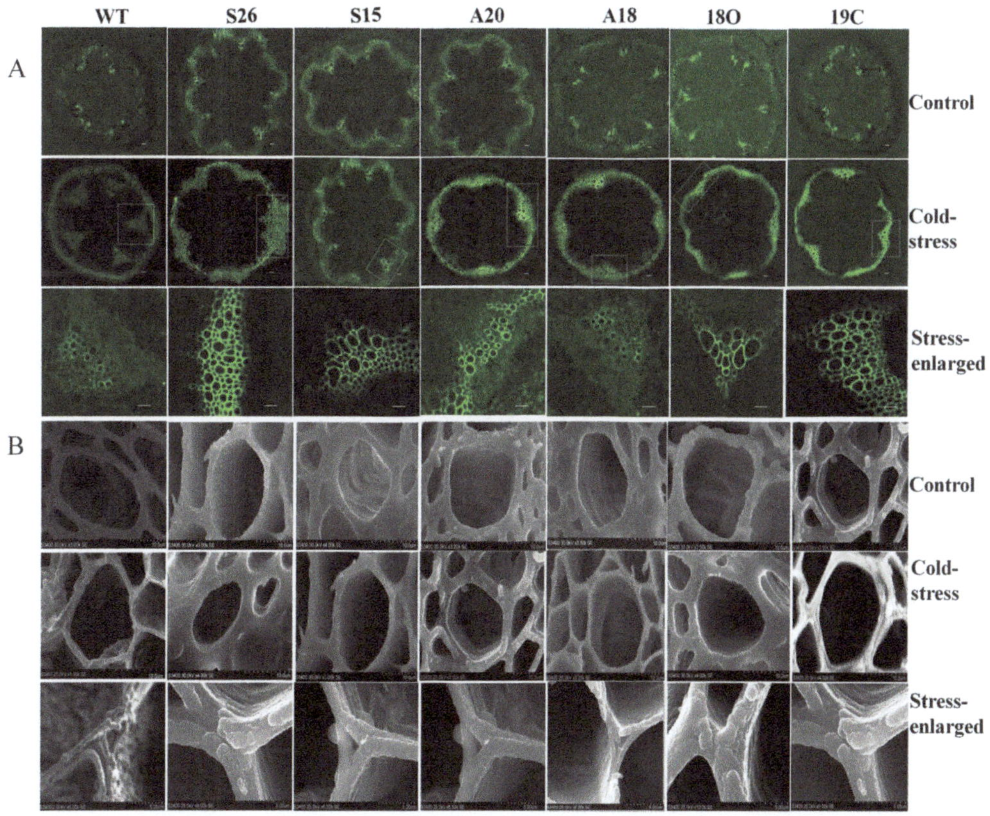

Figure 9. Pattern of lignin accumulation in transgenic *Arabidopsis* plants during cold stress. A) Transverse section of inflorescence stem of WT and transgenic lines (S26, S15, A20, A18, 18O and 19C) showing lignin accumulation pattern in WT and transgenic plants under cold stress. Close up view of the single vascular bundle of stress tissue is shown in the lower most panels (close view). Samples were collected after 192 h of the cold treatment. Lignin auto fluorescence was collected by excitation/emission wave lengths 488/505 nm by Confocal Laser Scanning Microscope. B) SEM analysis of vascular bundle walls showing thickness of lignin during cold stress.

Table 2. Details of the increase in lignifications of vascular bundle cell wall's width during cold stress in transgenic *Arabidopsis* over expressing *PaSOD*, *RaAPX* and both.

Plant type	Width of the vascular bundle wall (μ) ±SE	
	Control (20°C)	Cold stress (4°C)
WT	3±0.15	2±0.15
PaSOD (S26)	3±0.12	8±0.70
PaSOD (S15)	3±0.21	8±0.29
RaAPX (A20)	3±0.12	4±0.27
RaAPX (A18)	3±0.22	4±0.15
PaSOD + *RaAPX* (18O)	3±0.17	12±0.17
PaSOD + *RaAPX* (19C)	3±0.25	14±0.52

Each value is a mean of three separate biological replicates.

safer molecules like H_2O and O_2. The alpine climate is very harsh with low temperatures, the length of the daily photoperiod, and the quality of the incident light influences the plant growth in these areas. Because of having morphological, physiological and genetical adaptations to these stresses, the alpine vegetation can grow even in such harsh condition. So it is expected that these plants are known to have better enzymatic system whose activities are enhanced or are better adapted than their counterparts from plains. High altitude plants with efficient and active protective systems such as autoclavable superoxide dismutase from Potentilla [6], cell wall hydrolases with broad temperature range activity from Podophyllum [37], make these plants sustain under very low temperature conditions. Dissection of these adaptations in these plants has provided insights towards unraveling of the possible mechanisms of stress tolerance [14,11]. Cold stress was shown to enhance the levels of different ROS-scavenging enzymes [38,39] and accumulation of H_2O_2 in *Arabidopsis* callus cells was reported [40]. Scebba et al. [5] reported the induced expression of antioxidant enzymes SOD and CAT in wheat under cold stress. Over expression of *PaSOD* has improved salt and copper stress in *Arabidopsis thaliana* [11,12].

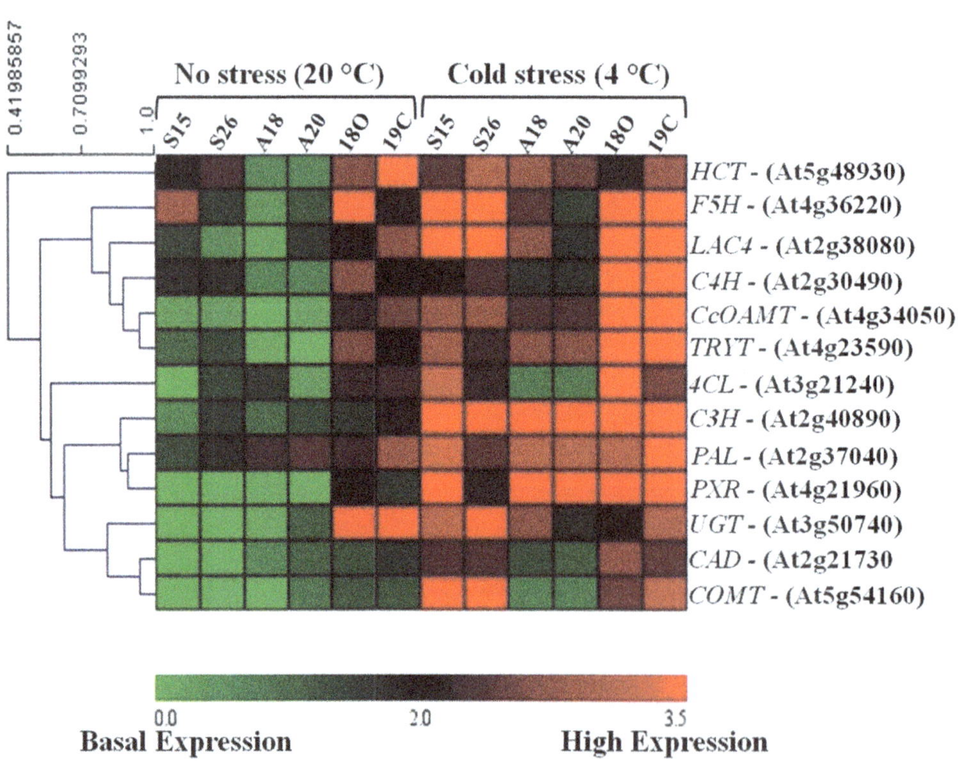

Figure 10. Heat Map of lignin biosynthesis pathway genes in transgenic *Arabidopsis* plants under cold stress (4°C). Phenylalanine ammonia lyase (*PAL1*), cinnamate 4-hydroxylase (*C4 H*), 4-coumarate:CoA ligase (*4CL-2*), ferulate 5-hydroxyase (*F5 H*), hydroxycinnamoyl-CoA transferase (*HCT*), *p*-coumarate 3-hydroxylase (*C3 H*), caffeoyl-CoA *O*-methyltransferase (*CCoAOMT1*), ferulate 5-hydroxylase (*F5 H*), caffeic acid *O*-methyltransferase (*COMT1*), cinnamyl alcohol dehydrogenase (*CAD2*), Laccase (*LAC4*), UDP-glucosyltransferase (*UGT 721 E1*) and Tyrosine ammonia lyase (*TRYT*) analysed in SOD, APX and double transgenic *Arabidopsis* plants.

Effects of low temperature stress includes reduced growth [41,42], wilting [43], chlorosis [44], impaired photosynthesis, increased electrolyte leakage across membranes [45] and impaired reproductive development which ultimately reduces the yield [46,47,48,49]. Reports were also available on successful manipulation of abiotic stresses with the simultaneous over expression of different antioxidant enzymes in Fescue plants, sweet potato and tobacco [50,51,52].

Present investigation dealt with the improvement of cold stress tolerance and studied the possible mechanism in *Arabidopsis* by the simultaneous over expression of antioxidant genes/enzymes *PaSOD* and *RaAPX* from the cold niche area plants. The present results showed that the WT plants were 'chilling sensitive' with stunted growth, reduced metabolism and with low biomass accumulation during cold stress. But the transgenic plants because of the manipulation of ROS detoxification mechanism by the over expression of antioxidant enzymes *PaSOD* and *RaAPX* performed better showing normal growth, more biomass, well developed root system, and higher yields in terms of more number of seeds per silique and thus have become 'chilling tolerant'. In the present results are in agreement with the work of Yuanyuan et al. [53], where he shows that increased soluble sugars play an important role during cold acclimation process, in transgenic *Arabidopsis* plants and also accumulation of soluble sugars and proline accumulation was recorded in response to cold stress. Soluble sugars exert their positive effects to protect plant cells from damage caused by cold stress through several ways, including serving as osmo-protectants, nutrient as well as interacting with the lipid bi-layer. These changes might also be responsible for maintaining the membrane stability and thus leading to reduced electrolyte leakage in the transgenic plants. Superoxide dismutase induced membrane damage tolerance was also reported by Kwon et al. [54] in transgenic tobacco plants. The photosynthetic performance of transgenic *Arabidopsis* plants was commendable with higher chlorophyll content throughout the cold stress whereas the WT plants were not able to sustain the lesser chlorophyll. Perhaps, this might be the reason in slower/stunted growth of WT during stress.

Lignins are the secondary metabolites, which maintain the structural integrity and provide mechanical strength to the stem and also impart hydrophobicity to vasculature allowing water and nutrient transportation [55]. The ROS catalysed by SOD and gets reduced into H_2O_2 which has a versatile role in the plant metabolic system. These molecules at low concentration acts as mediator of signalling pathways leading to stress acclimation at higher concentration it orchestrates the cellular damage and death [40]. Hydrogen peroxide acts as secondary messenger in stress signalling pathways due to its long half life and high permeability across membranes [56,57,58]. In addition to this the elevated levels of H_2O_2 have been shown to induce polymerization of lignin monomers (H-, S- and G- lignin) in to lignin [59].

Lignin induction has been correlated with cold, drought, or light stresses as well as mechanical injuries in a large number of different plant species, such as poplar, rice, pine, *Arabidopsis*, and soybean [60]. In our earlier study also, over expression of *PaSOD* induced lignin accumulation in the vascular bundles under NaCl stress which has been demonstrated to be the molecular basis for improved stress tolerance [14]. In the present study, anatomical investigation of vascular structures using Confocal and electron microscopy clearly showed that accumulation of lignin in the walls of the vascular strands of transgenic plants with cold stress. Transgenic plant expressing *PaSOD* alone and double transgenics were more lignified than *RaAPX* and WT plants. This high deposition of lignin in *PaSOD* transgenics, is probably due to the over-accumulation of H_2O_2 produced by scavenging of over produced superoxide ion ($O2^{-}$) under cold stress. Over-accumulation of H_2O_2 is toxic [40] and subsequent increased lignification may lead to choking of vascular system, thus cause damage and cell death in WT. In double transgenics, detoxification mechanism of increased levels of H_2O_2 can be achieved by the over expression of extra copy of downstream enzyme i.e. APX. This reduces the over-accumulated H_2O_2 levels and also probably distributes towards cell-signaling and lignifications and hence better adaptation to withstand the stress conditions. In the present study, the real time transcript analysis using qPCR also confirmed the up regulation of lignin biosynthetic pathway. A clear up regulation the core enzymes of the phenylpropanoid pathway namely PAL, C4 H and 4CL, a clear up regulation was observed in transgenics under cold stress. After that the pathway divides in to three directions, each is responsible for the production of individual components of monomers of lignin polymer. It is possible that the up regulated genes might have a significant role in antioxidant induced lignification during cold stress.

In conclusion, the simultaneous over expression of *PaSOD* and *RaAPX* from high altitude plants improved cold stress tolerance in *Arabidopsis*. This stress tolerance might be because of over expression of highly evolved and cold adapted genes, namely *PaSOD* and *RaAPX* as evidenced by the enzymes assays, which also detoxifies ROS as evidenced by the *in situ* ROS staining of leaves. Increased accumulation of TSS, proline in transgenic plants during cold stress, probably might be involved in protecting the membrane integrity as a result low electrolyte leakage was observed. Transgenic *Arabidopsis* plants had long roots with higher biomass without compromising yield, which is consistent with improved stress tolerance. No significant increase in all these parameters under normal growth condition (20°C) was also additional evidence to state that the cold stress tolerance is because of the over-expression of *PaSOD* and *RaAPX* only. This work clearly demonstrates that the modulation of endogenous ROS scavenging capacity against abiotic stresses can be successfully engineered by the simultaneous over expression of *PaSOD* and *RaAPX* in *Arabidopsis*. In addition, the results outlined the importance of the cytosolic antioxidant machinery in the cross-protection from multiple stresses in agriculturally important plants.

Supporting Information

Figure S1 Generalised scheme of lignin biosynthesis pathway. Each arrow shows a reaction in the pathway, and next to each arrow is the name of the enzyme that catalyzes the associated reaction, Abbreviations are listed as per pathway. CM, CHORISMATE MUTASE; PAT, PREPHENATE AMINOTRANSFERASE; AGT, AROGENATE DEHYDRATASE; PAL, PHENYLALANINE AMMONIA LYASE; C4H, CINNAMATE 4-HYDROXYLASE; 4CL, COUMARATE CoA LIGASE; CCR, CINNAMOYL-CoA REDUCTASE; CAD, CINNAMYL ALCOHOL DEHYDROGENASE; HCT, HYDROXYCINNAMOYL-CoA TRANSFERASE; C3H, *P*-COUMARATE 3-HYDROXYLASE; F5H, FERULATE 5-HYDROXYLASE; COMT, CAFFEIC ACID *O*-METHYLTRANSFERASE; UGT, UDP-GLUCOSYLTRANSFERASE; LAC, LACCASES; PXR, PEROXIDASES; CCoAOMT, CAFFEOYL-CoA 3-O-METHYLTRANSFERASE; H lignin, p-hydroxy phenyl lignin; G lignin, guaiacyl lignin monomers; S lignin, syringyl lignin monomers.

Figure S2 Growth of transgenic *Arabidopsis* plants under normal conditions (20°C) without cold stress. Five weeks old seedlings were transferred and allowed to grow at 20°C for 192 h and different stress tolerance parameters were estimated from the samples of these plants.

Figure S3 Root growth of WT and transgenic plants under normal conditions. A) Photograph showing root growth of transgenic plants without cold stress (192 h). B). Graphical representation of root length of transgenic plants without cold stress (192 h). Values are the representation of Mean±SE of three biological replicates. Transgenic plants developed better root system under normal growth conditions.

Figure S4 Estimation of total enzyme activities under normal growth conditions without cold stress both in WT and transgenics. A) superoxide dismutase and B) ascorbate peroxidase activity. Error bars represents ± SE of mean of three biological replicates.

Figure S5 Change in relative electrolyte conductivity (electrolyte leakage) in WT and transgenic plants under normal growth conditions. Error bars represents ± SE of mean of three biological replicates.

Table S1 Primer Sequence and PCR conditions for all the genes (including gene id and accession numbers) used for Semi-quantitative and Real Time expression analysis.

Acknowledgments

A.S., T.G. and V.D. acknowledge Fellowship by the CSIR, New Delhi, India. Authors thank Dr. Sanjay Kumar for providing cDNA of *RaAPX*. The manuscript represents CSIR-IHBT communication number 3576.

Author Contributions

Conceived and designed the experiments: PSA YS. Performed the experiments: AS VD TG. Analyzed the data: PSA YS. Contributed reagents/materials/analysis tools: AS VD TG PSA YS. Wrote the paper: PSA YS.

References

1. Scandalios JG (2005) Oxidative stress: molecular perception and transduction of signals triggering antioxidant gene defenses. Brazilian J. Med. Biol. Res. 38: 995–1014.
2. Hu D, Klann E, Theils E (2007) Superoxide dismutase and hippocampal function: age and isozymes matter. Antioxidant Redox. Signal. 9: 201–210.
3. Noctor G, Foyer CH (1998). Ascorbate and Glutathione: keeping active oxygen under control. Ann. Rev. Plant Physiol. Plant Mol. Biol. 49: 249–279.
4. Raven EL (2000) Peroxidase-catalyzed oxidation of ascorbate. Structural, spectroscopic and mechanistic correlations in ascorbate peroxidase. Subcell. Biochem. 35: 317–349.
5. Scebba F, Sebastiani L, Vitagliano C (1999) Protective enzymes against activated oxygen species in wheat (*Triticum aestivum* L.) seedlings: Responses to cold acclimation. J. Plant Physiol. 155: 762–768.
6. Sahoo R, Kumar S, Ahuja PS (2001) Induction of a new isozyme superoxide dismutase at low temperature in *Potentilla astrisanguinea* Lodd. variety *argyrophylla* (Wall.ex. Lehm) Griers. J. Plant Physiol.158: 1093–1097.
7. Yogavel M, Mishra PC, Gill J, Bhardwaj PK, Dutt S, et al. (2008) Structure of a superoxide dismutase and implications for copper-ion chelation. Acta Cryst. Biol. Cryst. D64: 892–901.
8. Kumar A, Dutt S, Bagler G, Ahuja PS, Kumar S (2012) Engineering a thermo-stable superoxide dismutase functional at sub-zero to >50°C, which also tolerates autoclaving. Sci. Rep. 2: 387: 1–8.
9. Wang Y, Ying Y, Chen J, Wang X (2004) Transgenic *Arabidopsis* overexpressing Mn-SOD enhanced salt-tolerance. Plant Sci. 167: 671–677.
10. Badawi GH, Kawano N, Yamauchi Y, Shimada E, Sasaki R, et al. (2004) Over-expression of ascorbate peroxidase in tobacco chloroplasts enhances the tolerance to salt stresses and water deficit. Plant Physiol. 121: 231–238.
11. Gill T, Kumar S, Ahuja PS, Sreenivasulu Y (2010b) Over-expression of *Potentilla* superoxide dismutase improves salt stress tolerance during germination and growth in *Arabidopsis thaliana*. J. Plant Genet. Transgen. 1: 1–10.
12. Gill T, Dogra V, Kumar S, Ahuja PS, Sreenivasulu Y (2012) Protein dynamics during seed germination under copper stress in *Arabidopsis* over-expressing *Potentilla* superoxide dismutase. J. Plant Res. 125: 165–172.
13. Pal AK, Acharya K, Vats SK, Kumar S, Ahuja PS (2013) Over-expression of *PaSOD* in transgenic potato enhances photosynthetic performance under drought. Biol. Planta. 57: 359–364.
14. Gill T, Sreenivasulu Y, Kumar S, Ahuja PS (2010a) Over-expression of superoxide dismutase exhibits lignifications of vascular structures in *Arabidopsis thaliana*. J. Plant Physiol. 167: 757–760.
15. Bechtold N, Ellis J, Pelletier GC (1993) *In planta* Agrobacterium mediated gene transfer by infiltration of adult *Arabidopsis* plants. C. R. Acad. Sci. Paris. 316: 1194–1199.
16. Murashige T, Skoog F (1962) A revised medium for rapid growth and bioassays with tobacco tissue cultures. Physiol. Plant. 15: 473–476.
17. Rozen S, Skaletsky H (2000) Primer3 on the WWW for general users and for biologist programmers. Methods Mol. Biol. 132: 365–386.
18. Shi LZ, Wang R, Huang G, Vogel P, Neale G, et al. (2011) HIF1α–dependent glycolytic pathway orchestrates a metabolic checkpoint for the differentiation of T_H17 and T_{reg} cells. J. Exp. Med. 208: 1367–1376.
19. Pfaffl MW, Horgan GW, Dempfle L (2002) Relative expression software tool (REST@) for group-wise comparison and statistical analysis of relative expression results in real-time PCR.Nucleic Acids Res. 30: e36.
20. Nakano Y, Asada K (1981) Hydrogen peroxide is scavenged by ascorbate-specific peroxidase in spinach chloroplasts. Plant Cell Physiol. 22: 867–880.
21. Bradford M (1976) A rapid and sensitive method for the quantitation of microgram quantities of protein utilizing the principles of protein dye binding. Analytical Biochem. 72: 248–254.
22. Beyer WF, Fridovich I (1987). Assaying for superoxide dismutase activity: some large consequences of minor changes in conditions. Anal. Biochem. 161: 559–566.
23. Lutts S, Kinet J, Bouharmont J (1996). NaCl induced senescence in leaves of rice (*Oryza sativa*) cultivars differing in salinity resistance. Annal. Bot. 78: 389–398.
24. Irigoyen JJ, Emerich DW, Sanchez-Diaz M (1992) Water stress induced changes in concentrations of proline and total soluble sugars in nodulated alfalfa (*Medicago sativa*) plants. Physiol. Plant. 84: 55–60.
25. Bates L, Waldren RP, Teare ID (1973) Rapid determination of free proline for water-stress studies. Plant and Soil. 39: 205–207.
26. Arnon DI (1949) Copper enzymes in isolated chloroplasts, polyphenoxidase in *Beta vulgaris*. Plant Physiol. 24: 1–15.
27. Jensen WA (1962) Botanical Histochemistry. p. 408. W. H. Freeman and Co., S. F. and London.
28. Koster KL, Lynch DV (1992) Solute accumulation and compartmentation during the cold acclimation of puma rye. Plant Physiol. 98: 108–113.
29. McKown R, Kuroki G, Warren G (1996) Cold responses of *Arabidopsis* mutants impaired in freezing tolerance. J. Exp. Bot. 47: 1919–1925.
30. Wanner LA, Junttila O (1999) Cold-induced freezing tolerance in *Arabidopsis*. Plant Physiol. 120: 391–399.
31. Gajewska E, Sklodowska M, Slaba M, Mazur J (2006) Effect of nickel on antioxidant enzyme activities, praline and chlorophyll contents in wheat shoots. Biol. Plant. 50: 653–659.
32. Munns R, Tester M (2008) Mechanisms of salinity tolerance. Ann. Rev. Plant Biol. 59: 651–681.
33. Fowler S, Thomashow MF (2002) *Arabidopsis* transcriptome profiling indicates that multiple regulatory pathways are activated during cold acclimation in addition to the CBF cold response pathway. Plant Cell 14: 1675–1690.
34. Zhu JK (2001) Plant salt tolerance. Trends Plant Sci. 6: 666–671.
35. Xiong L, Shumaker KS, Zhu JK (2002) Cell signaling during cold, drought and salt stresses. Plant Cell 14: 165–183.
36. Fridovich I (1975) Superoxide Dismutase. Ann. Rev. Biochem. 44: 147–159.
37. Dogra V, Ahuja PS, Sreenivasulu Y (2013) Change in protein content during seed germination of a high altitude plant *Podophyllum hexandrum* Royle. J. Proteom. 78: 26–38.
38. Prasad TK, Anderson MD, Martin BA, Stewart CR (1994) Evidence for chilling-induced oxidative stress in Maize seedlings and a regulatory role for hydrogen peroxide. Plant Cell 6: 65–74.
39. Sato Y, Murakami T, Funatsuki H, Matsuba S, Saruyama H, et al. (2001) Heat shock-mediated APX gene expression and protection against chilling injury in rice seedlings. J. Exp. Bot. 52: 145–151.
40. O'Kane D, Gill V, Boyd P, Burdon R (1996) Chilling, oxidative stress and antioxidant responses in *Arabidopsis thaliana* callus. Planta 198: 371–377.
41. Sowinski P, Rudzinska-Langwald A, Adamczyk J, Kubica I, Fronk J (2005) Recovery of maize seedling growth, development and photosynthetic efficiency after initial growth at low temperature. J. Plant Physiol. 162: 67–80.
42. Rymen B, Fiorani F, Kartal F, Vandepoele K, Inze D, et al. (2007) Cold nights impair leaf growth and cell cycle progression in maize through transcriptional changes of cell cycle genes. Plant Physiol. 143: 1429–1438.
43. Bagnall D, Wolfe JOE, King RW (1983) Chill-induced wilting and hydraulic recovery in mung bean plants. Plant Cell Environ. 6: 457–464.

44. Yoshida R, Kanno A, Kameya T (1996) Cool temperature-induced chlorosis in rice plants (II. Effects of cool temperature on the expression of plastid-encoded genes during shoot growth in darkness). Plant Physiol. 112: 585–590.
45. Nayyar H, Chander K, Kumar S, Bains T (2005a) Glycine betaine mitigates cold stress damage in Chickpea. Agro. Sustain. Deve. 25: 381–388.
46. Kaur G, Kumar S, Nayyar H, Upadhyaya HD (2008) Cold stress injury during the pod- filling phase in chickpea (*Cicer arietinum* L.): effects on quantitative and qualitative components of seeds. J. Agro. Crop Sci. 194: 457–464.
47. Ohnishi S, Miyoshi T, Shirai S (2010) Low temperature stress at different flower developmental stages affects pollen development, pollination, and pod set in soybean. Environ. Exp. Bot. 69: 56–62.
48. Kumar S, Malik J, Thakur P, Kaistha S, Sharma KD, et al. (2011) Growth and metabolic responses of contrasting chickpea (*Cicer arietinum* L.) genotypes to chilling stress at reproductive phase. Acta Physiol. Plant. 33: 779–787.
49. Suzuki K, Nagasuga K, Okada M (2008). The chilling injury induced by high root temperature in the leaves of rice seedlings. Plant Cell Physiol. 49: 433–442.
50. Lee HS, Ahsan N, Lee KW, Kim DH, Lee DG, et al. (2007) Simultaneous overexpression of both Cu-Zn superoxide dismutase and ascorbate peroxidase in transgenic tall fescue plants confers increased tolerance to a wide range of abiotic stresses. J. Plant Physiol. 164: 1626–1638.
51. Lu YY, Deng XP, Kwak SS (2010) Over expression of CuZn superoxide dismutase (CuZn SOD) and ascorbate peroxidase (APX) in transgenic sweet potato enhances tolerance and recovery from drought stress. African J. Biotech. 9: 8378–8391.
52. Faize M, Burgos L, Faize L, Piqueras A, Nicolas E, et al. (2011) Involvement of cytosolic ascorbate peroxidase and Cu/Zn-superoxide dismutase for improved tolerance against drought stress. J. Exp. Bot. 62: 2599–2613.
53. Yuanyuan M, Yali Z, Jiang L, Hongbo S (2009) Roles of plant soluble sugars and their responses to plant cold stress. African J. Biotech. 8: 2004–2010.
54. Kwon SY, Jeong YJ, Lee HS, Kim JS, Cho KY, et al. (2002) Enhanced tolerances of transgenic tobacco plants expressing both superoxide dismutase and Ascorbate peroxidase in chloroplasts against methyl viologen-mediated oxidative stress. Plant, Cell Environ. 25: 873–882.
55. Barriere Y, Riboulet C, Mechin V, Pichon M, Cardinal AJ, et al. (2007) Genetics and genomics of lignification in grass cell walls based on maize as a model system. G3-Genes Genome Genet. 1: 133–156.
56. Neill SJ, Desikan R, Clarke A, Hurst RD, Hancock JT (2002) Hydrogen peroxide and nitric oxide as signalling molecules in plants. J. Exp. Bot. 53: 1237–1247.
57. Huang C, Li J, Ke Q, Leonard SS, Jiang BH, et al. (2002) Ultraviolet-induced phosphorylation of p70(S6 K) at Thr(389) and Thr(421)/Ser(424) involves hydrogen peroxide and mammalian target of rapamycin but not Akt and atypical protein kinase C. Cancer Res. 62: 689–697.
58. Yang T, Poovaiah BW (2002) Hydrogen peroxide homeostasis: activation of plant catalase by calcium/calmodulin. Proc. Nat. Acad. Sci. USA 99: 4097–4102.
59. Denness L, McKenna JF, Segonzac C, Wormit A, Madhou P, et al. (2011) Cell wall damage-induced lignin biosynthesis is regulated by a reactive oxygen species- and jasmonic acid-dependent process in *Arabidopsis*. Plant Physiol. 156: 1364–1374.
60. Moura JCMS, Bonine CAV, de Oliveira Fernandes Viana J, Dornelas MC, Mazzafera P (2010) Abiotic and biotic stresses and changes in the lignin content and composition in plants. J. Integrat. Plant Biol. 52: 360–376.

12

The *SbMT-2* Gene from a Halophyte Confers Abiotic Stress Tolerance and Modulates ROS Scavenging in Transgenic Tobacco

Amit Kumar Chaturvedi, Manish Kumar Patel, Avinash Mishra*, Vivekanand Tiwari, Bhavanath Jha*

Discipline of Marine Biotechnology and Ecology, CSIR-Central Salt and Marine Chemicals Research Institute, Bhavnagar, Gujarat, India

Abstract

Heavy metals are common pollutants of the coastal saline area and *Salicornia brachiata* an extreme halophyte is frequently exposed to various abiotic stresses including heavy metals. The *SbMT-2* gene was cloned and transformed to tobacco for the functional validation. Transgenic tobacco lines (L2, L4, L6 and L13) showed significantly enhanced salt (NaCl), osmotic (PEG) and metals (Zn^{++}, Cu^{++} and Cd^{++}) tolerance compared to WT plants. Transgenic lines did not show any morphological variation and had enhanced growth parameters viz. shoot length, root length, fresh weight and dry weight. High seed germination percentage, chlorophyll content, relative water content, electrolytic leakage and membrane stability index confirmed that transgenic lines performed better under salt (NaCl), osmotic (PEG) and metals (Zn^{++}, Cu^{++} and Cd^{++}) stress conditions compared to WT plants. Proline, H_2O_2 and lipid peroxidation (MDA) analyses suggested the role of *Sb*MT-2 in cellular homeostasis and H_2O_2 detoxification. Furthermore *in vivo* localization of H_2O_2 and O_2^-; and elevated expression of key antioxidant enzyme encoding genes, *SOD*, *POD* and *APX* evident the possible role of *SbMT-2* in ROS scavenging/detoxification mechanism. Transgenic lines showed accumulation of Cu^{++} and Cd^{++} in root while Zn^{++} in stem under stress condition. Under control (unstressed) condition, Zn^{++} was accumulated more in root but accumulation of Zn^{++} in stem under stress condition suggested that *Sb*MT-2 may involve in the selective translocation of Zn^{++} from root to stem. This observation was further supported by the up-regulation of zinc transporter encoding genes *NtZIP1* and *NtHMA-A* under metal ion stress condition. The study suggested that *SbMT-2* modulates ROS scavenging and is a potential candidate to be used for phytoremediation and imparting stress tolerance.

Editor: Keqiang Wu, National Taiwan University, Taiwan

Funding: This study was supported by the Council of Scientific and Industrial Research (CSIR, www.csir.res.in), Government of India, New Delhi (BSC0109–SIMPLE; Senior Research Fellowship to AKC and VT). The funders had no role in study design, data collection and analysis, decision to publish, or preparation of the manuscript.

Competing Interests: The authors have declared that no competing interests exist.

* Email: avinash@csmcri.org (AM); bjha@csmcri.org (BJ)

Introduction

Abiotic stresses such as salinity, drought, temperature and heavy metals have significant effect on the agricultural production over the years [1]. Most often, plants may encounter these abiotic stresses simultaneously, resulting in to the substantial loss in agricultural productivity. Over the past two centuries, increased industrial and anthropogenic activities viz. mining, irrigation with sewage effluents/waste waters, use of phosphate fertilizers are the major sources of metal contamination to soil [2–3]. Salinity, drought and heavy metals have similar consequences ensuing oxidative stress, disturbances in ionic homeostasis and generation of reactive oxygen species (ROS) [1,4].

Metallothioneins (MTs) are the group of polypeptides which can bind with heavy metals through their thiols group via chelation and involved in the homeostasis of essential metals (Cu and Zn) and cellular detoxification of nonessential metals (Cd and Hg) [5,6]. Characteristics of metallothioneins are the cysteine (Cys) residue which is the basis of its classification. Based on the arrangement of Cys residues MTs are divided in two classes [7]. Plant MTs belong to Class II and further classified into four types (type 1 to 4) based on the position and allocation of cysteine residues [7]. Spatial expression of all four MT type has been reported which is localized to root, stem, leaves and developing seeds [7,8] and show differential tolerance to different metals. Although there are several reports regarding the possible role of MTs in plants but its physiological role is still not fully conclusive because of difficulties in its isolation and stability [7,9]. Over-expression of MTs in various model systems like Arabidopsis, tobacco, yeast and *E. coli* established its functional role in homeostasis and tolerance to Cu, Zn and Cd [10,11,12,13], high salinity, drought, low temperature, heavy metal ions, abscisic acid (ABA) and ethylene [12,14]. Besides above, MTs are reported in the inhibition of root elongation [15], fruit ripening, seed development [7] and provide disease resistance against pathogen attack [16].

Heavy metals affect the plant at cellular, biochemical and molecular level causing the oxidative stress. These toxic metals generate free radicals and reactive oxygen species (ROS) which damages the cell membrane, nucleic acids and photosynthetic pigments [17]. The equilibrium between ROS generation and quenching is prerequisite for the cell survival which is maintained by the intricate anti-oxidative system comprising of enzymatic [superoxide dismutase (SOD), catalase (CAT), peroxidase (POD)]

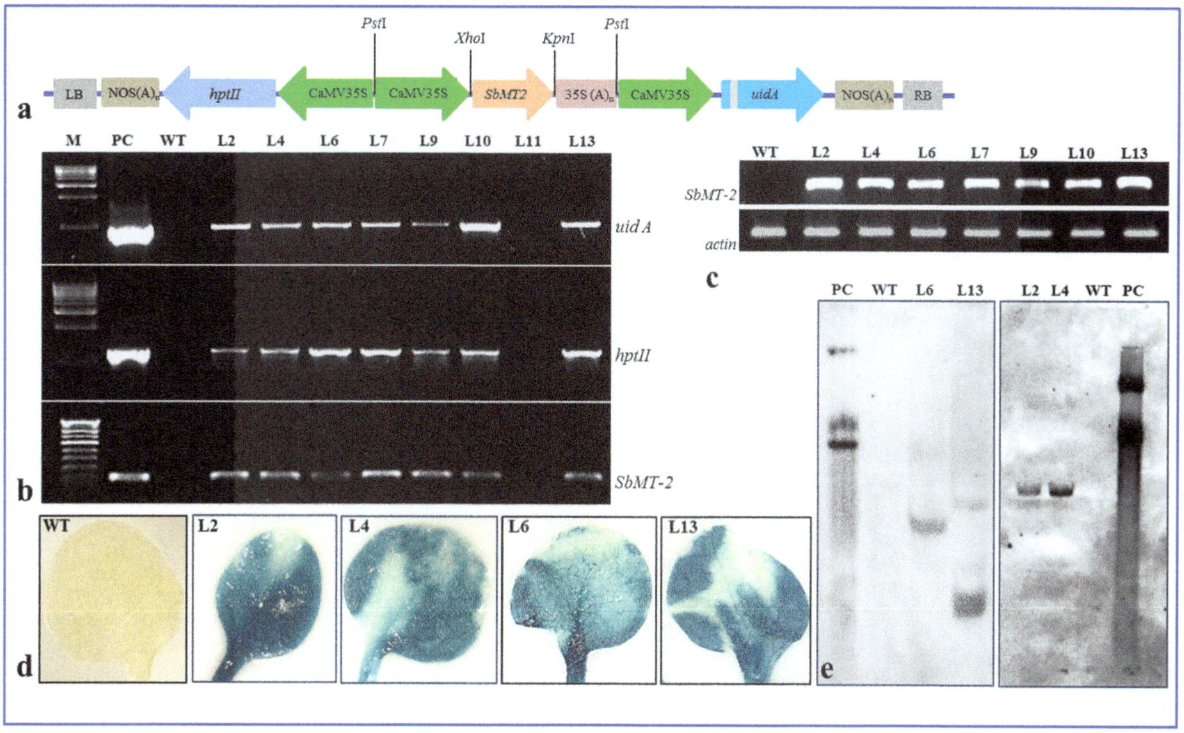

Figure 1. Analyses of transgenic tobacco plants. (a) Schematic map of plant expression gene construct pCAMBIA1301-*SbMT-2*. (b) Confirmation of transgenic lines by PCR amplification of *uidA*, *hptII* and *SbMT-2* genes (Lane M is molecular weight markers, Lane PC is positive control of PCR using plasmid *pCAMBIA1301-SbMT-2* as template, Lane WT is wild type control plant (non-transgenic) and Lane L is transgenic lines). (c) Over-expression of the *SbMT-2* gene in transgenic lines compared to wild-type control plants (Lane WT is wild type control plant and Lane L is transgenic lines) analyzed by semi-quantitative Rt-PCR, where *actin* was used as internal gene control. (d) Histochemical GUS staining of WT and transgenic plants. (e) Southern blot analysis (Lane PC is positive control plasmid *pCAMBIA1301-SbMT-2*, Lane WT is wild type control plant (non-transgenic) and Lane L is transgenic lines).

and non-enzymatic (ascorbate, glutathione and phenolic compounds) systems [4,12,17,18]. Under stress conditions the activity of this antioxidative system gets increased due to the increase of free radical formation [17].

Though MTs are known to be involved in abiotic/metal stresses and have been reported from several plant species, there are a few reports on halophytes under heavy metal stress [19]. *Salicornia brachiata* is an extreme halophyte and frequently exposed to heavy metals in coastal areas. The plant has nutritional value, unique oligosaccharide profile and requires NaCl for *in-vitro* regeneration [20–22]. The plant is considered as a model for the study of tolerance mechanism and several stress responsive genes have been isolated and characterized [13,23–28]. In our previous study, *SbMT*-2 gene was isolated and physiological role was determined in *E. coli* which showed its role in the homeostasis and detoxification of Zn, Cu and Cd ions [13]. The gene (*SbMT-2*) was considered as a potential candidate to be utilized for the genetic engineering of plants for phytoremediation of heavy metals and stress tolerance. Therefore in the present study *SbMT*-2 gene was transformed into tobacco for the functional validation. Biochemical, physiological and morphological responses of transgenic plants over-expressing *SbMT-2* gene under different abiotic stresses (metals- Zn, Cu & Cd; NaCl and osmotic stresses) were studied.

Results

Over-expression of the *SbMT*-2 gene

A plant expression vector (Figure 1a), harboring *SbMT*-2 gene driven by CaMV35S promoter was constructed and transformed to tobacco plants. Fourteen independent transgenic lines (T_0) were raised and preliminarily screened by PCR using gene specific primers (data not shown). These lines were grown in containment facility and seeds (T_1) were collected. T_1 Seeds were germinated on hygromycin containing media and selected T_1 transgenic lines were confirmed by PCR using gene specific primers (Figure 1b). Expected size of amplicon was found in all lines except L11. Based on *SbMT*-2 (Figure 1c) gene expression level, four independent T_1 transgenic lines; L2, L4, L6 and L13 were selected for the further morpho-physio-biochemical analyses. All transgenic lines showed high *gus* and *SbMT*-2 gene expression (Figure 1c and 1d). Southern blot confirmed single and double copy gene integration to L2, L4, L6 and L13 lines, respectively (Figure 1e).

Growth parameters in T_1 transgenic lines

Transgenic lines L2, L4, L6 and L13 showed high percentage of seed germination in all stress treatments (metals- Zn, Cu and Cd; NaCl and PEG) compared to wild type (WT) plants (Figure S1). Transgenic lines grown under different stress treatments were comparatively healthier than WT plants (Figure 2) and showed enhanced growth parameters under stress conditions (Figure 3). Shoot and root lengths were found significantly higher in transgenic lines compared to WT plants in all stress treatments. Similarly, fresh and dry weights of transgenic lines were significantly higher than WT plants. Among all stress treatments studied, transgenic lines showed better growth in osmotic stress followed by metal and NaCl stress.

Figure 2. Comparative morphology of WT and T_1 transgenic lines under different stress condition. WT and four independent T_1 transgenic tobacco lines (L2, L4, L6 and L13) grown under control, salt (200 mM NaCl), osmotic (10% PEG), Zn (5 mM), Cu (0.2 mM) and Cd (0.2 mM) stress for 3 weeks.

Leaf senescence assay and chlorophyll content

Stress tolerance of T_1 transgenic lines was studied by leaf disc senescence assay and chlorophyll estimation. In leaf discs assay, stress-induced necrosis resulted in the decrease of chlorophyll content was lower in the *SbMT-2* over-expressing lines compared to WT plants (Figure 4). The damage caused by stress treatments was visualized by the degree of bleaching of leaf tissues and it was evident that the transgenic plants (L2, L4, L6 and L13) had a better ability to tolerate osmotic stress followed by metal and NaCl stress (Figure 4). The chlorophyll content in the WT plants reduced significantly with stress treatments while the transgenic lines (L2, L4, L6 and L13) retained higher chlorophyll contents than WT plants (Figure 5a).

Relative water content, Electrolyte leakage and membrane stability index

The relative water content (RWC) was found almost similar (insignificant difference) in transgenic and WT plants under unstressed condition (control), however RWC was significantly higher in transgenic lines compared to WT plants under all stress treatments (Figure 5b). Membrane permeability measured by electrolyte leakage (EL), was found significantly stable in transgenic lines, which showed reduced electrolyte leakage compared to WT plants under stress condition (Figure 5c). Similarly, membrane stability index (MSI) of transgenic lines was significantly higher compared to WT under stress treatments (Figure 5d). RWC, EL and MSI evident that transgenic lines are thriving well in stress conditions compared to WT plants.

Localization of O_2^- & H_2O_2 and H_2O_2, proline & MDA content analysis

Leaves of transgenic lines and WT plants, subjected to different stress treatments showed *in vivo* localization of O_2^- and H_2O_2 (Figure 6). It was observed that leaves of WT plants showed more accumulation of O_2^- and H_2O_2 content compared to transgenic lines under various stress conditions. Transgenic lines showed significantly lower accumulation of H_2O_2, proline and MDA contents under stress condition compared to WT plants (Figure 7). Lower content of H_2O_2 exhibited the better ROS system in transgenic lines compared to WT plants. Lower accumulation of proline and MDA revealed that transgenic lines have higher osmoprotectants and lower lipid peroxidation, respectively compared to WT plants under stress condition.

Ion content analysis

Metallothioneins play a vital role in ion homeostasis and heavy metal binding. Metal ion sequestration was analyzed by ICP, which revealed high accumulation of metal ions in transgenic lines compared to WT plants (Figure 8). It was observed that metal ion contents were approximately same (insignificant difference) in WT and transgenic lines at control (unstressed) condition (Figure 8a). Under stress condition, metal ion accumulation was significantly higher in transgenic lines compared to WT plants (Figure 8b). Among different tissues, Zn accumulation was higher in shoot followed by root and leaves. However, high Cu and Cd contents were detected in roots followed by stem and leaves. Furthermore high affinity of *Sb*MT-2 was observed with Zn ions. Compared to control condition, Na^+ contents increased in transgenic as well as WT plants under NaCl stress treatment, however K^+ contents were decreased (Figure S2a). Similarly, transgenic lines showed higher K^+/Na^+ ratio compared to WT plants under NaCl stress condition (Figure S2b).

Transcript analysis of genes encoding metal transporters and antioxidative enzymes

Ion content analysis suggested that *Sb*MT-2 may involve in the selective translocation of Zn^{++} from root to stem. To further support this observation, expression analysis of zinc and heavy metal transporters encoding genes (*NtZIP1* and *NtHMA-A*) were studied under the metal stress (Zn, Cu and Cd) treatments. Expression of genes *NtZIP1* and *NtHMA-A* involved in heavy metal (Zn) translocation were up-regulated in transgenic lines compared to WT plants (Figure 9). Among different metal stress, the *NtZIP1* gene showed maximum up-regulation under Zn stress (Figure 9a). Transgenic lines, L2, L4, L6 and L13 showed about 10-, 13-, 8- and 6-fold expression compared to their respective control plants. However, *NtZIP1* expression was also up-regulated in WT plant. Similarly, the *NtHMA-A* gene was also up-regulated under metal stress compared to WT and control condition (Figure 9b). In contrast to other metal stress, inconsistent expression was observed among transgenic lines and WT under Cu stress treatment.

Common antioxidant enzymes involved in ROS scavenging mechanism are superoxide dismutase (SODs), peroxidase (POD) and ascorbate peroxidase (APX). Transcript analysis reveals that expression of the *NtSOD* gene increased concomitantly with stress treatments (except NaCl) in transgenic lines compared to WT plants (Figure 10a). About 3.5-fold expression was observed in L4

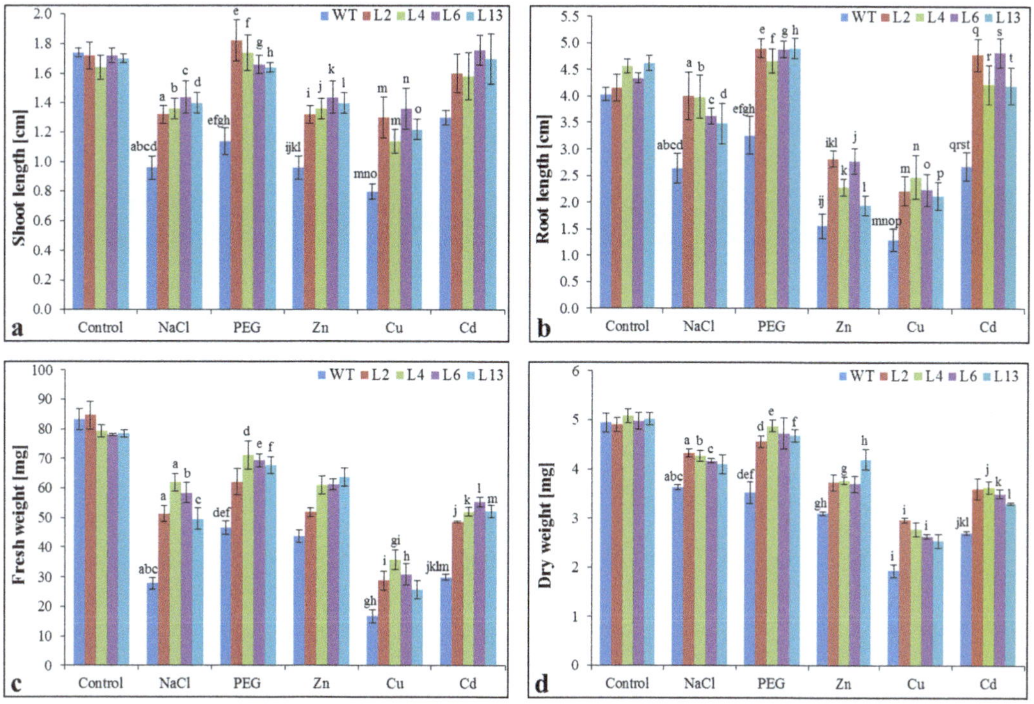

Figure 3. Analysis of *SbMT-2* transgenic tobacco lines under different stress condition. Comparison of (a) shoot length (cm), (b) root length (cm), (c) fresh weight (mg) and (d) dry weight (mg) of wild type (WT) and T_1 transgenic lines (L2, L4, L6 and L13) grown under control, salt (200 mM NaCl), osmotic (10% PEG), Zn (5 mM), Cu (0.2 mM) and Cd (0.2 mM) stress for 3 weeks. Graph represents the mean ± SE (of three replicates; n = 3) followed by similar letters are significantly different according to Tukey HSD at *P<0.05*.

line under Zn and osmotic stress, while about 2-fold expression was detected under Cu and Cd stress compared to control condition (untreated plants). Remaining L2, L6 and L13 lines showed higher relative expression of *NtSOD* gene under different stress conditions compare to WT plants. Surprisingly down-regulation of gene was found under NaCl stress in transgenic as well as WT plants compared to control (untreated) plants. Relative fold expression of *NtPOD* gene was found higher in transgenic lines compared to WT plants under metal stress treatments (Figure 10b). Transgenic line L4 showed maximum gene expression about 10 -fold under Zn, Cu and Cd stress. However, *NtPOD* gene expression was down-regulated in WT and transgenic lines under NaCl and PEG stress treatments. Elevated *NtAPX* gene expression was observed in transgenic lines (except L2 under NaCl and PEG) compared to WT plants under NaCl, osmotic and metal stress conditions (Figure 10c).

Discussion

Our previous study revealed that heterologous expression of *SbMT–2* gene not only augments zinc and copper tolerance but also increases metal ion sequestration in *E. coli* cells [13]. Further, in this study, *in planta* functional validation of *SbMT–2* gene has been elucidated. The *SbMT–2* gene was over-expressed in tobacco and its functional role was studied in different abiotic stresses *viz.* salt (200 mM NaCl), osmotic (10% PEG) and metals; Zn (1 mM), Cu (0.5 mM) and Cd (0.5 mM). Morphological variation was not observed in transgenic lines over-expressing *SbMT*-2 gene compared to WT plants (Figure 2). Transgenic lines showed higher seed germination percentage, increased root length, shoot length, fresh weight (FW), dry weight (DW) and chlorophyll content under different stress conditions compared to WT plants (Figure 3 and 5a) which reveal that *SbMT-2* leads to overcome the

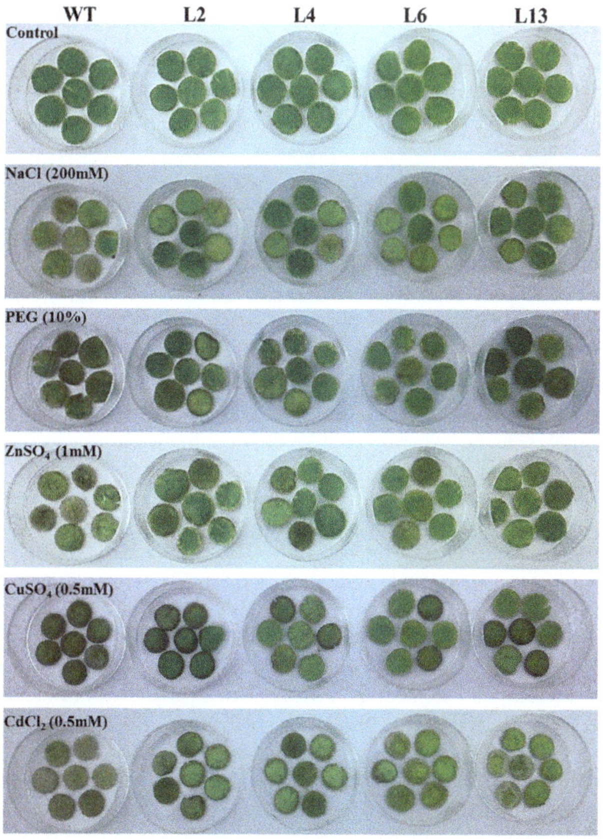

Figure 4. Leaf disc assay of transgenic tobacco lines (T_1) for different stress tolerance. Leaf discs of WT, L2, L4, L6 and L13 transgenic lines respectively were floated in different stress solution for 8 days.

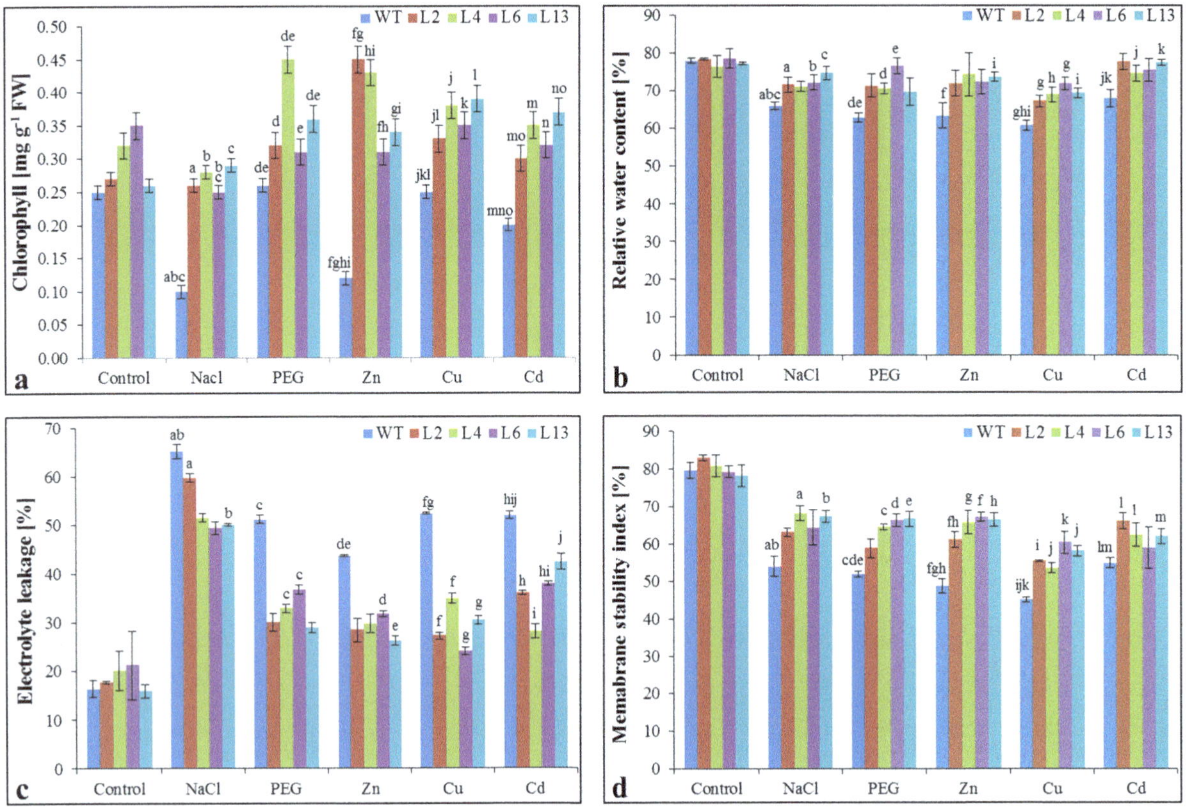

Figure 5. Analysis of *SbMT-2* transgenic tobacco lines under different stress condition. Comparison of (a) Chlorophyll content, (b) Relative water content, (c) Electrolyte leakage and (d) Membrane stability index of wild type (WT) and T_1 transgenic lines (L2, L4, L6 and L13) grown in hydroponic (control) and treated with salt (200 mM NaCl), osmotic (10% PEG), Zn (1 mM), Cu (0.5 mM) and Cd (0.5 mM) stress. Graph represents the mean ± SD (of three replicates; n = 3) followed by similar letters are significantly different according to Tukey HSD at $P<0.05$.

deleterious effect of salt, drought and metal stresses. Previously, ectopic expression of *BjMT2* gene conferred the increase in percentage seed germination, fresh weight (FW), dry weight (DW) and chlorophyll content in *Arabidopsis* under Cu and Cd stress [15]. Similar results were also observed under salt, drought, Zn and Cu stress in transgenic tobacco plants over-expressing clustered *OsMT1e-P* metallothionein gene [29].

Salt, drought and heavy metal stress are multigenic in nature, causing osmotic stress and thus create physiological drought conditions for plants. Transgenic lines displayed increase in relative water content under stress (Figure 5b) which indicates that *SbMT-2* gene may help in water retention to counteract the osmotic shock. Abiotic stress causes the perturbation in metabolic balance of the cell, resulting in the enhanced production of ROS, which in-turn damages cell membranes, nucleic acids and chloroplast pigments [15,30]. Electrolyte leakage (EL) and membrane stability index (MSI) is the indicator of cell membrane stability. In the present study electrolyte leakage was lower in transgenic lines while MSI was higher under stress condition (Figure 5c-d) which suggests the role of *SbMT-2* gene in cell membrane protection. It was further supported by lipid peroxidation analysis, H_2O_2 and proline quantification, in which high MDA, H_2O_2 and proline contents were found in WT plants compared to transgenic lines under stress condition (Figure 7). These result evident the possible role of *Sb*MT-2 in ROS scavenging/detoxification and maintaining the cellular homeostasis during stress condition. Furthermore, *in vivo* localization of H_2O_2 and O_2^- under different abiotic stress (Figure 6), confirmed the role of *SbMT*-2 gene in ROS scavenging.

Although role of metallothionein genes in ROS scavenging and detoxification have studied [12,14,16,29,31] but the molecular mechanism of ROS detoxification/scavenging is still unknown. ROS, being a signaling molecule play an important role in development and regulation of different metabolic processes whereas ROS toxicity resulted into the oxidative damage to the cell membrane and its components [30]. Plant cell maintains a stringent regulation over its production and scavenging. In *Arabidopsis* about 150 genes are involved in the homeostasis of ROS which comprised of ROS scavenging enzymes and ROS producing proteins [30]. Superoxide dismutase (SOD), peroxidase (POD) and ascorbate peroxidase (APX) are important ROS scavenging enzymes activated under different stress conditions to maintain the ROS homeostasis. Expression of genes encoding these ROS scavenging enzymes was studied in transgenic lines and compared with WT plants (Figure 10) to confirm the role of the *SbMT-2* gene in ROS scavenging and detoxification. Higher relative fold expression of *NtSOD* and *NtPOD* genes were found in transgenic lines under stress conditions, however similar result was also observed with the *NtAPX* gene except under salt stress. Moreover, high expression was detected in metal stress compared to salt and osmotic stress, which may be because of availability of cofactors of the enzymes.

Metallothioniens are involved in essential-metal homeostasis and impart protection against heavy metal toxicity by sequestra-

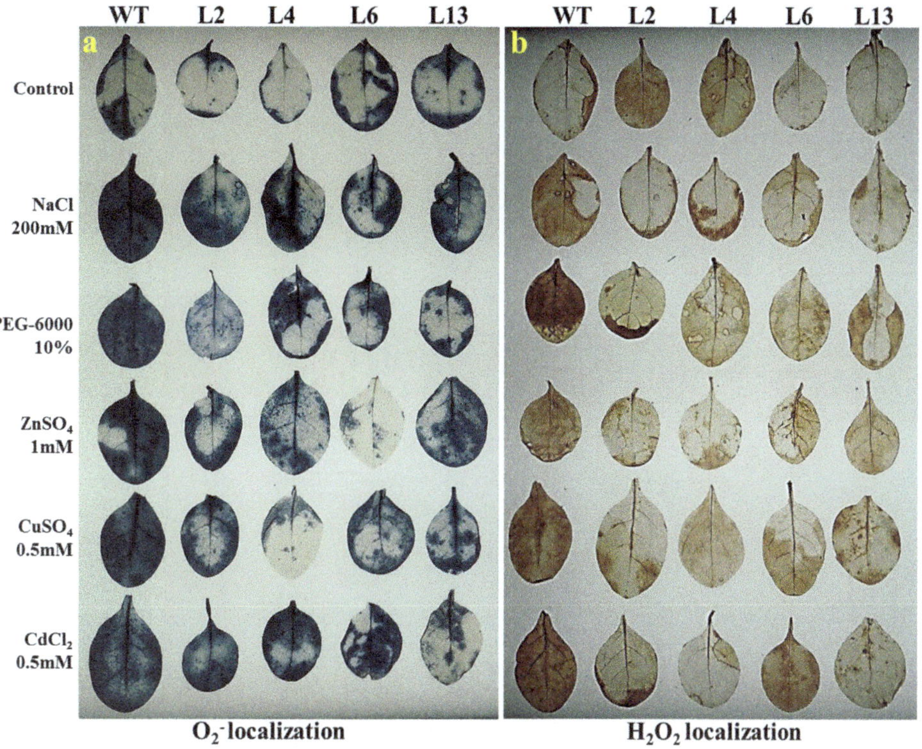

Figure 6. *In vivo* localization of O_2^- and H_2O_2 in WT and transgenic lines under different stress condition. Comparison of (a) O_2^- and (b) H_2O_2 localization in wild type (WT) and T_1 transgenic lines (L2, L4, L6 and L13) grown in hydroponic (control) and treated with salt (200 mM NaCl), osmotic (10% PEG), Zn (1 mM), Cu (0.5 mM) and Cd (0.5 mM) stress.

tion [6,9]. Plants maintain a high K^+/Na^+ ratio to combat with the deleterious effect of salinity. In present study transgenic lines over-expressing *SbMT-2* gene showed a high K^+/Na^+ ratio under stress compared to WT plants and control condition (Figure S2). It provides further evidence that *Sb*MT-2 may have role in ionic homeostasis and detoxification of H_2O_2 and thus impart salt tolerance. Transgenic lines showed accumulation of Cu^{++} and Cd^{++} in root compared to stem and leaves under stress condition. However, the *SbMT-2* gene over-expression leads to the accumulation of Zn^{++} in stem compared to control condition where it was accumulated more in roots (Figure 8). Zinc ion accumulation and tolerance varies among plant species and depends on MT type. The result exhibits *Sb*MT-2 may involve in the selective translocation of Zn^{++} from root to stem. Furthermore, the result also suggests that *Sb*MT-2 is involved in Cu and Cd binding and accumulation rather than translocation, as observed with Zn ions. Previously, it was reported that *Sb*MT2 protein exhibited high binding affinity and sequestration for Zn^{++} compared to Cu and Cd ions [13].

In order to support this observation, expression of type 1B heavy metal–transporting P-type ATPases (P_{1B} ATPase) transporter and zinc specific transporter encoding gene *NtHMA-A* and *NtZIP1* were analyzed under metal stress treatments (Figure 9). Up-regulation of genes in transgenic lines under metal stress especially, zinc compared to control and WT plants provides a supporting evidence that expression of *SbMT-2* gene may influence metal transporters and thus translocation of Zn was observed in the study. The qPCR analysis revealed higher up-regulation of these genes in transgenic lines compared to WT plants (about 6 to 10 fold up-regulation of *NtZIP1* gene in transgenic lines compared to about 2.5 fold of WT plants) under metal stress condition and it might be due to introgression of *SbMT-2* gene. Furthermore, higher down-regulation of these genes (*NtZIP1*and *NtHMA-A*) in transgenic plants compared to WT was observed under de-stress treatments, performed by re-culturing the plants in un-stressed conditions (Table S1). Results further suggest that the *SbMT2* gene may influence the expression of these genes.

It was observed that expression of MT type 2 gene *PsMT(A1)* of *Pisum sativum* enhanced metal tolerance in white poplar and accumulation of zinc and copper (in leaves and roots) respectively [11]. Expression of seed specific MT gene *MT4a* in *Arabidopsis* increased Cu^{++} accumulation but did not show any response under Zn stress [10]. The *OsMT1a*, a type 1 metallothionein gene, involved in the Zn^{++} accumulation and thus provides tolerance to the transgenic rice [12]. *Elsholtzia haichowensis* metallothionein 1 (*Eh*MT1) over-expression in tobacco plants enhances copper tolerance and accumulation in root cytoplasm [32] however, expression of *BjMT2* gene in *A. thaliana* showed copper and cadmium tolerance [15]. Sequestration, translocation and thereby accumulation are the important mechanism used by plants for the phytoextraction [33] and therefore *SbMT-2* gene may be utilized for the phytoremediation.

Conclusion

In conclusion, the present study provides an useful insight that *Sb*MT-2 may involve in maintaining the cellular homeostasis by modulating ROS scavenging/detoxification during stress conditions and thus impart tolerance to salt and osmotic stress. It was observed that *Sb*MT-2 provides protection against heavy metal toxicity by metal ions accumulation and may be involved in the

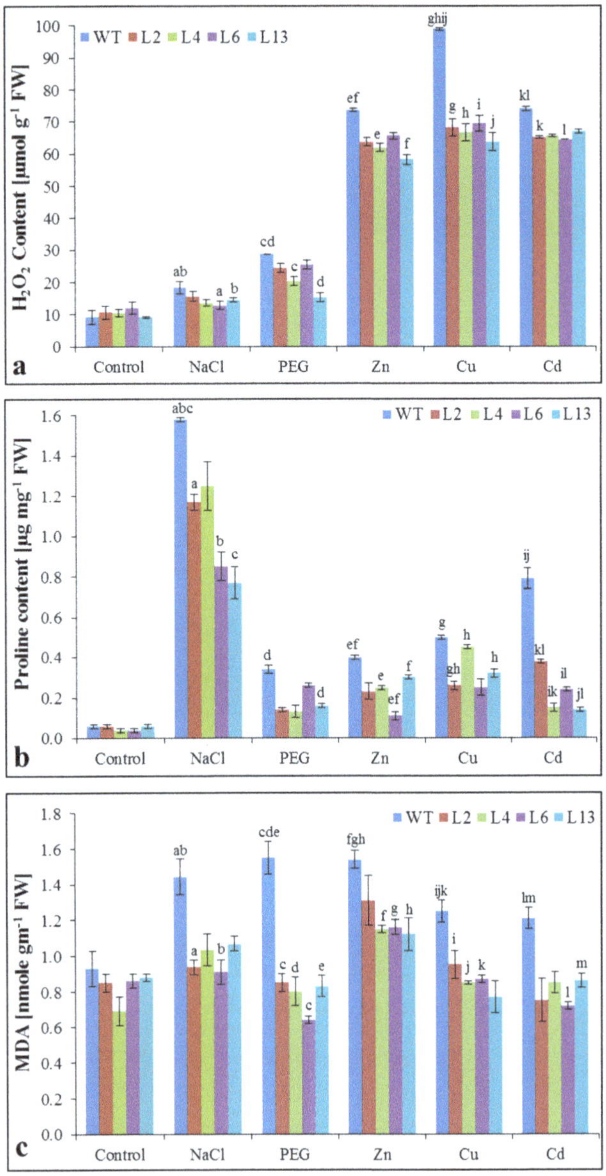

Figure 7. Estimation of H_2O_2, proline and MDA in transgenic tobacco lines under different stress condition. Comparison of (a) H_2O_2, (b) Proline and (c) MDA content of wild type (WT) and T_1 transgenic lines (L2, L4, L6 and L13) grown in hydroponic (control) and treated with salt (200 mM NaCl), osmotic (10% PEG), Zn (1 mM), Cu (0.5 mM) and Cd (0.5 mM) stress. Graph represents the mean ± SD (of three replicates; n = 3) followed by similar letters are significantly different according to Tukey HSD at *P<0.05*.

selective translocation of Zn^{++} from roots to stem (Figure 11). It is speculated that *SbMT-2* gene is a potential candidate for introgression to crop plants for imparting stress tolerance and phytoremediation.

Methods

Construction of plant transformation vector and tobacco transformation

The *SbMT-2* cDNA was amplified using forward (MT-2F: 5′-CTCGAGATGTCTCTTGCTTGTGGTGGTAAC-3′) and reverse MT-2R: 5′- GGTACCTCATTTTGCAAGTG-CAAGGGTTG -3′) primers containing *Xho*I and *Kpn*I sites respectively. The amplicon was digested with *Xho*I/*Kpn*I and cloned in pRT101 vector [34]. Thereafter, the gene cassette "CaMV35S-*SbMT-2*" was excised with *Pst*I and cloned into the pCAMBIA1301 vector. The resulting vector was mobilised into *Agrobacterium tumefaciens* (EHA105) and further transformed to tobacco (*Nicotiana tabacum* cv. xanthii) plants according to the standard protocol [35].

Confirmation of transgenic lines and expression of transgene

Total genomic DNA was isolated from leaves of transgenic and wild type (WT; control/untransformed) tobacco plants (T_1) using Qiagen DNeasy plant mini kit and quantified by Nanodrop Spectrophotometer (ND1000, Wilmington, USA). The integration of transgene in different lines was confirmed by PCR analyses using the *SbMT-2* gene, reporter gene *gus* and selection marker gene *hptII* specific primers (Table 1). Presence and expression of the *SbMT-2* gene in selected lines (L2, L4, L6 and L13) were confirmed by Southern blot and semi-quantitative Rt-PCR, respectively.

Genomic DNA (20 µg) from WT and transgenic lines L2, L4, L6 and L13 was digested with *Hind*III, separated on agarose (0.7%) by electrophoresis and transferred to a Hybond N^+ membrane (Amersham Pharmacia, UK) using alkaline transfer buffer (0.4 N NaOH with 1 M NaCl). DNA blot was hybridized with PCR-generated probe for the *uidA* gene labeled with DIG-11-dUTP following pre-hybridization and hybridization carried out at 42°C overnight in DIG EasyHyb buffer solution [36–39]. The hybridized membrane was detected by using CDP-Star chemiluminescent as substrate, following manufacturer user guide (Roche, Germany) and signals were visualized on X-ray film after 30 min.

Total RNA was extracted from transgenic tobacco (T_1) and wild-type plants using Qiagen RNeasy plant mini kit and the cDNA was made by reverse transcription with Superscript II RT (Invitrogen, USA). The actin gene was used as an internal control and both genes (*SbMT-2* and actin) were amplified using gene specific primers (Table 1). Histochemical GUS assay was performed with leaves as described by Jefferson [40].

Analyses of transgenic plants under different abiotic stress

Transgenic and wild type tobacco lines were maintained under controlled containment facility. Further morphological and physio-biochemical analyses were performed with T_1 transgenic lines under different abiotic stress treatments.

Growth parameters

Seeds of transgenic lines (L2, L4, L6 and L13) and WT plants were germinated on MS medium supplemented with 200 mM NaCl, 10% PEG 6000, 5 mM $ZnSO_4$, 0.2 mM $CuSO_4$ or 0.2 mM $CdCl_2$ under culture room conditions and the percentage of seed germination was calculated after 20 days.

Different growth parameters were measured under same stress conditions using T_1 seedlings. Seeds of transgenic lines were germinated on the MS media supplemented with 20 mg/l hygromycin, while seeds of WT plant were germinated on MS media only. WT and hygromycin positive T_1 seedlings were transferred to MS medium supplemented with 200 mM NaCl, 10% PEG, 5 mM $ZnSO_4$, 0.2 mM $CuSO_4$ or 0.2 mM $CdCl_2$ after eight days and grown for further 21 days. Growth

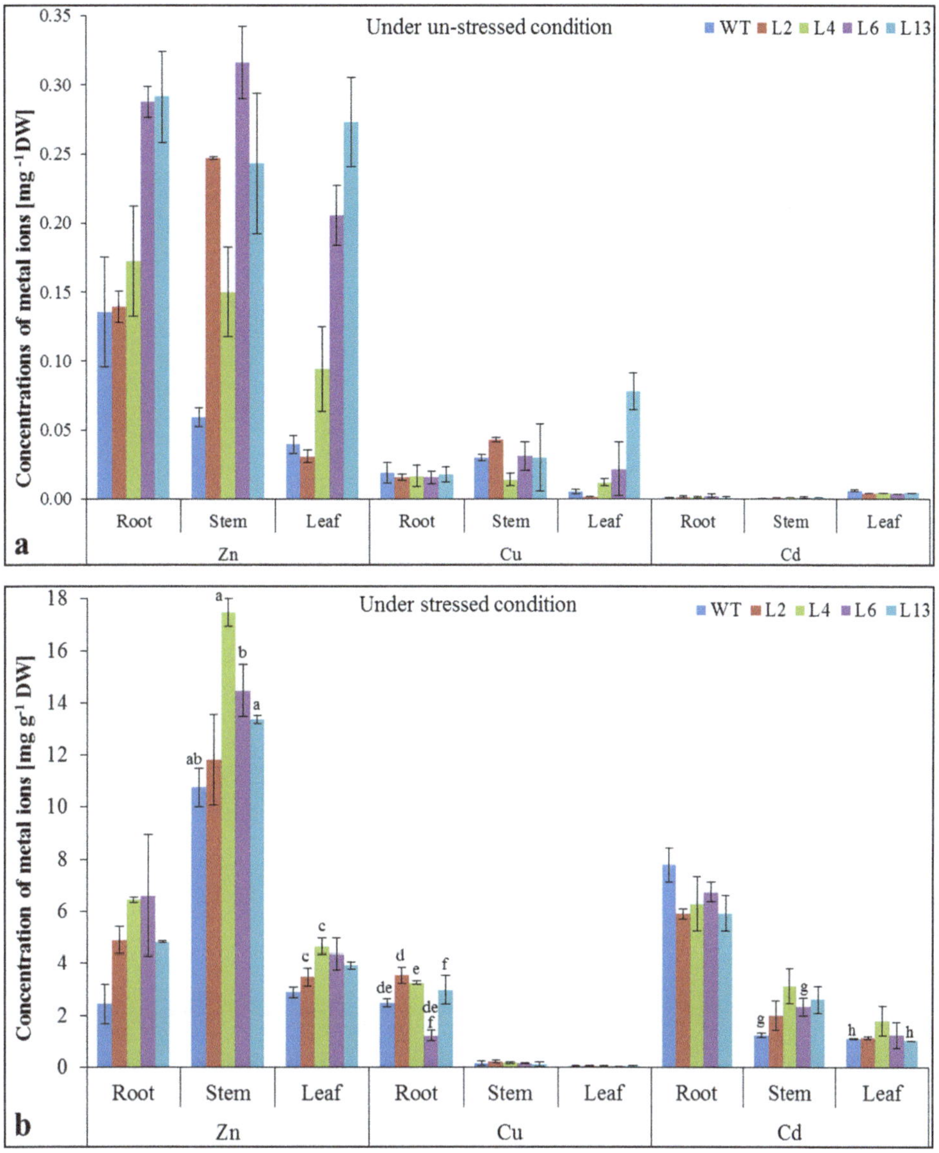

Figure 8. Metal ions content in transgenic tobacco lines under different stress condition. Comparison of Zn^{++}, Cu^{++} and Cd^{++} content in WT and transgenic lines in root, stem and leaf under (a) control and (b) stress condition (1 mM Zn, 0.5 mM Cu and Cd). Graph represents the mean ± SD (of three replicates; n = 3) followed by similar letters are significantly different according to Tukey HSD at *P<0.05*.

parameters; shoot length, root length, fresh weight and dry weight were recorded and compared with wild type plants.

Leaf senescence assay and chlorophyll estimation

Leaf discs from 30 days old WT and transgenic plants (L2, L4, L6 and L13) were used for the stress tolerance assay. Healthy leaves of similar age were detached and leaf discs (of 5 mm in diameter) were punched out. About 8 discs of each plant were floated in 5 ml ½ Hoagland media (control) supplemented with 200 mM NaCl, 10% PEG, 1 mM $ZnSO_4$, 0.5 mM $CuSO_4$ or 0.5 mM $CdCl_2$ for 7 days. The effect of these treatments on leaf discs were assessed by observing phenotypic changes.

Leaf discs (control and treated) were further subjected to chlorophyll isolation. Leaf discs were thoroughly homogenized in chilled N, N-dimethylformamide (DMF) at 4°C and thereafter centrifuged at 10,000 g for 10 min. Supernatant was aspirated out and O.D. was recorded at 664 and 647 nm. Total chlorophyll content was calculated per gram fresh weight of tissue according to Porra et al. [41].

Relative water content, Electrolyte leakage and Membrane stability index

WT and T_1 seedlings (L2, L4, L6 and L13) were transferred to ½ Hoagland hydroponics culture and maintained for 20 days. Healthy young plants of same age and size were collected for each treatment (24 h for 200 mM NaCl and 0.5 mM $CuSO_4$; 12 h for 10% PEG, 1 mM $ZnSO_4$ and 0.5 mM $CdCl_2$).

About 100 mg (FW) leaves of control and treated plants were submerged into the deionised water and after 12 h turgid weight (TW) were recorded. Samples were further kept at 80°C for 48 h

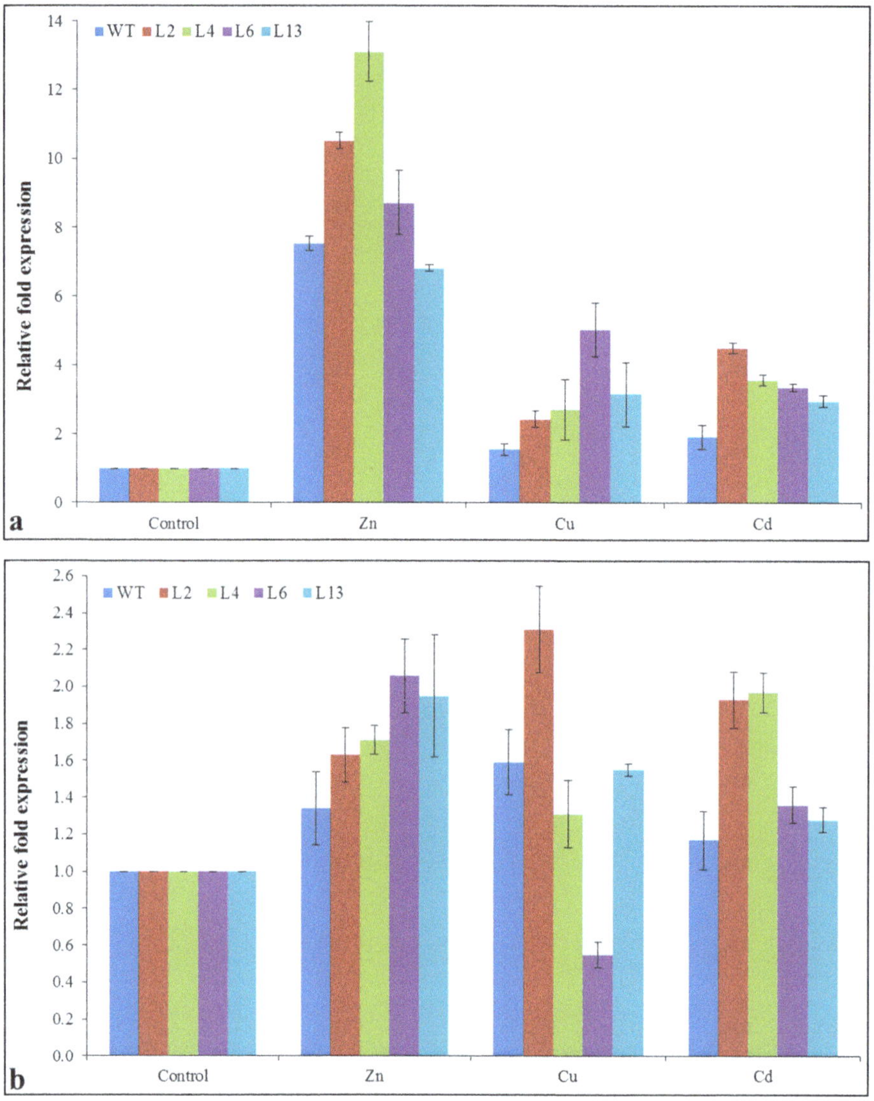

Figure 9. Expression analysis of zinc transporter encoding genes in transgenic tobacco lines under different metal ion stress condition. Comparison of relative fold expression of (a) *NtZIP1* and (b) *NtHMA-A* gene under Zn (1 mM), Cu (0.5 mM) and Cd (0.5 mM) stress.

to record the dry weight (DW). The relative water content was calculated as: *RWC (%) = (FW-DW/TW-DW) ×100.*

Collected leaves were washed thoroughly with deionized water to remove surface-adhered electrolytes. Samples were kept in closed vials containing 10 ml deionised water and incubated at 25°C on a rotary shaker for 24 h. Subsequently, the electrical conductivity (EC) of the solution (L_t) was determined using conductivity meter (SevenEasy, Mettler Toledo AG 8603, Switzerland). Samples were autoclaved at 120°C for 20 min, cooled up to 25°C and electrical conductivity (L_0) was determined. The electrolyte leakage was determined: *Electrolyte leakage (%)* $= (L_t/L_0) \times 100$.

Healthy young leaves of stress-treated plants were taken and membrane stability index (MSI) was determined [42]. Leaves (200 mg) were kept in close vials containing 10 ml deionized water. A set of vials was incubated at 40°C for 30 min while second set of vials was incubated at 100°C for 10 min. Electrical conductivity was recorded for both sets (L1 for 40°C while L2 for 100°C) and MSI was calculated using the formula as: *MSI = [1-(L1/L2)] ×100.*

In vivo localization of O_2^- and H_2O_2

Histochemical staining was performed for the *in vivo* detection of O_2^- and H_2O_2 using nitro-blue tetrazolium (NBT) and 3, 3-diaminobenzidine (DAB), respectively [43]. The presence of O_2^- in transgenic and WT leaves exposed to different stresses (as described above) was detected by immersing the leaf samples in NBT solution (1 mg ml^{-1} in 10 mM phosphate buffer; pH 7.8) at room temperature for 2 h and then illuminatedfor 12 h in light until blue spots appeared. For the localization of H_2O_2, treated leaves were incubated in DAB solution (1 mg ml^{-1} in 10 mM phosphate buffer; pH 3.8) at room temperature for 6 h in dark, thereafter exposed to the light until brown spots were appeared. Leaf samples were treated with absolute ethanol (for bleaching chlorophyll contents) before the documentation.

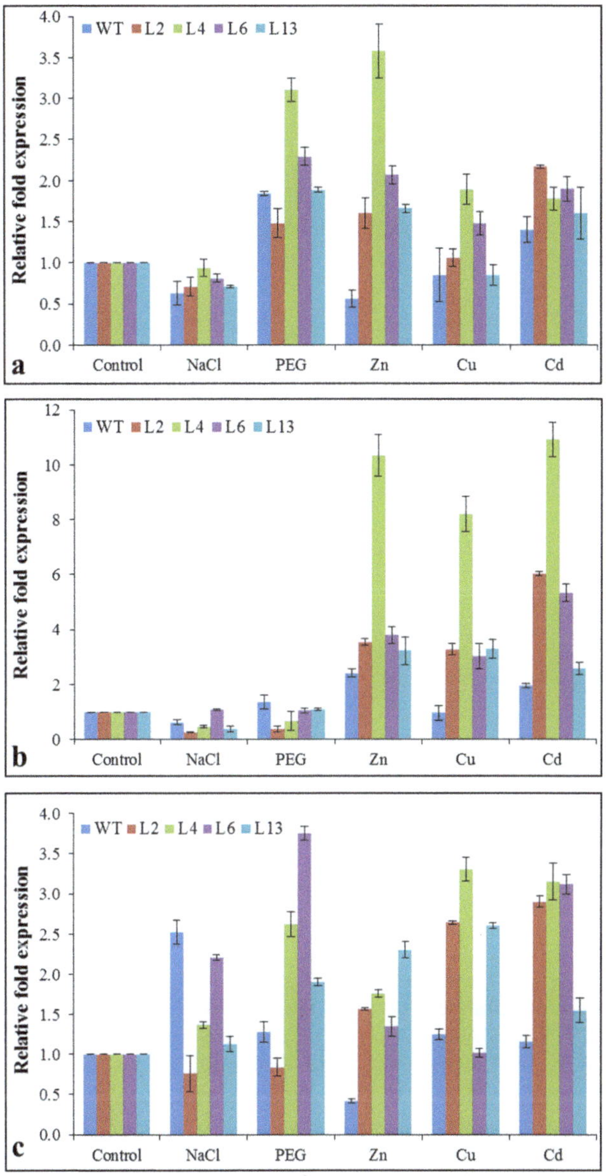

Figure 10. Expression analysis of antioxidant enzyme encoding genes in transgenic tobacco lines under different stress condition. Comparison of relative fold expression of (a) *NtSOD* gene, (b) *NtPOD* gene and (c) *NtAPX* gene under salt (200 mM NaCl), osmotic (10% PEG), Zn (1mM), Cu (0.5 mM) and Cd (0.5 mM) stress.

Quantification of H_2O_2, proline and lipid peroxidation

The H_2O_2 and free proline content in leaf samples (WT and transgenic lines under different stress conditions) were measured [44,45]. Proline and H_2O_2 levels were calculated by the standard curve, prepared against known concentration of proline or H_2O_2 measured at 520 and 560 nm absorbance respectively. Lipid peroxidation was estimated by determining the concentration of malondialdehyde (MDA) produced by thiobarbituric acid (TBA) reaction [46]. Leaf samples were extracted in 2 ml 0.1% trichloroacetic acid (TCA) and 0.5 ml extract was reacted with 2.0 ml of TBA reagent followed by boiling at 95°C for 30 min. Samples were cooled at ice, centrifuged at 10000 g for 5 min and absorbance of the supernatants was measured at 440 nm, 532 nm and 600 nm.

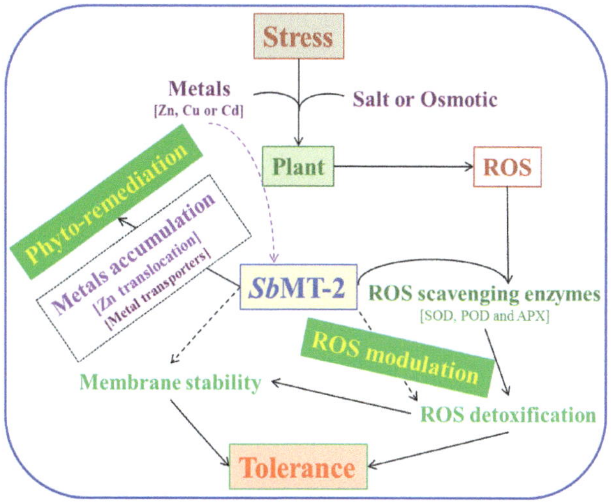

Figure 11. A hypothetical model for the role of *Sb*MT-2 in phyto-remediation and ROS modulation under abiotic stress.

Ion content analysis

For ion content analysis, tissues (root, shoot and leaves) from 4-week-old plants grown in hydroponic medium and treated with 200 mM NaCl, 10% PEG 6000, 1 mM $ZnSO_4$, 0.5 mM $CuSO_4$ or 0.5 mM $CdCl_2$ were washed with deionised water, dried in hot air oven for 48 h at 70°C and digested with 4 ml perchloric acid-nitric acid solution (3:1). The solution was heated to dry and further diluted to 25 ml with deionised water. Ion contents were measured by inductively coupled plasma optical emission spectrometer (Optima2000DV, PerkinElmer, Germany).

Transcriptional regulation of metal transporters and antioxidative enzymes

Total RNA was isolated from the stressed and control plants by using RNeasy plant mini kit (Qiagen) following the manufacturer's instructions and quantified with Nanodrop spectrophotometer (NanoDrop, USA). The cDNA was prepared using 2 µg of total RNA with a SuperScript RT III first-strand cDNA synthesis kit (Invitrogen, USA). The expression pattern of genes; *NtZIP1*, *NtHMA-A*, encoding metal transporters [47] and *NtSOD*, *NtPOX* and *NtAPX*, encoding antioxidant enzymes [48] which involved in heavy metal transportation and ROS scavenging, respectively were analyzed under different stress by using gene-specific primer pairs (Table 1), while the gene actin was used as an internal control. The quantitative real time PCR (qRT-PCR) was performed in a Bio-Rad IQ5 detection system (Bio-Rad, USA) with QuantiFast Kit (Qiagen, USA). Fold expression was calculated by the method described by Livak and Schmittgen [49] and specificity of qRT-PCR was monitored by melt curve analysis.

Statistical analysis

Each experiment was carried out in three replicates. All data was expressed as mean ± SD and subjected to analysis of variance (ANOVA) to determine the significance of difference between the means of WT and transgenic plants of each treatment group. A Tukey HSD multiple comparison of mean test was used, significant differences were considered at $P<0.05$ and indicated by similar letters.

Table 1. Primer sets used for the confirmation and transcript analysis of transgenic lines.

Genes	Primer sequences (5′ to 3′)	PCR conditions
uidA	F: GATCGCGAAAACTGTGGAAT	95°C, 5min; 34 Cycles: 95°C, 1min; 60°C, 1min; 72°C, 2 min; Extension: 72°C, 7 min
	R: TGAGCGTCGCAGAACATTAC	
hptII	F: TTCTTTGCCCTCGGACGAGTG	95°C, 5min; 34 Cycles: 95°C, 1min; 55°C, 1min; 72°C, 2 min; Extension: 72°C, 7 min
	R: ACAGCGTCTCCGACCTGATG	
SbMT-2	F: AGGTTGCAAGATGTTCC TG	95°C, 5min; 35 Cycles: 95°C, 30sec; 60°C, 45sec; 72°C, min; Melt curve: 60–95°C
	R: TCCATCGTTCTCAGCTGCTAC	with 0.5°C/cycle increment
Actin	F: CGTTTGGATCTTGCTGGTCGT	
	R: CAGCAATGCCAGGGAACATAG	
NtZIP1	F: TGGTGGCTCAGTCTGGAGAT	
	R: CGAAGGAGCTCAGAACTGGAA	
NtHMA-A	F: ACAAAGTGCTCGGACACCAA	
	R: CTTCTCGGTTGCAGAGTCCT	
NtSOD	F: AGCTACATGACGCCATTTCC	
	R: CCCTGTAAAGCAGCACCTTC	
NtPOD	F: CTTGGAACACGACGTTCCTT	
	R: TCGCTATCGCCATTCTTTCT	
NtAPX	F: CAAATGTAAGAGGAAACTCAGAGGA	
	R: CAGCCTTGAGCCTCATGGTACCG	

Supporting Information

Figure S1 Percentage of seed germination under different stress condition. Seeds of WT and transgenic lines were germinated on MS media supplemented with 200 mM NaCl, 10% PEG, 5 mM Zn, 0.2 mM Cu or 0.2 mM Cd. Graph represents the mean ± SD (of three replicates; n = 3) followed by similar letters are significantly different according to Tukey HSD at *P< 0.05*.

Figure S2 Sodium and potassium ion content in transgenic tobacco lines under different stress condition. Comparison of (a) Na^+ and K^+ content and (b) K^+/Na^+ ratio in WT and transgenic lines in root, stem and leaf under stressed (200 mM NaCl) and un-stressed conditions. Graph represents the mean ± SD (of three replicates; n = 3) followed by similar letters are significantly different according to Tukey HSD at $P<0.05$.

Table S1 Comparison of relative fold expression (down–regulation) of zinc transporter encoding genes (*NtZIP1* and *NtHMA-A*) in transgenic tobacco lines and wild type plants under de–stress condition.

Acknowledgments

CSIR-CSMCRI Communication No. PRIS-144/2013.

Author Contributions

Conceived and designed the experiments: AM BJ. Performed the experiments: AKC MKP VT. Analyzed the data: AM AKC. Wrote the paper: AKC AM.

References

1. Sreenivasulu N, Sopory SK, Kavi Kishor PB (2007) Deciphering the regulatory mechanisms of abiotic stress tolerance in plants by genomic approaches. Gene 388: 1–13.
2. Seth CS, Remans T, Keunen E, Jozefczak M, Gielen H, et al. (2012) Phytoextraction of toxic metals: a central role for glutathione. Plant Cell Environ 35: 334–346.
3. Chary NS, Kamala CT, Raj DSS (2008) Assessing risk of heavy metals from consuming food grown on sewage irrigated soils and food chain transfer. Ecotoxicol Environ Safety 69: 513–524.
4. Tewari RK, Kumar P, Sharma PN, Bisht S (2002) Modulation of oxidative stress responsive enzymes by excess cobalt. Plant Sci 162: 381–388.
5. Hamer DH (1986) Metallothionein. Annu Rev Biochem 55: 913–951.
6. Huang GY, Wang YS (2010) Expression and characterization analysis of type 2 metallothionein from grey mangrove species (*Avicennia marina*) in response to metal stress. Aquat Toxicol 99: 86–92.
7. Cobbett C, Goldsbrough P (2002) Phytochelatins and metallothioneins: Roles in heavy metal detoxification and homeostasis. Annu Rev Plant Biol 53: 159–182.
8. Zhou JM, Goldsbrough PB (1994) Functional homologs of fungal metallothionein genes from *Arabidopsis*. Plant Cell 6: 875–884.
9. Hassinen VH, Tuomainen M, Peräniemi S, Schat H, Kärenlampi SO, et al. (2009) Metallothioneins 2 and 3 contribute to the metal-adapted phenotype but are not directly linked to Zn accumulation in the metal hyperaccumulator, *Thlaspi caerulescens*. J Exp Bot 60: 187–196.
10. Rodriguez Lloriente ID, Perez-Palacios P, Doukkali B, Caviedes MA, Pajuelo E (2010) Expression of the seed-specific metallothionein *MT4a* in plant vegetative tissues increase Cu and Zn tolerance. Plant Sci 178: 327–332.
11. Turchia A, Tamantini I, Camussia AM, Racchi ML (2012) Expression of a metallothionein A1 gene of *Pisum sativum* in white poplar enhances tolerance and accumulation of zinc and copper. Plant Sci 183: 50–56.
12. Yang Z, Wu Y, Li Y, Ling HQ, Chu C (2009) *OsMT1a*, a type 1 metallothionein, plays the pivotal role in zinc homeostasis and drought tolerance in rice. Plant Mol Biol 70: 219–229.
13. Chaturvedi AK, Mishra A, Tiwari V, Jha B (2012) Cloning and transcript analysis of type 2 metallothionein gene (*SbMT-2*) from extreme halophyte

Salicornia brachiata and its heterologous expression in *E. coli*. Gene 499: 280–287.
14. Xue T, Li X, Zhu W, Wu C, Yang G, et al. (2009) Cotton metallothionein *GhMT3a*, a reactive oxygen species scavenger, increased tolerance against abiotic stress in transgenic tobacco and yeast. J Exp Bot 60: 339–349.
15. Zhigang A, Cuijie L, Yuangang Z, Yejie D, Wachter A, et al. (2006) Expression of *BjMT2*, a metallothionein 2 from *Brassica juncea*, increases copper and cadmium tolerance in *Escherichia coli* and *Arabidopsis thaliana*, but inhibits root elongation in *Arabidopsis thaliana* seedlings. J Exp Bot 57: 3575–3582.
16. Wong HL, Sakamoto T, Kawasaki T, Umemura K, Shimamoto K (2004) Down-regulation of metallothionein, a reactive oxygen scavenger, by the small GTPase *OsRac1* in rice. Plant Physiol 135: 1447–1456.
17. Zhang FQ, Wang YS, Lou ZP, Dong JD (2007) Effect of heavy metal stress on antioxidative enzymes and lipid peroxidation in leaves and roots of two mangrove plant seedlings (*Kandelia candel* and *Bruguiera gymnorrhiza*). Chemosphere 67: 44–50.
18. Schutzenduble A, Polle A (2002) Plant responses to abiotic stresses: heavy metal-induced oxidative stress and protection by mycorrhization. J Exp Bot 53: 1351–1365.
19. Kholodova V, Volkov K, Kuznetsov V (2010) Plants under heavy metal stress in saline environments, in: Sherameti, I., Varma, A. (Eds.), Soil heavy metals, Soil Biology vol. 19. Springer Berlin Heidelberg. pp. 163–183.
20. Joshi M, Mishra A, Jha B (2012) NaCl plays a key role for *in vitro* micropropagation of *Salicornia brachiata*, an extreme halophyte. Ind Crops Prod 35: 313–316.
21. Jha B, Singh N, Mishra A (2012a) Proteome profiling of seed storage proteins reveals the nutritional potential of *Salicornia brachiata* Roxb., an extreme halophyte. J Agric Food Chem 60: 4320–4326.
22. Mishra A, Joshi M, Jha B (2013) Oligosaccharide mass profiling of nutritionally important *Salicornia brachiata*, an extreme halophyte. Carbohydr Polym 92: 1942–1945.
23. Jha B, Sharma A, Mishra A (2011) Expression of *SbGSTU* (tau class glutathione S-transferase) gene isolated from *Salicornia brachiata* in tobacco for salt tolerance. Mol Biol Rep 38: 4823–4832.
24. Singh N, Mishra A, Jha B (2014). Ectopic over-expression of peroxisomal ascorbate peroxidase (*SbpAPX*) gene confers salt stress tolerance in transgenic peanut (*Arachis hypogaea*). Gene On-line: 10.1016/j.gene.2014.06.037.
25. Joshi M, Jha A, Mishra A, Jha B (2013) Developing transgenic *Jatropha* using the *SbNHX1* gene from an extreme halophyte for cultivation in saline wasteland. PLoS ONE 8: e71136.
26. Singh N, Mishra A, Jha B (2014) Over-expression of the peroxisomal ascorbate peroxidase (*SbpAPX*) gene cloned from halophyte *Salicornia brachiata* confers salt and drought stress tolerance in transgenic tobacco. Mar Biotechnol 16: 321–332.
27. Udawat P, Mishra A, Jha B (2014) Heterologous expression of an uncharacterized universal stress protein gene (*SbUSP*) from the extreme halophyte, *Salicornia brachiata*, which confers salt and osmotic tolerance to *E. coli*. Gene 536: 163–170.
28. Tiwari V, Chaturvedi AK, Mishra A Jha B (2014) The transcriptional regulatory mechanism of the peroxisomal ascorbate peroxidase (*pAPX*) gene cloned from an extreme halophyte, *Salicornia brachiata*. Plant Cell Physiol 55: 201–217.
29. Kumar G, Ritturaj H, Kushwaha V, Sabharwal P, Kumari S, et al. (2012) Clustered metallothionein genes are co-regulated in rice and ectopic expression of *OsMT1e-P* confers multiple abiotic stress tolerance in tobacco via ROS scavenging. BMC Plant Biol 12: 107.
30. Mittler R (2002) Oxidative stress, antioxidants and stress tolerance. Trends Plant Sci 7: 405–410.
31. Zhu J, Zhang Q, Wua R, Zhang Z (2010) *Hb*MT2, an ethephon-induced metallothionein gene from *Hevea brasiliensis* responds to H_2O_2 stress. Plant Physiol Biochem 48: 710–715.
32. Xia Y, Qic Y, Yuana Y, Wanga G, Cuia J, et al. (2012) Overexpression of *Elsholtzia haichowensis* metallothionein 1 (*Eh*MT1) in tobacco plants enhances copper tolerance and accumulation in root cytoplasm and decreases hydrogen peroxide production. J Hazard Mater 233–234: 65–71.
33. Cherian S, Oliveira MM (2005) Transgenic Plants in Phytoremediation: Recent Advances and New Possibilities. Environ Sci Technol 39: 9377–9390.
34. Töpfer R, Matzeit V, Gronenborn B, Schell J, Steinbiss HH (1987) A set of plant expression vectors for transcriptional and translational fusions. Nucleic Acids Res 15: 5890.
35. Horsch RB, Fry JE, Hoffmann NL, Eichholtz D, Rogers SG, et al. (1985) A simple and general method for transferring genes into plants. Science 227: 1229–1231.
36. Joshi M, Mishra A, Jha B (2011) Efficient genetic transformation of *Jatropha curcas* L. by microprojectile bombardment using embryo axes. Ind Crop Prod 33: 67–77.
37. Singh N, Mishra A, Joshi M, Jha B (2010) Microprojectile bombardment mediated genetic transformation of embryo axes and plant regeneration in cumin (*Cuminum cyminum* L.). Plant Cell Tiss Organ Cult 103: 1–6.
38. Pandey S, Mishra A, Patel MK, Jha B (2013) An efficient method for *Agrobacterium*-mediated genetic transformation and plant regeneration in cumin (*Cuminum cyminum* L.). Appl Biochem Biotechnol 171: 1–9.
39. Tiwari V, Chaturvedi AK, Mishra A, Jha B (2014) An efficient method of *Agrobacterium*-mediated genetic transformation and regeneration in local Indian cultivar of groundnut (*Arachis hypogaea*) using grafting. Appl Biochem Biotechnol In-press, doi: 10.1007/s12010-014-1286-3.
40. Jefferson RA (1987) Assaying chimeric genes in plants: the GUS fusion system. Plant Mol Biol Rep 5: 387–405.
41. Porra RJ, Thompson WA, Kriedemann PE (1989) Determination of accurate extinction coefficients and simultaneous equations for assaying chlorophylls a and b extracted with four different solvents: verification of the concentration of chlorophyll standards by atomic absorption spectroscopy. Biochemica et Biophysica Acta 975: 384–391.
42. Sairam RK (1994) Effect of homobrassinolide application on metabolism and grain yield under irrigated and moisture stress conditions of two wheat varieties. Plant Growth Regul 14: 173–181.
43. Shi J, Fu XZ, Peng T, Huang XS, Fan QJ, et al. (2010) Spermine pretreatment confers dehydration tolerance of citrus in vitro plants via modulation of antioxidative capacity and stomatal response. Tree Physiol 30: 914–922.
44. He Z, Wang ZY, Li J, Zhu Q, Lamb C, et al. (2000) Perception of brassinosteroids by the extracellular domain of the receptor kinase BRI. Science 288: 2360–2363.
45. Bates LS, Waldern R, Teare ID (1973) Rapid determination of free proline for water stress studies. Plant Soil 39: 205–207.
46. Draper HH, Hadley M (1990) Malondialdehyde determination as index of lipid peroxidation. Methods in Enzymology 186: 421–431.
47. Siemianowski O, Barabasz A, Kendziorek M, Ruszczyńska A, Bulska E, et al. (2014) *HMA4* expression in tobacco reduces Cd accumulation due to the induction of the apoplastic barrier. J Exp Bot 65: 1125–1139.
48. Huang XS, Luo T, Fu XZ, Fan QJ, Liu JH (2011) Cloning and molecular characterization of a mitogen-activated protein kinase gene from *Poncirus trifoliata* whose ectopic expression confers dehydration/drought tolerance in transgenic tobacco. J Exp Bot 62: 5191–5206.
49. Livak KJ, Schmittgen TD (2001) Analysis of relative gene expression data using real-time quantitative PCR and the $2^{-\Delta\Delta C_T}$ method. Methods 25: 402–408.

13

Overexpression of a Soybean Ariadne-Like Ubiquitin Ligase Gene *GmARI1* Enhances Aluminum Tolerance in Arabidopsis

Xiaolian Zhang, Ning Wang, Pei Chen, Mengmeng Gao, Juge Liu, Yufeng Wang, Tuanjie Zhao, Yan Li*, Junyi Gai*

National Key Laboratory of Crop Genetics and Germplasm Enhancement, National Center for Soybean Improvement, Key Laboratory for Biology and Genetic Improvement of Soybean (General, Ministry of Agriculture), Nanjing Agricultural University, Nanjing, Jiangsu, China

Abstract

Ariadne (ARI) subfamily of RBR (Ring Between Ring fingers) proteins have been found as a group of putative E3 ubiquitin ligases containing RING (Really Interesting New Gene) finger domains in fruitfly, mouse, human and Arabidopsis. Recent studies showed several RING-type E3 ubiquitin ligases play important roles in plant response to abiotic stresses, but the function of ARI in plants is largely unknown. In this study, an ariadne-like E3 ubiquitin ligase gene was isolated from soybean, *Glycine max* (L.) Merr., and designated as *GmARI1*. It encodes a predicted protein of 586 amino acids with a RBR supra-domain. Subcellular localization studies using Arabidopsis protoplast cells indicated GmARI protein was located in nucleus. The expression of *GmARI1* in soybean roots was induced as early as 2–4 h after simulated stress treatments such as aluminum, which coincided with the fact of aluminum toxicity firstly and mainly acting on plant roots. In vitro ubiquitination assay showed GmARI1 protein has E3 ligase activity. Overexpression of *GmARI1* significantly enhanced the aluminum tolerance of transgenic Arabidopsis. These findings suggest that *GmARI1* encodes a RBR type E3 ligase, which may play important roles in plant tolerance to aluminum stress.

Editor: Ji-Hong Liu, Key Laboratory of Horticultural Plant Biology (MOE), China

Funding: This work was supported by the National Hightech R & D Program of China (2013AA102602), the National Key Basic Research Program of China (2011CB1093), the Program for Changjiang Scholars and Innovative Research Team in University (PCSIRT13073), the Program for New Century Excellent Talents in University (NCET-12-0891), the MOE 111 Project (B08025), the Program for High-level Innovative and Entrepreneurial Talents in Jiangsu Province, and the Jiangsu Higher Education PAPD Program. The funders had no role in study design, data collection and analysis, decision or preparation of the manuscript.

Competing Interests: The authors have declared that no competing interests exist.

* Email: yanli1@njau.edu.cn (YL); sri@njau.edu.cn (JG)

Introduction

Ubiquitination is an enzymatic, protein post-translational modification by which proteins are selectively targeted for a variety of cellular processes including DNA transcription and repair, cell cycle and division, response to stresses and many others [1]. This process is carried out by three types of enzyme, including an ubiquitin-activating enzyme (E1), an ubiquitin-conjugating enzyme (E2), and an ubiquitin protein ligase (E3) [2]. Encoded by a large gene family of widely divergent isoforms [3], E3 ligases play important roles in governing the ubiquitin signaling pathway by transferring ubiquitin from E2 conjugation to specific protein substrates. E3 ligases are generally divided into two families, with either a HECT or RING-finger domain(s) [1], [4]. The RING-type E3 ubiquitin ligases are generally identified by the presence of conserved cysteine- and histidine- rich RING finger motifs that coordinate zinc atoms [5]. Recently several RING-type E3 ubiquitin ligases were found to play important roles in plant responses to abiotic and biotic stresses. The pepper E3 ubiquitin ligase RING1 gene, *CaRING1*, is required for cell death and the salicylic acid (SA)-dependent defense response [6]. *AtAIRP1* and *AtAIRP2* play roles in abscisic acid (ABA)-mediated drought stress responses in Arabidopsis [7]. In soybean, a RING-finger protein encoded by *GmRFP1* was identified and shown to be involved in ABA signaling and stress responses through the ubiquitin-proteasome pathway [8].

RBR (Ring Between Ring fingers) proteins are characterized by the presence of their RING1 – IBR – RING2 supra-domain, which is composed of two RING finger domains plus an IBR (In Between Rings) domain [9]. Many RBR proteins are known to have E3 ubiquitin ligase activity [10]. ARIADNE (ARI) proteins, a subclass of RBRs, have been identified in fruitfly [11], mouse [12], [13], human [14–16], and Arabidopsis [17], [18]. ARI proteins are characterized by the presence of an N-terminal acid-rich cluster, followed by a C3HC4 RING-finger motif, a central IBR or B-box, a second C3HC4 RING-finger structure, and Leu-rich domain at the C terminus. ARI proteins share their RBR domain with PARKIN, a protein involved in autosomal recessive familial Parkinson's disease [9]. PARKIN functions as E2-dependent ubiquitin-protein ligase [17]. Recent studies suggest that the ARI/PARKIN proteins define a new class of RING-finger E3 ligases [19]. There are only few studies on ARI proteins in plants.

Figure 1. The protein structure of GmARI1. Acid: acid-rich cluster; **Leu**: Leu-rich cluster; **RING1**: a C3HC4 RING-finger; **IBR**: In Between RING fingers (IBR); **RING2**: a second C3HC4 RING finger; **Coile~**: Coiled coil.

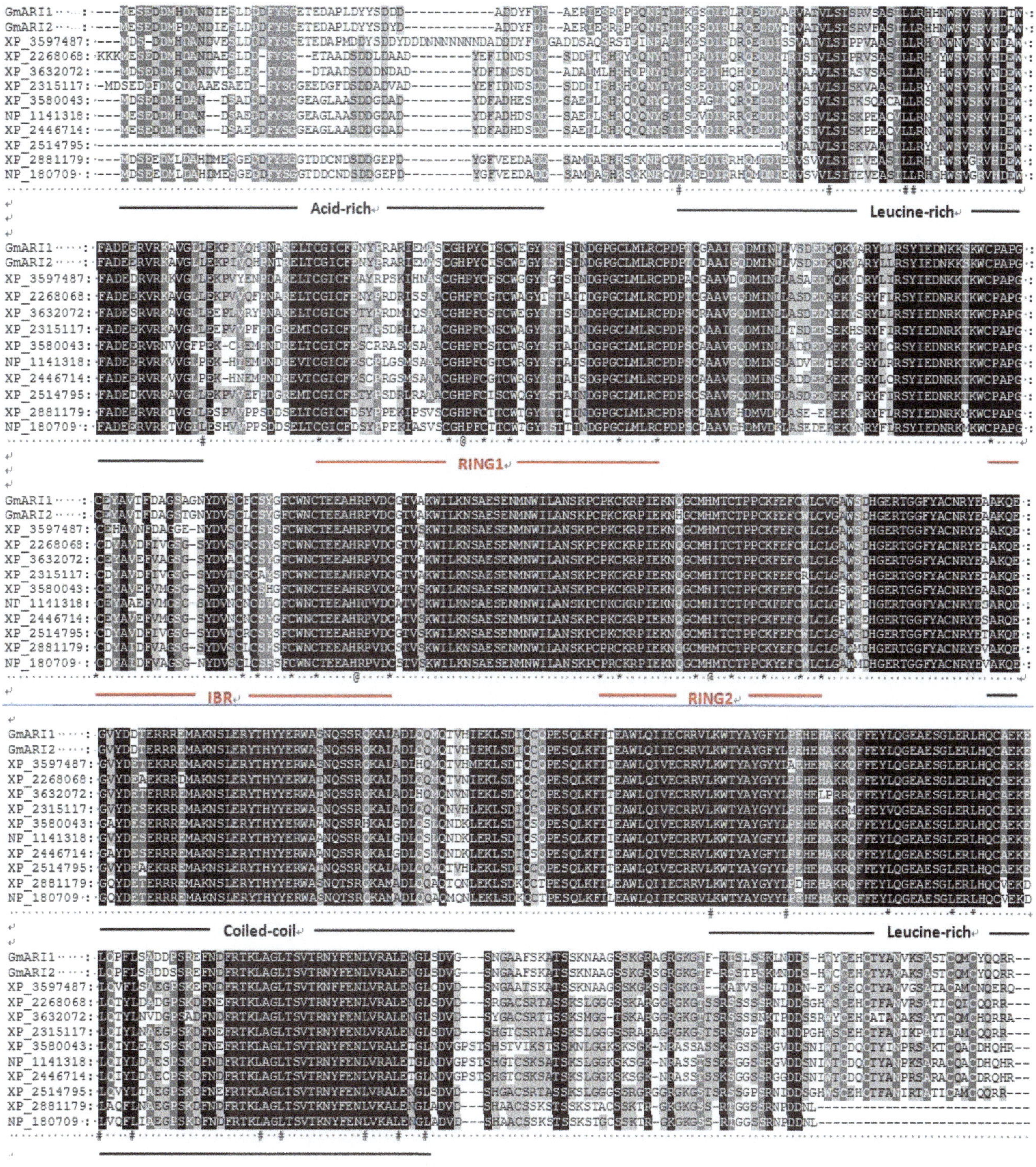

Figure 2. Multiple sequence alignment of GmARI1 with other RBR domain containing proteins. XP_3597487: *Medicago truncatula*; XP_2268068 and XP_3632072: *Vitis vinifera*; XP_2315117: *Populus trichocarp*; XP_3580043: *Brachypodium distachyon*; NP_1141318: *Zea mays*; XP_2446714: *Sorghum bicolor*; XP_2514795: *Ricinus communis*; XP_2881179: *Arabidopsis lyrata*; NP_180709: *Arabidopsis thaliana*; *: Cys; @: His; #: Leu and Ile.

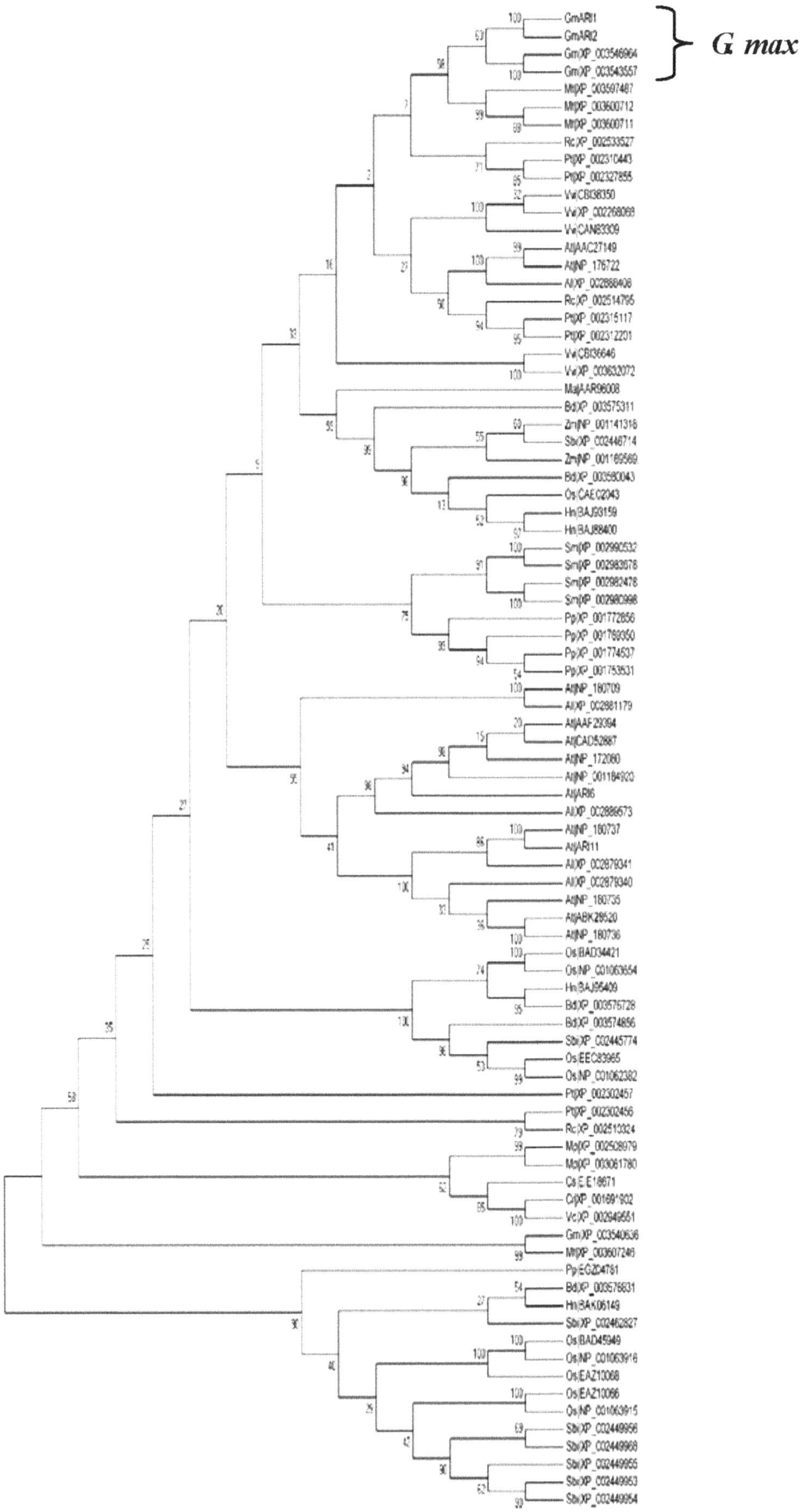
G. max

Figure 3. Phylogenetic relationships among soybean ARI proteins and other RBR domain containing proteins. The numbers on the tree indicate percent bootstrap values. The species abbreviations are listed as follows; *Al: Arabidopsis lyrata; At: Arabidopsis thaliana; Bd: Brachypodium distachyon; Cr: Chlamydomonas reinhardtii; Cs: Coccomyxa subellipsoidea; Hn: Hordeum vulgare; Ma: Musa acuminata; Mp: Micromonas pusilla; Mt: Medicago truncatula; Nt: Nicotana plumbaginifolia; Os: Oryza sativa subsp. Japonica; Pp: Physcomitrella patens; Pt: Populus trichocarpa; Rc: Ricinus communis; Sbi: Sorghum bicolor; Sm: Selaginella moellendorffii; Vc: Volvox carteri, Vvi: Vitis vinifera.*

Recently *AtARI12* in Arabidopsis was identified to be involved in UV-B signaling pathway [20].

Soybean (*Glycine max* [L.] Merr.) is widely grown as a major source of vegetable oil and protein. Soybean quality and yield are affected by various abiotic and biotic stresses. Soybean is also an important crop grown in South China, where acid soils comprise approximately 21% of the total land area [21], [22]. Aluminum (Al) toxicity is a major limiting factor of plant growth and crop production on acidic soils. There is large variation in Al tolerance among soybean varieties, and most of the Al tolerant varieties are from China [23]. Another study showed many Al tolerant varieties are from South China [24]. However, the genes underlying the Al tolerance in soybean remain largely unknown, except recently a soybean malate transporter gene *GmALMT1* which was shown to mediate root malate efflux which underlies soybean Al tolerance [25].

Increasing evidence indicates that RING-type E3 ubiquitin ligases play important roles in plant response to abiotic stresses. However, to date, there are no reports on the functions of soybean ARI proteins. Therefore, in this study, an ariadne-like E3 ubiquitin ligase gene *GmARI1* was cloned from soybean, and its gene expression patterns in different soybean tissues were studied. The transcriptional changes of *GmARI1* in response to various stress such as aluminum (Al) and plant hormone treatments were investigated using real-time quantitative PCR (qRT-PCR). We further characterized the *GmARI1* gene function by its subcellular location, in vitro ubiquitination assay, and performance of the transgenic Arabidopsis overexpressing *GmARI1* under Al stress. The possible mechanisms and signal pathways involved in soybean response to Al are also discussed.

Materials and Methods

Plant materials

Seeds of the soybean (*Glycine max* [L.] Merr.) cultivar Nannong 1138-2, provided by the National Center for Soybean Improvement (Nanjing, China), were germinated in sand under 25°C, 60% relative humidity (RH) and a photoperiod of 16 h/8 h (light/dark) cycle (light intensity was about 110 µmol photons. $m^{-2}s^{-1}$). Nannong 1138-2 is a released cultivar adapted to South China, which has good agronomic traits and moderate Al tolerance. The soybean plants at VE stage (emergence) were transferred to the 'standard' nutrient solution [26], and grown for another ten days before various stress and hormone treatments. The nutrient solution was renewed every five days.

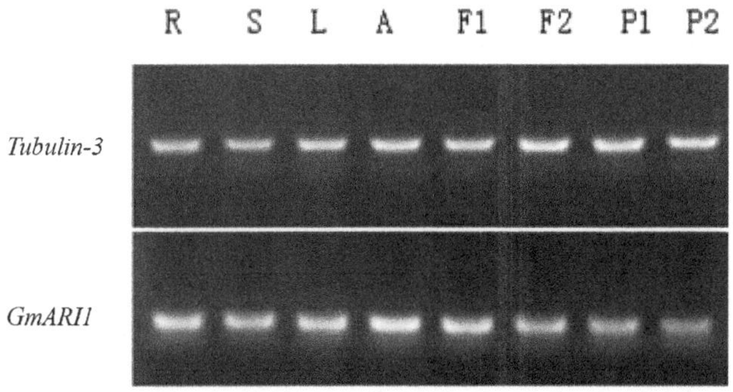

Figure 4. Tissue expression patterns of *GmARI1* in soybean. R: Roots; S: Stems; L: Leaves; A: Apex (SAMs); F1, F2: Flowers; P1, P2: Pods.

Isolation of the *GmARI1* gene from soybean

The full-length opening reading frame of the *GmARI1* gene was obtained by RT-PCR using soybean RNA. Total RNA was isolated using Trizol Reagent (Invitrogen, USA) according to the user's manual. 0.2 µg of the purified total RNA was used to synthesize first-strand cDNA by the MMLV-reverse transcriptase (TaKaRa). The primers: 5′-TCCCAATTCTTCTTCTGCCCTAG-3′ and 5′-GCAACCTTTCTTCCAAG CCTTAC -3′ were designed to amplify the *GmARI1* gene located on Chromosome 11. The PCR products were cloned into the pGEM-T vector (Promega) and sequenced (Invitrogen). The sequencing results showed that two *ARI* genes, *GmARI1* and *GmARI2*, were isolated using above primers, due to their high similarity of 97%.

Sequence analysis

Protein domains were analyzed by the SMART (Simple Modular Architecture Research Tool) (http://smart.embl-heidelberg.de/) and Pfam (Protein families database of alignments and HMMs). The molecular mass, isoelectric point and secondary structure were predicted using ProtParam and SOPMA on the ExPASY(http://www.expasy.org/tools/). The BLASTP program at GenBank (http://www.ncbi.nlm.nih.gov/blast) was used to search the homologous sequences of GmARI1/GmARI2 from Non-Redundant (NR) database. Alignment was performed with ClustalW2 (http://www.ebi.ac.uk/Tools/msa/clustalw2/) and MUSCLE (http://www.ebi.ac.uk/Tools/msa/muscle/) using the default settings. The phylogenetic tree was constructed by the neighbor-joining algorithm (NJ) using MEGA version 5 with 1000 bootstraps.

Semi-quantitative RT-PCR

To study the tissue expression pattern of *GmARI1*, soybean roots, stems, leaves, and shoot apical meristem (SAM) were collected from 15-day-old plants, flowers and pods were collected from plants at 20 days after flowering (DAF). All tissues were frozen immediately in liquid nitrogen and stored at −80°C. The semi-quantitative RT-PCR assay was performed with 0.1 µg RNA as template for cDNA synthesis. Primers 5′-CTCCATTCTCCATTCTCCTCCTTTGC-3′ and 5′-GTCGTCGTCGCTGTAGTAGT CC -3′ were used for *GmARI1*. Primers:5′-ATCTCATTCCCTTCCCTCGTCTG-3′ and 5′-CTGCCTCTGTGAACTCCATCTCG -3′ were used for *Tubulin-3* (GeneBank accession No. U12286) as the internal control. The PCR products were examined by electrophoresis in 2.0% agarose gel.

Transient expression of the GmARI1-GFP fusion protein

GmARI1-GFP was cloned into pMDC83 vector, with the expression driven by the cauliflower mosaic virus 35S promoter. The ORF of *GmARI1* was amplified by PCR using primers:5′-GGGGACAAGTTTGTACAAAAAA GCAGGCTTCCCAATTCTTCTTCTGCCCTAG-3′ and 5′-GGGGACCACTTTG-

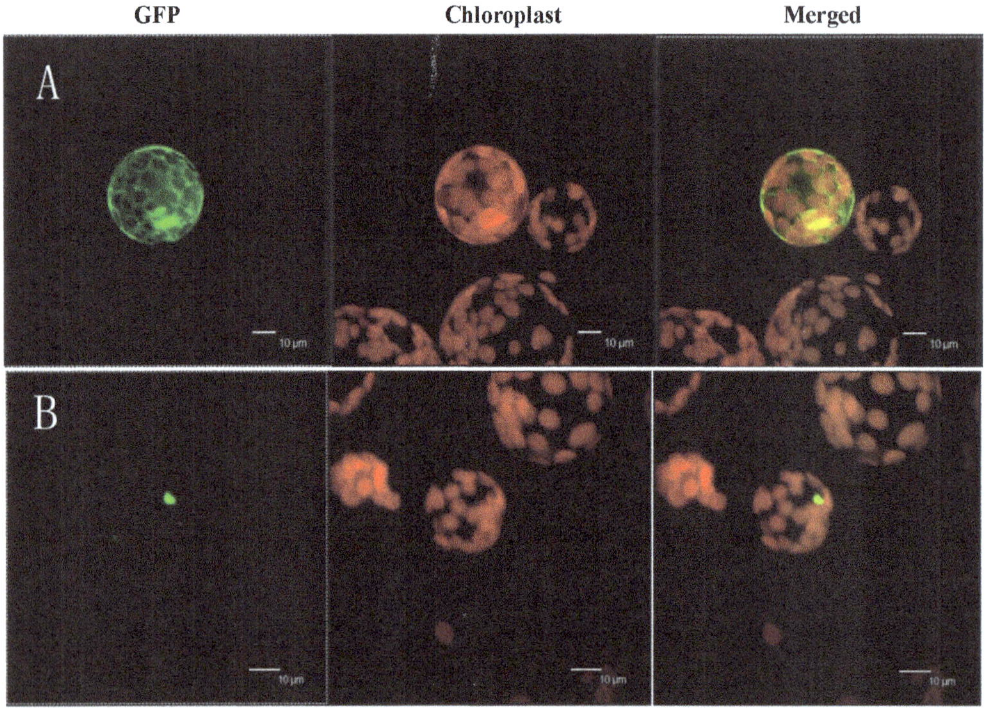

Figure 5. Sub-cellular localizations of GmARI1 protein. **A**: 35S-GFP; **B**: 35S-GmARI1-GFP.

TACA AGAAAGCTGGGTCTCGACGTTGTTGATAGCA-CATCTG -3′, without the stop codon. The *35S-GmARI1-GFP* in-frame fusion construct and control vector of *35S-GFP* were introduced into the Arabidopsis protoplasts cells by PEG-mediated protocol [27], [28]. After culturing in dark at 23°C, the localization of GFP was observed with a confocal microscope the next day (Leica TCS SP2).

Stress treatments

The plants were put in the 'standard' nutrient solution [26] with one of the following chemicals for various treatments: Al stress (10 μM $Al(NO_3)_3$, pH 4.3), drought (osmotic) stress (400 mM mannitol), salt stress (200 mM NaCl), absicisic acid (ABA,100 μM), indoleacetic acid (IAA, 100 μM), jasmonic acid (JA, 100 μM), and salicylic acid (SA, 150 μM), with 'standard' nutrient solution (Mg^{2+} was withdrew for Al treatment) as control. The leaves and roots were harvested at 0, 0.5, 1, 2, 4, 6, 8, 12, 24, and 48 h after each treatment. Each sample was the mixture of three seedlings and each treatment was repeated three times. All samples were immediately frozen in liquid nitrogen and stored at −80°C for later use.

Real-time quantitative PCR

RNAs of different treatments were extracted using Trizol reagent (Invitrogen) and purified with RNase-free DNase I. The cDNA was synthesized from 0.2 μg RNA in a 10 μl reaction volume using PrimeScript R 1st Strand cDNA Synthesis kit (TaKaRa). Primers for *Tubulin-3* (GeneBank accession No. U12286) were 5′- TCATTCCCTTCCCTCGTCTGC-3′ and 5′-CCTCCTTGGTGCTCATCTTGC-3′. Primers for *GmARI1* were 5′-CGCTGGTTCCTGAATTTCCCTTG-3′ and 5′- GTCGTCGTCGCTGTAGTAGTCC-3′. Quantitative real-time PCR was performed with SYBR Green method on ABI 7500 Fast Real-Time PCP system. The following procedure was used for qPCR: 95°C for 5 min; 40 cycles of 95°C for 3 sec; 60°C for 30 sec; 72°C for 30 sec. Data was analyzed using the $2^{-\Delta\Delta CT}$ method as described by Livak and Schmittgen [29].

In vitro E3 ubiquitin ligase activity assay of GmARI1 protein

The full-length cDNA of GmARI1 with SalI/XhoI restriction enzyme sites was amplified by PCR, using primers ARI1-F: 5′-ACGCGTCGACATGGAGTCAGAGGATATGCAC-3′ and ARI1-R: 5′-CCGCTCGAGTCGACGTTGTTGATAGCA-CATCTG -3′. The fragment was cloned into the vector pET28a via the SalI/XhoI restriction sites, with 6×His tag fused to GmARI1 at the N-terminal. The expression construct (pET28a-His_6-GmARI1) was transformed into *E.coli* BL21 (DE3) cells to produce a recombinant His_6-GmARI1 fusion protein with an expected mass of about 66 kDa. The transformed cells harboring pET28a-His_6-GmARI1 were grown at 37°C with vigorous shaking until an OD600 of 0.4–0.6 is reached and induced with 0.1 mM isopropylthio-b-galactoside (IPTG) for 12 h at 16°C. The overexpressed His_6-GmARI1 was purified using Ni–NTA resin according to the supplier's instructions (GE life sciences). The protein concentration was determined as described by Bradford [30] using BSA as a standard.

For the autoubiquitination assay, each reaction (30 μl final volume) contained 10 μg of recombinant ubiquitin (Ub, Sigma), 0.1 μg rabbit E1 (Boston Biochemicals), 0.2 μg human E2 (UbcH5b, Boston Biochemicals), 2 mM ATP, 50 mM Tris–HCl (pH 7.5), 5 mM MgCl2, and 2 mM DTT contained 500 ng purified His6-GmARI1 [31]. After incubation at 30°C for 2–3 h, the reactions were stopped with sodium dodecyl sulfate-polyacrylamide gel electrophoresis (SDS-PAGE) loading buffer at 95°C for 5 min. The reaction samples were electrophoretically separated on 12% SDS-PAGE gels and transferred on two PVDF membranes separately. The two membranes were blocked and thereafter

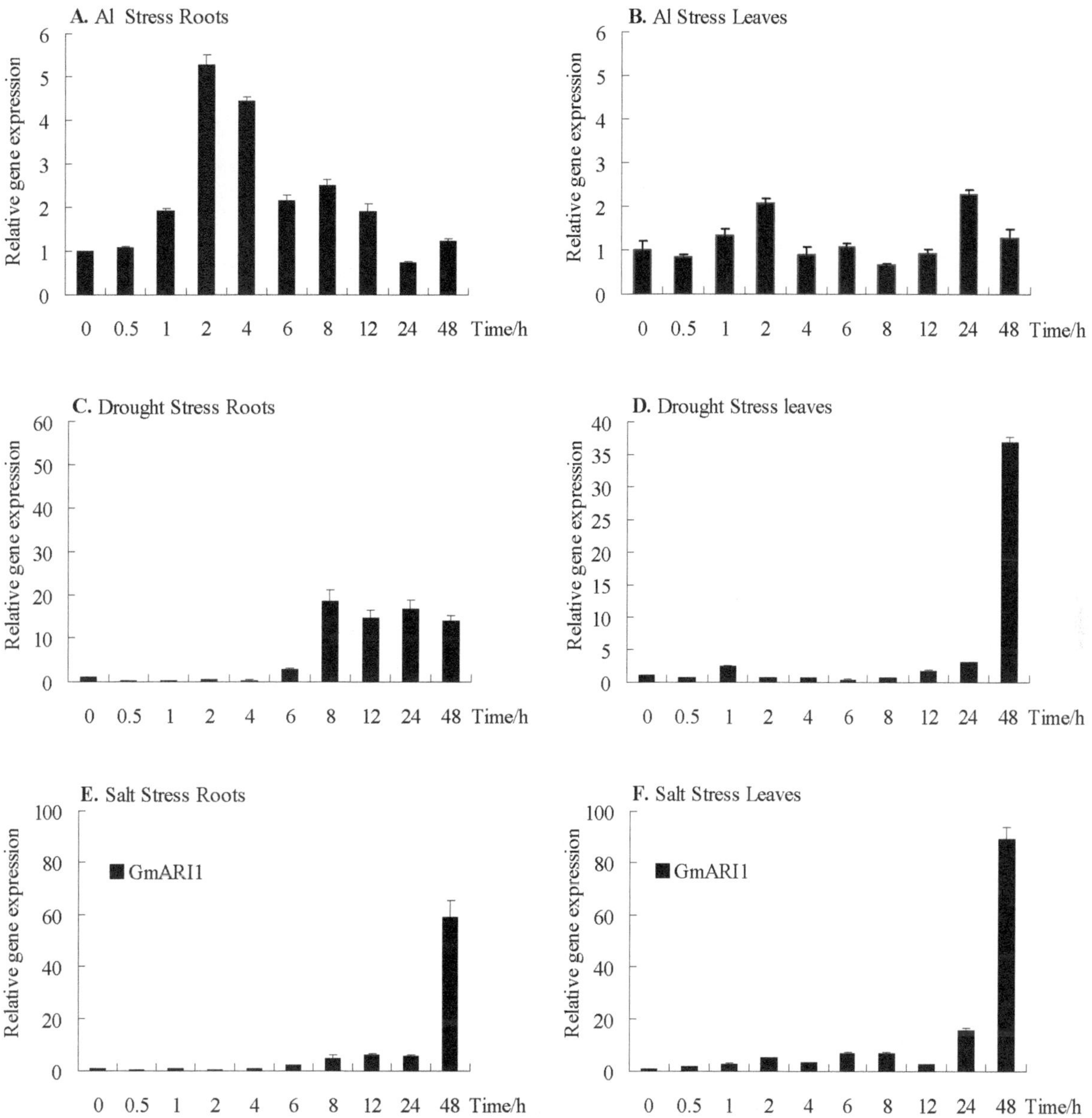

Figure 6. Relative gene expression levels of *GmARI1* in soybean under simulated stress treatments. A. relative gene expression in roots under Al stress (10 μM $Al(NO_3)_3$, pH4.3) **B**. relative gene expression in leaves under Al stress (10 μM $Al(NO_3)_3$, pH4.3) **C**. relative gene expression in roots under drought stress (400 mM mannitol) **D**. relative gene expression in leaves under drought stress (400 mM mannitol) **E**. relative gene expression in roots under salt stress (200 mM NaCl) **F**. relative gene expression in leaves under salt stress (200 mM NaCl). Error bars represent the standard error of three replicates.

blotted with an anti-ubiquitin monoclonal antibody (Santa Cruz Biotechnology) and an anti-His_6 monoclonal antibody (Sigma, USA) for 6 h at a 1:3000 dilution, respectively. After extensive washing, each of the bound primary antibody was detected with a horseradish peroxidase-conjugated goat anti-rabit IgG secondary antibody using the 3,3′-diaminobenzidine (DAB) development kit according to the manufacturer's protocol (Bio Basic Inc, Canada).

Generation of *GmARI1* transgenic Arabidopsis

The *GmARI1* gene was amplified by RT-PCR as described above and cloned into the plant expression vector pMDC83 under the control of CaMV 35S promoter by Gateway technology (Invitrogen), and the recombined plasmid was transferred into *A. tumefaciens* strain EHA105. Arabidopsis plants (Col-0 ecotype) were transformed using the floral dip method [32]. Twenty transgenic lines of *GmARI1* were obtained. Eight T3 lines of the transgenic *GmARI1* Arabidopsis were examined by RT-PCR to select positive transgenic lines for further analyses.

Al-tolerance of the *GmARI1* transgenic Arabidopsis

Homozygous T_3 seeds of the transgenic lines and wild type plants were used for Al-tolerance analysis. Seeds were surfaced-sterilized as described before [33] and germinated on 1/2 MS medium for 7 days. The seedlings were transferred to 1/2 MS without Mg^{2+} but with 0 or 15 μM $Al(NO_3)_3$ and 8% Agar, pH 4.3, and then put on the medium with the plates placed

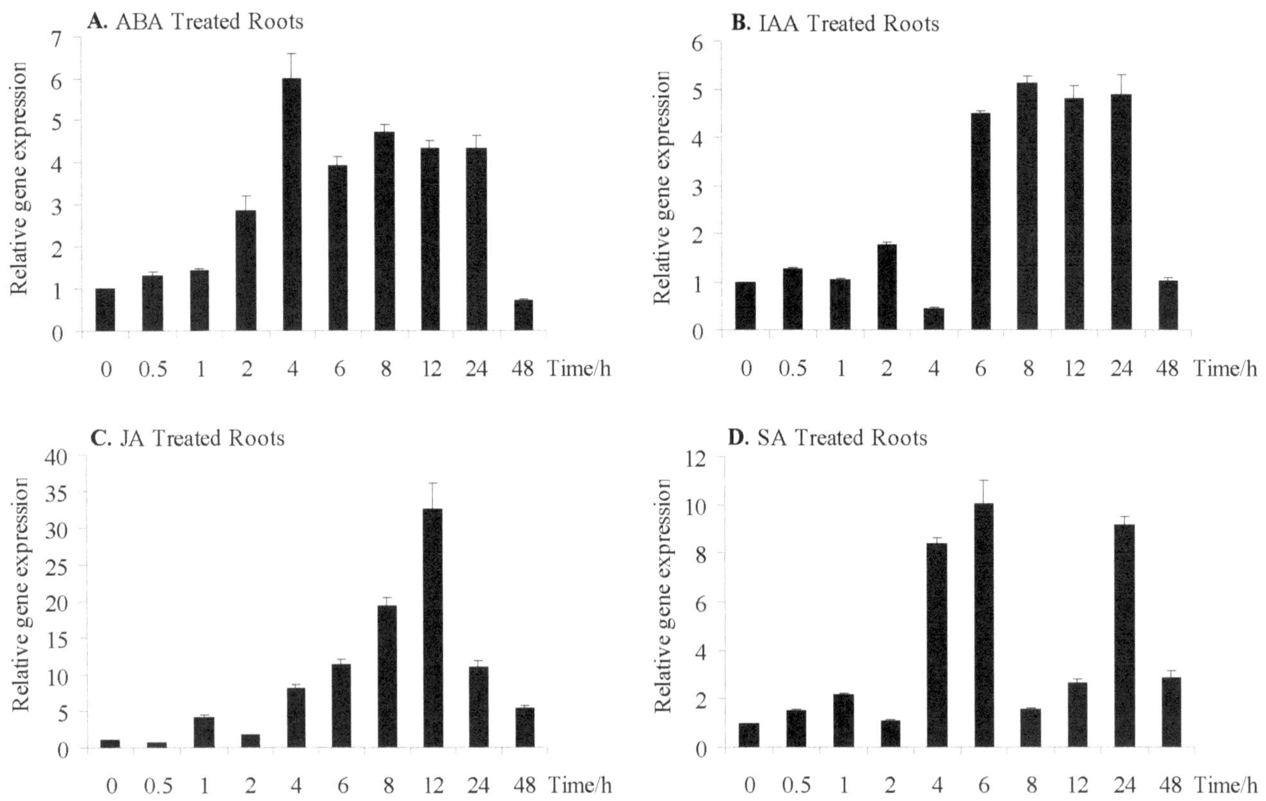

Figure 7. Relative gene expression levels of *GmARI1* in the roots of soybean after exogenous application of plant hormones. A. 100 μM ABA treatment **B.** 100 μM IAA treatment **C.** 100 μM JA treatment **D.** 150 μM SA treatment. Error bars represent the standard error of three replicates.

vertically. After 15 days, the lengths of the roots were measured, and the Relative Root Growth (RRG) in each line was calculated as: RRG (%) = $(RL_{Alt}- RL_{Al0})/(RL_{ct}- RL_{c0})$, in which RL_{Al0} represent the root length before Al treatment, RL_{Alt} represent the root length after 15 days of Al treatment (15 μM $Al(NO_3)_3$, pH 4.3), RL_{c0} represent the root length before growth on the control medium, and RL_{ct} represent the root length after growth on the control medium (0 μM $Al(NO_3)_3$, pH 4.3). The statistical analysis of the experimental data was conducted by t-test with SPSS [34].

Results

Isolation and sequence analysis of *GmARI1* gene

The cDNA of *GmARI1* gene is 2043 bp in length, containing an open reading frame (ORF) of 1761 bp, a 112 bp 5′-untranslated region (UTR), and a 170 bp 3′-UTR. Its homologous gene, *GmARI2*, shares 97% of similarity in the nucleotide sequences of the ORF region. The *GmARI1* and *GmARI2* nucleotide sequence and the predicted amino acid sequence have been deposited in GenBank (accession number JX392390 and JX392391). The genomic sequence of *GmARI1* and *GmARI2* variants in cultivar Williams 82 have 15 exons and 14 introns (http://www.phytozome.net/soybean.php), which located on chromosome 11 and 12, respectively.

The deduced protein of *GmARI1* comprises 586 amino acids with the predicted molecular mass of 66.99 kDa and isoelectric point of 5.37. GmARI1 protein has a RBR domain, which contains an IBR (C6HC) domain flanked by RING1 and RING2 (C3HC4) (Fig. 1). The secondary structure of GmARI1 protein is predicted to be composed of 45.56% alpha helix, 11.09% extended strand, 41.64% random coil, and 1.71% beta turn.

In order to determine the relationship of GmARI1 and other RBR proteins, Blastp on Uniprot was used to search the homologues proteins of GmARI1. The top 87 amino acid sequences with RBR conserved domain were selected from different plant species, including *Arabidopsis thaliana*, *Medicago truncatula*, *Ricinus communis*, and *Zea mays L.* Multiple sequence alignment of these 87 amino acid sequences was performed. In addition to the ARI proteins from soybean which showed high similarity with *GmARI1*, the protein from *M. truncatula* showed 80% similarity with *GmARI1* (Fig. 2) The phylogenetic tree was drawn using MEGA 5.0 program based on Neighbor-Joining (NJ) with 1000 bootstrap replications (Fig. 3), which also showed ARI proteins from *G. max* was closely related to the protein from *M. truncatula*.

Tissue expression pattern of *GmARI1* and the subcellular localization of its protein

GmARI1 gene was expressed in roots, stems, leaves, SAMs, flowers, and pods (Fig. 4). We determined the subcellular localization of the GmARI1 protein by transient expression of *35S-GmARI1-GFP* in Arabidopsis protoplasts. GmARI1 was located in the nucleus, while the GFP control was mainly located in the cytoplasm (Fig. 5).

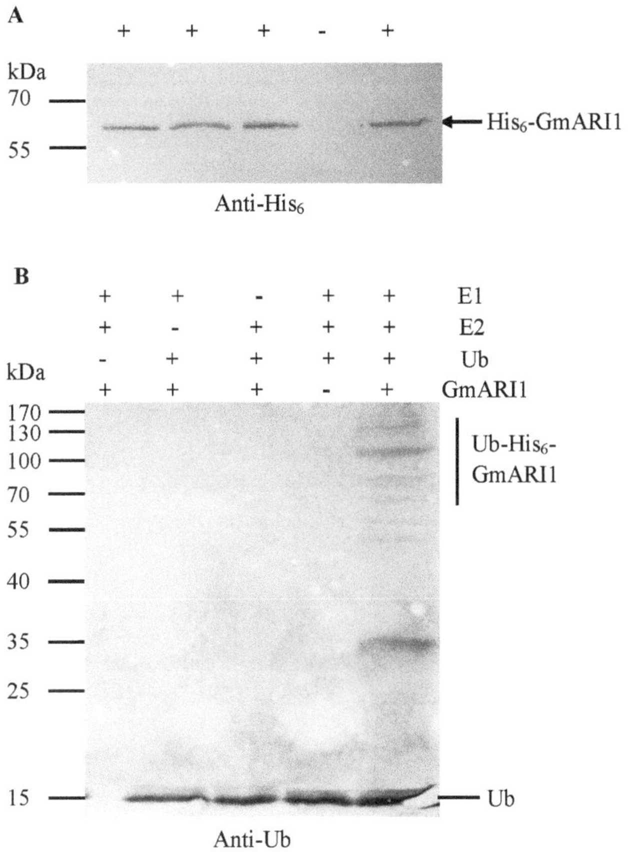

Figure 8. Western blot analysis of GmARI1 protein. A. Western blot using anti-His6 monoclonal antibody in the presence (+) or absence (−) of the purified His6 tag-GmARI1 proteins. **B**. In vitro E3 ubiquitin ligase activity assay of GmARI1 protein. His6 tag-GmARI1 fusion proteins were assayed for E3 activity in the presence of rabbit E1, human E2 (UbcH5b), and ubiquitin (Ub). The left numbers denote the molecular masses of marker proteins in kilodaltons.

Expression pattern of *GmARI1* under stress and plant hormone treatments

Real-time quantitative PCR was carried out to examine the expression pattern of *GmARI1* gene in soybean under different treatments including various abiotic stresses and plant hormones (Fig. 6 and 7). Under 10 μM Al (pH4.3) stress, the transcripts of *GmARI1* in soybean roots increased and peaked during 2 h to 4 h, but the transcripts in leaves did not change significantly (Fig. 6A, B). When treated with 400 mM mannitol (osmotic stress), the *GmARI1* gene expression was induced from 6 h to 48 h in roots, while showed delayed induction in leaves (Fig. 6C, D). Under 200 mM NaCl stress, the *GmARI1* gene expression showed a high induction after 24 h in both roots and leaves (Fig. 6E, F).

The transcriptional changes of *GmARI1* in response to ABA, IAA, JA and SA were also investigated in soybean roots. The transcripts of *GmARI1* were up-regulated during 2-24 h after ABA treatment (Fig. 7A). *GmARI1* was induced after 6 h IAA treatment and its transcript abundance was maintained at a higher level until 24 h (Fig.7B). The expression of *GmARI1* showed continual increase from 4 h and peaked around 12 h after JA treatment (Fig. 7C). Increased transcripts of *GmARI1* were detected at 4 h, 6 h and 24 h after SA treatment (Fig. 7D).

E3 ubiquitin ligase activity of GmARI1 protein

To test if GmARI1 has the E3 ubiquitin ligase activity, a full-length GmARI1 protein with maltose binding protein (6 × His tag) was expressed in *E. coli* and subsequently affinity-purified (His$_6$-GmARI1) from the soluble fraction (Fig. S1). The purified recombinant His$_6$-GmARI1 protein was about 66 kDa as expected, and the western blotting with anti-His$_6$ monoclonal antibody also showed the purified target recombinant protein had the right molecular weight of 66 kDa (Fig. 8A). In vitro self-ubiquitination assays were performed in the presence of rabbit E1, human E2 (UbcH5b), and Ub (Fig. 8B). Polyubiquitination was detected only in the presence of E1, E2, Ub and His$_6$-GmARI1. A negative result was observed if either E1, E2, Ub or His$_6$-GmARI1 was omitted in the reaction. These results indicate that GmARI1 has E3 ubiquitin ligase activity.

Performance of transgenic plants over-expressing *GmARI1* under Al stress

The gene expression of *GmARI1* was induced early by Al stress, therefore we further studied its function in transgenic plants. An expression plasmid vector of *pMDC83-GmARI1* was constructed and introduced into Arabidopsis plants using floral dip method. Transgenic T$_3$ Arabidopsis over-expressing *GmARI1* were generated and the positive transgenic lines were identified by RT-PCR (Fig. S2). Seeds of three T$_3$ homozygous transgenic lines (*GmARI1-1, GmARI1-2*, and *GmARI1-3*) and wild type Col-0 were germinated on 1/2 MS medium. After 10 days, the seedlings were transferred to 1/2 MS (pH4.3) containing 15 μM Al or 0 μM Al as a control. Fifteen days later, the root growth of the wild type Col-0 was severely inhibited by 15 μM Al as compared with the control medium (0 μM), while the transgenic lines were little affected by Al (Fig. 9 A). The relative root growth (RRG) of the transgenic lines was significantly ($p<0.01$) longer than the wild type under Al treatment (Fig. 9B). The relative abundance of *GmARI1* in the transgenic lines of *GmARI1-1* and *GmARI1-2* is higher than *GmARI1-3*, which coincided with RRG result (Fig. S3 and Fig. 9B).

Discussion

RING-type E3 ubiquitin ligases play important roles in plant responses to abiotic stresses [7], [8], [35]. RBR subclass of RING-containing E3 ligases were recently shown as an important group of proteins since the discovery of *parkin*, a mutation causing the familial autosomal-recessive juvenile parkinsonism (AR-JP) [36]. However, there is little research on the plant RBR family. The RBR family was classified into 14 subfamilies, including Ariadne (ARI), ARA54, Dorfin, parkin, PlantI, PlantII, and XAP3 [9], but the function of ARI class is largely unknown. In this study, *GmARI1* gene was isolated and characterized from soybean. Based on the analysis of predicted protein domains, *GmARI1* gene belongs to the Ariadne subfamily, characterized by the presence of an N-terminal acid-rich cluster, followed by a RBR domain and a Coiled coil region at the C-terminus [17]. The origin of Ariadne can be traced back earlier in time, as it was found not only in animal, fungal, and plant, but also in some protist [37]. The blastp result indicated among the RBR domain containing protein from other species, *Medicago truncatula* had the highest similarity (80%) with *GmARI1* and other ARI proteins from soybean.

GmARI1 gene was expressed ubiquitously in roots, stems, SAMs, leaves, flowers, and pods of soybean plants. The GmARI1 protein was located in the nucleus of the cell. Real-time quantitative PCR showed that the expression level of *GmARI1* in soybean root under Al (10 μM, pH 4.3) treatment reached the

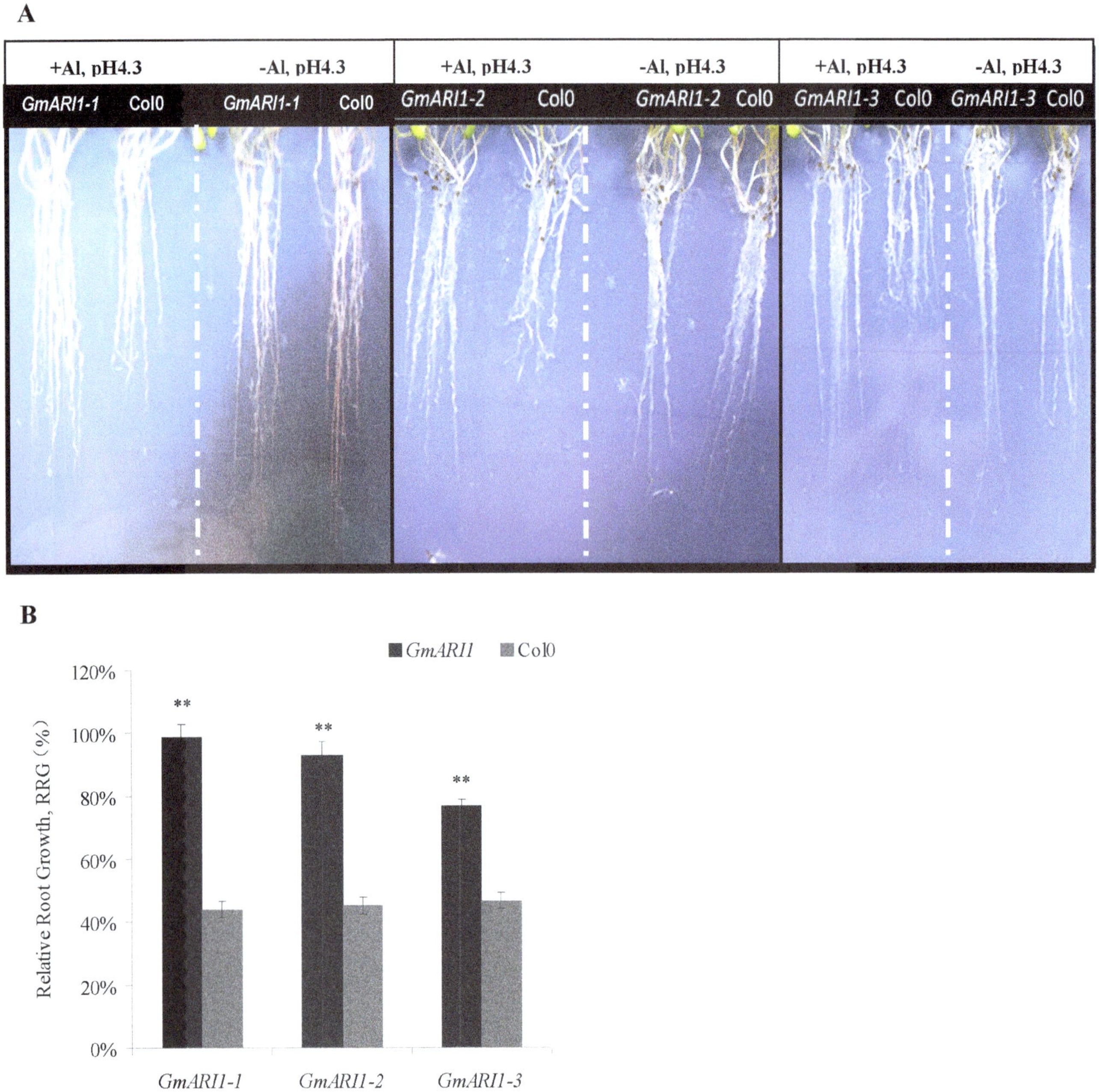

Figure 9. Performance of the transgenic Arabidopsis under 15 μM Al treatment. A. Root phenotypes of the *35S: GmARI1* overexpression lines *GmARI1-3* and the wild type Col-0 grown on 15 μM Al in 1/2 MS (+Al, pH4.3), and 0 μM Al in 1/2 MS (-Al, pH4.3). **B**. Relative root growth (RRG, %) of the transgenic Arabidopsis lines (*GmARI1-1, GmARI1-2, GmARI1-3*) and the wild type (Col-0). RRG was calculated by the root growth length under Al treatment (15 μM Al in 1/2 MS, pH4.3) divided by the root growth length under control (0 μM Al in 1/2 MS, pH4.3). Error bars represent the standard error (SE), ** indicate the significance level of 0.01 by t-tests.

peak during 2 to 4 hours, but no significant change was detected in the leaves. This coincides with the fact that Al toxicity was first and mainly acting on plant roots [38], limiting water and nutrition absorption [39], which further inhibiting the development of whole plant and reducing yield [40], [41]. Therefore in the early stage (less than 4 hours) of Al stress, induced gene expression of *GmARI1* in soybean roots may play important roles to trigger downstream signaling pathways to protect root cells from Al toxicity.

It has been reported that Al induces oxidative stress and DNA damage in plant cells [42–44]. Reactive oxygen species (ROS) constantly attack DNA, leading to oxidative DNA damage [45]. The cell cycle checkpoint regulators could detect and respond to such damage, leading to inhibition of root growth [46]. Protein ubiquitination is involved in DNA transcription and repair, cell cycle and division [1], and is emerging as a critical regulatory mechanism of DNA damage response [47]. Several RING domain-containing E3 ubiquitin ligases play an essential role in response to DNA damage [48]. In vitro ubiquitination assay showed GmARI1 has E3 ligase activity (Fig. 8), therefore we hypothesize that GmARI might be involved in the oxidative DNA damage repair to confer Al tolerance. We investigated the co-expression pattern of *GmARI* (http://bioinformatics.cau.edu.cn/SFGD/), and found Glyma02g15070 was on the top list of coexpression genes with *GmARI*. The homolog gene of Glyma02g15070 in arabidopsis is AT1G49670, which was shown to be involved in oxidative stress tolerance (http://www.arabidopsis.org/). Suppression of oxidative stress might help plants reduce the damage or root growth inhibition [49], [50]. Therefore, the Al tolerance observed in *GmARI* overexpression lines might be due to the improved tolerance to oxidative stress, or/and other signaling cascades.

The activities of cell wall-bound peroxidases in the annual legume, *Cassia tora*, significantly increased with Al concentrations, and were regulated by JA [51]. Another study showed that the SA-signaling and SA-dependent expression of a respiratory burst oxidase homolog gene is involved in Al responsive oxidative burst in Arabidopsis [52]. Here in this study, the expression of *GmARI1* was induced by Al stress, as well as JA and SA treatments in soybean roots. These suggested *GmARI1* might mediate soybean response to Al through oxidative species signals, which may overlap with plant hormone signaling pathways. The T_3 transgenic Arabidopsis plants over-expressing the *GmARI1* gene showed significant improvement in Al tolerance comparing with wild type plants, which further support the important role of *GmARI1* gene in plant response to Al stress.

Supporting Information

Figure S1 Expression and purification of the recombinant GmARI1 proteins. The recombinant His6-GmARI1 proteins were expressed in *E.coli* BL21 (DE3) and analyzed by SDS–PAGE. Lane 1, total proteins from *E. coli* cells before IPTG induction; lane 2, total proteins containing pET28a-GmARI1 from *E. coli* cells after induction by IPTG; lane 3, purified recombinant His6-GmARI1 protein.

Figure S2 RT-PCR confirmation of the transgenic Arabidopsis T3 lines GmARI 1 to 8. (−): Arabidopsis wild ecotype Col-0; (+): plasmid pMDC83-GmARI1.

Figure S3 Expression of the *GmARI1* gene in 2-week-old Arabidopsis plants quantified by qRT-PCR using actin (*AtACT2*) as the reference gene. The Arabidopsis plants were germinated and grown on 1/2 MS medium (pH5.8) for two weeks and then transferred to 1/2 MS medium with 25 μM $AlCl_3$ (pH4.3). Two hours later, tissues were sampled from the wild type Arabidopsis Col0 and homozygous transgenic lines separately (each sample was the mixture of four plants). Error bars are the standard errors from three replications.

Acknowledgments

We thank the staff at JiangPu Experimental Station for their assistance with the soybean seed production, and PLoS One editors and reviewers for their valuable comments and suggestions.

Author Contributions

Conceived and designed the experiments: JG YL XZ NW PC YW. Performed the experiments: XZ NW MG JL. Analyzed the data: XZ NW MG PC YL JG. Contributed reagents/materials/analysis tools: JG TZ YL. Wrote the paper: XZ NW YL JG.

References

1. Hershko A, Ciechanover A (1998) The ubiquitin system. Annu Rev Biochem 67: 425–479.
2. Pickart CM (2001) Mechanisms underlying ubiquitination. Annu Rev Biochem 70: 503–533.
3. Stone SL, Hauksdottir H, Troy A, Herschleb J, Kraft E, et al. (2005) Functional analysis of the RING-type ubiquitin ligase family Arabidopsis. Plant Physiol 137: 13–30.
4. Moon J, Parry G, Estelle M (2004) The ubiquitin-proteasome pathway and plant development. Plant Cell 16: 3181–3195.
5. Barlow PN, Luisi B, Milner A, Elliott M, Elliott R (1994) Structure of the C3H4 domain by 1H-nuclear magnetic resonance spectroscopy: a new structural class of zinc-finger. J Mol Biol 237: 201–211.
6. Lee DH, Choi HW, Hwang BK (2011) The Pepper E3 ubiquitin ligase RING gene *CaRING1* is required for cell death and the Salicylic Acid-dependent defense response. Plant Physiol 156: 2011–2025.
7. Cho SK, Ryu MY, Seo DH, Kang BG, Kim WT (2011) The Arabidopsis RING E3 ubiquitin ligase AtAIRP2 plays combinatory roles with AtAIRP1 in Abscisic Acid-mediated drought stress responses. Plant Physiol 157: 2240–2257.
8. Du QL, Cui WZ, Zhang CH, Yu DY (2010) *GmRFP1* encodes a previously unknown RING-type E3 ubiquitin ligase in soybean (*Glycine max*). Mol Biol Rep 37: 685–693.
9. Marin I, Ferrus A (2002) Comparative genomics of the RBR family, including the Parkinson's disease-related gene *parkin* and the genes to the Ariadne subfamily. Mol Biol Evol 19: 2039–2050.
10. Eisenhaber B, Chumak N, Eisenhaber F, Hauser MT (2007) The ring between ring fingers (RBR) protein family. Genome Biol 8: 209.
11. Aguilera M, Oliveros M, Matinez-Padron M, Barbas JA, Ferrus A (2000) Ariadne-1: A vital drosophila gene is required in development and defines a new conserved family of RING-finger proteins. Genetics 155: 1231–1244.
12. Itier JM, Ibanez P, Mena MA, Abbas N, Cohen-Salmon C, et al. (2003) Parkin gene inactivation alters behaviour and dopamine neuro-transmission in the mouse. Hum Mol Genet 12: 2277–2291.
13. Bromann PA, Weiner JA, Apel ED, Lewis RM, Sanes JR (2004) A putative ariadne-like E3 ubiquitin ligase (PAUL) that interacts with the muscle-specific kinase(MuSK). Gene Expr Patterns 4: 77–84.
14. Moynihan TP, Ardley HC, Nuber U, Rose SA, Jones PF, et al. (1999) The ubiquitin-conjugating enzymes UbcH7 and UbcH8 interact with RING finger/IBR motif-containing domains of HHARI and H7-AP1. J Biol Chem 274(43): 30963–30968.
15. Ardley HC, Tan NG, Rose SA, Markham AF, Robinson PA (2001) Features of the parkin/ariadne-like ubiquitin ligase, HHARI, that regulate its interaction with the ubiquitin-conjugating enzyme, Ubch7. J Biol Chem 276(22): 19640–19647.
16. Marteijn JA, van Emst L, Erpelinck-Verschueren CA, Nikoloski G, Menke A, et al. (2005) The E3 ubiquitin-protein ligase *Triad1* inhibits clonogenic growth of primary myeloid progenitor cells. Blood 106: 4114–4123.
17. Mladek C, Guger K, Hauser MT (2003) Identification and characterization of the *ARIADNE* gene family in Arabidopsis. A group of puptative E3 ligase. Plant Physiol 131: 27–40.
18. Marin I (2010) Diversification and specialization of plant RBR ubiquitin ligase. PLoS ONE 5(7): e11579.
19. Wenzel DM, Klevit RE (2012) Following Ariadne's thread: a new perspective on RBR ubiquitin ligases. BMC Biol 10: 24.
20. Lang-Mladek C, Xie L, Nigam N, Chumak N, Binkert M, et al. (2012) UV-B signaling pathways and fluence rate dependent transcriptional regulation of ARIADNE12. Physiol Plant 145: 527–539.
21. Huang BQ, Bai JH, Xue XQ (2001) Advances in studies on aluminum toxicity and tolerance in plants. Chinese Bulletin of Botany 18(4): 385–395.
22. Liu Q, Zheng SJ, Lin XY (2004) Plant physiological and molecular biological mechanism in response to aluminium toxicity. Chinese Journal of Applied Ecology 15: 1641–1649.
23. Sapra VT, Mebrahtu T, Mugwira LM (1982) Soybean germplasm and cultivar aluminum tolerance in nutrient solution and bladen clay loam soil. Agron J 74: 687–690.
24. Nian H, Huang H, Yan XL, Lu YG (1998) Studies on adaptability of soybean to acidic and aluminum toxicity I Study on the screening and identification for acidic and aluminum-resistance soybeans. Soybean Science 16(3): 191–196.
25. Liang CY, Pineros M, Tian J, Yao ZF, Sun LL, et al. (2013) Low pH, aluminum and phosphorus coordinately regulate malate exudation through GmALMT1 to improve soybean adaptation to acid soils. Plant Physiol 161: 1347–61.
26. Tocquin P, Corbesier L, Havelange A, Pieltain A, Kurtem E, et al. (2003) A novel high efficiency, low maintenance, hydroponic system for synchronous growth and flowering of Arabidopsis thaliana. BMC Plant Biol 3: 2.
27. Yoo SD, Cho YH, Sheen J (2007) Arabidopsis mesophyll protoplasts: a versatile cell system for transient gene expression analysis. Nat Protoc 2: 1565–1572.
28. Wu FH, Shen SC, Lee LY, Lee SH, Chan MT, et al. (2009) Tape-Arabidopsis Sandwich - a simpler Arabidopsis protoplast isolation method. Plant Methods 5: 16.

29. Livak KJ, Schmittgen TD (2001) Analysis of Relative Gene Expression Data Using Real- Time Quantitative PCR and the $2^{-\Delta\Delta CT}$ method. Methods 25: 402–408.
30. Bradford MM (1976) A rapid and sensitive method for the quantification of microgram quantities of protein utilizing the principal of protein-dye binding. Anal. Biochem 72: 248–254.
31. Zhang Y, Yang C, Li Y, Zheng N, Chen H, et al. (2007) SDIR1 is a RING finger E3 ligase that positively regulates stress-responsive abscisic acid signaling in Arabidopsis. Plant Cell 19(6): 1912–1929.
32. Clough SJ, Bent AF (1998) Floral dip: a simplified method for Agrobacterium-mediated transformation of Arabidopsis thaliana. Plant J 16: 735–743.
33. Ardie SW, Xie L, Takahashi R, Liu S, Tahano T (2009) Cloning of a high-affinity K+ transporter gene PutHKT2; 1 from Puccinellia tenuiflora and its functional comparison with OsHKT2; 1 from rice in yeast and Arabidopsis. J Exp Bot 60: 3491–3502.
34. Norusis MJ (1990) SPSS advanced statistics user's guide: SPSS Chicago.
35. Cheng MC, Hsieh EJ, Chen JH, Chen HY, Lin TP (2012) Arabidopsis RGLG2, Functioning as a RING E3 Ligase, Interacts with AtERF53 and Negatively Regulates the Plant Drought Stress Response. Plant Physiol 158: 363–375.
36. Beasley SA, Hristova VA, Shaw GS (2007) Structure of the Parkin in-between-ring domain provides insights for E3-ligase dysfunction in autosomal recessive Parkinson's disease. Proc Natl Acad Sci USA 104: 3095–3100.
37. Marin I, Lucas JI, Gradilla AC, Ferrus A (2004) Parkin and relatives: the RBR family of ubiquitin ligases. Physiol Genomics 17: 253–263.
38. Taylor GJ (1988) The physiology of aluminum tolerance in higher plants. Commun. Soil Sci. Plant Anal 19: 1179–1194.
39. Barcelo J, Poschenrieder C (2002) Fast root growth responses, root exudates, and internal detoxification as clues to the mechanisms of aluminium toxicity and resistance: A review. Environ Exp Bot 48: 75–92.
40. Foy CD (1996) Tolerance of barley cultivars to an acid, aluminum-toxic subsoil related to mineral element concentrations in their shoots. J Plant Nutr 19: 1361–1380.
41. Ma JF, Furukawa J (2003) Recent progress in the research of external Al detoxification in higher plants: a minireview. J Inorg Biochem 97: 46–51.
42. Achary VM, Jena S, Panda KK, Panda BB (2008) Aluminium induced oxidative stress and DNA damage in root cells of Allium cepa L. Ecotoxicol. Environ. Saf 70: 300–310.
43. Achary VM, Panda BB (2010) Aluminium-induced DNA-damage and adaptive response to genotoxic stress in plant cells are mediated through reactive oxygen intermediates. Mutagenesis 25: 201–209.
44. Yamamoto Y, Kobayashi Y, Devi SR, Rikiishi S, Matsumot H (2002) Aluminum toxicity is associated with mitochondrial dysfunction and the production of reactive oxygen species in plant cells. Plant Physiol 128: 63–72.
45. Markkanen E, Hübscher U, van Loon B (2012) Regulation of oxidative DNA damage repair: the adenine: 8-oxo-guanine problem. Cell Cycle 11: 1070–1075.
46. Nezames CD, Sjogren CA, Barajas JF, Larsen PB (2012) The Arabidopsis cell cycle checkpoint regulators TANMEI/ALT2 and ATR mediate the active process of aluminum-dependent root growth inhibition. Plant Cell 24(2): 608–21.
47. Pinder JB, Attwood KM, Dellaire G (2013) Reading, writing, and repair: the role of ubiquitin and the ubiquitin-like proteins in DNA damage signaling and repair. Front Genet 4: 45.
48. Bartocci C, Denchi EL (2013) Put a RING on it: regulation and inhibition of RNF8 and RNF168 RING finger E3 ligases at DNA damage sites. Front Genet 4: 128.
49. Panda SK, Sahoo L, Katsuhara M, Matsumoto H (2013) Overexpression of alternative oxidase gene confers aluminum tolerance by altering the respiratory capacity and the response to oxidative stress in Tobacco Cells. Mol Biotechnol 54: 551–63.
50. Yin L, Mano J, Wang S, Tsuji W, Tanaka K (2010) The involvement of lipid peroxide-derived aldehydes in aluminum toxicity of tobacco roots. Plant Physiol 152: 1406–17.
51. Xue YJ, Tao L, Yang ZM (2008) Aluminum-induced cell wall peroxidase activity and lignin synthesis are differentially regulated by jasmonate and nitric oxide. J Agric Food Chem 56 (20): 9676–9684.
52. Kunihiro S, Hiramatsu T, Kawano T (2011) Involvement of salicylic acid signal transduction in aluminum-responsive oxidative burst in Arabidopsis thaliana cell suspension culture. Plant Signal Behav 6: 611–616.

14

A Soybean C2H2-Type Zinc Finger Gene *GmZF1* Enhanced Cold Tolerance in Transgenic *Arabidopsis*

Guo-Hong Yu[1]☯, Lin-Lin Jiang[1]☯, Xue-Feng Ma[1,2], Zhao-Shi Xu[3], Meng-Meng Liu[1], Shu-Guang Shan[1], Xian-Guo Cheng[1]*

1 Key Lab. of Plant Nutrition and Fertilizer, Ministry of Agriculture, Institute of Agricultural Resources and Regional Planning, Chinese Academy of Agricultural Sciences, Beijing, China, **2** Institute of Agro-Products Processing Science and Technology, Chinese Academy of Agricultural Sciences, Beijing, China, **3** Institute of Crop Science, Chinese Academy of Agricultural Sciences (CAAS)/National Key Facility for Crop Gene Resources and Genetic Improvement, Key Laboratory of Biology and Genetic Improvement of Triticeae Crops, Ministry of Agriculture, Beijing, China

Abstract

Zinc finger proteins were involved in response to different environmental stresses in plant species. A typical Cys2/His2-type (C2H2-type) zinc finger gene *GmZF1* from soybean was isolated and was composed of 172 amino acids containing two conserved C2H2-type zinc finger domains. Phylogenetic analysis showed that GmZF1 was clustered on the same branch with six C2H2-type ZFPs from dicotyledonous plants excepting for GsZFP1, and distinguished those from monocotyledon species. The GmZF1 protein was localized at the nucleus, and has specific binding activity with EP1S core sequence, and nucleotide mutation in the core sequence of *EPSPS* promoter changed the binding ability between GmZF1 protein and core DNA element, implying that two amino acid residues, G and C boxed in core sequence TGACAGTGTCA possibly play positive regulation role in recognizing DNA-binding sites in GmZF1 proteins. High accumulation of *GmZF1* mRNA induced by exogenous ABA suggested that *GmZF1* was involved in an ABA-dependent signal transduction pathway. Over-expression of *GmZF1* significantly improved the contents of proline and soluble sugar and decreased the MDA contents in the transgenic lines exposed to cold stress, indicating that transgenic *Arabidopsis* carrying *GmZF1* gene have adaptive mechanisms to cold stress. Over-expression of *GmZF1* also increased the expression of cold-regulated *cor6.6* gene by probably recognizing protein-DNA binding sites, suggesting that *GmZF1* from soybean could enhance the tolerance of *Arabidopsis* to cold stress by regulating expression of cold-regulation gene in the transgenic *Arabidopsis*.

Editor: Fan Chen, Institute of Genetics and Developmental Biology, Chinese Academy of Sciences, China

Funding: This work was supported by the National Key Project for Cultivation of New Varieties of Genetically Modified Organisms (2013ZX08002-005). The funders had no role in study design, data collection and analysis, decision to publish, or preparation of the manuscript.

Competing Interests: The authors have declared that no competing interests exist.

* Email: chengxianguo@caas.cn

☯ These authors contributed equally to this work.

Introduction

Plants are usually exposed to various environmental stress factors affecting plant growth and crop productivity, such as drought, high salt and low temperature. As an adaptive response, the plants have developed a resistance mechanism to abiotic stresses to achieve an optimal adaptation [1]. It was reported that a number of transcription factor gene from different plants was induced or repressed during the responses and acclimations [2]. For example, some typical transcription factors, AP2/ERF, bZIP, NAC, MYB, MYC, WRKY and zinc finger proteins have been well characterized, and confirmed to be involved in stress response via transcriptional regulation modulation [3,4,5], and these transcription factors usually have specific structure and/or conserved domains functioning in the adaptation to abiotic stresses [6,7].

The first zinc finger transcription factor IIIA (TFIIIA) was recognized as a repeated zinc-binding motif in *Xenopus*, and is a classical Cys2/His2-type (C2H2-type) zinc finger gene [6]. The C2H2-type zinc finger has been well characterized in the eukaryotic transcription factors having specific $CX_{2\text{-}4}CX_3FX_5LX_2HX_{3\text{-}5}H$ structure (X: any amino acid; number: amino acid amounts), and this conserved motif contains two pairs of specific Cys and His residues which form a tetrahedral structure with a zinc ion [7]. In plants, the C2H2-type zinc finger has a highly conserved QALGGH motif in a putative DNA-contacting surface, which is unique in zinc-finger proteins from plants [8,9]. *In vitro* analysis revealed that the conserved QALGGH motif in plants plays a critical role in DNA binding activity, and each amino acid residue in the motif seems to be essential for maintaining effective binding between DNA and C2H2-type zinc finger proteins [10,11], and these specific domains conferred that zinc-finger proteins play important regulatory role by recognizing the target sequences and regulating expression of target gene in a plant-specific manner [1,8,12].

As a first C2H2-type zinc finger protein in plants, ZPT2-1 was found in petunia (*Petunia hybrid*), and identified to interact with the specific DNA sequences in the promoter region of 5-enolpyruvylshikimate-3-phosphate synthase (EPSPS) [1,13]. Thereafter, several TFIIIA-type zinc finger proteins have been consecutively reported [1,8], and these C2H2-type zinc finger genes paly important regulation roles in responses to various abiotic stresses [8,9,14]. In petunia, the TFIIIA-type zinc finger genes, *ZPT2-2* and *ZPT2-3*, were regulated by cold and/or drought, and over-expression of *ZPT2-3* gene in the transgenic petunia increased the tolerance of plants to drought stress [15,16]. *STZ* is one of the *ZPT2*-related genes in *Arabidopsis*, and was isolated by complementation of the salt-sensitive phenotype with a yeast calcineurin mutant [17], and the expression of *STZ* gene has been identified to be responsive to drought, salt, cold and abscisic acid (ABA), and that constitutive expression of *STZ* resulted in a suppression in growth, and accompanied an enhancement of plants adaptation to drought and osmotic stresses [18,19,20].

To date, few zinc finger proteins from cultivar soybean (*Glycine max*) have been isolated and characterized. *SCOF-1,* a typical C2H2-type zinc finger gene, was isolated from soybean and was confirmed to have a positive role in regulating the expression of cold-regulated gene and enhancing cold tolerance in transgenic plants [21]. *GsZFP1*, a zinc finger transcription factor lacking typical QALGGH motif was isolated from wild soybean (*Glycine soja* L. G07256), and transformed into *Arabidopsis*, and confirmed to play a crucial role in withstanding cold and drought stresses [22], indicating that the presence of QALGGH motifs in the C2H2 zinc finger *GsZFP1* gene from wild soybean seems to be not a crucial element in functioning during adaptation to abitotic stress. In our study, a novel C2H2-type zinc finger gene, *GmZF1* was isolated from cultivar soybean, and identified to have two typical conserved QALGGH motifs, and is significantly different from *GsZFP1* gene in wild soybean [22]. To our knowledge, the understanding on the function of zinc finger C2H2 from cultivar soybean in responses to cold stress is still limited. Therefore, based on the structure analysis and *in vitro* identification of *GmZF1* gene, we transformed the *GmZF1* gene into *Arabidopsis* plants and identified three homozygous lines which were subsequently used to observe the function effects of *GmZF1* gene in responses to cold stress and exogenous ABA. Our data showed that *GmZF1* from cultivar soybean was induced by ABA, and over-expression of *GmZF1* gene significantly enhanced the tolerance of the transgenic *Arabidopsis* to low temperature stress.

Results

Characteristics of the *GmZF1* gene

The *GmZF1* cDNA is composed of 765-bp nucleotides, and eocodes a predicted protein of 172 amino acids with a calculated molecular mass of 19.2 kDa, and flanked by a 97 bp fragment at 5′ end and a 149 bp fragment at 3′ end at two untranslating regions, respectively (accession number DQ055134). Electrophoresis analysis showed that PCR product from the genomic DNA of soybean seedlings has the same size as that from the cDNA length generated by RT-PCR using total RNA as remplate, and sequencing also showed that the generated *GmZF1* clones from gDNA and cDNA have identical sequence, indicating that *GmZF1* gene has no intron in the genomic DNA. A homology search against the GenBank database showed that GmZF1 is a homolog of C2H2-type zinc finger proteins (ZFPs), and has two typical C2/H2 type zinc finger domains ($CX_2CX_3FX_3QALGGHX_3H$) (Fig. 1a). GmZF1 has 57%, 48%, 47% and 46% homology with the C2H2-type ZFPs from *Medicago truncatula*, Petunia, *Arabidopsis* and rice, respectively, and no highly homology was observed between these genes from different plants, suggesting that *GmZF1* belongs to a novel subfamily member of C2H2 zinc finger. Like C2H2 zinc finger genes from four kinds of plants mentioned above, the GmZF1 protein also has plant-specific QALGGH motifs as well as a conserved motif L-box with more Leu residues, and contains a short hydrophobic region with a highly conserved DLN box near the C-terminus of GmZF1 (Fig. 1a), which may function as a transcription repression domain [16,18]. It is noteworthy that GmZF1 have two conserved domains with QALGGH motifs, and is significantly different from *GsZFP1* isolated from wild soybean, the later only has one C2H2 domain lacking QALGGH motif, and the number of amino acid residues between two His (H) residues is composed of four amino acids, which is different from GmZF1 having three amino acids between two His amino acids (as indicated by # and * respectively in Fig. 1a), and no DLN box was found at the C-terminus of GsZFP1 [22].

For profiling the differences in genetic characteristics between C2H2 family members from different plants, a systematic phylogenetic analysis was carried out. The results showed that GmZF1 was clustered on the same branch with six C2H2-type ZFPs from dicotyledonous plants, and distinguished them from monocotyledon species (Fig. 1b), indicating that dicotyledonous plants and monocotyledon plants have significant difference in genetic characteristics of C2H2-type genes, and probably play different regulation roles in plant responses to abiotic stresses. What's worthy reminding is that *GmZF1* from cultivar soybean and *GsZFP1* from wild soybean respectively belong to two different subfamily members in C2H2 zinc finger super family.

Analysis for SDS-PAGE and DNA-binding activity of GmZF1

Analysis of SDS-PAGE indicated that the GmZF1::GST fusion protein with 45.2 kD molecular weight was successfully expressed in *E. coli* strain (Fig. 2a), and the size of expressed protein is identical with the predicted molecular weight of target protein in the fusion vector vector pGEX-4 T-1. As a DNA-binding motif, a C2H2-type zinc-finger domain in the GmZF1 has been identified in many transcription factors. In our study, we observed that the transcription factors containing the zinc finger motif could specifically bind to EP1S core sequence (TGACAGTGTCA), which was originally identified as a *cis*-element within the *EPSPS* gene promoter in petunia [13,23]. For identifying DNA-binding ability of GmZF1, a gel-shift assay was performed *in vitro* using the procedure as described in the method. Data showed that GmZF1 protein was expressed as fusion proteins with GST in *E. coli.* To better understand the molecular mechanism of specific-binding between the GmZF1 proteins and EP1S core sequences, a series of synthesis probes including wild type and mutants was prepared as the following descriptions. Briefly, the EP1S core sequence (wild-type, E1) and two probes (E2: bases substitution, E3: with four bases substitution) in the repeated sequences were prepared (Fig. 2b), and fusion protein was purified for gel shift assay. Assay showed that the GmZF1::GST fusion protein strongly bound to E1 and E2 probes, but weakly binding to E3 probe (Fig. 2c), no complex was found in reaction solution containing GST protein and probes, suggesting that GmZF1 protein could bind specifically to the EP1S core sequence *in vitro*, and two amino acid residues, G and C boxed in core sequence of TGACAGTGTCA probably play key role in recognizing DNA-binding sites in GmZF1 proteins.

Subcellular localization of GmZF1

To identify the subcellular localization of GmZF1, we fused the full length of GmZF1 to GFP vector under control of the constitutive 35S promoter. Both the recombinant DNA constructs encoding a GmZF1::GFP fusion protein and a GFP protein were respectively introduced into *Arabidopsis* protoplast cells.

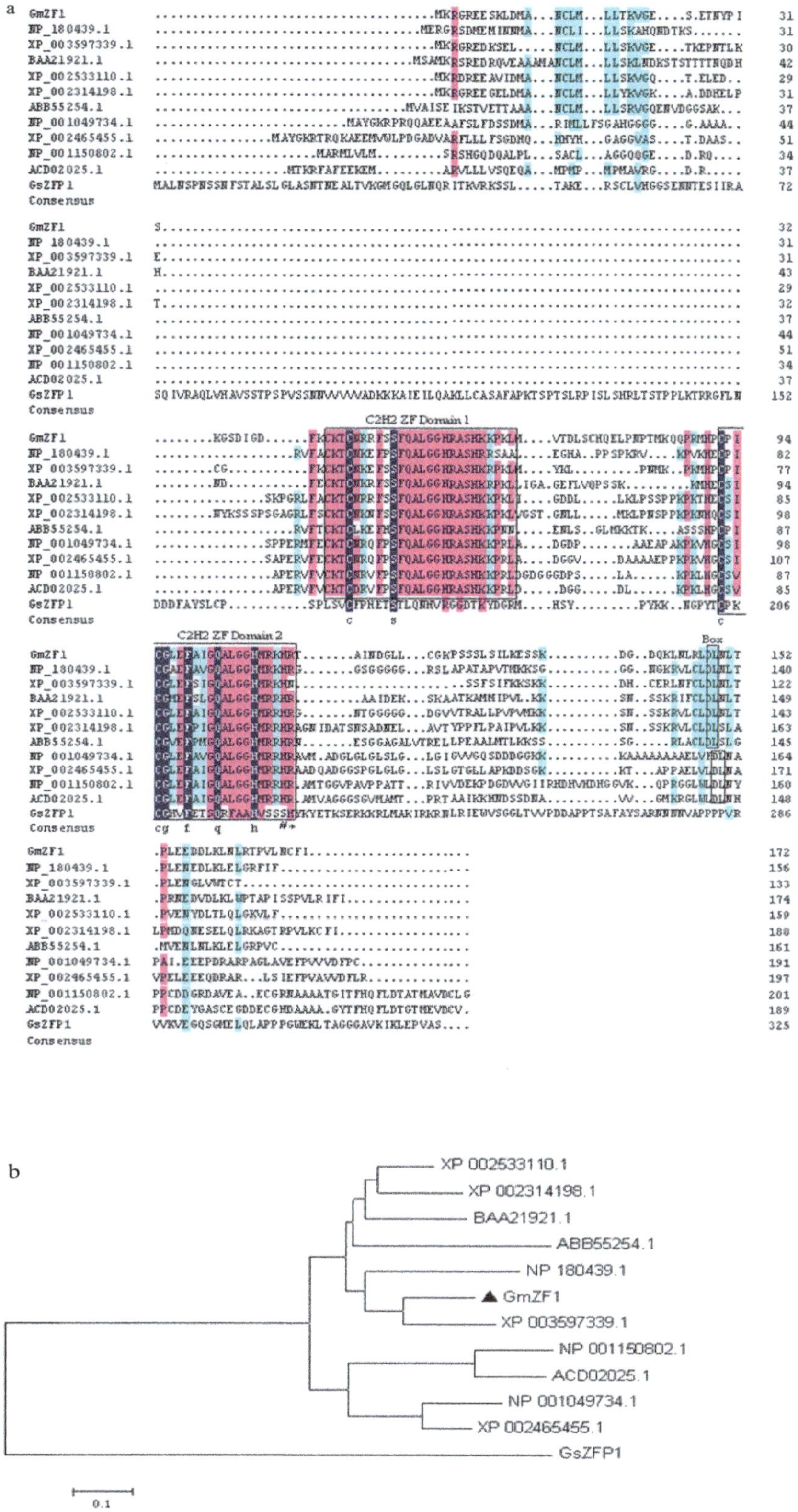

Figure 1. Sequence characteristic of GmZF1 deduced amino acids. a: Multiple alignment of GmZF1 amino acid sequences with XP_003597339.1 (*Medicago truncatula*), BAA21921.1 (*Petunia hybrid*)), NP_180439.1 (*Arabidopsis*) and NP_001049734.1 (*Oryza sativa*). The two zinc finger domains and the conserved DLN amino residues are boxed. b: Phylogenetic analysis of GmZF1 and related proteins. A mid-point rooted neighbor-joining phylogeny was constructed by 12 amino acid sequences from diverse organisms. Excepting for GmZF1, the other C2H2 type zinc finger proteins were demonstrated by XP_002533110.1 (*Ricinus communis*), XP_003597339.1 (*M. truncatula*), BAA21921.1 (*Petunia*), XP_002314198.1 (*Populus*), NP_180439.1 (*Arabidopsis*), ABB55254.1 (*Brassica*), NP_001049734.1 (*Oryza sativa*), XP_002465455.*1* (*Broomcorn*), NP_001150802.1 (*Zea mays*), *ACD02025.1* (*Triticum aestivum*) and GsZFP1 (*Glycine sojia.* L).

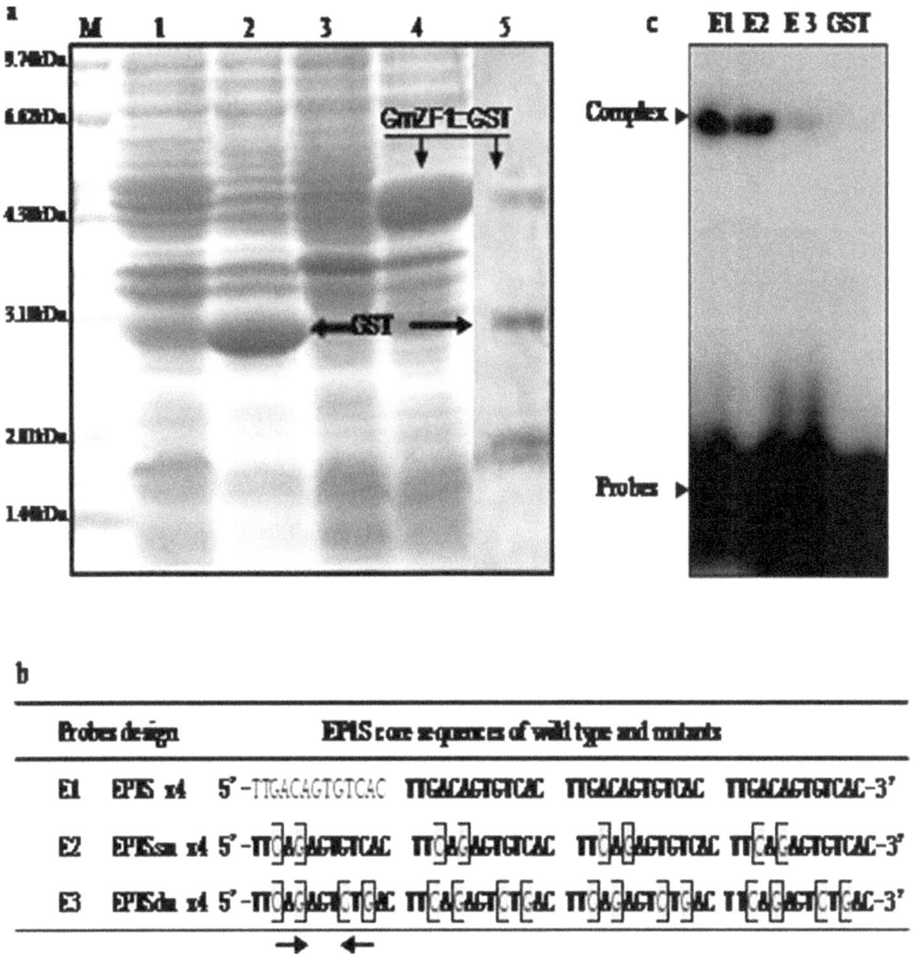

Figure 2. SDS PAGE induction expression and gel shift assay for DNA-binding activities *in vitro.* a: Induction expression and purification of target proteins, GmZF1::GST fusion protein and GST protein. Lane 1-uninduced strains containing pGEX4T-1 vector; Lane 2-induced strains containing pGEX4T-1; Lane-3-uninduced strains containing GmZF1::GST fusion; Lane 4-induced strains containing GmZF1::GST fusion; Lane 5-purified target protein as indicated by arrows. b: Probe design and composition. Nucleotide sequences of EP1S and mutated EP1S (E1, E2 and E3) probes. The nucleotide mutations in the EP1S core motif for each probe are boxed. c: The gel shift assay was performed in a solution containing 0.2 μg GmZF1::GST proteins or GST proteins and the probes ^{32}P-labeled EP1S (E1) or mutant EP1S (E2 and E3), respectively. The GmZF1::EP1S complex and free probes are indicated by arrows.

Localization of the GmZF1::GFP fusion protein was visualized exclusively in the nucleus (Fig. 3b), whereas the control GFP (35S::GFP) was distributed throughout the protoplast cells (Fig. 3a), demonstrating that GmZF1 is a nuclear protein possibly functioning as a transcriptional activator.

Expression of *GmZF1* in response to stress

To examine the transcription profile of *GmZF1* in soybean under cold stress, the expression of *GmZF1* in soybean was detected using semi-quantitative RT-PCR. The result showed that *GmZF1* was weakly expressed in young seedlings, and induced by low temperature and reached a maximum at 5 h after 4°C stress (Fig. 3c). Data also showed that *GmZF1* gene could be induced by exogenous ABA, and seems to have similar expression pattern under cold stress, and maintained at a high expression level when the seedlings were kept at ABA treatment (Fig. 3d), suggesting that *GmZF1* might be involved in plant responses to cold stress via an ABA-dependent pathway.

Phenotype changes of plants exposed to cold stress

To observe the difference in germination and survival rates between the wild type and the transgenic *Arabidopsis*, two independent culture experiments was performed at 22°C and at 4°C, respectively (Fig. 4a, b) Fig. 4a showed that both wild type and transgenic lines almost exhibited identical germination rates under favorable condition, but the survival rate of transgenic line OE1 was significantly higher than that of wild types under cold stress when wild type and transgenic lines of one-week-old seedlings were exposed to 4°C for one week (Fig. 4e). For observing the phenotypic changes of wild type and transgenic *Arabidopsis* under cold stress, the seedlings of two-week-old were transferred onto the MS medium, and cultured at 22°C (Fig. 4c) and 4°C (Fig. 4d) for one week, respectively (note: white bacterial plaques appeared in the mediums because of possible contaminant from culture processes). Data showed that transgenic lines showed a better phenotypic characteristic in withstanding cold stress compared to wild types (Fig. 4d), because both fresh weight and root lengths from the transgenic lines were significantly increased comparing to the wild types (Fig. 4f, g). These data suggested that *GmZF1* in the transgenic *Arabidopsis* could enhance the tolerance

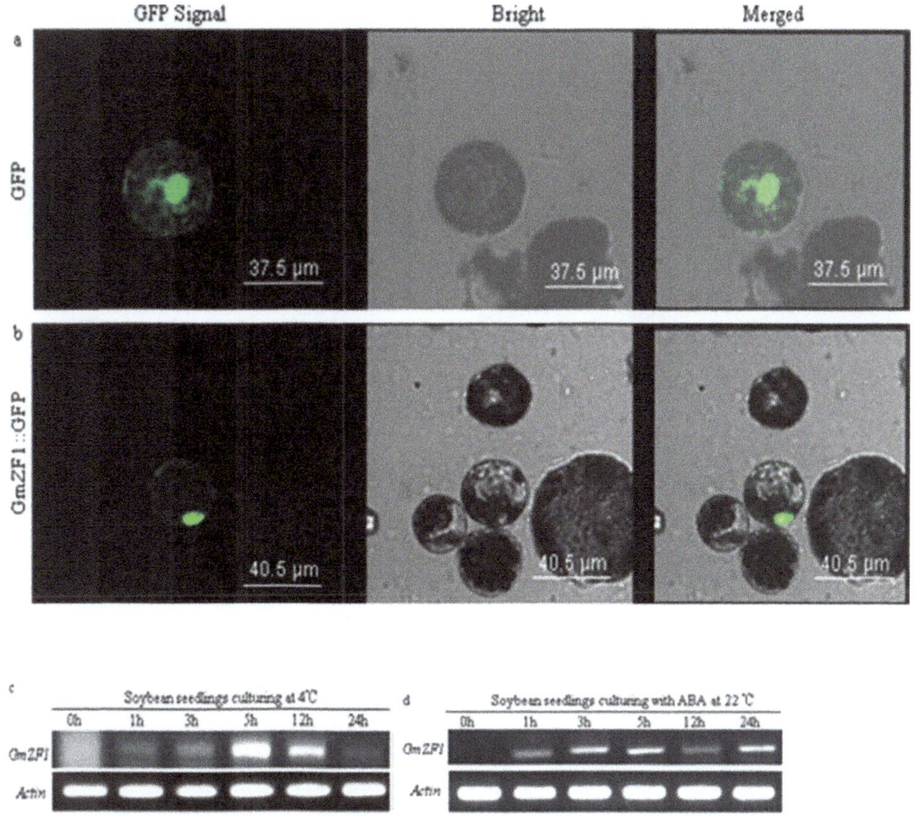

Figure 3. Subcellular localization of the GmZF1 proteins and mRNA accumulation of *GmZF1* gene in soybean seedlings respectively exposed to cold and ABA. a: Images expressing h16318-GFP in protoplast cells; b: images expressing the GmZF1::GFP fusion protein, fluorescent-field illumination was used to examine GFP signal (left); followed by bright-field illumination (middle) and confocal microscopy (right) for an merged image of bright and fluorescent illumination. c: the relative accumulation of *GmZF1* mRNA in soybean seedlings of 10-day-old exposed to 4°C at different time points; d: the accumulation of GmZF1 mRNA in soybean seedlings of 10-day-old exposed to ABA of 200 μM at different time points.

plants to the cold stress. *GmZF1* and *GsZFP1* [22] seems to play similar roles in acclimating the tolerance of plants to cold stress, although these two genes have differences in the structure, especially in the conserved domains of C2H2 zinc finger gene, suggesting that different zinc finger genes in the same super family could demonstrate an uniformity in gene functions even though they have the difference in the genetic characteristics, and this seems to imply that the diversity genetic of genes is an essential factor in regulating the expression of target genes related to abiotic stresses. Therefore, to better elucidate the molecular regulation mechanism of *GmZF1* from cultivar and *GsZFP1* from wild soybean in responding to cold stress, further investigations is necessary.

Over-expression of *GmZF1* led to physiological changes in the transgenic lines under cold stress

To further characterize the function of *GmZF1*, we generated *Arabidopsis* transgenic plants carrying *35S::GmZF1*, in which *GmZF1* was driven by the CaMV35S promoter. Total 22 transformed *Arabidopsis* plants with kanamycin-resistant were obtained, and three independent homozygous transgenic lines (OE1, OE4 and OE7) carrying *GmZF1* were continuously selected by the PCR analysis until T3 transgenic *Arabidopsis* lines. For cold stress, 5-week-old wild-type and transgenic lines were transferred to the incubator at 4°C for two days, and morphological observation showed that the transgenic lines were not significantly inhibited in growth comparing to wild type in spite of being a short term cold stress, and the growth of the transgenic lines carrying *GmZF1* gene were slightly damaged comparing to wild type (Fig. 5a, b). However, three transgenic lines have remarkable physiological changes in accumulation of the proline, soluble sugar and MDA in the leaves comparing to the wild type after cold stress. Cold stress enhanced accumulation of proline and soluble sugar in the wild type and transgenic lines, but the increase times in the contents of proline in the transgenic lines (OE1, 1.3-fold; OE4, 1.3-fold and OE7, 1.6-fold) were significantly higher than that in the wild type (1.1-fold) (Fig. 5c). Similarly, the increase times in the content of soluble sugar in the leaves from OE1, OE4 and OE7 (1.7-fold, 1.9-fold and 2.2-fold) were also higher than that in the wild type (1.4-fold) after cold stress for two days (Fig. 5d). In contrast, the contents of MDA in the transgenic lines (OE1, 0.012 $\mu mol{\cdot}g^{-1}$; OE4, 0.006 $\mu mol{\cdot}g^{-1}$ and OE7, 0.015 $\mu mol{\cdot}g^{-1}$) were decreased comparing to the wild type (0.019 $\mu mol{\cdot}g^{-1}$) after cold stress (Fig. 5e), demonstrating that over-expression of *GmZF1* reduced oxidative damage of membrane lipid peroxidation, and further maintained osmotic balance in the plant cells.

To observe the physiological effects of exogenous ABA supply on the transgenic *Arabidopsis* lines, the leaves from the wild type and the transgenic *Arabidopsis* lines exposed to ABA solution was sampled, and used for the determination in the content changes of praline, soluble sugar and MDA. Data showed that the contents of proline in the wild type were almost identical with that in the transgenic lines before ABA treatment. However, the content of

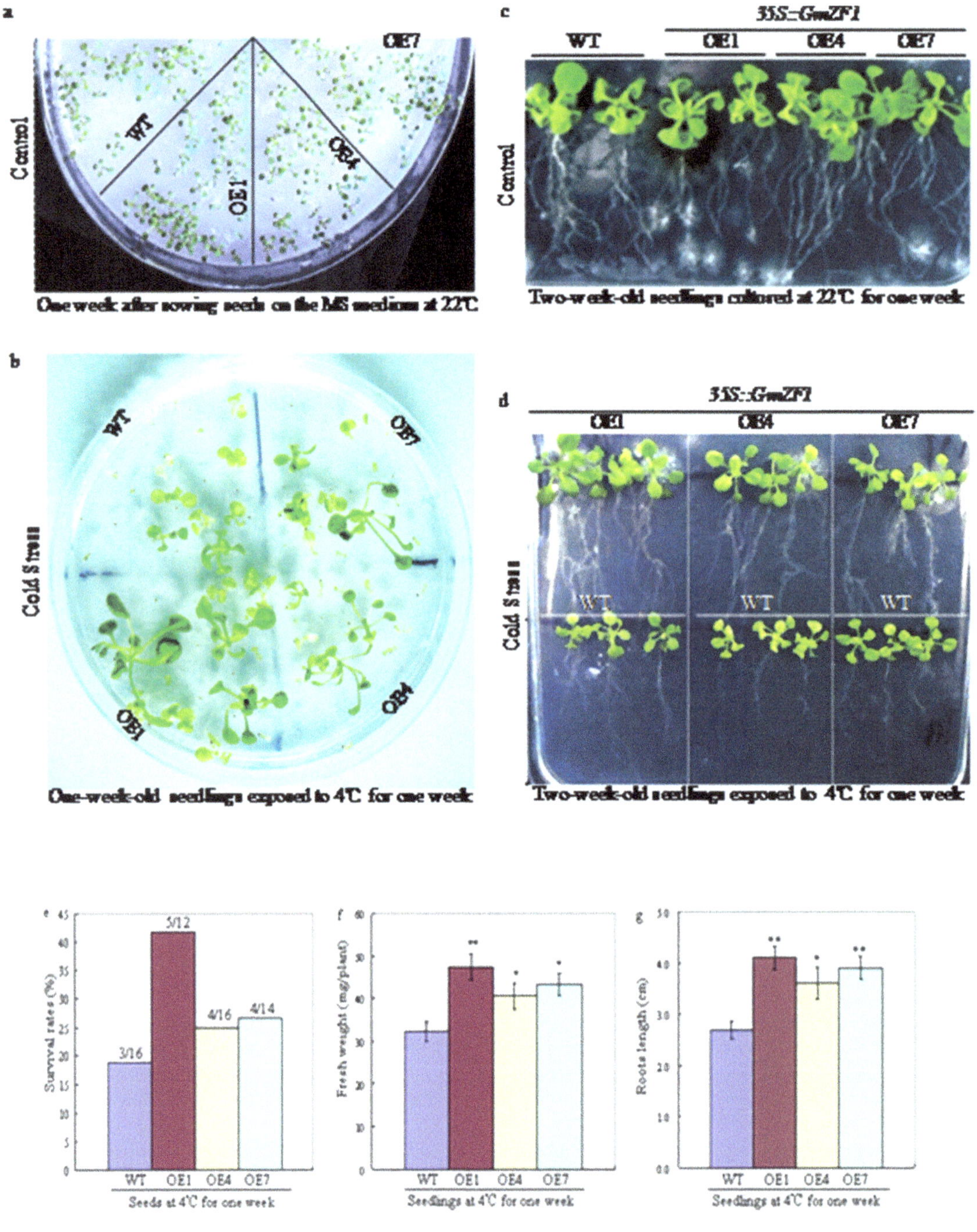

Figure 4. Morphological identification on wild type and transgenic *Arabidopsis* exposed to cold stress at 4°C on the MS medium. a: *Arabidopsis* seedlings of one-week-old after culturing at 22°C; c: *Arabidopsis* seedlings of two-week-old cultured at 22°C for one week; b and d: *Arabidopsis* seedlings of one-week-old cultured at 4°C for one week or two weeks, respectively; e: the survival rates of *Arabidopsis* seedlings exposed to 4°C for one week; f: the fresh weight per plant of two-week-old *Arabidopsis* after culturing at 4°C for one week; g: the roots length of two-week-old *Arabidopsis* seedlings after culturing at 4°C. WT-wild type; OE1, OE4 and OE7- transgenic lines.

proline in the transgenic lines (OE1, 142.4 ng·mg^{-1}; OE4, 427.2 ng·mg^{-1} and OE7, 412.9 ng·mg^{-1}) were significantly higher than that in the wild type (125.7 ng·mg^{-1}) after application of ABA (Fig. 6c). In addition, ABA treatment also enhanced the levels of soluble sugar in the wild type and the transgenic lines, but the increase times in the content of soluble sugar in the transgenic lines (OE1, 1.5-fold; OE4, 2.1-fold and OE7, 2.0-fold) were higher than that in the wild type (1.4-fold) after two days by ABA treatment for (Fig. 6d). The contents of MDA in the transgenic lines (OE1, 0.012 μmol·g^{-1}; OE4, 0.009 μmol·g^{-1} and OE7, 0.011 μmol·g^{-1}) were reduced comparing to the wild type (0.013 μmol·g^{-1}) after ABA treatment, whereas the contents of MDA in the transgenic lines (OE1, 0.009 μmol·g^{-1}; OE4, 0.009 μmol·g^{-1} and OE7, 0.008 μmol·g^{-1}) were slightly higher than that in the wild type (0.007 μmol·g^{-1}) before ABA treatment (Fig. 6e).

The induction of numerous stress-responsive genes is a hallmark of stress acclimation in plants. To elucidate the molecular

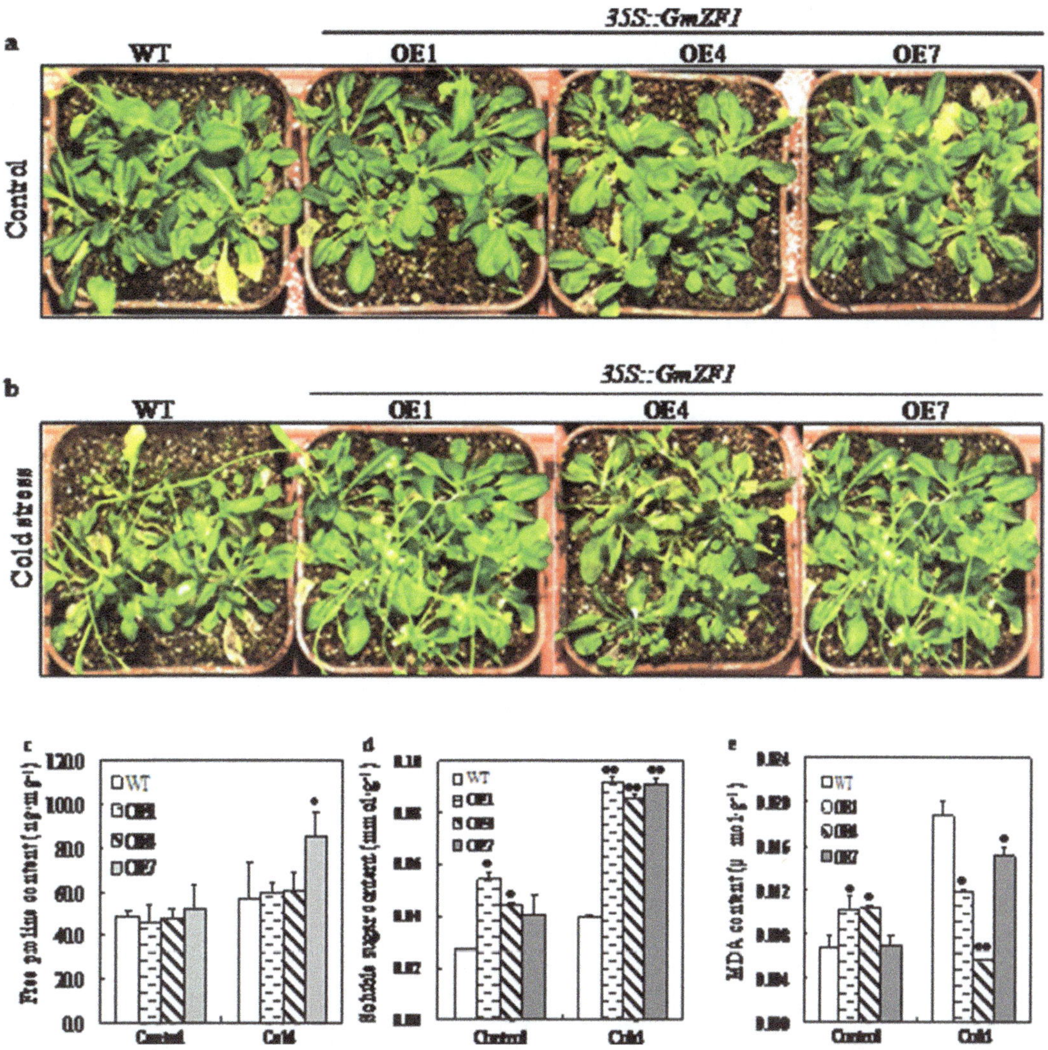

Figure 5. Phenotypic and physiological changes in wild type and transgenic *Arabidopsis* lines under cold stress. a: Phenotypes of wild type and transgenic lines of 5-week-old seedlings before cold stress (Control). b: Phenotypes of wild type and transgenic lines of 5-week-old seedlings exposed to 4°C for one week (Cold stress). c, d and e respectively represent the content changes of free proline, soluble sugar and MDA in the leaves of wild type and in transgenic lines before (Control) and after cold stress (Cold). WT-wild type; OE1, OE4 and OE7-transgenic lines; single * and duble ** respectively means significant difference at 0.05 and 0.01 levels.

mechanism of *GmZF1* in responding to cold, both wild type and transgenic seedlings of *Arabidopsis* were incubated at 4°C or by ABA for 48 hours to examine the expression of cold-responsive genes *cor6.6* at different time points by real-time PCR analysis. As shown in Fig. 7, both wild type and transgenic lines (OE1, OE4 and OE7) exhibited a similar levels in the expression of *cor6.6* before cold treatment, and the OE1 line began to appear significant up-regulations in *cor6.6* expressions in the roots after 12 hours cold treatment, while the OE7 line significantly increased the expression of *cor6.6* gene in the leaves after 12 h cold stress (Fig. 7a, b) When the seedlings of *Arabidopsis* were exposed to exogenous ABA, the expressions of *cor6.6* gene in the roots of the OE1 and OE7 lines were significantly induced after 24 hours (Fig. 7c), and the all transgenic lines (OE1, OE4 and OE7) exhibited remarkable increase in the expression of *cor6.6* gene in the leaves after 12 hours (Fig. 7d). Although both the wild type and transgenic lines at some time points did not exhibit significant differences in the expression of *cor6.6*, but the transgenic lines, OE1, OE4 and OE7 still showed higher transcript levels of *cor6.6* gene than the wild type under cold stress or exogenous ABA supply (Fig. 7b, c and d), suggesting that *GmZF1* may play an important role in activating cold-resistance genes in the transgenic *Arabidopsis* responding to cold stress through an signal pathway depending ABA.

Fig. 7e, f showed that the expressions of *GmZF1* gene in the roots and in the leaves of transgenic lines significantly increased after exposing the plants to cold stress for 6 hours, and this result seems to be identical with that in soybean seedlings exposed to cold stress or ABA (Fig. 3c, d). Fig. 7e showed that the expression of *GmZF1* gene in the roots of OE4 and OE7 was almost identical, but the OE7 line exhibited obvious increase in the expression of *GmZF1* gene in the leaves after 6 hours cold stress relative to other transgenic lines, and then maintained at relative stable levels at other time points (Fig. 7f). In addition, accumulation pattern of *GmZF1* gene in the transgenic *Arabidopsis* lines treated by exogenous ABA also showed that *GmZF1* gene in the leaves of the transgenic *Arabidopsis* lines was strongly induced by ABA, and the accumulation levels *GmZF1* gene in the roots of OE1 line were

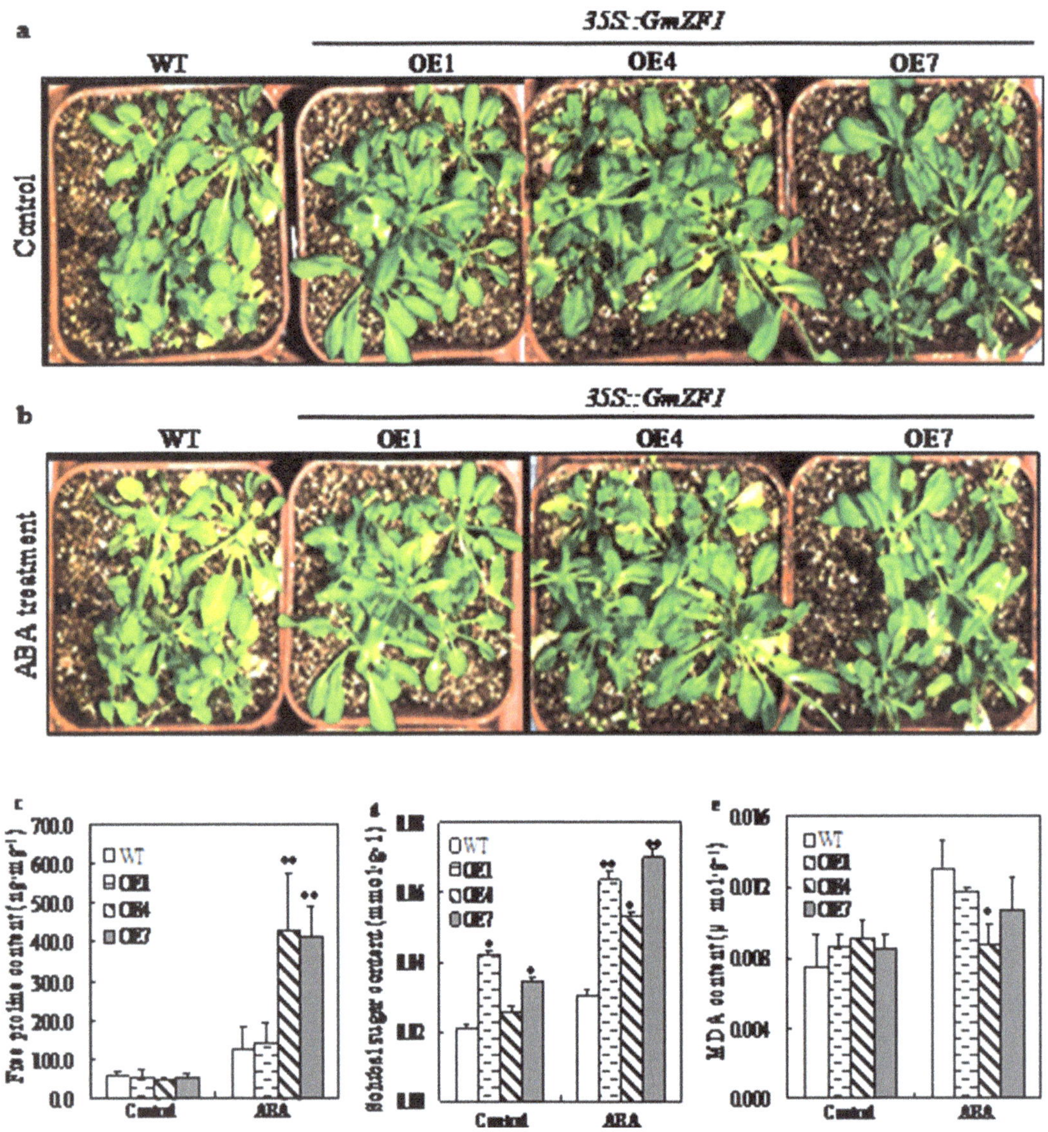

Figure 6. Phenotypic and physiological changes in wild type and transgenic *Arabidopsis* lines exposed to ABA. a: Phenotypes of wild type and transgenic lines of 5-week-old seedlings before ABA treatment. b: Phenotypes of wild type and transgenic lines of 5-week-old seedlings exposed to 200 μM ABA for one week. c, d and e respectively represent the content changes of free proline, soluble sugar and MDA in the leaves of wild type and transgenic lines before (Control) and after ABA treatment (ABA). WT-wild type; OE1, OE4 and OE7-transgenic lines; single * and duble ** respectively means significant difference at 0.05 and at 0.01 levels.

increased significantlyafter exposing to ABA for 24 hours (Fig. 7g) hours, and the accumulation of *GmZF1* mRNA in the leaves of OE the transgenic lines exhibited strong induction by ABA after 6 hours, and only OE7 line has almost identical induction accumulation patterns of *GmZF1* mRNA by ABA, and the induction accumulation seems to be inhibited after 48 hours (Fig. 7h). However, the accumulation of *GmZF1* mRNA in the OE7 transgenic line was little after exposing to ABA supply compared to the other transgenic lines, indicating that different transgenic lines exhibited different accumulation patterns in expressions of *GmZF1* gene in the plant organs, and seems to imply that accumulation patterns of *GmZF1* mRNA in the transgenic plants probably are related to the positions and copy numbers of *GmZF1* gene inserting in the genome of *Arabidopsis*.

Discussion

The C2H2 zinc-finger transcription factors were usually thought to be involved in plant development and have various adaptive responses to the environment stress [10,24]. Although the roles of some C2H2 zinc-finger transcription factors have been identified to be related to stress and developmental processes, the functions of C2H2 ZFPs from soybean involved in stress response are largely unknown [22].

In this study, as a novel C2H2 zinc finger protein gene, *GmZF1* from soybean was characterized. Sequence analysis revealed that the GmZF1 had high identity with other C2H2-type ZFPs, and shared two zinc finger motifs containing a conserved plant-specific QALGGH amino acid sequence which was proved to be critical for DNA-binding activity [11]. Based on the present data, we predicted that the binding activity of GmZF1 to EP1S core elements was probably mediated by the QALGGH sequence.

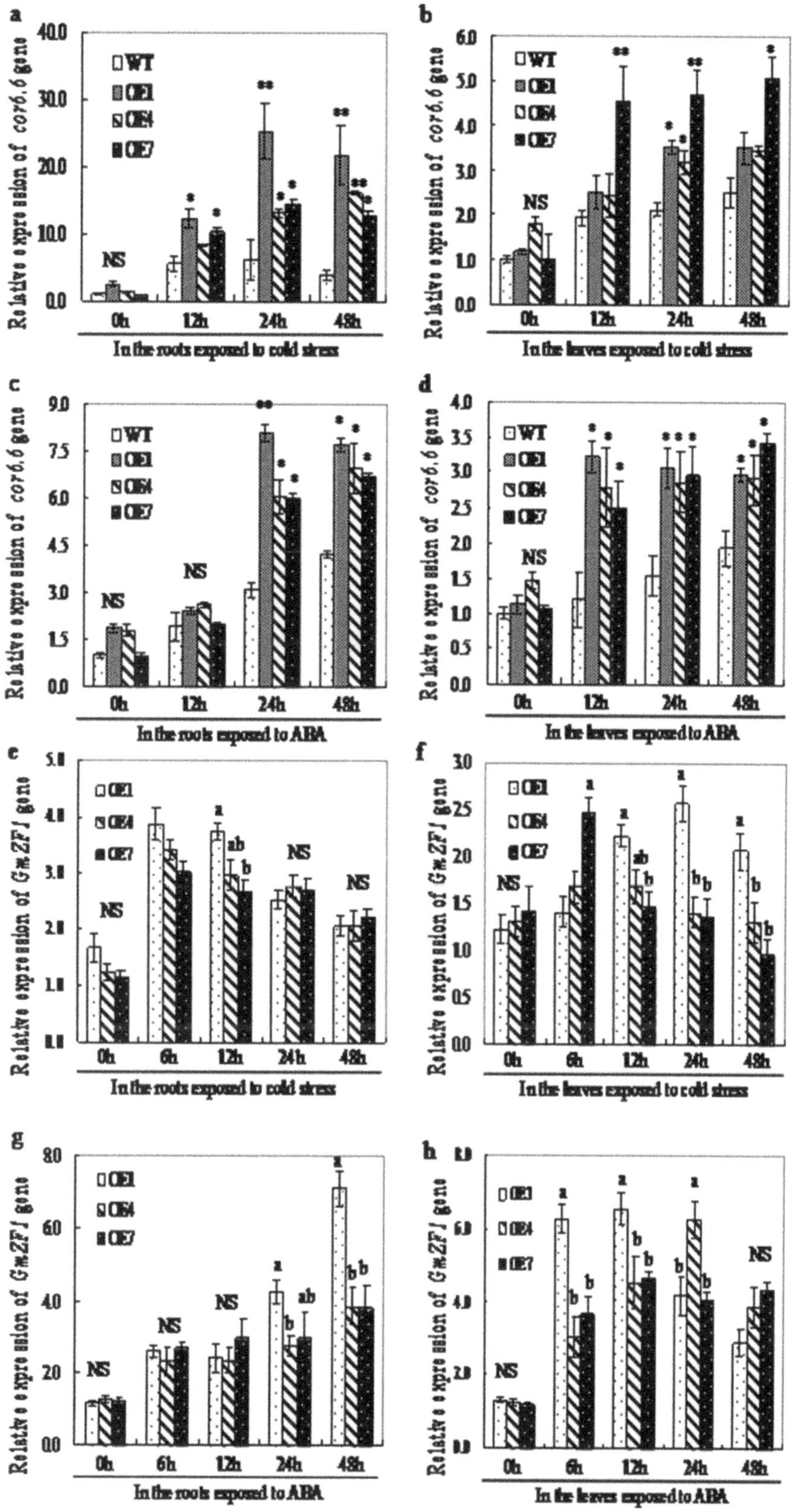

Figure 7. Responsive expression of *cor6.6* gene and relative accumulation of *GmZF1* mRNA in wild type and transgenic lines both under cold stress and ABA supply. a and b respectively represent the relative expressions of *cor6.6* gene in the roots and leaves of plants exposed to cold stress; c and d respectively represent the relative expressions of *cor6.6* gene in the roots and leaves of plants exposed to ABA, and the expression level of *cor6.6* gene in unstressed wild type was set at 1.0 in the transgenic *Arabidopsis* lines. e: the relative expressions of *GmZF1* gene in the roots; f: the relative expressions of *GmZF1* gene in the leaves of transgenic plants exposed to cold stress; g and h respectively represent the relative expressions of *GmZF1* gene in the roots and leaves of transgenic plants exposed to ABA. Error bars indicate the standard deviation; NS means

no significant differences; lowercase letters labeled on the columns mean statistical difference at 0.05 level ce at 0.05 level; single * and duble ** respectively means significant difference at 0.05 and at 0.01 levels; WT-wild type; OE1, OE4 and OE7-transgenic lines.

Additionally, the C-terminus of *GmZF1* gene contains typical Leu-rich L-box and DLN-box which play roles in protein interactions or in maintaining the folded structure [19]. The DLN-box was thought to function in transcriptional repression. As reported previously, the DLN-boxes in the zinc finger proteins ZPT2–3 from petunia [16] and STZ/ZAT10 from *Arabidopsis* exhibited repression roles in transcription activities [18]. However, some zinc finger proteins containing the DLN-box were involved in transcriptional activation, such as ThZF1 from *Thellungiella halophila* [25] and CaZF from chickpea [26]. Subcellular localization analysis revealed that GmZF1 localized at nuclei (Fig. 3), implying that GmZF1, like other ZFPs from plant TFIIIA-type, functions as a transcription factor in plant cells and may play an important role in signaling pathway in soybean under abiotic stress.

As a key regulator, ABA plays an important role in signaling pathways under stresses, such as drought, low temperature and osmotic stress, and induces the expressions of a number of genes that respond to abiotic stress [30,31]. For example, the expression of *AZF2* gene from *Arabidopsis* was strongly induced following ABA treatment under drought and salt stress [1,12,18]. However, it was reported that some C2H2-type zinc finger genes are induced by dehydration and cold stress, but do not respond to exogenous ABA [2,32,33], suggesting that two kinds of signaling transduction pathways, ABA-independent pathway and ABA-dependent pathway play different regulation roles in responding to abiotic stresses, repectively, and that the initial stress signal was converted into cellular responses [34]. In this study, our data proved that *GmZF1* was involved in plant responses to cold stress through an ABA-dependent signal transduction pathway since the expression of *GmZF1* in soybean seedlings was clearly induced by ABA (Fig. 4). In addition, accumulation pattern of *GmZF1* gene in the transgenic *Arabidopsis* lines treated by exogenous ABA also showed that over-expression of *GmZF1* gene in *Arabidopsis* was strongly induced by ABA.

Some of the C2H2-type ZFPs from different plant species are confirmed to play regulatory roles in stress responses, such as AZF2 and STZ in Arabidopsis [12,17,27], SCOF-1 in soybean [21], GsZFP1 in wild soybean [22], StZFP1 in potato [9], DST in rice [28] and TaCHP in wheat [29]. Expression analysis revealed that GmZF1 was clearly induced by cold stress (Fig. 4), suggesting that *GmZF1* might be involved in plant responses to cold stress in soybean. Transgenic *Arabidopsis* plants over-expressing *GmZF1* were evaluated for the involvement of *GmZF1* gene in cold tolerance of plants. Although no morphological differences between the transgenic lines and the wild type was observed, the remarkable differences in accumulation of free proline, soluble sugar and MDA were confirmed after cold and ABA treatment (Fig. 5c–e and Fig. 6c–e). Accumulation of proline by stress-induced has been observed in many plant species, and functions as an osmo-protectant coping with stress [32,35,36]. For example, the transgenic tobacco plants over-expressing *GmDREB3* could accumulate much free proline than the wild-type plants after drought stress treatment for 16 d [37]. In plants, soluble sugar has been shown to fulfill a dual role as both metabolites and as signaling molecules [38,39] that may play important roles in the mechanisms of plant responding to the stress [40,41]. In our study, the accumulations of proline and soluble sugar in the wild type and transgenic lines were improved, and increase times in the contents of proline and soluble sugar in the transgenic lines were higher than that in wild type during cold and ABA treatment (Fig. 5 and Fig. 6), proving that the accumulation of soluble sugars has been occurred in many plant species during cold acclimation [42]. Based on our investigations, both wild type and transgenic *Arabidopsis* lines basically have no significant differences in germination rates. However, the survival rates from these two phenotypes was remarkable when these two seedlings were exposed to cold stress, the survival rates of transgenic *Arabidopsis* lines cultured at 4°C significantly increased after one week comparing to the wild types (Fig. 4 a,b), indicating that *GmZF1* gene could play regulation role during the plant growth and development. Our study showed that over-expression of *GmZF1* gene significantly enhanced the transcription levels of *cor6.6* gene responding to cold stress and ABA supply, and increased the accumulation of proline, soluble sugar in the transgenic *Arabidopsis*, and reduced the content of MDA in the transgenic lines comparing to the wild type, and significantly improved the tolerance of the transgenic lines exposed to cold stress and exogenous ABA (Fig. 5e; Fig. 6e). In conclusion, over-expression of *GmZF1* resulted in an enhancement in cold tolerance of the transgenic *Arabidopsis* by activating transcription of *cor6.6* and/or cold resistance-genes (Fig. 7) as well as accumulation changes of proline, soluble sugar and MDA [43–45]. Bioinformatics analysis fund that the promoter with 1955 bp of *cor6.6* gene contains 7×GACA and 6×GTCA repeat in core element TTGA-CAGTGTCAC, respectively (NCBI Reference Sequence in *Arbidopsis thaliana* chromosome 5: NC_003076.8-AT5G15950/NM_121600.3/NP_197099.1), and these two core elements possibly play key role in recognizing DNA-binding sites in target proteins, because the mutation of G and C in the two boxed core elements could change the ability of GmZf1 protein binding to DNA, especially the mutation of G and C in the GCTA box (Fig. 3c). Therefore, *GmZF1* could play regulation role by activating the transcription of *cor6.6* cold-regulated gene in *Arabidopsis,* and led to an enhancement in the tolerance of transgenic plants to cold stress.

Materials and Methods

Plant materials and growth conditions

Soybean seedlings (*Glycine max*) were grown in a growth chamber at 24°C with 60% relative humidity under 16 h light and 8 h darkness. *Arabidopsis* plants (genotype Colombia) were grown in a controlled environmental chamber at 22°C and 70% humidity with a 14 h light/10 h darkness cycle under normal light intensity (150 $Em^{-2}\cdot s^{-1}$). T_1 seeds were sterilized and planted on the MS medium containing kanamycin of 50 $\mu g\cdot mL^{-1}$ for the selection of transgenic plants. After continuously screening on the MS medium containing kanamycin of 50 $\mu g\cdot mL^{-1}$ for two times and the generated T3 seedlings of transgenic *Arabidopsis* lines were transplanted to pots or MS medium for further investigations.

Isolation and sequence analysis of *GmZF1*

The zinc finger gene *GmZF1* was cloned from soybean based on the expressed sequence tag (EST) database of soybean. Total RNA was extracted from soybean seedlings using RNA Prep. Pure plant kit (Tiangen Biological Company, Beijing), and cDNA synthesis was performed by a reverse transcription kit (TaKaRa Dalian BioCompany). The entire *GmZF1* cDNA of was obtained by PCR using the specific primers (F: 5′-AGAGGAAACTAGCTAGGG-CACTTC-3′ and R: 5′-CCCGAGAACTAAGAAGTTTCG-TATT-3′). The deduced amino acid sequences of soybean GmZF1 zinc finger protein were matched by a protein blast procedure (http://www.ncbi.nlm.nih.gov/BLAST/). BioEdit version 7.0 software was used to multiple sequence alignments. Phylogenetic analysis was carried out by MEGA version 4.0 with adopting position correction distance using a bootstrap replicate number of 1000.

Induction expression of *GmZF1* fusion protein and electrophoretic mobility shift assay

An entire 516-bp *GmZF1* fragment containing the DNA binding domain was amplified by PCR using a pair of primers F and R (F: 5′-ACAACTCGAGATGAAGAGAGGCAGAGAA-3′ and R: 5′-AGACGAATTCAATGAAACAATTGAGCAC-3′), and was subcloned onto the pGEX4 T-1 vector by inserting at two specific sites, *EcoR*I and *Xho*I (Amersham Biosciences), and the recombinant pGEX-4 T-1 plasmids were identified by sequencing, and transferred into *Escherichia coli* BL21 cells (Amersham Biosciences). The induction of target fusion proteins was performed by adding 0.8 mM IPTG (Isopropyl-β-d-thiogalactopyranoside) into the culture of *E. coli* strains, and the cultures carrying GmZF1::GST fusion proteins was incubated for 5 hours at 37°C with 250 rpm The bacterial cells were shattered by ultrasonication in phosphate-buffered saline (PBS) and centrifuged at 11000 *g* for 10 min to remove insoluble cell debris, and then the supernates were collected, and applied on the gel containing 12% ployacrylamide for electrophoresis detection of fusion protein by the method of SDS-PAGE described in the book [46]. The GST::GmZF1 fusion protein was purified using a glutathione-Sepharose 4B column (Amersham Biosciences) according to the manufacturer's instructions. The 52 bp DNA fragment containing four copies of wild-type or mutant EP1S core sequences were labeled by 25 $\mu Ci \cdot \mu L^{-1}$ of γ-^{32}P-dATP (Amersham Biosciences). The DNA-binding reaction was performed in a 20 μL binding buffer containing ^{32}P-labelled probe, glycerine and purified fusion protein [24], and were subjected to electrophoresis with the gel containing 0.53 Tris-borate-EDTA, 6% polyacrylamide, and the gels were dried and visualized by autoradiography.

Subcellular localization of *GmZF1* gene

In brief, an entire *GmZF1* cDNA fragment was amplified by PCR procedure with a pair of specific primers (F: 5′-AACACTG-CAGATGAAGAGAGGCAGAGA-3′), and R: 5′-CGGGATC-CAATGAAACAATTGAGCAC-3′.). Based on the specific sites in the MCS of GFP vector, two specific sites, a *Pst*I and *BamH*I were respectively inserted into the 5′ ends of the primers for subcloning requirement. The digested PCR fragment was inserted into the multiple cloning sites (MCS) in the GFP vector carrying a GFP protein driven by a 35 S promoter, and this construct carrying 35 S::GFP or 35 S::GmZF1::GFP were introduced into *Arabidopsis* protoplast cells, which was prepared by young seedlings of *Arabidopsis,* and the transformation procedures were performed by the method as described previously [47,48]. The images were captured using a Nikon Eclipse TE2000-U microscope (Nikon).

Transformation of *Arabidopsis*

A pair of specific primers, the *GmZF1* cDNA fragment was generated by PCR procedure using the specific primers (F: 5′-AACACCATGGCCATGAAGAGAGGCAGAGA-3′, and R: 5′-AGACACTAGTAATGAAACAATTGAGCAC-3′. The generated PCR product directly was inserted onto the pCAMBIA1304 vector using a pair of specific restriction sites (*Nco*I and *Spe*I), and an over-expression vector p35 S::GmZF1 was constructed, and introduced into *Agrobacterium tumefaciens* strain C58C1. The flowering *Arabidopsis* seedlings were used for genetic transformation at 25°C by the method of floral-dip. T3 transgenic *Arabidopsis* plants were continuously cultured in the medium containing kanamycin of 45 $mg \cdot L^{-1}$ and three positive transgenic *Arabidopsis* were identified and the T3 seeds of transgenic *Arabidopsis* lines and wild type *Arabidopsis* were sowed on MS medium based on the requirements of experiment, and the resulting seedlings were accordingly transferred into the pots and MS medium for cold stress and ABA treatment.

Stress treatments

Cold stress was performed by incubating soybean seedlings of 10-day-old at 4°C for 24 hours and *Arabidopsis* seedlings of two-week-old seedlings were incubated at 4°C for 48 hours. ABA treatment was respectively carried out by completely spraying 200 $\mu mol \cdot L^{-1}$ ABA onto the 10-day-old soybean seedlings and two-week-old *Arabidopsis* seedlings Soybean leaves from the same position were respectively sampled at 0, 1, 3, 5, 12 and 24 hours, and *Arabidopsis* leaves were collected respectively at 0, 6, 12, 24 and 48 hours. Equal amounts of seeds from wild type and transgenic lines were sowed on MS medium at 22°C (Control), and the generated seedlings of one-week-old *Arabidopsis* were cultured at 22°C and at 4°C (Cold stress), respectively, for one week in two independent growth chambers, and used for investigation in survival rates. For evaluating the changes of Arabidopsis biomass and root development under cold stress, two-week-old seedlings of *Arabidopsis* from control culture were transplanted on the MS medium, and cultured respectively at 22°C (Control) and at 4°C (Cold stress) for one week in two independent growth chambers, and biological fresh weights per plant and root lengths from two treatments were statistically investigated based on the three replicates.

Gene expression analysis

Based on the equal amounts of leaves RNA from each time point, the expression pattern of *GmZF1* gene in soybean was characterized using semi-quantitative RT-PCR by GmF and GmR primers, and accumulation analysis of *GmZF1* mRNA in the transgenic *Arabidopsis* lines was performed using the specific primers (GmF: 5′-ATGAAGAGAGGCAGAGAA-3′ and GmR: 5′-AATGAAACAATTGAGCAC-3′) by quantitative real-time PCR (qPCR) For understanding the expression profile of GmZF1 gene and cold-responsive marker gene *cor6.6* in the transgenic *Arabidopsis* under low temperature stress and exogenous ABA supply, the wild type Arabidopsis and transgenic *Arabidopsis* lines carrying GmZF1 gene were incubated at 4°C for 48 hours in darkness. All samples including the roots and leaves were collected with three biological replicates at the designated time intervals after cold and ABA treatment, and were quickly frozen in liquid nitrogen and stored at 80°C for total RNA isolation. As a reference gene in *Arabidopsis*, actin was used as an internal maker, and amplified by the specific primers [(F 5′-AAGTATCCTATT-GAGCATGGTGTTG-3′; R 5′- CTGGCGTACAAGGAGAG-A-3′), (accession number: AEE76148)], and the *cor6.6* gene was amplified by the primers (F 5′- ATGTCAGAGACCAACAA-GAATG-3′; R5′- CTTGTTCAGGCCGGTCTTG-3′) (accession number: CAA38894) with qPCR, which was performed using a SYBR premix Ex Taq kit (TaKaRa) according to the manufacturer's instructions on a 7900HT Real-time PCR system.

Accumulation analysis of free proline, soluble sugar and malondialdehyde (MDA)

0.5 g fresh leaves from each treatment of the wild-type and the transgenic *Arabidopsis* lines were respectively collected at different time point, and respectively were treated according to requirements of experiment and used for measurements of free praline, soluble sugar and MDA. The contents of free proline were measured by the previous procedure [40,49,50]. Measurement of MDA content in *Arabidopsis* leaves was determined as described previously [51,52], and the content of soluble sugar was measured with the method as described by [53].

Author Contributions

Conceived and designed the experiments: XGC ZSX. Performed the experiments: GHY LLJ XFM MML SGS. Analyzed the data: ZSX. Contributed reagents/materials/analysis tools: GHY. Wrote the paper: XGC GHY ZSX.

References

1. Kodaira KS, Qin F, Tran LSP, Maruyama K, Kidokoro S, et al. (2011) *Arabidopsis* Cys2/His2 zinc-finger proteins AZF1 and AZF2 negatively regulate abscisic acid-repressive and auxin-inducible genes under abiotic stress conditions. Plant Physiol 157: 742–756.
2. Yamaguchi-Shinozaki K, Shinozaki K (2006) Transcriptional regulatory networks in cellular responses and tolerance to dehydration and cold stresses. Annu Rev Plant Biol 57: 781–803.
3. Umezawa T, Fujita M, Fujita Y, Yamaguchi-Shinozaki K, Shinozaki K (2006) Engineering drought tolerance in plants: discovering and tailoring genes to unlock the future. Curr Opin Biotechnol 17: 113–122.
4. Zhang G, Chen M, Chen X, Xu Z, Guan S, et al. (2008) Phylogeny, gene structures, and expression patterns of the ERF gene family in soybean (*Glycine max*). J Exp Bot 59: 4095–4107.
5. Xu ZS, Chen M, Li LC, Ma YZ (2011) Functions and application of the AP2/ERF transcription factor family in crop improvement. J Integr Plant Biol 53: 570–585.
6. Laity JH, Lee BM and Wright PE (2001) Zinc finger proteins: new insights into structural and functional diversity. Curr Opin Struct Biol 11: 39–46.
7. CO Pabo, Peisach E, Grant RA (2001) Design and selection of novel Cys2His2 zinc finger proteins. Annu Rev Biochem 70: 313–340.
8. Takatsuji H (1999) Zinc-finger proteins: the classical zinc finger emerges in contemporary plant science. Plant Mol Biol 39: 1073–1078.
9. Tian ZD, Zhang Y, Liu J, Xie CH (2010) Novel potato C2H2-type zinc finger protein gene, *StZFP1*, which responds to biotic and abiotic stress, plays a role in salt tolerance. Plant Biol 12: 689–697.
10. Ciftci-Yilmaz S, Mittler R (2008) The zinc finger network of plants. Cell Mol Life Sci 65: 1150–1160.
11. Kubo K, Sakamoto A, Kobayashi A, Rybka Z, Kanno Y, et al. (1998) Cys2/His2 zinc-finger protein family of petunia: evolution and general mechanism of target-sequence recognition. Nucleic Acids Res 26: 608–615.
12. Chen Y, Sun A, Wang M, Zhu Z, Ouwerkerk PB (2013) Functions of the CCCH type zinc finger protein OsGZF1 in regulation of the seed storage protein GluB-1 from rice. Plant Mol Biol DOI 10.1007/s11103-013-0158-5.
13. Takatsuji H, Mori M, Benfey PN, Ren L, Chua NH (1992) Characterization of a zinc finger DNA-binding protein expressed specifically in Petunia petals and seedlings. EMBO J 11: 241–249.
14. Liu P, Xu ZS, Pan-Pan L, Hu D, Chen M, et al. (2013) A wheat plasma membrane-localized *PI4K* gene possessing threonine autophophorylation activity confers tolerance to drought and salt in *Arabidopsis*. J Exp Bot 64: 2915–2927.
15. Van Der Krol AR, Van Poecke RMP, Vorst OFJ, Voogd C, Van Leeuwen W, et al. (1999) Developmental and wound-, cold-, desiccation-, ultraviolet-b-stress-induced modulations in the expression of the Petunia Zinc Finger Transcription Factor Gene *ZPT2–2*. Plant Physiol 121: 1153–1162.
16. Sugano S, Kaminaka H, Rybka Z, Catala R, Salinas J, et al. (2003) Stress responsive zinc finger gene *ZPT2–3* plays a role in drought tolerance in petunia. Plant J 36: 830–841.
17. Lippuner V, Cyert MS, Gasser CS (1996) Two classes of plant cDNA clones differentially complement yeast calcineurin mutants and increase salt tolerance of wild-type yeast. J Biol Chem 271: 12859–12866.
18. Sakamoto H, Maruyama K, Sakuma Y, Meshi T, Iwabuchi M, et al. (2004) *Arabidopsis* Cys2/His2-type zinc-finger proteins function as transcription repressors under drought, cold, and high-salinity stress conditions. Plant Physiol 136: 2734–2746.
19. Sakamoto H, Araki T, Meshi T, Iwabuchi M (2000) Expression of a subset of the Arabidopsis Cys (2)/His (2)-type zinc-finger protein gene family under water stress. Gene 248: 23–32.
20. Mittler R, Kim YS, Song L, Coutu J, Coutu A, et al. (2006) Gain-and loss-of-function mutations in Zat10 enhance the tolerance of plants to abiotic stress. FEBS Lett 580: 6537–6542.
21. Kim JC, Lee SH, Cheong YH, Yoo CM, Lee SI, et al. (2001) A novel cold-inducible zinc finger protein from soybean, SCOF-1, enhances cold tolerance in transgenic plants. Plant J 25: 247–259.
22. Luo X, Bai X, Zhu D, Li Y, Ji W, et al. (2012) GsZFP1, a new Cys2/His2-type zinc-finger protein, is a positive regulator of plant tolerance to cold and drought stress. Planta 235: 1141–1155.
23. Takatsuji H, Nakamura N, Katsumoto Y (1994) A new family of zinc finger proteins in petunia: structure, DNA sequence recognition, and floral organ-specific expression. Plant Cell 6: 947–958.
24. Lu X, Li Y, Su Y, Liang Q, Meng H, et al. (2012) An *Arabidopsis* gene encoding a C2H2-domain protein with alternatively spliced transcripts is essential for endosperm development. J Exp Bot 63: 5935–5944.
25. Xu SM, Wang XC, Chen J (2007) Zinc finger protein 1 (ThZF1) from salt cress (*Thellungiella halophila*) is a Cys-2/His-2-type transcription factor involved in drought and salt stress. Plant Cell Rep 26: 497–506.
26. Jain D, Roy N, Chattopadhyay D (2009) *CaZF*, a plant transcription factor functions through and parallel to HOG and calcineurin pathways in *Saccharomyces cerevisiae* to provide osmotolerance. PLoS ONE 4: e5154.
27. Nakai Y, Nakahira Y, Sumida H, Takebayashi K, Nagasawa Y, et al. (2013) Vascular plant one-zinc-finger protein 1/2 transcription factors regulate abiotic and biotic stress responses in Arabidopsis. Plant J 73: 761–775.
28. Huang XY, Chao DY, Gao JP, Zhu MZ, Shi M, et al. (2009) A previously unknown zinc finger protein, DST, regulates drought and salt tolerance in rice via stomatal aperture control. Genes & Development 23: 1805–1817.
29. Li C, Lv J, Zhao X, Ai X, Zhu X, et al. (2010) *TaCHP*: a wheat zinc finger protein gene down-regulated by abscisic acid and salinity stress plays a positive role in stress tolerance. Plant Physiol 154: 211–221.
30. Xiong L, Schumaker KS, Zhu JK (2002) Cell signaling during cold, drought, and salt stress. Plant Cell 14: 165–183.
31. Finkelstein RR, Gampala SSL, Rock CD (2002) Abscisic acid signaling in seeds and seedlings. Plant Cell 14: 15–45.
32. Davletova S, Schlauch K, Coutu J, Mittler R (2005) The zinc-finger protein Zat12 plays a central role in reactive oxygen and abiotic stress signaling in *Arabidopsis*. Plant Physiol 139: 847–856.
33. Zhu JK (2002) Salt and drought stress signal transduction in plants. Annu Rev Plant Biol 53: 247–273.
34. Shinozaki K, Yamaguchi-Shinozaki K, Seki M (2003) Regulatory network of gene expression in the drought and cold stress responses. Curr Opin Plant Biol 6: 410–417.
35. Igarashi Y, Yoshiba Y, Sanada Y, Yamaguchi-Shinozaki K, Wada K, et al. (1997) Characterization of the gene for 1-pyrroline-5-carboxylate synthetase and correlation between the expression of the gene and salt tolerance in Oryza sativa L. Plant Mol Biol 33: 857–865.
36. Dobrá J, Vanková R, Havlová M, Burman AJ, Libus J, et al. (2011) Tobacco leaves and roots differ in the expression of proline metabolism-related genes in the course of drought stress and subsequent recovery. J Plant Physiol 168: 1588–1597.
37. Chen M, Xu Z, Xia L, Li L, Cheng X, et al. (2009) Cold-induced modulation and functional analyses of the DRE-binding transcription factor gene, *GmDREB3*, in soybean (*Glycine max*). J Exp Bot 60: 121–135.
38. Bolouri Moghaddam MR, Le Roy K, Xiang L, Rolland F, Van den Ende W (2010) Sugar signaling and antioxidant network connections in plant cells. FEBS J 277: 2022–2037.
39. Smeekens S, Ma J, Hanson J, Rolland F (2010) Sugar signals and molecular networks controlling plant growth. Curr Opin Plant Biol 13: 273–278.
40. Sperdouli I, Moustakas M (2012) Interaction of proline, sugars, and anthocyanins during photosynthetic acclimation of *Arabidopsis thaliana* to drought stress. J Plant Physiol 169: 577–585.
41. Ramel F, Sulmon C, Gouesbet G, Couéel I (2009) Natural variation reveals relationships between pre-stress carbohydrate nutritional status and subsequent responses to xenobiotic and oxidative stress in *Arabidopsis thaliana*. Anna Bot 104: 1323–1337.
42. Guy CL (1990) Cold Acclimation and freezing stress tolerance: Role of protein metabolism. Annu Rev Plant Biol 41: 187–223.
43. Giannakoula A, Moustakas M, Mylona P, Papadakis I, Yupsanis T (2007) Aluminium tolerance in maize is correlated with increased levels of mineral nutrients, carbohydrates and proline and decreased levels of lipid peroxidation and Al accumulation. J Plant Physiol 165: 385–396.
44. Zhan X, Wang B, Li H, Liu R, Kalia RK, et al. (2012) *Arabidopsis* proline-rich protein important for development and abiotic stress tolerance is involved in microRNA biogenesis. Proc Natl Acad Sci USA 109: 18198–203.
45. Xu ZS, Xia LQ, Chen M, Cheng XG, Zhang RY, et al. (2007) Isolation and molecular characterization of the *Triticum aestivum* L. Ethylene-responsive

factor 1 (*TaERF1*) that increases multiple stress tolerance. Plant Mol Biol 65: 719–732.
46. Davis LG, Kuehl WM, Battey JF (1994) Section, 11-3: Acidic guanidine isothiocyanate/phenol/chlororm extraction for isolation of RNA; Section, 17-1: Electrophoresis of protein on sodium dodecy sulfate polyacrylamide gels. In Basic methods in Molecular Biology, 2nd ed.1994, 335–338, 661–668, Appleton & Lange, USA.
47. Han MJ, Jung KH, Yi G, Lee DY, An G (2006) Rice Immature Pollen 1 (RIP1) is a regulator of late pollen development. Plant Cell Physiol. 47: 1457–1472.
48. Liu XM, Nguyen XC, Kim KE, Han HJ, Yoo J, et al. (2013) Phosphorylation of the zinc finger transcriptional regulator *ZAT6* by *MPK6* regulates *Arabidopsis* seed germination under salt and osmotic stress. Biochem Biophys Res Commun 430: 1054–1059.
49. Giannakoula A, Moustakas M, Syros T, Yupsanis T (2012) Aluminium stress induces upregulation of an efficient antioxidant system in the Al-tolerant maize line but not in the Al-sensitive line. Environ Exp Bot 67: 487–494.
50. Zhan X, Wang B, Li H, Liu R, Kalia RK, et al. (2012) Arabidopsis proline-rich protein important for development and abiotic stress tolerance is involved in microRNA biogenesis. Proc Natl Acad Sci USA 109: 18198–203.
51. Yue Y, Zhang M, Zhang J, Tian XL, Duan LS, et al. (2012) Overexpression of the *AtLOS5* gene increased abscisic acid level and drought tolerance in transgenic cotton. J Exp Bot 63: 3741–3748.
52. Ai L, Li ZH, Xie ZX, Tian XL, Eneji AE, et al. (2008) Coronatine alleviates polyethylene Glycol-induced water stress in Two Rice (*Oryza sativa L.*) cultivars. J Agronomy and Crop Sci 194: 360–368.
53. Taji T, Ohsumi C, Iuchi S, Seki M, Kasuga M, et al. (2002) Important roles of drought-and cold-inducible genes for galactinol synthase in stress tolerance in *Arabidopsis thaliana*. Plant J 29: 417–426.

15

The Positive Regulatory Roles of the TIFY10 Proteins in Plant Responses to Alkaline Stress

Dan Zhu[1,3], Rongtian Li[2], Xin Liu[1], Mingzhe Sun[3], Jing Wu[3], Ning Zhang[3], Yanming Zhu[3]*

1 College of Life Science, Qingdao Agricultural University, Qingdao, P.R. China, **2** Key Laboratory of Molecular Biology, College of Heilongjiang Province, Heilongjiang University, Harbin, P.R. China, **3** Plant Bioengineering Laboratory, Northeast Agricultural University, Harbin, P.R. China

Abstract

The TIFY family is a novel plant-specific protein family, and is characterized by a conserved TIFY motif (TIFF/YXG). Our previous studies indicated the potential roles of TIFY10/11 proteins in plant responses to alkaline stress. In the current study, we focused on the regulatory roles and possible physiological and molecular basis of the TIFY10 proteins in plant responses to alkaline stress. We demonstrated the positive function of TIFY10s in alkaline responses by using the *AtTIFY10a* and *AtTIFY10b* knockout Arabidopsis, as evidenced by the relatively lower germination rates of *attify10a* and *attify10b* mutant seeds under alkaline stress. We also revealed that ectopic expression of *GsTIFY10a* in *Medicago sativa* promoted plant growth, and increased the NADP-ME activity, citric acid content and free proline content but decreased the MDA content of transgenic plants under alkaline stress. Furthermore, expression levels of the stress responsive genes including *NADP-ME*, *CS*, *H^+-ppase* and *P5CS* were also up-regulated in *GsTIFY10a* transgenic plants under alkaline stress. Interestingly, *GsTIFY10a* overexpression increased the jasmonate content of the transgenic alfalfa. In addition, we showed that neither GsTIFY10a nor GsTIFY10e exhibited transcriptional activity in yeast cells. However, through Y2H and BiFc assays, we demonstrated that GsTIFY10a, not GsTIFY10e, could form homodimers in yeast cells and in living plant cells. As expected, we also demonstrated that GsTIFY10a and GsTIFY10e could heterodimerize with each other in both yeast and plant cells. Taken together, our results provided direct evidence supporting the positive regulatory roles of the TIFY10 proteins in plant responses to alkaline stress.

Editor: Keqiang Wu, National Taiwan University, Taiwan

Funding: This work was supported by Heilongjiang Provincial Higher School Science and Technology Innovation Team Building Program (grant no. 2011TD005 to YMZ), National Natural Science Foundation of China (grant no. 31171578 to YMZ), National Major Project for Cultivation of Transgenic Crops (grant no. 2011ZX08004-002 to YMZ). The funders had no role in study design, data collection and analysis, decision to publish, or preparation of the manuscript.

Competing Interests: The authors have declared that no competing interests exist.

* Email: ymzhu@neau.edu.cn

Introduction

Salt-alkaline stress is one of the most severe environmental challenges and affects all aspects of plant physiological and metabolic processes [1,2]. Compared with neutral salt stress, soil alkalization leads to high pH stress, poor fertility, dispersed physical property and low water content, and thereby causes much stronger inhibition of plant growth and development [3–5]. In recent years, research on plant responses to salt stress have identified the molecular basis of stress signal transduction pathways and salt tolerance mechanisms, and have been at the forefront of plant stress biology [6]. Unfortunately, until now, little attention has been paid on the molecular mechanisms of plant responses to alkaline stress [7,8].

Alfalfa (*Medicago sativa L.*) is an important worldwide leguminous forage crop and distributes over a wide range of climatic conditions [9,10]. It has become one of the most important plants due to its high productivity, high feed value and potential roles in soil improvement and soil conservation [11,12]. However, alfalfa yield and symbiotic nitrogen-fixation capacity were severely restricted by adverse environmental stresses, especially soil salinity and alkalinity [10,13]. With the global climate change and the global shrinkage of arable lands, a grimmer reality of soil salinity and alkalinity is painted. Therefore, it is of fundamental importance to explore salt/alkaline-tolerant alfalfa through rational breeding and genetic engineering strategies.

The TIFY family, a novel plant-specific protein family, is characterized by a conserved TIFY motif (TIFF/YXG), and comprises 18 members in *Arabidopsis* and 20 members in rice [14,15]. It has been well suggested that TIFY genes play important roles in the jasmonate (JA) signaling pathway [16,17], plant growth and development[18–21], and pathogen responses [22–24]. For example, in *Arabidopsis*, TIFY genes are suggested to negatively regulate the key transcriptional activator of JA responses [25], such as MYC2 [26], MYC3, MYC4 [26,27], MYB21, MYB24 [28], bHLH017 and bHLH003 [29]. Furthermore, *AtTIFY4a* and *AtTIFY4b* regulate lamina size and curvature [30], whereas *AtTIFY1* plays a role in petiole and hypocotyl elongation [31]. Recent research in tomato and tobacco suggested that JAZ proteins regulated the progression of cell death during host and nonhost interactions [23].

Recently, several lines of direct evidence supported that TIFY genes also fulfilled important function in plant responses to

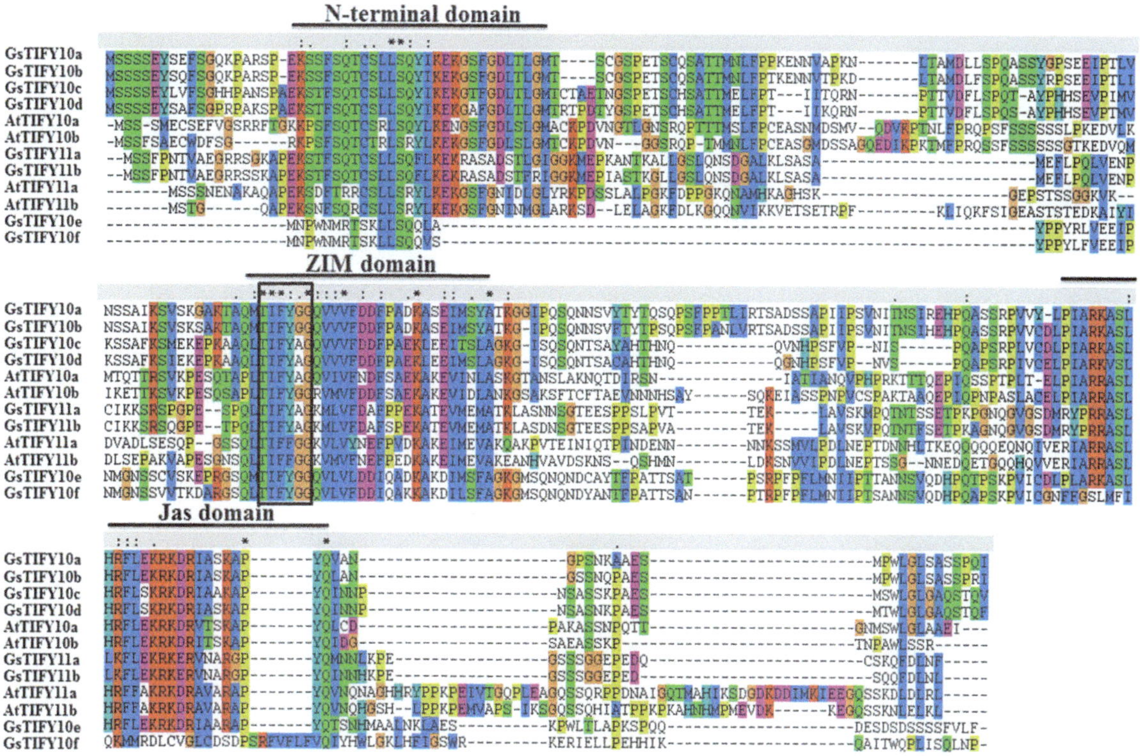

Figure 1. Sequence alignment of the Arabidopsis and wild soybean TIFY10/11 subgroup members based on the full-length amino acid sequences. The conserved N-terminal domain, ZIM domain and Jas domain were marked as solid lines. The TIFY motif (TIFF/YXG) was marked as a black solid box. Sequences were aligned by using ClustalX, and gaps were introduced to maximize alignment.

environmental challenges [32,33]. Overexpression of *OsTIFY11a* resulted in increased tolerance to salt and dehydration stresses [15]. Furthermore, OsTIFY3 acted as a transcriptional regulator of the OsbHLH148-mediated JA signaling pathway leading to drought tolerance [34]. Although a series of studies have demonstrated the biological function of TIFYs in salt and drought tolerance, little evidence is given on their roles in alkaline stress responses.

Glycine soja, is a wild soybean species and belongs to the same family *Leguminosae* with *Medicago sativa*. *Glycine soja* has extreme excellent tolerance to salt-alkaline stress [35], which makes it as an ideal candidate for exploring resistant genes and breeding of transgenic legume crops with superior salt-alkaline tolerance. In previous studies, we constructed a transcriptional profile of *Glycine soja* (G07256) roots in response to alkaline stress (50 mM $NaHCO_3$, pH8.5) [36], and identified three TIFY genes *GsTIFY10a*, *GsTIFY10e* (also named as *GsJAZ2*) and *GsTIFY11b* as alkaline stress responsive genes. We further demonstrated that overexpression of *GsTIFY10a* and *GsTIFY10e* in *Arabidopsis* improved plant alkaline tolerance [37,38]. In contrast, *GsTIFY11b* overexpression led to decreased salt tolerance [39]. A genomic analysis revealed 34 TIFY genes in *Glycine soja* genome and these GsTIFY proteins were clustered into two groups [40]. Group I comprises 9 members containing a GATA zinc-finger domain (GsTIFY1a, 1b, 1c, 1d, 2a, 2b, 2c, 2d, and 2e), and group II consists of 25 members without GATA zinc-finger domains. Among the group II TIFY proteins, GsTIFY10s (10a, 10b, 10c, 10d, 10e, and 10f) and GsTIFY11s (11a and 11b) were clustered together into one subgroup, here designated as GsTIFY10/11 subgroup. Transcriptional profiles revealed that all GsTIFY10/11 members were dramatically up-regulated at the early stage of alkaline stress, indicating potential roles of GsTIFY10/11s in alkaline stress responses.

In this study, we aimed to identify the regulatory roles and possible physiological and molecular basis of the TIFY10 proteins in plant responses to alkaline stress. We verified the positive function of TIFY10s in alkaline responses by using the *AtTIFY10a* and *AtTIFY10b* knockout Arabidopsis. We also demonstrated the increased alkaline tolerance of transgenic alfalfa ectopically expressing *GsTIFY10a*, and investigated the physiological basis by which *GsTIFY10a* overexpression conferred to increased alkaline tolerance. Finally, we determined the transcriptional activity and dimerization characteristics of GsTIFY10a and GsTIFY10e in yeast and plant cells. Taken together, our results provided direct evidence that TIFY10 proteins positively regulated plant alkaline stress responses.

Results

The TIFY10/11 subgroup in Glycine soja and *Arabidopsis thaliana*

As shown in Fig. 1, the TIFY10/11 subgroup comprises 8 members in *Glycine soja* and 4 members in *Arabidopsis*. Protein sequence analyses showed that all TIFY10/11 proteins contained two highly conserved domains: a ZIM/TIFY domain which mediated homo- and hetero-dimerization, and a Jas domain which played a critical role in repression of JA signaling. It is worth noted that except for GsTIFY10e and GsTIFY10f, all other TIFY10/11 proteins included an N-terminal domain (Fig. 1, Fig. 2C). We further examined the phylogenetic relationship of TIFY10/11s in *Glycine soja* and *Arabidopsis thaliana*. As shown in Fig. 2A, TIFY10/11 proteins were divided into two branches (TIFY10 and

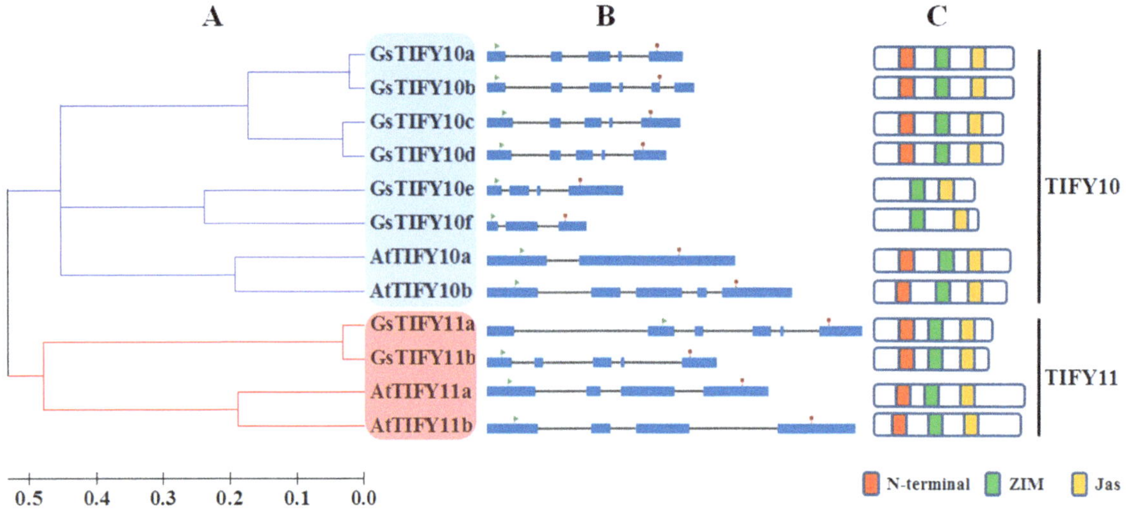

Figure 2. The Arabidopsis and wild soybean TIFY10/11 subgroup proteins. a. Phylogenetic analysis of the Arabidopsis and wild soybean TIFY10/11 subgroup proteins. A neighbor-joining tree was constructed with the full-length TIFY10/11 protein sequences by using MEGA 5.0. **b**. Exon/intron structures of the Arabidopsis and wild soybean TIFY10/11 genes. Exons were represented by blue boxes, and grey lines connecting two exons represented introns. Both the exons and introns were drawn to scale. **c**. The distribution of conserved domains within Arabidopsis and wild soybean TIFY10/11 proteins. The relative positions of each conserved domain within each protein were shown in color.

TIFY11). GsTIFY10a and GsTIFY10b, GsTIFY10c and GsTIFY10d, GsTIFY10e and GsTIFY10f were closely related with each other, respectively. Among the TIFY10 proteins, GsTIFY10a/b/c/d exhibited high similarity in exon distribution (Fig. 2B), amino acid sequence and domain architecture (Fig. 2C).

To get better understanding of TIFY10/11s sequence diversity, we analyzed the species and numbers of the cis-regulatory elements in their promoter regions. A series of typical elements related to environmental stress, hormone responsiveness and transcription factor (TF) binding sites were identified (Table 1), indicating the involvement of TIFY10/11s in hormone-dependent environmental stress responses and the regulation of TFs on TIFY10/11 expression or activity. Among the 33 elements listed in Table 1, several elements, including ABRELATERD1, ABRERATCAL, ACGTATERD1, ARR1AT, CACGTGMOTIF, DPBFCOREDCDC3, GAREAT, GT1GMSCAM4 and MYCCONSENSUSAT, were found in all of the TIFY10/11s promoter regions. We also noticed several elements only existed in specific TIFY10/11 promoters, for example ABREATRD22, ABREOSRAB21, ABREMOTIFAOSOSEM, ACGTABREMOTIFA2OSEM, ACGTCBOX and ACGTTBOX.

We further checked the responses of GsTIFY10/11s expression to alkaline stress based on the RNA-seq data. Our results revealed that expression of all the GsTIFY10/11 genes was rapidly and greatly induced by alkaline stress. They showed similar alkaline induced expression patterns, with a maximum point at 1 h after $NaHCO_3$ treatment (Fig. S1A). Among them, *GsTIFY10e* and *GsTIFY10f* exhibited the greatest alkaline stress induction, while *GsTIFY10a* and *GsTIFY10b* showed the highest expression levels (Fig. S1B). Taken together, these results strongly suggested the potential roles of GsTIFY10/11 proteins in alkaline stress responses.

The *AtTIFY10a/b* knockout decreased alkaline tolerance at the seed germination stage

To confirm the regulatory roles of TIFY10 proteins in alkaline stress responses, we adopted the T-DNA insertion mutant *Arabidopsis* of *AtTIFY10a* (*attify10a*, SALK_011957) and *AtTIFY10b* (*attify10a*, SALK_025279) (Fig. 3A). PCR-based analysis demonstrated the homozygous T-DNA insertion in the *attify10a* and *attify10b* mutants, and RT-PCR results confirmed that the *AtTIFY10a* and *AtTIFY10b* genes did not express in the *attify10a* and *attify10b* mutants, respectively (Fig. 3B).

The wild-type (WT) and mutant Arabidopsis seeds were germinated and grown on 1/2MS solid medium at pH5.8 (Control) or pH8.5 (Alkaline stress), respectively. As shown in Fig. 3C, WT and *AtTIFY10a/b* knockout Arabidopsis seedlings showed similar growth on 1/2MS solid medium at pH5.8, but growth of mutant seedlings was inhibited more severely than that of WT on 1/2MS solid medium at pH8.5. Under alkaline stress, WT seeds maintained relatively high germination rates (86.5%) on the 3rd day, but the germination rates of *attify10a* and *attify10b* seeds dropped to 54.2% and 77.0%, respectively (Fig. 3D). These results suggested that alkaline tolerance of *AtTIFY10a/b* knockout decreased at the seed germination stage, and further confirmed the positive roles of TIFY10 proteins in alkaline stress responses.

GsTIFY10a overexpression in alfalfa enhanced plant alkaline tolerance

In previous studies, we have demonstrated the involvement of three GsTIFY genes in salt-alkaline stress, among which *GsTIFY10a* could dramatically improve the alkaline tolerance. In an attempt to generate transgenic alfalfa with superior alkaline tolerance, we ectopically expressed *GsTIFY10a* in the wild type *Medicago sativa* through the *Agrobacterium tumefaciens*-mediated transformation strategy. The *GsTIFY10a* gene was under the control of the cauliflower mosaic virus (CaMV) 35S promoter, with the binding enhancers E12 and omega (Fig. 4A). After glufosinate selection, the regenerated alfalfa seedlings were analyzed by PCR and semi-quantitative RT-PCR assays. We identified a total of six transgenic lines (Fig. 4B), and three of them, with different expression levels (#12, #13 and #28), were used to examine the responses to alkaline stress.

We first compared the growth performance of WT and *GsTIFY10a* transgenic alfalfa plants under alkaline stress. As shown in Fig. 4C, under control conditions, transgenic lines

Table 1. Sequence diversity and cis-regulatory elements of soybean and Arabidopsis TIFYs promoters.

No.	Site Name	Signal Sequence	Description	GsTIFY10/11s								AtTIFY10/11s			
				10a	10b	10c	10d	10e	10f	11a	11b	10a	10b	11a	11b
1	ABREATRD22	RYACGTGGYR	ABRE in Arabidopsis dehydration-responsive gene rd22			1	1								
2	ABRELATERD1	ACGTG	ABRE-like sequence required for etiolation-induced expression of erd1 in *Arabidopsis*	3	4	2	3	4	5	2	2	5	7	5	6
3	ABREOSRAB21	ACGTSSSC	ABRE of wheat Em and rice rab21 genes		1								1		
4	ABREMOTIFAOSOSEM	TACGTGTC	ABRE-like sequence found in rice Osem gene promoter											1	
5	ABRERATCAL	MACGYGB	ABRE-related sequence in the upstream regions of Ca^{2+}-responsive upregulated genes	2	4	2	2	3	3	1	2	7	4	3	5
6	ACGTABREMOTIFA2OSEM	ACGTGKC	Sequence requirement of ACGT-core of motif A in ABRE of the rice gene, OSEM										1	1	2
7	ACGTATERD1	ACGT	ACGT sequence required for etiolation-induced expression of erd1 in *Arabidopsis*	4	6	6	4	6	8	4	2	12	18	12	8
8	ACGTCBOX	GACGTC	"C-box" according to the nomenclature of ACGT elements								2				
9	ACGTTBOX	AACGTT	"T-box" according to the nomenclature of ACGT elements											2	
10	ARFAT	TGTCTC	ARF binding site found in the promoters of primary/early auxin response genes; AuxRE; enriched in the 5′-flanking region of genes up-regulated by both IAA and BL	1	1							2	1		
11	ARR1AT	NGATT	"ARR1-binding element" found in Arabidopsis (CK)	11	4	8	5	6	5	7	11	13	13	10	6
12	ASF1MOTIFCAMV	TGACG	ASF-1 binding site; are found in many promoters and are involved in transcriptional activation of several genes by auxin and/or salicylic acid		1		1		2	1		3	1	1	2
13	BIHD1OS	TGTCA	Binding site of OsBIHD1, a rice BELL homeodomain transcription factor (disease)	4	2	3	1	4	4	1			1	2	5
14	CACGTGMOTIF	CACGTG	"CACGTG motif"; "G-box"; Binding site of Arabidopsis GBF4	2	2	2	2	2	2	2	2	2	4	2	4
15	CACGCAATGMGH3	CACGCAAT	Sequence found in D4 element in Soybean GH3 gene promoter; Confers auxin inducibility			1									
16	CATATGGMSAUR	CATATG	Sequence found in soybean SAUR15A promoter; Involved in auxin responses				2					2			
17	CBFHV	RYCGAC	Binding site of barley CBF1 and CBF2	1	1			1	1	2		2		1	1
18	CGCGBOXAT	VCGCGB	"CGCG box" recognized by AtSR1-6 (Arabidopsis thaliana signal-responsive genes)		2							4	2	2	
19	CPBCSPOR	TATTAG	The sequence critical for Cytokinin-enhanced Protein Binding in vitro		1	3	3		2	1	2	2			3
20	CRTDREHVCBF2	GTCGAC	Preferred sequence for AP2 transcriptional activator HvCBF2; Core CRT/DRE motif							2		2			
21	DPBFCOREDCDC3	ACACNNG	A novel class of bZIP transcription factors binding core sequence (ABA)	1	1	1	1	1	1	1	2	3	5	2	3
22	DRE2COREZMRAB17	ACCGAC	"DRE2" core found in maize rab17 gene promoter; rab17 is induced by ABA	1								1		1	
23	DRECRTCOREAT	RCCGAC	Core motif of DRE/CRT cis-acting element in many genes in *Arabidopsis* and rice	1	1									1	1

Table 1. Cont.

No.	Site Name	Signal Sequence	Description	GsTIFY10/11s								AtTIFY10/11s			
				10a	10b	10c	10d	10e	10f	11a	11b	10a	10b	11a	11b
24	ERELEE4	AWTTCAAA	"ERE (ethylene responsive element)" of tomato E4 and carnation GST1 genes; ERE motifs mediate ethylene-induced activation of the U3 promoter region	1	1		1					1		2	
25	GADOWNAT	ACGTGTC	Sequence present in 24 genes in the GA-down regulated d1 cluster (106 genes) found in Arabidopsis seed germination; This motif is similar to ABRE										1	1	2
26	GARE1OSREP1	TAACAGA	"Gibberellin-responsive element (GARE)" found in rice cystein proteinase promoter			1						1	2		1
27	GAREAT	TAACAAR	GARE (GA-responsive element)	1	2	3	1	2	1	1	2	1	3	1	1
28	GT1GMSCAM4	GAAAAA	"GT-1 motif" found in the promoter of soybean CaM isoform, SCaM-4; Plays a role in pathogen- and salt-induced SCaM-4 gene expression	5	4	1	3	3	4	3	3	1	2	3	3
29	LTRECOREATCOR15	CCGAC	Core of low temperature responsive element of cor15a gene in Arabidopsis	1	1	1								2	3
30	MYB1AT	WAACCA	MYB recognition site found in the promoters of the dehydration-responsive gene rd22	2	3	1	3	3	2		1		3	2	2
31	MYB2CONSENSUSAT	YAACKG	MYB recognition site found in the promoters of the dehydration-responsive gene rd22	1				1	1	1			6	2	3
32	MYBCORE	CNGTTR	Binding site plant MYB proteins ATMYB1 and ATMYB2 from *Arabidopsis*	1		1	3	4	4	1		1	8	4	6
33	MYCCONSENSUSAT	CANNTG	MYC recognition site found in the promoters of the dehydration-responsive gene rd22	8	8	6	6	10	14	2	4	6	14	4	8

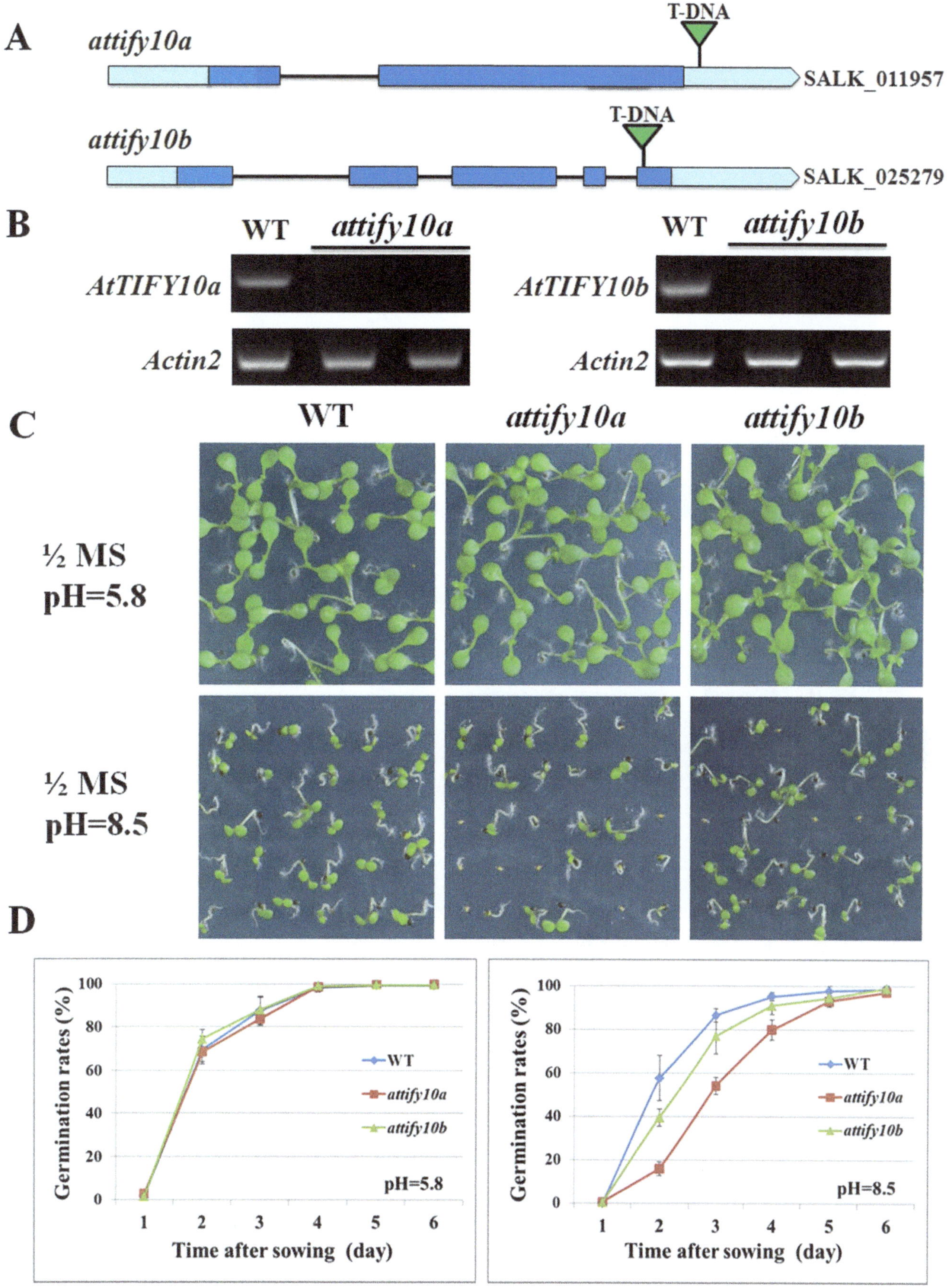
A
attify10a
T-DNA
SALK_011957
attify10b
T-DNA
SALK_025279
B
WT
attify10a
AtTIFY10a
Actin2
WT
attify10b
AtTIFY10b
Actin2
C
WT
attify10a
attify10b
½ MS
pH=5.8
½ MS
pH=8.5
D
Germination rates (%)
Time after sowing (day)
WT
attify10a
attify10b
pH=5.8
pH=8.5

Figure 3. The *AtTIFY10a/b* knockout mutant *Arabidopsis* showed decreased alkaline tolerance at the seed germination stage. a. Schematic representation of the *AtTIFY10a/b* T-DNA insertion mutant lines. The exons and introns of the *AtTIFY10a/b* genes were showed as boxes and lines, and the T-DNA insertion sites were marked as triangles. **b**. RT-PCR analyses showing that *AtTIFY10a/b* did not expressed in the *attify10a/b* mutants. **c**. The growth performance of WT, *attify10a* and *attify10b* mutant Arabidopsis under alkaline stress. *Arabidopsis* seeds were germinated and grown on 1/2MS medium at pH5.8 or pH8.5. Photographs were taken 6 days after germination. **d**. Seed germination rates of WT and mutant lines. Seeds were considered to be germinated when the radicles completely penetrated the seed coats. A total of 90 seeds from each line were used for each experiment. Data are means (±S.E.) of three replicates.

showed no obvious differences in seedling growth compared with WT. After 100 or 150 mM $NaHCO_3$ treatment for 14 d, both WT and transgenic lines showed growth retardation in a dose-dependent manner. However, the growth inhibition of transgenic lines was less severe than that of WT. After 150 mM $NaHCO_3$ treatment, all the transgenic lines could maintain continous growth, whereas WT plants showed severe chlorosis (Fig. 4C). In details, the shoot length, ground fresh weight and dry weight of

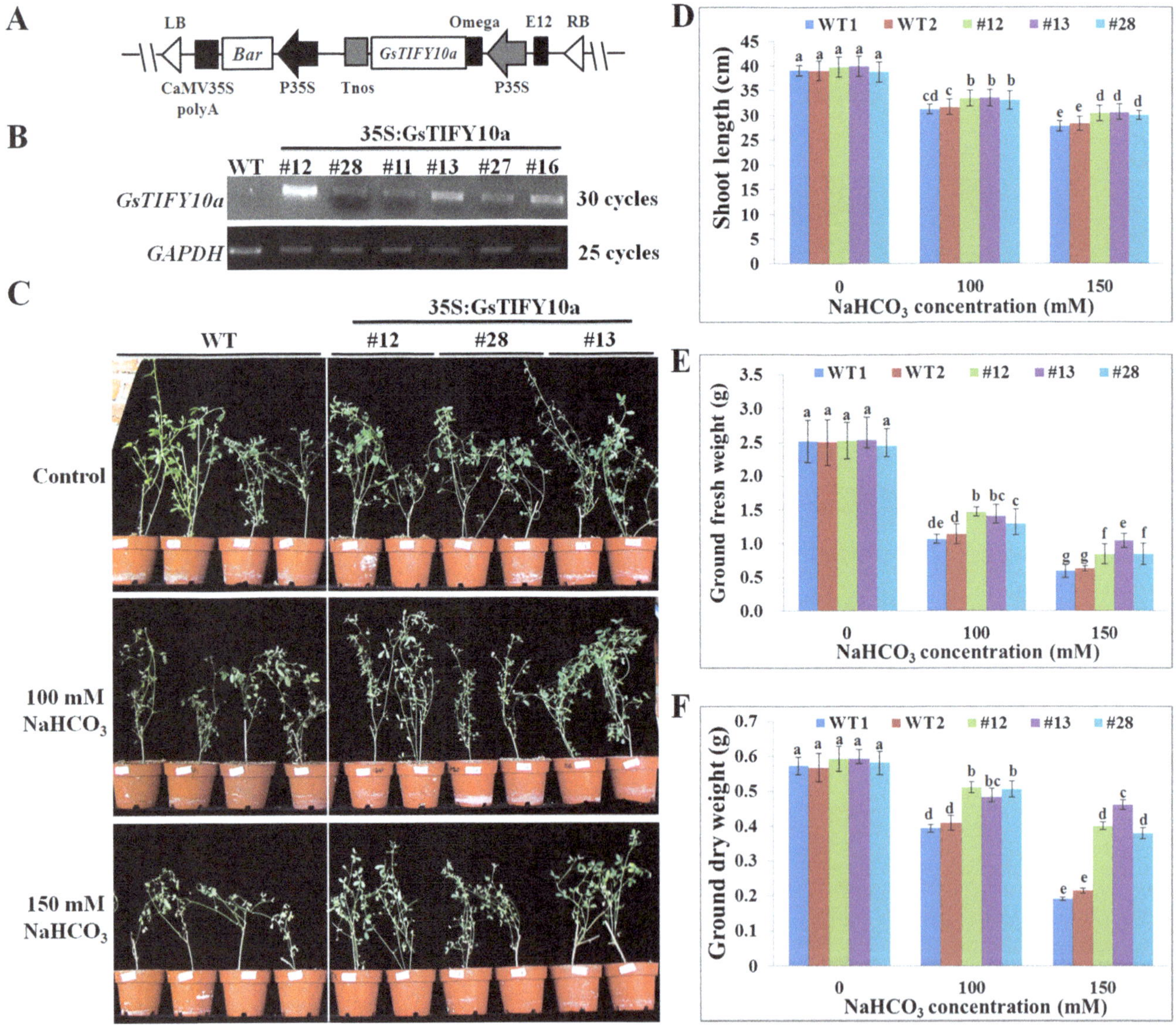

Figure 4. Overexpression of *GsTIFY10a* in alfalfa promoted plant growth under alkaline stress. a. Schematic representation of expression constructs to ectopically express *GsTIFY10a* in Medicago *sativa*. **b**. Semi-quantitative RT-PCR analysis showing the transcript levels of *GsTIFY10a* in transgenic alfalfa lines. **c**. Growth performance of WT and transgenic lines under control conditions or $NaHCO_3$ treatments. **d**. The shoot length of WT and transgenic plants. **e**. The ground fresh weight of WT and transgenic plants. **f**. The ground dry weight of WT and transgenic plants. For phenotypic analysis under alkaline stress, the propagated WT and *GsTIFY10a* transgenic plants with similar sizes (approximately 25 cm high) were treated with 1/8 Hoagland nutrient solution containing either 0, or 100, or 150 mM $NaHCO_3$ every 3 days for a total of 12 days. Photographs were taken 12 days after initial treatment. Thirty plants of each line were used for each experiment. Data are means (±SE) of three replicates. Significant differences were determined by one-way ANOVA ($P<0.0001$) statistical analysis. Different letters show significant differences between groups as indicated by Dunnett's posttests ($P<0.05$).

both WT and transgenic alfalfa plants were decreased gradually with increased $NaHCO_3$ concentration, but transgenic lines were much taller (Fig. 4D) and displayed more biomass accumulation (Fig. 4E and 4F) than WT. These results suggested that overexpression of *GsTIFY10a* in alfalfa promoted plant growth under alkaline stress.

To further elucidate the influence of *GsTIFY10a* overexpression in alfalfa, several alkaline stress-related physiological and biochemical parameters were analyzed in both WT and transgenic plants, respectively. It is generally accepted that alkaline stress is characterized by high pH value, and always causes much stronger inhibition of plant growth than salt stress [41]. Previous studies showed that NADP-ME could help to maintain the cytosolic pH homeostasis[42], and citric acid was an indicator of plant responses to pH challenge [43]. Therefore, we investigated the NADP-ME activity and citrate acid content of WT and transgenic plants, in an attempt to understand the physiological mechanisms responsible for the increased alkaline tolerance of *GsTIFY10a* transgenic alfalfa. As shown in Fig. 5A and 5B, alkaline stress obviously improved the NADP-ME activity and citrate acid content in both WT and transgenic lines, however, an obvious up-regulation was observed in the transgenic lines. These results indicated that the alleviation of high pH damage in *GsTIFY10a* transgenic alfalfa might partially result from the ability to maintain the cytosolic pH homeostasis through increased NADP-ME activity and citrate acid content.

As a type of compatible osmolyte, proline plays a critical role in protecting plants from environmental stresses [44]. Our results revealed that transgenic alfalfa plants accumulated more free proline than WT in the presence of 100 or 150 mM $NaHCO_3$ (Fig. 5C). To further test the cell membrane stability, we further determined the malon dialdehyde (MDA) content of WT and transgenic plants. Under control conditions, the MDA content of transgenic lines was similar to that of WT (Fig. 5D). After $NaHCO_3$ treatment for 14 d, the MDA content of WT were significantly higher than that of transgenic alfalfa. Collectively, these results demonstrated that the increased alkaline tolerance of *GsTIFY10a* transgenic alfalfa might be related to the elevated levels of NADP-ME activity, citrate content and proline content, as well as reduced MDA content.

GsTIFY10a overexpression up-regulated the expression levels of stress responsive genes

As described above, *GsTIFY10a* overexpression increased the NADP-ME activity, citrate content and proline content of transgenic alfalfa. To explore the molecular basis of *GsTIFY10a* in the cytoplasmic pH regulation and osmotic regulation under alkaline stress, we examined the expression levels of *NADP-ME*, *CS*, *H^+-Ppase*, and *P5CS* in WT and two transgenic lines (#12 and #13). The real-time PCR results showed that their expression was greatly induced by alkaline stress in both WT and transgenic lines (Fig. 6). Expectedly, their expression levels in transgenic plants were significantly higher than that in WT, which explained

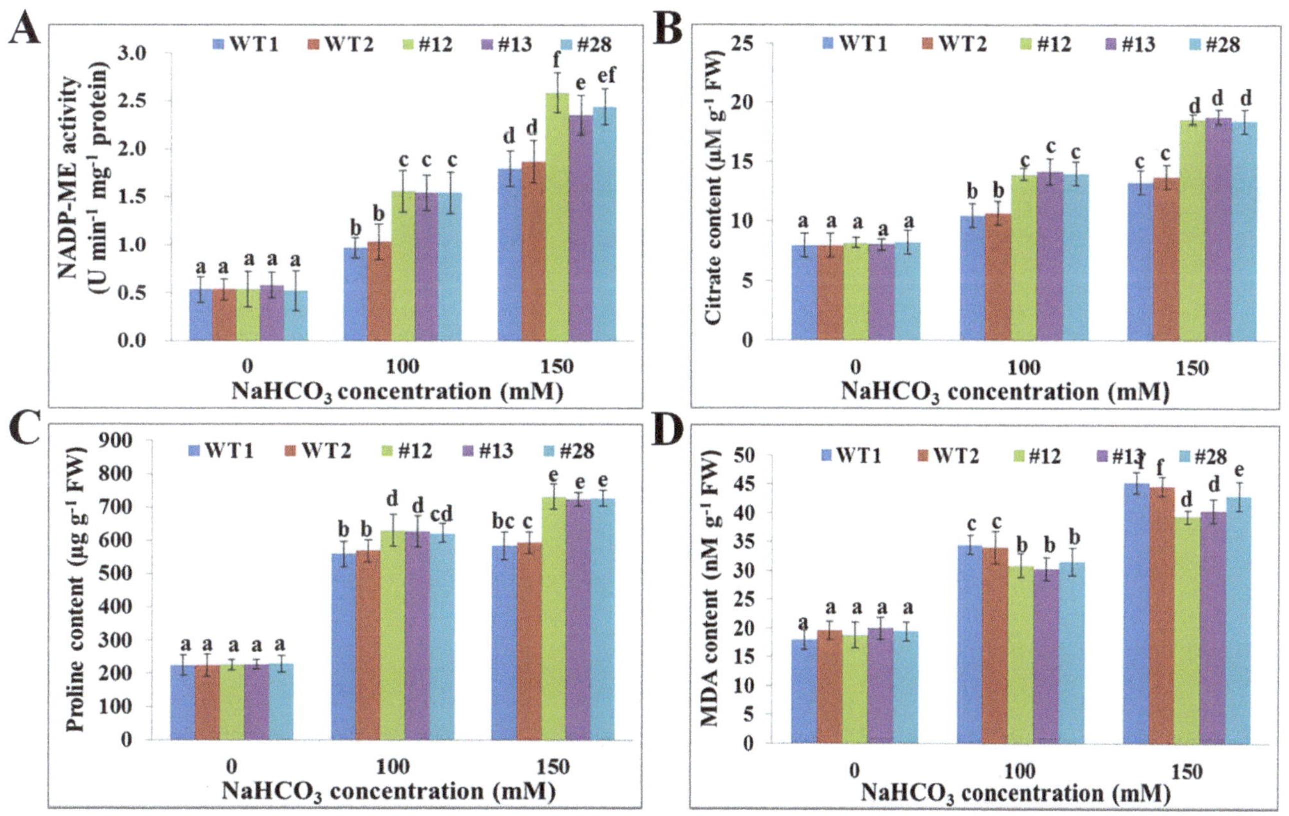

Figure 5. *GsTIFY10a* overexpression altered several physiological indices of transgenic plants under alkaline stress. a. The NADP-ME activity of WT and transgenic lines. **b**. The citric acid content of WT and transgenic lines. **c**. The free proline content of WT and transgenic lines. **d**. The MDA content of WT and transgenic lines. Thirty plants of each line were used for each experiment. Data are means (±SE) of three replicates. Significant differences were determined by one-way ANOVA ($P<0.0001$) statistical analysis. Different letters show significant differences between groups as indicated by Dunnett's posttests ($P<0.05$).

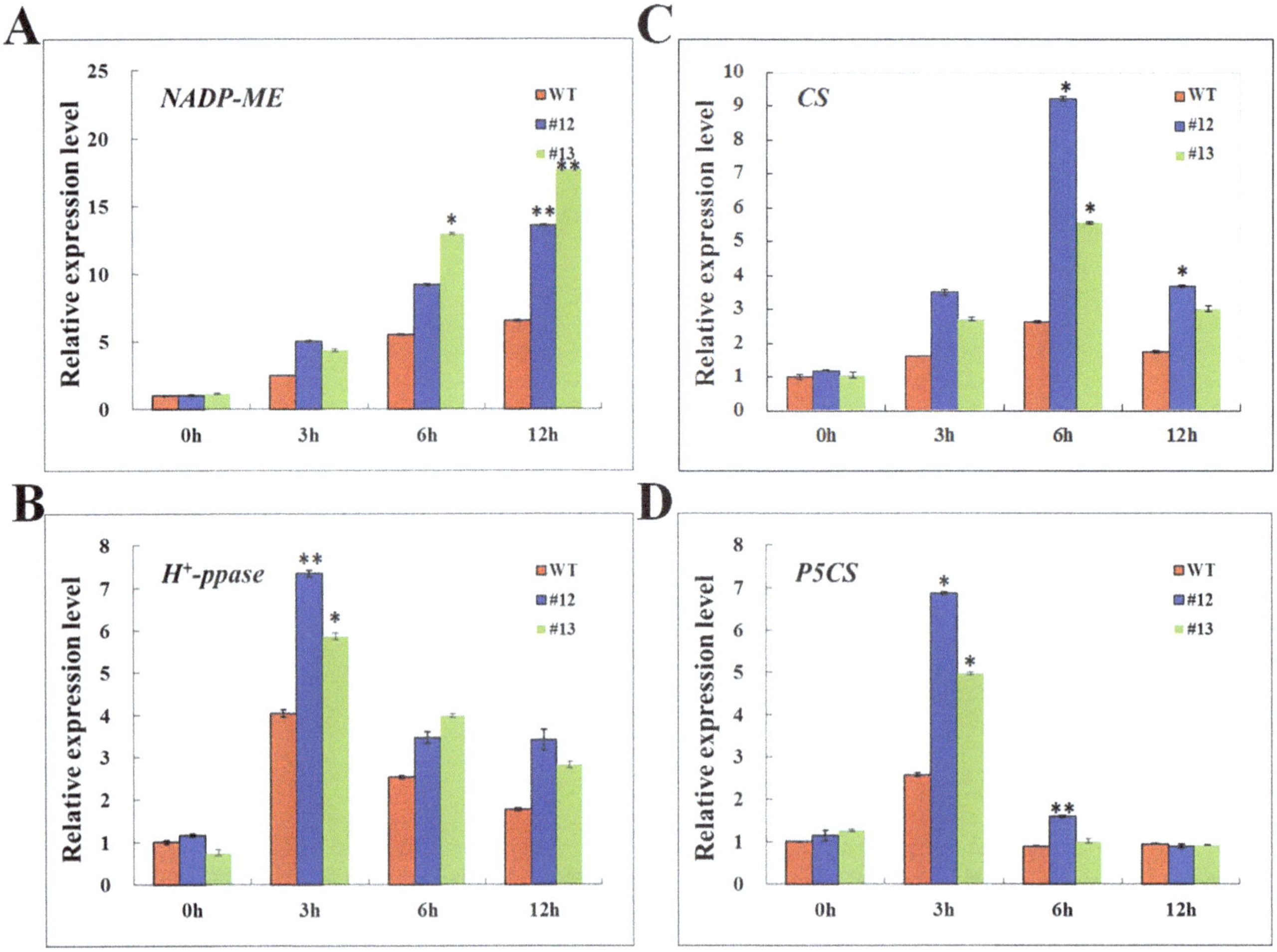

Figure 6. *GsTIFY10a* overexpression up-regulated the expression levels of several stress responsive genes. a. Increased expression levels of *NADP-ME* in transgenic plants under alkaline stress. **b**. Increased expression levels of *CS* in transgenic plants. **c**. Increased expression levels of *H+-ppase* in transgenic plants. **d**. Increased expression levels of *P5CS* in transgenic plants. To explore the expression patterns of stress-responsive genes, the 4-week-old WT and transgenic seedlings (line #12 and #13) after shoot cottage were treated with 1/8 Hoagland solution containing 50 mM $NaHCO_3$ (pH 8.5) for 0, 3, 6, and 12 h, respectively. Relative transcript levels were determined by quantitative real-time PCR with the *MtGAPDH* gene as an internal reference, and were normalized to WT at 0 h. Values represented the means of three independent biological replicates, and three technological replicates for each. *$P<0.05$; **$P<0.01$ by Student's t-test.

the up-regulation of the NADP-ME activity, citrate acid content and free proline content in transgenic lines. These results also implied that *GsTIFY10a* overexpression promoted the transcript accumulation levels of the stress responsive genes, which might be helpful for the intracellular pH homeostasis and osmotic regulation under alkaline stress.

Increased JA content in *GsTIFY10a* transgenic lines

TIFY/JAZ proteins are repressors of JA signaling in plants. Our previous studies revealed that *GsTIFY10a* overexpression in Arabidopsis repressed the transcription of JA responsive genes [38]. Hence, in this study, we determined the JA content in the transgenic alfalfa lines. As shown in Fig.7, JA contents were significantly increased in the transgenic alfalfa lines ($F_{3,8} = 581.042$, $P = 1.07*10^{-9}$).

Transcriptional activity assays of GsTIFY10a and GsTIFY10e

Our previous studies demonstrated that two of the *Glycine soja* TIFY10 proteins GsTIFY10a and GsTIFY10e positively regulated

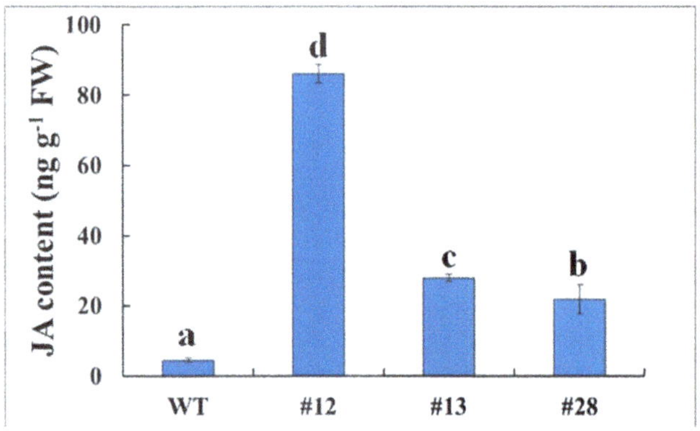

Figure 7. Increased JA content in *GsTIFY10a* transgenic plants. Leaves of WT and transgenic plants were harvested for JA extraction, and subjected for HPLC analysis to determine the content of endogenous JA. Each data point represents the mean (±SE) of three samples from independent sets of plants. Significant differences were found using one-way ANOVA analysis ($P = 1.07*10^{-9}$).

plant tolerance to alkaline stress. Both of them were found to localize at the nuclei of plant cells, indicating that they might act as transcriptional regulators. To further investigate the molecular basis of the TIFY10 proteins in alkaline responses, we identified the transcriptional activity of the GsTIFY10a and GsTIFY10e proteins. The full-length *GsTIFY10a* and *GsTIFY10e* genes were fused to the GAL4 DNA-binding domain in the pGBKT7 vector, and then introduced to the yeast reporter strain AH109. The AtbZIP1 transcription factor, which showed transcriptional activity in yeast cells, was used as a positive control [45,46]. LacZ activity was assessed by using the β-galactosidase filter lift assays. As shown in Fig.8, only the recombinant yeast cells carrying the AtbZIP1-BD vector displayed LacZ activity. These results implied that neither GsTIFY10a nor GsTIFY10e showed transcriptional activity in yeast cells.

Dimerization analyses between GsTIFY10a and GsTIFY10e

It has been suggested that the ZIM domains mediated homo- and heterodimerization of the TIFY proteins [47]. To verify if GsTIFY10a and GsTIFY10e could form homodimers or heterodimers with each other, we performed the yeast two hybrid analyses. The AtbZIP63 transcription factor, which formed homodimers in plant cells [48], was used as a positive control. As shown in Fig.9A, the yeast cells carrying GsTIFY10a-BD/GsTIFY10a-AD, GsTIFY10a-BD/GsTIFY10e-AD, GsTIFY10e-BD/GsTIFY10a-AD and AtbZIP63-BD/AtbZIP63-AD (positive control) were capable of growth on the both SD/-Trp-Leu and SD/-Trp-Leu-Ade-His medium. However, the yeast cells harboring GsTIFY10e-BD/GsTIFY10e-AD, GsTIFY10a-BD/AD (negative control), GsTIFY10e-BD/AD (negative control) could not grow on the SD/-Trp-Leu-Ade-His medium. These results demonstrated that GsTIFY10a, not GsTIFY10e, could form homodimers, and GsTIFY10a and GsTIFY10e could heterodimerize with each other in yeast cells.

To further verify their physical interaction in living plant cells, we performed the bimolecular fluorescence complementation (BiFC) assays. To this end, we fused GsTIFY10a to the N-terminal YFP fragment and GsTIFY10a/e to the C-terminal YFP fragment, to generate GsTIFY10a-YFPN and GsTIFY10a/e-YFPC constructs, respectively. GsTIFY10a-YFPN/GsTIFY10a-YFPC and GsTIFY10a-YFPN/GsTIFY10e-YFPC were co-transformed into *Arabidopsis* protoplasts, respectively, and the empty YFPN/YFPC was used as a negative control. As shown in Fig. 9B, YFP fluorescence was observed from protoplasts co-transformed with GsTIFY10a-YFPN/GsTIFY10a-YFPC and GsTIFY10a-YFPN/GsTIFY10e-YFPC, but not from YFPN/YFPC. The BiFc results confirmed the homodimerization of GsTIFY10a, as well as the heterodimerization between GsTIFY10a and GsTIFY10e in plant cells.

Discussions

The TIFY protein family, characterized by a highly conserved TIFY motif, constitutes a particular class of plant-specific transcription factors with a broad range of biological functions. According to their distinct domain architectures, TIFY proteins could be classified into four subfamilies: the TIFY subfamily containing only the TIFY domain, the JAZ subfamily containing the TIFY domain and the Jas domain, the PPD subfamily containing the PPD, TIFY and a truncated Jas domain, and the ZML subfamily containing the TIFY, CCT and ZML domains[14]. Among them, the JAZ subfamily proteins could be further clustered into five groups (group I-V) [14]. The first group (group I) is composed of the TIFY10 and TIFY11 proteins. In this study, we focused on the group I TIFY proteins in Arabidopsis and wild soybean. All of the group I TIFY proteins contained the conserved TIFY and Jas domains (Fig. 1, 2), which was the canonical characteristic of JAZ subfamily. It is noteworthy that except for GsTIFY10e and GsTIFY10f, all other TIFY10/11s included an N-terminal domain (Fig. 1, 2).

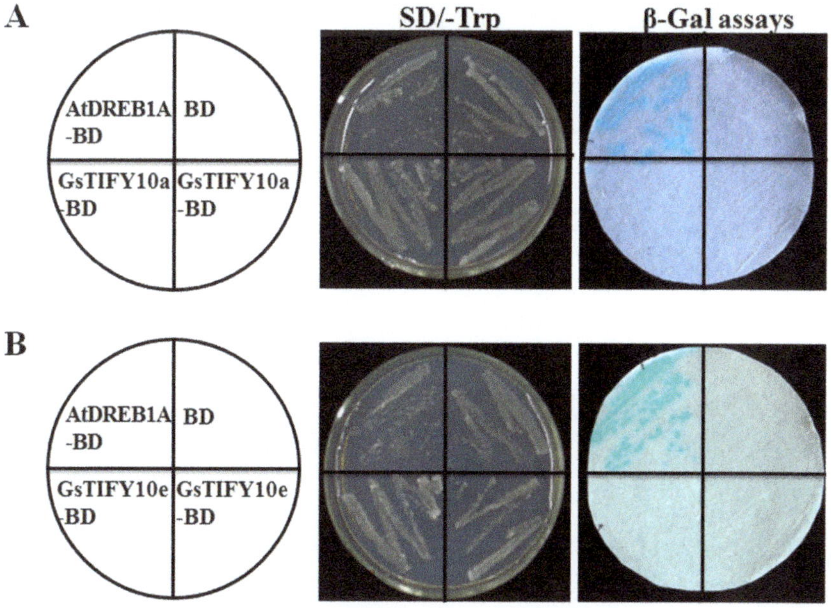

Figure 8. Transcriptional activity analysis of the GsTIFY10a and GsTIFY10e proteins. a. Transcriptional activity analysis of GsTIFY10a. **b.** Transcriptional activity analysis of GsTIFY10e. The pGBKT7-GsTIFY10a and pGBKT7-GsTIFY10e vectors were transformed into the yeast reporter strain AH109, and LacZ activity was assessed by using the β-galactosidase filter lift assays. The AtDREB1A transcription factor was used as a positive control.

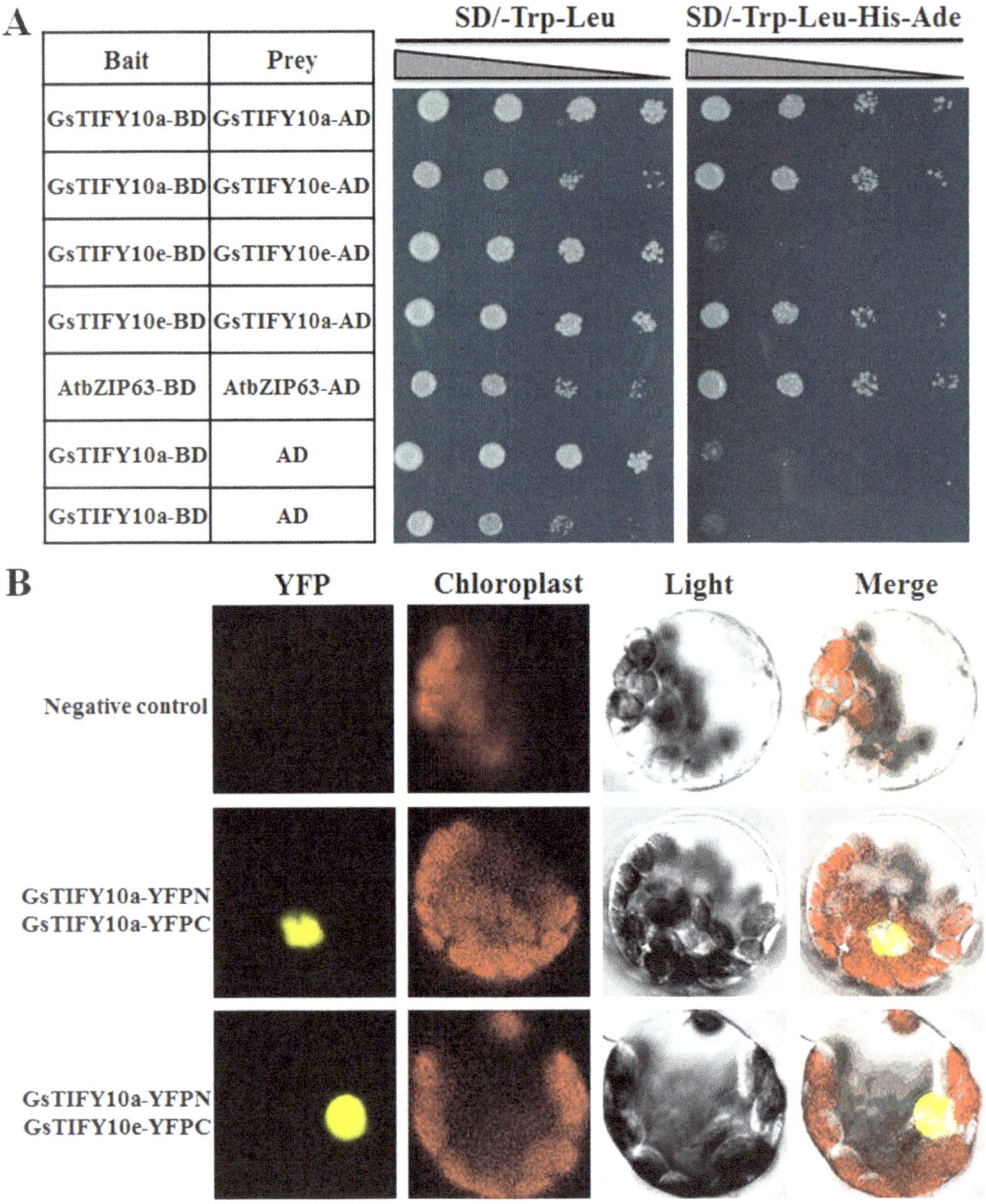

Figure 9. Dimerization assays of the GsTIFY10a and GsTIFY10e proteins. a. Dimerization analyses between GsTIFY10a and GsTIFY10e in yeast cells. Pictures showed the growth performance of recombinant yeast cells harboring different plasmids on SD/-Trp-Leu and SD/-Trp-Leu-Ade-His medium. The AtbZIP63-BD/AtbZIP63-AD combination was used as a positive control, and the GsTIFY10a-BD/AD and GsTIFY10e-BD/AD combinations were used as negative controls. **b**. Dimerization analyses between GsTIFY10a and GsTIFY10e in living plant cells. The YFPN/YFPC combination was used as a negative control. Pictures showed the YFP fluorescence, chlorophyll auto-fluorescence, light and overlay visions.

The best-characterized TIFY10 protein is the Arabidopsis TIFY10a/JAZ1, which was found to play a critical role in JA signaling. In the absence of JAs, JAZ1 interacts with and represses the downstream transcriptional activators, such as MYC2 [26], MYC3, MYC4 [27], MYB21, MYB24 [28], bHLH017 and bHLH003 [29], which control the expression of JA-responsive genes. In the presence of JAs, AtTIFY10a protein is recognized and degraded by the SCF (COI1) E3 ubiquitin ligase, releasing the downstream transcription factors [49]. Jasmonoyl isoleucine (JA-Ile) and coronatine could promote the physical interaction between JAZ1 and COI1, and the C-terminal Jas domain of JAZ1 is critical for JA-Ile/coronatine-dependent interaction with COI1 [49,50]. Recently, a transacting factor AtBBD1 was suggested to interact with AtTIFY10a and bind to the JARE element upstream of the JA responsive gene AtJMT [51]. In addition to the negative regulatory role in JA signaling pathway, AtTIFY10a was also reported to be involved in phytochrome A [52] and auxin signaling [53]. Furthermore, AtTIFY10a physically interacted with ICE1 and ICE2 transcription factors, repressed the transcriptional function of ICE1, and thereby affected plant freezing stress responses [33]. However, no direct evidence supporting its role in alkaline stress responses was reported until now.

Recently, the systemic transcriptional analyses and several lines of genetic evidence suggested the potential regulatory roles of TIFY10/11s in plant responses to environmental challenges

[15,33]. Our previous research also revealed the alkaline stress induced expression of *Glycine soja* TIFY10/11 genes [36,40], which was further confirmed by the RNA-seq data (Fig. S1). Consistently, we also observed some cis-regulatory elements related to environmental stress response in the TIFY10/11s promoters (Table 1). In addition, we also found elements involved in hormone responsiveness, suggesting the crosstalk between environmental stress and hormone signal. Several elements were common to all TIFY10/11s, implying that TIFY10/11 proteins were involved in the same signal transduction pathway. Remarkably, several elements only existed in specific TIFY10/11 promoters, which might indicate the specific response of TIFY10/11s to environmental stress.

Among the eight GsTIFY10/11 genes, overexpression of *GsTIFY10a* and *GsTIFY10e* in *Arabidopsis* obviously improved plant alkaline tolerance [37,38], while *GsTIFY11b* overexpression led to decreased salt tolerance [39]. In this study, we suggested the positive roles of AtTIFY10a/b in plant tolerance to alkaline stress by using the plate germination assays (Fig. 3), further supporting the important roles of TIFY10s in plant alkaline stress responses. It is worth to note that, under the same alkaline stress treatment, *attify10a* (54.2% on the 3rd day) exhibited relatively lower germination rates than *attify10b* (77.0%), indicating a more important role of *AtTIFY10a* than *AtTIFY10b* in alkaline stress responses.

In the current study, we also transformed the *GsTIFY10a* gene into *Medicago sativa* in an attempt to obtain transgenic alfalfa with superior alkaline tolerance. Our results demonstrated that *GsTIFY10a* overexpression dramatically promoted growth of transgenic plants under alkaline stress. *GsTIFY10a* transgenic lines displayed much better at shoot height, ground fresh weight and dry weight than WT under alkaline stress (Fig. 4), which was in line with the better growth performance of *GsTIFY10a* transgenic Arabidopsis [38]. Furthermore, we also investigated the potential physiological and molecular basis of *GsTIFY10a* in response to alkaline stress (Fig. 5, 6). Firstly, *GsTIFY10a* overexpression could help plant to deal with the high pH damage by up-regulating the NADP-ME activity and citrate acid content (Fig. 5). The up-regulation of *NADP-ME* and *CS* gene expression in transgenic plants might be helpful to explain the increase of NADP-ME activity and citrate acid content (Fig. 6). Meanwhile, we also observed an obvious increase of *H^{+}-ppase* expression in transgenic lines under alkaline stress (Fig. 6), which was also helpful for maintaining the cytosolic pH homeostasis. These results were consistent with our previous observation that *GsTIFY10a* overexpression up-regulated the expression of *NADP-ME* and *H^{+}-ppase* in Arabidopsis [38]. Secondly, *GsTIFY10a* overexpression led to greater proline accumulation (Fig. 5) and up-regulated expression of the *P5CS* gene (Fig. 6), which encodes a key enzyme in the proline biosynthesis process. Proline serves as a compatible osmolyte, molecular chaperone and ROS scavenger in plant responses to environmental stress [54]. The increased accumulation of free proline in *GsTIFY10a* transgenic lines might be of great importance for the effective osmo-regulation and ROS scavenging of plant cells under alkaline stress. In addition, *GsTIFY10a* transgenic lines displayed lower levels of MDA content under alkaline stress (Fig. 5D). MDA is widely recognized as an indicator for lipid peroxidation resulted from the elevated ROS accumulation in plant cells [55]. The decreased MDA accumulation also indicated the more effective adaptation of transgenic plants to ROS damage caused by alkaline stress. Taken together, we speculated that *GsTIFY10a* overexpression is of fundamental importance for the cytosolic pH regulation, osmo-regulation and ROS scavenging by regulating the related gene expression and enzyme activity, and thereby promoted plant growth under alkaline stress.

A great number of studies have demonstrated that JA signaling is critical for plant stress responses and TIFY proteins are involved in plant tolerance to both biotic and abiotic stresses such as wounding [52,56], pathogen [23,57], drought stress [15,34], and salt stress [20,32]. Our studies suggested the involvement of TIFYs in alkaline stress responses and the crosstalk between alkaline stress and JA signaling [37,38,40]. All of the *GsTIFY10/11* genes were dramatically up-regulated at the early stage of alkaline stress treatment (Fig. S1). Overexpression of GsTIFY10e/JAZ2 significantly improved plant alkaline stress tolerance [37]. Knockout of AtTIFY10a and AtTIFY10b also inhibited seed germination under alkaline stress (Fig. 3). Furthermore, *GsTIFY10a* expression was greatly induced by both alkaline stress and MeJA treatment [38].

In addition to the increased alkaline tolerance, *GsTIFY10a* overexpression in Arabidopsis conferred MeJA insensitivity. *GsTIFY10a* repressed transcription of the JA responsive genes such as *PDF1.2* (Plant Defensin 1.2), *VSP2* (Vegetative Storage Protein 2), *AOS* (Allelen Oxide Synthase) and *LOX2* (Lipoxygenase 2), whose expression was also induced by alkaline stress. However, in the present study, we found that JA contents were significantly increased in the transgenic alfalfa lines (Fig.7). As we know, environmental stress could stimulate the biosynthesis of JAs. The elevated JAs levels promoted the interaction of TIFY/JAZ with SCF^{Col1}, which led to degradation of TIFY/JAZ proteins, and subsequently release of the targeted TFs. Activation of TFs induced transcription of JA-responsive genes, and the newly synthesized TIFY/JAZ proteins could restore the repression of TFs. The *GsTIFY10a* transgenic lines always kept high levels of GsTIFY10a proteins, which repressed the TFs activity and JA signaling. Hence, plants need to synthesize more JAs to degrade the GsTIFY10a proteins. Collectively, these results strongly suggested the crosstalk between alkaline stress and JA signaling. Molecular basis of the hypothesis that TIFY10 proteins mediate alkaline stress responses through JA signaling need be further studied.

The JAZ proteins were suggested to act as transcriptional repressors in JA signaling because they do not contain a known DNA binding domain. Due to the nuclear localization of several JAZ subfamily members [25,58], it is proposed that they might exert their effect on gene expression through protein interaction with transcription factors. Our previous revealed that, like other JAZs, GsTIFY10a and GsTIFY10e localized in the nuclei of plant cells. In this study, we found that GsTIFY10a and GsTIFY10e showed no transcriptional activity in yeast cells (Fig.8), indicating them as transcriptional repressors. On the other hand, it has been well demonstrated that the ZIM domains of JAZ/TIFYs mediate formation of homo- and heteromeric complexes [59]. For example, the Arabidopsis JAZ1/TIFY10a and JAZ2/TIFY10b could form homodimers and heterodimers [59]. In this study, by using the Y2H and BiFc technologies, we found that GsTIFY10a could form homodimers, as well as heterodimers with GsTIFY10e, while GsTIFY10e could not homodimerize (Fig.9). Similarly, in case of Arabidopsis JAZs, JAZ7, JAZ8, JAZ9, JAZ11, and JAZ12 could not homodimerize, but JAZ9 and JAZ12 could form heterodimers with other JAZs [59].

In summary, here we provide novel insights into the regulatory roles and potential molecular basis of TIFY10 proteins in plant responses to alkaline stress. We obtained the transgenic alfalfa with superior alkaline tolerance by ectopically expressing *GsTIFY10a*, and investigated the possible physiological and molecular mechanism by which *GsTIFY10a* regulated plant alkaline tolerance.

For now, the *GsTIFY10a* transgenic alfalfa is at the stage of the "biosafety evaluation of genetically modified organisms", and will provide a bearing on future aims at salt-alkaline soil management.

Materials and Methods

Plant materials and growth conditions

Seeds of the wild type *Arabidopsis thaliana* (Columbia ecotype), *attify10a* (SALK_011957) and *attif10b* (SALK_025279) mutant lines were kindly provided by the European Arabidopsis Stock Centre (UK). Both WT and mutant Arabidopsis seeds were germinated and grown in standard nutrient solution as described by Tocquin et al. (2003). The Arabidopsis seedlings were maintained in a greenhouse under controlled environmental conditions (21–23°C, 100 μmol m^{-2} s^{-1}, 60% relative humidity, 16 h light/8 h dark cycles).

Alfalfa (*Medicago sativa* L. cv. Nongjing No. 1, a local species) was used in this study, and was kindly obtained from Heilongjiang Academy of Agricultural Sciences (Harbin, China). Alfalfa was grown in a greenhouse under controlled environmental conditions (24–26°C, 600 μmol m^{-2} s^{-1}, 80% relative humidity, 16 h light/ 8 h dark cycles).

Seed germination assays of *AtTIFY10a/b* knockout Arabidopsis

Homozygous T-DNA insertion mutant Arabidopsis was obtained by using a PCR-based method as described previously [60]. Briefly, the gene specific primers across the T-DNA insertion sites of *attify10a* and *attif10b* mutant (5′-TGCGATCCAGCCAAAGCGTCT-3′ and 5′-GAGTATTTGATAGTATGGTTCGTCAACAA-3′ for *attify10a*, 5′-TGAGAAGAGGAAGGATAGGTAATGCAT-3′ and 5′-CAAGCTAATGTTGAGATCGGCAGAG-3′ for *attify10b*) were used for homozygous identification. T-DNA insertion was confirmed by PCR amplification with the following primers (5′-TGCGATCCAGCCAAAGCGTCT-3′ and 5′-GATTTGGGTGATGGTTCACGTAGTG-3′ for *attify10a*, 5′-GAGAAGAGGAAGGATAGGTAATGCAT-3′ and 5′-GATTTGGGTGATGGTTCACGTAGTG-3′ for *attify10b*).

RT-PCR analysis was used to confirm the silence of *AtTIFY10a* and *AtTIFY10b* genes in the T-DNA insertion mutants. The following gene specific primers were used (5′-TGCGATCCAGCCAAAGCGTCT-3′ and 5′-GAGTATTTGATAGTATGGTTCGTCAACAA-3′ for *AtTIFY10a*, 5′-TGAGAAGAGGAAGGATAGGATCACATC-3′ and 5′-CAAGCTAATGTTGAGATCGGCAGAG-3′ for *AtTIFY10b*).

For plate germination assays under alkaline stress, seeds of WT and mutant lines were surfaced-sterilized with 5% sodium hypochlorite (NaClO) for 6–8 min with shaking, washed with sterilized distilled water for 6–8 times, and kept at 4°C for 3 days to break seed dormancy. After that, WT and mutant *Arabidopsis* seeds were sown on 1/2 MS agar medium with 1% (w/v) sucrose and 0.8% (w/v) agar at pH5.8 (Control conditions) or pH8.5 (Alkaline stress). The germination percent was recorded for consecutive 6 days after sowing, and pictures were taken to show the growth performance of each line. Ninety seeds from each line were used for each experiment and the experiments were repeated for three times.

Generation of *GsTIFY10a* overexpression transgenic alfalfa

In order to investigate the effect of *GsTIFY10a* on alkaline stress tolerance, *GsTIFY10a* was cloned into the pBEOM plant expression vector, and was under the control of the cauliflower mosaic virus (CaMV) 35S promoter, with the binding enhancers E12 and omega. The *Bar* gene was used as the selectable marker. The recombinant vector pBEOM-GsTIFY10a was introduced into *Agrobacterium tumefaciens* strain EHA105, and then introduced into *Medicago sativa* by using the cotyledonary node method as described previously [13]. Briefly, alfalfa seeds were surface sterilized with 70% ethanol for 1 min, 0.1% (v/v) $HgCl_2$ for 15 min, and then washed with sterilized water for 3–5 times. Sterilized seeds were germinated and grown on 1/2MS medium for 8 days, and then seedlings were aseptically excised at the cotyledonary node position as explants. The explants were infected with *A. tumefaciens* EHA105 for 15 min, and placed vertically into MS medium (pH 5.2) containing 1 mg L^{-1} 6-benzylaminopurine (6-BA) and 100 mM L^{-1} acetosyringone for 3 days. After that, explants were washed in liquid MS medium (pH 5.8) containing 1 mg L^{-1} 6-BA, 0.5 mg L^{-1} glufosinate and 100 mg L^{-1} amoxicillin. Then explants were transferred onto solid MS medium (pH 5.8) containing 1 mg L^{-1} 6-BA, 0.5 mg L^{-1} glufosinate and 100 mg L^{-1} amoxicillin, and grown for another 14 days to form the glufosinate-resistant regenerated shoots. The regnerated shoots were then transferred to new solid medium as described above, and grown for another 14 days. Elongated shoots were then transferred to new 1/2MS medium only containing 100 mg L^{-1} amoxicillin, until roots appeared.

The regenerated alfalfa plants were then transplanted into soil under controlled conditions and confirmed by PCR analysis using CaMV35S promoter specific forward primer and *Bar* gene specific reverse primer (5′-TGCACCATCGTCAACCACTACATCG-3′ and 5′-CCAGCTGCCAGAAACCCACGTCATG-3′). To eliminate the potential existence of the chimeric transgenic plants, new-born leaves from different branches of each regenerated plants were used for PCR identification. The transcript levels of *GsTIFY10a* in the PCR-positive plants were further analyzed by semi-quantitative RT-PCR analyses with gene specific primers (5′-ACAGAGCCAGCCTTCATTTCC-3′ and 5′-CGAACCCGACTCACGAAGAAG-3′). The alfalfa glyceraldehyde-3-phosphate dehydrogenase gene (*MtGAPDH*, Accession: Medtr3g085850) was used as an internal control, and PCR amplified with the following primer pair (5′-GTGGTGCCAAGAAGGTTGTTAT-3′ and 5′-CTGGGAATGATGTTGAAGGAAG-3′).

Phenotypic analyses of transgenic alfalfa under alkaline stress

For phenotypic analyses under alkaline stress, the lignified WT and transgenic alfalfa plants were used for vegetative propagation through stem cuttings. The propagated seedlings were transplanted into plastic culture pots filled with a mixture of peat moss: soil (1:1; v/v), irrigated with 1/8 Hoagland nutrient solution and grown in a greenhouse under controlled conditions. To eliminate the potential existence of the chimeric transgenic plants, each propagated seedlings were subjected for PCR identification before stress treatment. PCR-positive plants with similar sizes (approximately 25 cm high) were then exposed to alkaline stress by irrigating with 1/8 Hoagland solution containing either 0, or 100, or 150 mM $NaHCO_3$ every 3 days for a total of 12 days. Photographs were taken on the 12th day. Thirty plants of each line were used for each experiment, and the experiments were repeated for at least three times.

The MDA content was determined according to the protocol described by Peever et al. [61]. The citric acid content was determined by using a spectrophotometer (UV-2550, Shimadzu, Japan) at the absorbance of 490 nm according to the method of Zhu [62]. Free proline content was measured according to the method of Bates et al. [63]. NADP-ME activity was measured as described by Geer et al. [64].

Table 2. Gene-specific primers used for quantitative real-time PCR assays.

Gene name	Gene ID	Primer Sequence (5′ to 3′)
GsGAPDH	DQ355800	Forward: GACTGGTATGGCATTCCGTGT Reverse: GCCCTCTGATTCCTCCTTGA
GsTIFY10a		Forward: TCCAGCGCAATTAAGTCTGTGAG Reverse: TTTGGTGGCATAGGACATGATCT
MtGAPDH	Medtr3g085850	Forward: GTGGTGCCAAGAAGGTTGTTAT Reverse: CTGGGAATGATGTTGAAGGAAG
P5CS	EU371644.1	Forward: TCGGGGTCCAGTAGGAGTTG Reverse: AGTAGTTAGGTCTTTGTGGGTGTAGG
H⁺-ppase	XM_003609415.1	Forward: TCTCCACCGACGCATCTATCA Reverse: GGCATTATCCCAAGCACCG
NADP-ME	XM_003630679.1	Forward: TGGCGGTGTTGAGGATGTCT Reverse: CTGACAGTGGGAGGTAAGAGGC
Citrate Synthase	HM030734.1	Forward: TTAAAGCCAGGAAACGAAAGC Reverse: AAGGCAACAGCAACCTCAATC

All of the above numerical data were subjected to statistical analyses using EXCEL 2010 and/or IBM SPSS statistics 19, and analyzed by Student's T-test and/or one-way ANOVA analysis.

Quantitative real-time PCR analyses

To investigate the expression patterns of stress-responsive genes, the 4-week-old seedlings after shoot cottage were treated with 1/8 Hoagland solution containing 50 mM $NaHCO_3$ (pH 8.5) for 0, 3, 6, and 12 h, respectively. Equal amounts of leaves were harvested and stored at −80°C after snap-frozen in liquid nitrogen.

Total RNA was extracted by using RNeasy Plant Mini Kit (Qiagen, Valencia, CA, USA), and subjected to cDNA synthesis by using SuperScript™ III Reverse Transcriptase kit (Invitrogen, Carlsbad, CA, USA). Quantitative real-time PCR was performed using a Stratagene MX3000P real-time PCR instrument and the SYBR Select Master Mix (Applied Biosystems, USA). The MtGAPDH gene was used as an internal reference. Expression levels of all candidate genes were calculated by using the $2^{-\Delta\Delta CT}$ method, and the relative intensities were calculated and normalized as described previously 65. Three independent biological replicates were carried out and subjected to real-time PCR in triplicate. Primers used for quantitative real-time PCR were designed by using Primer 5 software and listed in Table 2.

Transcriptional activation assays

The full-length coding regions of *GsTIFY10a* and *GsTIFY10e* were amplified by using the following primer pairs: 5′-CTTGAATTCTCGAGCTCATCGG-3′ and 5′-GAAGTCGACGATTTGAGGTGAA-3′ (for GsTIFY10a), 5′-CGGGAATTCAATCCATGGAAC-3′ and 5′-GCCGTCGACTAACACAAAGCTGG-3′ (for GsTIFY10e). The PCR products were cloned to the pGBKT7 vector to express the GsTIFY10a-BD and GsTIFY10e-BD fused proteins.

For the transcriptional activation assays, the corresponding vectors pGBKT7 (negative control), pGBKT7-GsTIFY10a, pGBKT7-GsTIFY10e and pGBKT7-AtbZIP1 (positive control) were transformed into yeast strain AH109. The transformants were selected on SD/-Trp medium. LacZ activity was analyzed by using the β-galactosidase filter lift assay.

Dimerization analyses of GsTIFY10a and GsTIFY10b in yeast cells

The full-length coding regions of *GsTIFY10a* and *GsTIFY10e* were amplified by using the following primer pairs: 5′-GCTCATATGTCGAGCTCATCGG-3′ and 5′-CTTGAATTCGATTTGAGGTGAA-3′ (for *GsTIFY10a*), 5′- GCCCATATGACCTCAGTGGAG-3′ and 5′-GTTGGATCCTAACACAAAGCTGG-3′ (for *GsTIFY10e*). The PCR products were cloned to the pGADT7 vector to express the GsTIFY10a-AD and GsTIFY10e-AD fused proteins.

For homodimerization analyses of GsTIFY10a and GsTIFY10e, the combinations pGBKT7-GsTIFY10a/pGADT7-GsTIFY10a and pGBKT7-GsTIFY10e/pGADT7-GsTIFY10e were co-transformed into yeast strain AH109. For heterodimerization analyses between GsTIFY10a and GsTIFY10e, the combinations pGBKT7-GsTIFY10a/pGADT7-GsTIFY10e and pGBKT7-GsTIFY10e/pGADT7-GsTIFY10a were co-transformed into yeast. The pGBKT7-GsTIFY10a/pGADT7 and pGBKT7-GsTIFY10e/pGADT7 combinations were used as negative controls, and the pGBKT7-AtbZIP63/pGADT7-AtbZIP63 combination was used as a positive control. The transformants were selected on SD/-Trp-Leu and SD/-Trp-Leu-His-Ade medium, respectively.

Bimolecular fluorescence complementation (BiFc) assays

The full-length coding region of *GsTIFY10a* was PCR amplified and fused into the N-terminus of the amino-terminal half of YFP protein (YFPN), to generate the GsTIFY10a-YFPN construct. The coding regions of *GsTIFY10a* and *GsTIFY10e* were fused into the N-terminus of the carboxyl-terminal half of YFP protein (YFPC), to generate the GsTIFY10a-YFPC and GsTIFY10e-YFPC construct.

BiFc assays were conducted by transient transformation of *Arabidopsis* protoplasts prepared as described [66]. Yellow fluorescence was measured by excitation at 514 nm and emission at 527 nm, by using confocal laser-scanning microscope Leica SP2 (Leica, Wetzlar, Germany). YFP fluorescence, chlorophyll autofluorescence and light visions were recorded in separate channels and then merged into an overlay image.

Supporting Information

Figure S1 Expression patterns of the Arabidopsis and wild soybean TIFY10/11 subgroup genes under 50 mM $NaHCO_3$ (pH 8.5) treatment based on the RNA-seq data. **a**. Relative expression levels of the TIFY10/11 genes under alkaline stress. The expression levels at 0 h were considered as 1. **b**. Expression levels of the TIFY10/11 genes under alkaline stress.

Acknowledgments

We thank members of the lab for discussions and comments on this manuscript. We gratefully acknowledge support from Key Laboratory of Agricultural Biological Functional Genes - College of Heilongjiang Province.

Author Contributions

Conceived and designed the experiments: DZ MS YZ. Performed the experiments: DZ JW NZ. Analyzed the data: MS RL. Contributed reagents/materials/analysis tools: YZ RL XL. Wrote the paper: DZ MS.

References

1. Xu J, Ji P, Wang B, Zhao L, Wang J, et al. (2013) Transcriptome sequencing and analysis of wild Amur Ide (*Leuciscus waleckii*) inhabiting an extreme alkaline-saline lake reveals insights into stress adaptation. PLoS ONE 8: e59703.
2. Sun Y, Wang F, Wang N, Dong Y, Liu Q, et al. (2013) Transcriptome exploration in Leymus chinensis under saline-alkaline treatment using 454 pyrosequencing. PLoS ONE 8: e53632.
3. Zhang WJ, Niu Y, Bu SH, Li M, Feng JY, et al. (2014) Epistatic association mapping for alkaline and salinity tolerance traits in the soybean germination stage. PLoS ONE 9: e84750.
4. Han X, Cheng Z, Meng H (2012) Soil properties, nutrient dynamics, and soil enzyme activities associated with garlic stalk decomposition under various conditions. PLoS ONE 7: e50868.
5. Wang H, Wu Z, Han J, Zheng W, Yang C (2012) Comparison of ion balance and nitrogen metabolism in old and young leaves of alkali-stressed rice plants. PLoS ONE 7: e37817.
6. Guan Q, Wu J, Yue X, Zhang Y, Zhu J (2013) A nuclear calcium-sensing pathway is critical for gene regulation and salt stress tolerance in Arabidopsis. PLoS Genet 9: e1003755.
7. Xu W, Jia L, Shi W, Baluska F, Kronzucker HJ, et al. (2013) The tomato 14-3-3 protein TFT4 modulates H^+ efflux, basipetal auxin transport, and the PKS5-J3 pathway in the root growth response to alkaline stress. Plant Physiol 163: 1817–1828.
8. Liu J, Guo Y (2011) The alkaline tolerance in Arabidopsis requires stabilizing microfilament partially through inactivation of PKS5 kinase. J Genet Genomics 38: 307–313.
9. Tong Z, Xie C, Ma L, Liu L, Jin Y, et al. (2014) Co-expression of bacterial aspartate kinase and adenylylsulfate reductase genes substantially increases sulfur amino acid levels in transgenic alfalfa (*Medicago sativa* L.). PLoS ONE 9: e88310.
10. Li W, Wei Z, Qiao Z, Wu Z, Cheng L, et al. (2013) Proteomics analysis of alfalfa response to heat stress. PLoS ONE 8: e82725.
11. Shi Y, Guo R, Wang X, Yuan D, Zhang S, et al. (2014) The regulation of alfalfa saponin extract on key genes involved in hepatic cholesterol metabolism in hyperlipidemic rats. PLoS ONE 9: e88282.
12. Bogino P, Abod A, Nievas F, Giordano W (2013) Water-limiting conditions alter the structure and biofilm-forming ability of bacterial multispecies communities in the alfalfa rhizosphere. PLoS ONE 8: e79614.
13. Sun M, Sun X, Zhao Y, Zhao C, Duanmu H, et al. (2014) Ectopic expression of GsPPCK3 and SCMRP in Medicago sativa enhances plant alkaline stress tolerance and methionine content. PLoS ONE 9: e89578.
14. Bai Y, Meng Y, Huang D, Qi Y, Chen M (2011) Origin and evolutionary analysis of the plant-specific TIFY transcription factor family. Genomics 98: 128–136.
15. Ye H, Du H, Tang N, Li X, Xiong L (2009) Identification and expression profiling analysis of TIFY family genes involved in stress and phytohormone responses in rice. Plant Mol Biol 71: 291–305.
16. Qi T, Huang H, Wu D, Yan J, Qi Y, et al. (2014) Arabidopsis DELLA and JAZ proteins bind the WD-repeat/bHLH/MYB complex to modulate gibberellin and jasmonate signaling synergy. Plant Cell 26: 16.
17. Van der Does D, Leon-Reyes A, Koornneef A, Van Verk MC, Rodenburg N, et al. (2013) Salicylic acid suppresses jasmonic acid signaling downstream of SCFCOI1-JAZ by targeting GCC promoter motifs via transcription factor ORA59. Plant Cell 25: 744–761.
18. Oh Y, Baldwin IT, Galis I (2013) A jasmonate ZIM-domain protein NaJAZd regulates floral jasmonic acid levels and counteracts flower abscission in Nicotiana attenuata plants. PLoS ONE 8: e57868.
19. Toda Y, Yoshida M, Hattori T, Takeda S (2013) RICE SALT SENSITIVE3 binding to bHLH and JAZ factors mediates control of cell wall plasticity in the root apex. Plant Signal Behav 8.
20. Toda Y, Tanaka M, Ogawa D, Kurata K, Kurotani K, et al. (2013) RICE SALT SENSITIVE3 forms a ternary complex with JAZ and class-C bHLH factors and regulates jasmonate-induced gene expression and root cell elongation. Plant Cell 25: 1709–1725.
21. Hakata M, Kuroda M, Ohsumi A, Hirose T, Nakamura H, et al. (2012) Overexpression of a rice TIFY gene increases grain size through enhanced accumulation of carbohydrates in the stem. Biosci Biotechnol Biochem 76: 2129–2134.
22. Song S, Qi T, Fan M, Zhang X, Gao H, et al. (2013) The bHLH subgroup IIId factors negatively regulate jasmonate-mediated plant defense and development. PLoS Genet 9: e1003653.
23. Ishiga Y, Ishiga T, Uppalapati SR, Mysore KS (2013) Jasmonate ZIM-domain (JAZ) protein regulates host and nonhost pathogen-induced cell death in tomato and Nicotiana benthamiana. PLoS ONE 8: e75728.
24. Demianski AJ, Chung KM, Kunkel BN (2012) Analysis of Arabidopsis JAZ gene expression during Pseudomonas syringae pathogenesis. Mol Plant Pathol 13: 46–57.
25. Thines B, Katsir L, Melotto M, Niu Y, Mandaokar A, et al. (2007) JAZ repressor proteins are targets of the SCF(COI1) complex during jasmonate signalling. Nature 448: 661–665.
26. Fernandez-Calvo P, Chini A, Fernandez-Barbero G, Chico JM, Gimenez-Ibanez S, et al. (2011) The Arabidopsis bHLH transcription factors MYC3 and MYC4 are targets of JAZ repressors and act additively with MYC2 in the activation of jasmonate responses. Plant Cell 23: 701–715.
27. Niu Y, Figueroa P, Browse J (2011) Characterization of JAZ-interacting bHLH transcription factors that regulate jasmonate responses in Arabidopsis. J Exp Bot 62: 2143–2154.
28. Song S, Qi T, Huang H, Ren Q, Wu D, et al. (2011) The Jasmonate-ZIM domain proteins interact with the R2R3-MYB transcription factors MYB21 and MYB24 to affect Jasmonate-regulated stamen development in Arabidopsis. Plant Cell 23: 1000–1013.
29. Fonseca S, Fernandez-Calvo P, Fernandez GM, Diez-Diaz M, Gimenez-Ibanez S, et al. (2014) bHLH003, bHLH013 and bHLH017 Are New Targets of JAZ Repressors Negatively Regulating JA Responses. PLoS ONE 9: e86182.
30. White DW (2006) PEAPOD regulates lamina size and curvature in Arabidopsis. Proc Natl Acad Sci U S A 103: 13238–13243.
31. Shikata M, Matsuda Y, Ando K, Nishii A, Takemura M, et al. (2004) Characterization of Arabidopsis ZIM, a member of a novel plant-specific GATA factor gene family. J Exp Bot 55: 631–639.
32. Ismail A, Riemann M, Nick P (2012) The jasmonate pathway mediates salt tolerance in grapevines. J Exp Bot 63: 2127–2139.
33. Hu Y, Jiang L, Wang F, Yu D (2013) Jasmonate regulates the inducer of cbf expression-C-repeat binding factor/DRE binding factor1 cascade and freezing tolerance in Arabidopsis. Plant Cell 25: 2907–2924.
34. Seo JS, Joo J, Kim MJ, Kim YK, Nahm BH, et al. (2011) OsbHLH148, a basic helix-loop-helix protein, interacts with OsJAZ proteins in a jasmonate signaling pathway leading to drought tolerance in rice. Plant J 65: 907–921.
35. Chen P, Yan K, Shao H, Zhao S (2013) Physiological Mechanisms for High Salt Tolerance in Wild Soybean (Glycine soja) from Yellow River Delta, China: Photosynthesis, Osmotic Regulation, Ion Flux and antioxidant Capacity. PLoS ONE 8: e83227.
36. Ge Y, Li Y, Zhu YM, Bai X, Lv DK, et al. (2010) Global transcriptome profiling of wild soybean (Glycine soja) roots under NaHCO3 treatment. BMC Plant Biol 10: 153.
37. Zhu D, Cai H, Luo X, Bai X, Deyholos MK, et al. (2012) Over-expression of a novel JAZ family gene from Glycine soja, increases salt and alkali stress tolerance. Biochem Biophys Res Commun 426: 273–279.
38. Zhu D, Bai X, Chen C, Chen Q, Cai H, et al. (2011) GsTIFY10, a novel positive regulator of plant tolerance to bicarbonate stress and a repressor of jasmonate signaling. Plant Mol Biol 77: 285–297.
39. Zhu D, Bai X, Zhu YM, Cai H, Li Y, et al. (2012) [Isolation and functional analysis of GsTIFY11b relevant to salt and alkaline stress from Glycine soja]. Yi Chuan 34: 230–239.
40. Zhu D, Bai X, Luo X, Chen Q, Cai H, et al. (2013) Identification of wild soybean (Glycine soja) TIFY family genes and their expression profiling analysis under bicarbonate stress. Plant Cell Rep 32: 263–272.
41. Yang CW, Wang P, Li CY, Shi DC, Wang DL (2008) Comparison of effects of salt and alkali stresses on the growth and photosynthesis of wheat. Photosynthetica 46: 107–144.
42. Edwards GE, Andreo CS (1992) NADP-malic enzyme from plants. Phytochemistry 31: 1845–1857.
43. Garcia J, Torres N (2011) Mathematical modelling and assessment of the pH homeostasis mechanisms in Aspergillus niger while in citric acid producing conditions. J Theor Biol 282: 23–35.
44. Xiong H, Li J, Liu P, Duan J, Zhao Y, et al. (2014) Overexpression of OsMYB48-1, a Novel MYB-Related Transcription Factor, Enhances Drought and Salinity Tolerance in Rice. PLoS ONE 9: e92913.
45. Sun X, Li Y, Cai H, Bai X, Ji W, et al. (2012) The Arabidopsis AtbZIP1 transcription factor is a positive regulator of plant tolerance to salt, osmotic and drought stresses. J Plant Res 125: 429–438.
46. Weltmeier F, Rahmani F, Ehlert A, Dietrich K, Schutze K, et al. (2009) Expression patterns within the Arabidopsis C/S1 bZIP transcription factor network: availability of heterodimerization partners controls gene expression during stress response and development. Plant Mol Biol 69: 107–119.
47. Chini A, Fonseca S, Chico JM, Fernandez-Calvo P, Solano R (2009) The ZIM domain mediates homo- and heteromeric interactions between Arabidopsis JAZ proteins. Plant Journal 59: 77–87.
48. Siberil Y, Doireau P, Gantet P (2001) Plant bZIP G-box binding factors. Modular structure and activation mechanisms. Eur J Biochem 268: 5655–5666.

49. Melotto M, Mecey C, Niu Y, Chung HS, Katsir L, et al. (2008) A critical role of two positively charged amino acids in the Jas motif of Arabidopsis JAZ proteins in mediating coronatine- and jasmonoyl isoleucine-dependent interactions with the COI1 F-box protein. Plant J 55: 979–988.
50. Withers J, Yao J, Mecey C, Howe GA, Melotto M, et al. (2012) Transcription factor-dependent nuclear localization of a transcriptional repressor in jasmonate hormone signaling. Proc Natl Acad Sci U S A 109: 20148–20153.
51. Seo JS, Koo YJ, Jung C, Yeu SY, Song JT, et al. (2013) Identification of a novel jasmonate-responsive element in the AtJMT promoter and its binding protein for AtJMT repression. PLoS ONE 8: e55482.
52. Robson F, Okamoto H, Patrick E, Harris SR, Wasternack C, et al. (2010) Jasmonate and phytochrome A signaling in Arabidopsis wound and shade responses are integrated through JAZ1 stability. Plant Cell 22: 1143–1160.
53. Grunewald W, Vanholme B, Pauwels L, Plovie E, Inze D, et al. (2009) Expression of the Arabidopsis jasmonate signalling repressor JAZ1/TIFY10A is stimulated by auxin. EMBO Rep 10: 923–928.
54. Liu J, Zhu JK (1997) Proline accumulation and salt-stress-induced gene expression in a salt-hypersensitive mutant of Arabidopsis. Plant Physiol 114: 591–596.
55. Mittler R (2002) Oxidative stress, antioxidants and stress tolerance. Trends Plant Sci 7: 405–410.
56. Chung HS, Koo AJ, Gao X, Jayanty S, Thines B, et al. (2008) Regulation and function of Arabidopsis JASMONATE ZIM-domain genes in response to wounding and herbivory. Plant Physiol 146: 952–964.
57. Jiang S, Yao J, Ma KW, Zhou H, Song J, et al. (2013) Bacterial effector activates jasmonate signaling by directly targeting JAZ transcriptional repressors. PLoS Pathog 9: e1003715.
58. Yan Y, Stolz S, Chetelat A, Reymond P, Pagni M, et al. (2007) A downstream mediator in the growth repression limb of the jasmonate pathway. Plant Cell 19: 2470–2483.
59. Chung HS, Howe GA (2009) A critical role for the TIFY motif in repression of jasmonate signaling by a stabilized splice variant of the JASMONATE ZIM-domain protein JAZ10 in Arabidopsis. Plant Cell 21: 131–145.
60. Alonso JM, Stepanova AN, Leisse TJ, Kim CJ, Chen H, et al. (2003) Genome-wide insertional mutagenesis of Arabidopsis thaliana. Science 301: 653–657.
61. Peever TL, Higgins VJ (1989) Electrolyte leakage, lipoxygenase, and lipid peroxidation induced in tomato leaf tissue by specific and nonspecific elicitors from *Cladosporium fulvum*. Plant Physiol 90: 867–875.
62. de la Fuente JM, Ramirez-Rodriguez V, Cabrera-Ponce JL, Herrera-Estrella L (1997) Aluminum tolerance in transgenic plants by alteration of citrate synthesis. Science 276: 1566–1568.
63. Bates LS, Waldren RP, Teare ID (1973) Rapid determination of free proline for water-stress studies. Plant and Soil 39: 3.
64. Geer BW, Krochko D, Williamson JH (1979) Ontogeny, cell distribution, and the physiological role of NADP-malic enxyme in Drosophila melanogaster. Biochem Genet 17: 867–879.
65. Willems E, Leyns L, Vandesompele J (2008) Standardization of real-time PCR gene expression data from independent biological replicates. Anal Biochem 379: 127–129.
66. Yoo SD, Cho YH, Sheen J (2007) Arabidopsis mesophyll protoplasts: a versatile cell system for transient gene expression analysis. Nat Protoc 2: 1565–1572.

De Novo Transcriptome and Small RNA Analysis of Two Chinese Willow Cultivars Reveals Stress Response Genes in *Salix matsudana*

Guodong Rao[2]⁹, Jinkai Sui[2]⁹, Yanfei Zeng[1,2], Caiyun He[1,2], Aiguo Duan[1,2], Jianguo Zhang[1,2]*

1 State Key Laboratory of Tree Genetics and Breeding, Research Institute of Forestry, Chinese Academy of Forestry, Beijing, Republic of China, **2** Key Laboratory of Tree Breeding and Cultivation, State Forestry Administration, Research Institute of Forestry, Chinese Academy of Forestry, Beijing, Republic of China

Abstract

Salix matsudana Koidz. is a deciduous, rapidly growing, and drought resistant tree and is one of the most widely distributed and commonly cultivated willow species in China. Currently little transcriptomic and small RNAomic data are available to reveal the genes involve in the stress resistant in *S. matsudana*. Here, we report the RNA-seq analysis results of both transcriptome and small RNAome data using Illumina deep sequencing of shoot tips from two willow variants(*Salix. matsudana* and *Salix matsudana* Koidz. cultivar 'Tortuosa'). *De novo* gene assembly was used to generate the consensus transcriptome and small RNAome, which contained 106,403 unique transcripts with an average length of 944 bp and a total length of 100.45 MB, and 166 known miRNAs representing 35 miRNA families. Comparison of transcriptomes and small RNAomes combined with quantitative real-time PCR from the two *Salix* libraries revealed a total of 292 different expressed genes(DEGs) and 36 different expressed miRNAs (DEMs). Among the DEGs and DEMs, 196 genes and 24 miRNAs were up regulated, 96 genes and 12 miRNA were down regulated in *S. matsudana*. Functional analysis of DEGs and miRNA targets showed that many genes were involved in stress resistance in *S. matsudana*. Our global gene expression profiling presents a comprehensive view of the transcriptome and small RNAome which provide valuable information and sequence resources for uncovering the stress response genes in *S. matsudana*. Moreover the transcriptome and small RNAome data provide a basis for future study of genetic resistance in *Salix*.

Editor: Matty Knight, George Washington University School of Medicine and Health Sciences, United States of America

Funding: The research was sponsored by grants from the National Natural Science Foundation of China (31400569), the Fundamental Research Funds for the Central Non-profit Research Institution of CAF (RIF2013-11), Beijing Co-building Plan for Scientific Research and Postgraduate Education (2013K0140, 2013K0141, 2014K0151, 2014K0152), and Collaborative Innovation Plan of Jiangsu Higher Education. The funders had no role in study design, data collection and analysis, decision to publish, or preparation of the manuscript.

Competing Interests: The authors have declared that no competing interests exist.

* Email: raoguodong02@163.com

⁹ These authors contributed equally to this work.

Introduction

The increasing concern about climate change and energy security has resulted in the focusing on the economic importance of *Salix* species, given their utility for bioenergy production. However, plants have to tolerate many abiotic and biotic stresses, which are serious threats to agriculture and forestry. These common environmental stresses include salinity, extreme temperatures, drought, chemical toxicity, oxidative stress, pests, and pathogen infection. To exploit their potential for renewable energy, willows need to be kept free of pests and diseases and the yield should be improved without significantly increasing the requirement for fertilizers and water [1].

Previous studies have reveled many *Salix* species have the capacity to tolerate abiotic and biotic stress. The willow species *S. caprea* and *S. cinera* are grown on nutrient- deficient and industrially- contaminated soils; thus, they sever as biological filters for waste water and sludge disposal [2]. Nine different clones of six species of *Salix* (*Salix cordata* Muhlenb. non Michaux, *S. fragilis* L., *S. caprea* L., *S. cinerea* L., *S. burjatica* Nazarov. and *S. viminalis* L.) and one hybrid (*S.* x *calodendron* Wimm.) were exposed to heavy metals in solution culture showed reduced phytotoxicity and increased metal resistance [2]. Two Russian willow species, *Salix viminalis* and *Salix dasyclados* have shown strong genetic control of freezing resistance at during autumn [3]. *Salix caprea*, *Salix repens* and their F1 hybrids had the ability to tolerate different insects and pathogens [4].

Many willow species are characterized by their biological property and ability to adapt to stressful conditions, including high biomass, tolerance to heavy metals, rapid growth, and tolerance of flooding [5]. *Salix matsudana* Koidz. is a deciduous, rapidly growing tree that is native to northeastern China, and is one of the most widely distributed and commonly cultivated willow species in China [6]. The Chinese name for *S. matsudana* is 'Han Liu', where 'Liu' means willow and 'Han' means that the willow is drought tolerant. *Salix matsudana* Koidz. cultivar 'Tortuosa' is widely propagated and planted as an ornamental tree in China because of its attractive twisted stem. This variant is also a water loving plant, which is often cultivated near the river.

The high-throughput next generation sequencing technology such as RNA-seq and Digital Gene Expression (DGE) is an efficient method to discover genes in an unbiased manner [7–10]. Measurement of mRNA and miRNA expression levels, and elucidation of the regulatory relationship between miRNA and their corresponding mRNA targets are crucial for understanding many pathways and biological systems including plant development, and biotic and abiotic stress responses [11].

However, there have been a few studies probing the ability of *S. matsudana* withstand drought stress. In this study, we performed a transcriptome and miRNA sequencing from one-year-old stem shoots of *Salix matsudana* Koidz. and *Salix matsudana* Koidz. cultivar 'Tortuosa' using deep sequencing technology to discover the differentially expressed genes and miRNAs in stem shoots of the two varieties of *Salix*. Our results showed that 196 transcripts, which are mainly related to plant abiotic and biotic stress responses, are commonly up-regulated in *S. matsudana*. These up-regulated genes in *S. matsudana* included a set of biotic and abiotic stress response genes, such as pathogen defense, insect resistance, antibiotics tolerance, antioxidant, and hormone related genes. Functional analysis of miRNA targets also showed that many genes were involved in stress resistance in *S. matsudana*. The transcriptome and small RNAome reported here on the two Chinese willow cultivars significantly enhance our knowledge of stress response genes in *Salix*.

Methods

Plant materials

Shoots from a natural population of *Salix matsudana* and *Salix matsudana* Koidz. cultivar 'Tortuosa' were harvested in spring 2013 at Beijing Botanical Garden of China. To ensure that there absolute minimal differences in the growth conditions, all shoots were sampled within the distribution of 30×30 m^2. Samples were frozen in liquid nitrogen immediately for the extraction of total RNA.

This study was carried out in strict accordance with the recommendations in the Guide for Observational and field studies (http://www.plosone.org/static/publication). All necessary permits were obtained for the described field studies. The sampling of all individuals of *Salix. matsudana* and *Salix matsudana* Koidz. cultivar 'Tortuosa' were approved by Xiping Zheng, director of Beijing Botanical Garden of China.

Libraries construction and deep sequencing

For each group of samples, equal numbers of shoots from five individual plants were pooled for the total RNA extraction with the TRIZOL reagent according to the manufacturer' instructions (Invitriogen). Small RNA and transcriptome sequencing, used the total RNA from the different pooled samples. After eliminating traces of genomic DNA by treatment with DNase I, an automated capillary gel electrophoresis was used to assess RNA quality and quantity using a Bioanalyzer 2100 with RNA 6000 Nano Labchips (Agilent Technologies Ireland, Dublin, Ireland). sRNA and Transcriptome sequencing libraries were prepared using an Illumina Small RNA Sample Prep Kitand an Illumina TruSeq RNA Sample Prep Kit, respectively. After the two libraries had been prepared, raw reads generated by using the Illumina Hiseq 2000 were initially processed to get clean reads. Then, all the clean reads were assembled using a de novo assembly program Trinity [12,13].

Gene validation and quantitative real-time PCR analysis

Ten selected genes specific PCR primers (Table S2) were designed corresponding to the conserved region of willow cDNA sequences from the sequenced database. PCR was performed in a total volume of 50 μl containing 2.5 mmol/L Mg^{2+}, 0.15 mmol/L dNTPs, 0.4 mmol/L of each primer, 15 ng cDNA and 0.5 U PFU DNA polymerase (NEB). The PCR products were purified and ligated into the PMD19-T vector (Takara Bio), and then transformed into *E. coli* Jm109. After sequencing the positive clones with ABI 3730 (Applied Biosystems, USA), the sequences were validated to the corresponding transcriptome seqences.

Quantitative real-time PCR of mature miRNA was performed following the high-stringency protocol in which a poly A polymerase was used to add a poly A tail. The selected genes real-time quantitative PCR were conducted using the SYBR Green Perfect (Takara) and StepOnePlus™ System (Applied Biosystems). All of the PCR products were sequenced and the dissociation curve was analyzed to verify the amplification specificity. The purified PCR products were used to make standard curve for establishing a quantitative correlation between the CT values and the transcript copy numbers [14]. Each qRT-PCR reaction was repeated at least three times, and each standard curve contains at least 5 points. 5.8 s rRNA and *Populus* actin gene were used for reference gene for miRNA and transcripome genes expression validation, respectively.

Small RNA prediction and identification of new mi RNAs in willow

Small interfering RNA (siRNA) has their own structure feature, which is a 20–25 nt long double-stranded RNA, and each strand of which is 2 nt longer than the other at the 3′ end. According to this, reads were aligned with each other in order to find potential siRNA candidate by using the alignment software Bowtie combined with Perl scripts. To identify new miRNA, sRNAs with non-annotation information were subjected to a secondary structure prediction by using the software RNAfold [15]. Further identification of new miRNA was conducted by Mireap (http://sourceforge.net/projects/mireap/).

Result

Transcriptome sequencing by RNA-seq and *de novo* assembly

Two RNA-seq libraries were prepared from the total RNA extracted from *Salix matsudana* ('SM') and *Salix matsudana* Koidz. cultivar 'Tortuosa' ('SMT') and were subjected to pair-end read (PE) sequencing using Illumina platform. We removed reads with adaptors, reads with unknown nucleotides larger than 5%, and low quality reads to obtain clean PE reads in SM and SMT samples. A total of 60,521,772 clean PE reads consisting of 6,506,875,453 nucleotides (nt) with an average GC content of 45.11% were obtained in the SM sample while 58,648,772 clean PE reads consisting of 6,292,916,298 nt with an average GC content of 45.14% were obtained in the SMT sample (Table 1). An overview of the raw reads data is given in Figure S1. All high-quality clean reads were assembled into 106,430 contigs with an average length of 944 bp having a minimum length of 201 bp and a maximum length of 17,226 bp (Figure 1A). The contigs were further joined into 48,817 unigenes with an average length of 773 bp, a N50 length of 1,468 bp and a N90 length of 294 bp (Figure 1B).

Table 1. Statistics of transcriptome output sequencing.

Samples	SM	SMT
Total raw reads	61,958,422	60,131,098
Total clean reads	60,521,772	58,648,772
Total clean nucleotides (nt)	6,506,875,453	6,292,916,298
Q20 percentage	97.55%	98.14%
N percentage	0.06%	0.10%
GC percentage	45.11%	45.14%

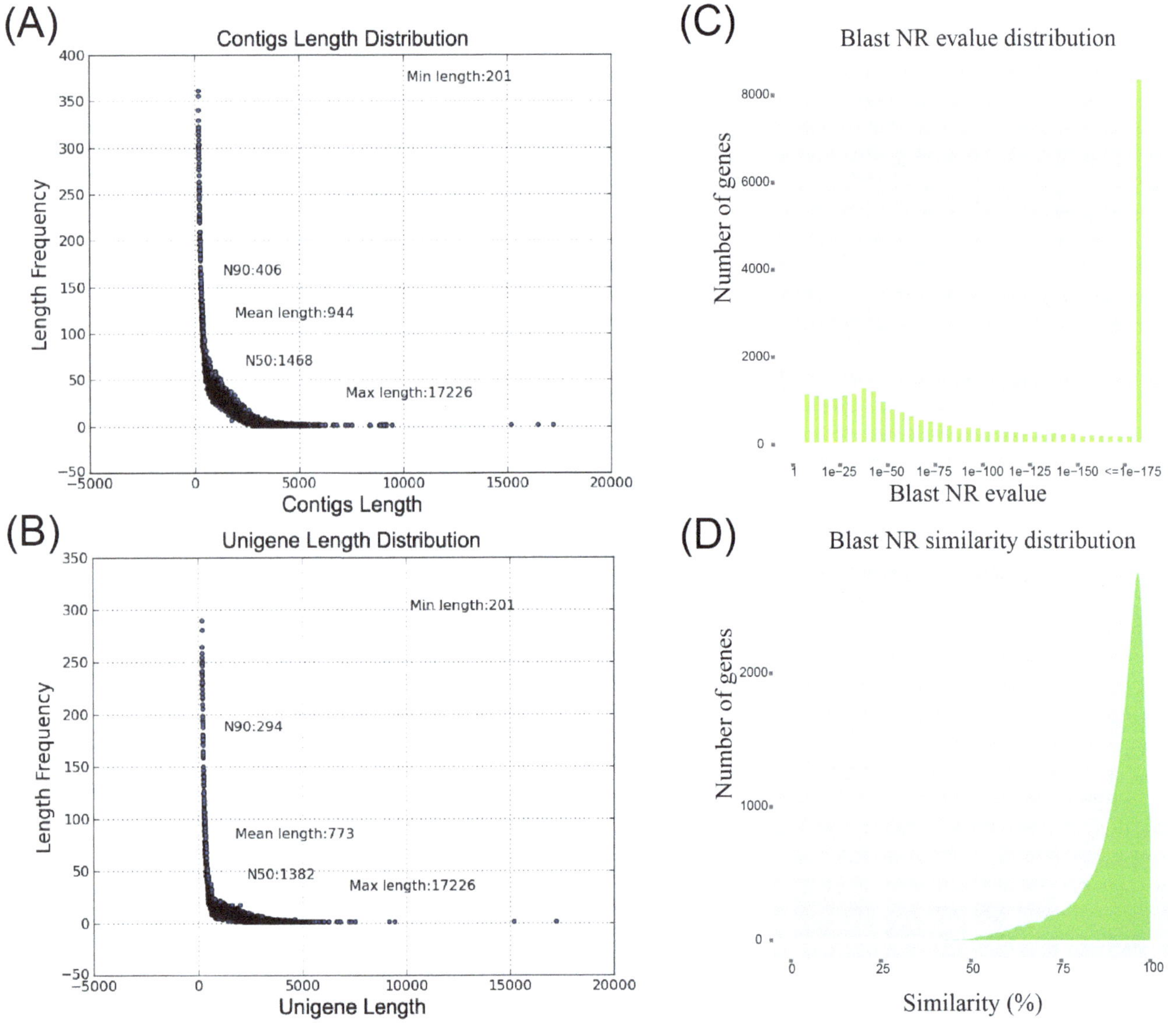

Figure 1.The length distribution of contigs and unigenes, and E-value distribution and similarity distribution of Blast NR hits. Minimum length, N90; mean length, N50; and maximum length of contigs and unigenes are shown in their corresponding positions. (**A**) Contigs length distribution. (**B**) Unigene length distribution. (**C**) E-value distribution of Blast NR hits for matched unigene sequences. (**D**) Similarity distribution of Blast NR hits for each unigene.

Functional annotation and classification of the assembled unigenes

Gene annotation was performed based on sequence homologies to the databases of NCBI non-redundant protein (Nr), NCBI nucleotide sequence (Nt), Kyoto Encyclopedia of Genes and Genomes (KEGG), Swiss-Prot protein, Protein family, Gene Ontology (GO), and Cluster of Orthologous Groups (COGs) using BLAST2GO. In total, 2,592 (5.2% of all unigenes) unigenes annotated in all databases, and 34,049 (69.74% of all unigenes) significantly matched a sequence in at least one of the public databases (Table 2). For the nr annotations, 29,657 of unigenes (60.75%) were found to have matches in the database. Further analysis of the BLAST data showed that 64.11% of the top hits had strong homology with the E-value $<1.0e^{-50}$ while 35.89% of the matched sequences showed moderate homology with the E-value between $1.0e^{-5}$ and $1.0e^{-50}$ (Figure 1C).

The identity distribution pattern showed that 79.44% of the sequences had a similarity higher than 80% (Figure 1D). The majority of the annotated sequences corresponded to the known nucleotide sequences of plant species with 83.14%, 6.46%, 2.87%, 0.63%, and 0.36% matches with *Populus trichocarpa*, *Ricinus communis*, *Vitis vinifera*, *Glycine max*, and *Medicago truncatula* respectively.

The size distribution of the BLAST-aligned coding sequence (CDS) and predicted proteins are shown in Figure S2A, B, respectively. The remaining 41.49% (20,256/48,817) of unigenes that did not match sequences in the databases were analyzed by ESTScan to predict coding regions. An additional 12,740 unigenes (62.89%) also showed orientation in the transcriptome coding sequence (Figure S2C, D). The sequences without a homologous hit may represent novel genes specifically expressed in willow stem shoots, or they could be attributed to other technical or biological biases such as assembly parameters. Furthermore, some cDNA are non-coding, lineage-specific or highly variable and need to be further verified.

GO annotation is an international classification system that can provide standardized vocabulary for assigning functions of the uncharacterized sequences [16]. A total of 23,249 unigenes (47.62% of all the assembled unigenes) were assigned at least one GO term. GO annotation assignments classified 21,577 unique contigs into 22 subcategories of the biological process category, 11 subcategories of the molecular function category, and 14 subcategories of the cellular component category at level 2 (Figure 2). Among biological processes, transcript sequences assigned to biological regulation, cellular process, metabolic process and single-organism were the most abundant. The subcategories with the most highly abundant transcripts in the cellular components category included cell, cell part, macromolecular complex, membrane, and organelle. Within the molecular function category, the majority of the GO terms were predominantly assigned to the binding and catalytic activity.

The database of COGs is a phylogenetic classification of protein encoded in completely sequenced genomes [17]. Overall, 10,073 of 48,817 (20.63%) unigenes were assigned to the COG classification (Table 2). Among the 26 COG categories, the cluster for 'general function prediction only' (1,663, 14.66%) associated with the basic physiological and metabolic functions represented the largest group, followed by post-translational modification, protein turnover, chaperons (1,376, 12.13%), signal transduction (840, 7.40%), 'translation (812, 7.15%), and intracellular trafficking and secretion (577, 5.08%). However, only a few unigenes were assigned to cell motility (5, 0.04%), extracellular structures (37, 0.33%), and nuclear structure (49, 0.43%). One unigene was annotated as 'unnamed protein' (Figure 3A).

KEGG is a database resource that integrates genomic, chemical and systemic functional information that can facilitate to understand the biological functions of genes in terms of networks [18]. The assembled unigenes were annotated with KEGG Orthology (KO) numbers using BLASTx alignments against KEGG with a cut-off E value of 10^{-5}. Overall, 5,444 of 48,817 (11.15%) unigenes were matched in the database and were assigned to 242 KEGG pathways. Among the five categories of pathway hierarchy 1, the 'metabolism' represented the largest group, which were sorted into 11 subcategories including carbohydrate metabolism, energy metabolism, amino acid metabolism, lipid metabolism, nucleotide metabolism, and some others (Figure 3B).

Detection of differentially expressed genes between two transcription libraries

A quality control test for the data assembled from each cDNA library confirmed that they were suitable for statistical analysis for differentially expressed genes identification (Figure S3). A total of 292 differentially expressed genes were detected between SM and SMT groups when $\log_2$ (fold change) >1 and $P<0.005$ were used as cutoff values. Further analysis showed that 196 genes involved in stress resistance, pathogen defense, antibiotics tolerance, insect resistance, antioxidant, hormone related and ribosomal RNAs

Table 2. statistics of functional annotation of unigenes in public databases.

Public database	Number of Unigenes	Percentage (%)
Annotated in NR	29,657	60.75
Annotated in NT	30,551	62.58
Annotated in KEGG	5,444	11.15
Annotated in SwissProt	19,354	39.64
Annotated in PFAM	17,869	36.6
Annotated in GO	23,249	47.62
Annotated in KOG/COG	10,073	20.63
Annotated in all Databases	2,592	5.3
Annotated in at least one Database	34,049	69.74
Total Unigenes	48,817	100

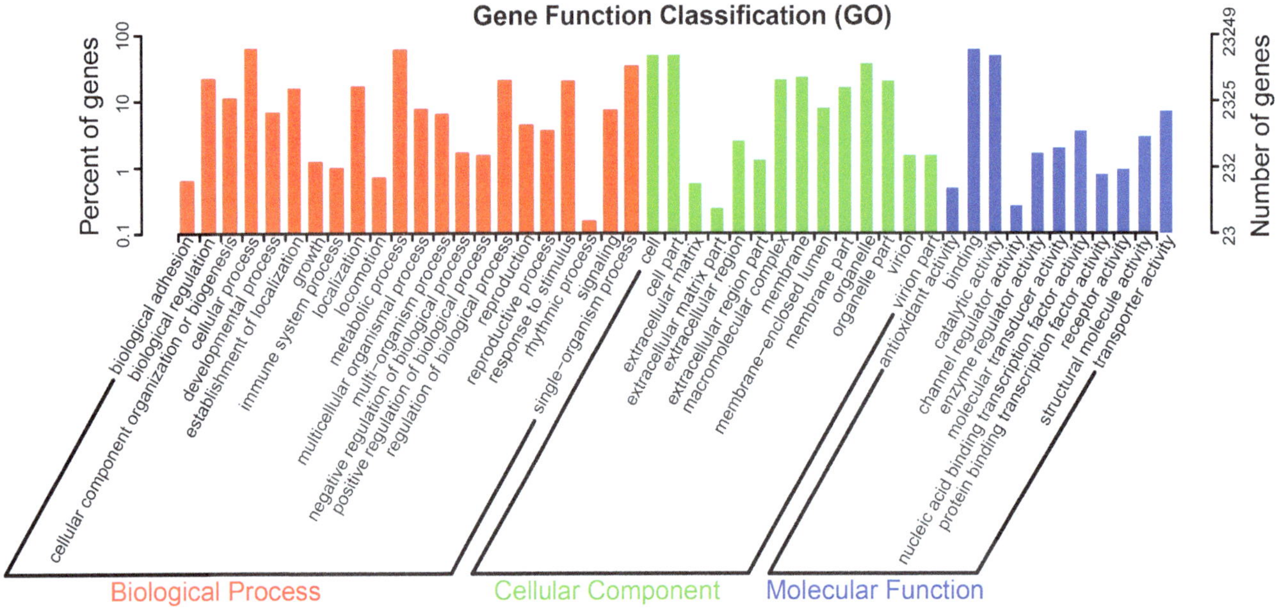

Figure 2. Number of unigenes in each functional category. Unigenes were classified into different functional groups based on a set of plant specific GO terms within biological process, cellular component, and molecular function.

were up regulated (Figure 4A), whereas 96 genes involved in cell wall synthesis, photosynthesis, stress resistance and antibiotics tolerance were down regulated in SM as compared to SMT. Most of the differentially expressed genes are related to plant adaption stress which includes drought, cold, salt, pathogen, insect, and antibiotics. Among the 196 up regulated genes, 133 genes (67.86%) are related to stress adaption indicating that *S. matsudana* has a better ability to tolerant harsh environment.

Deep sequencing of small RNA

Two libraries were constructed for small RNA deep sequencing, and we identified 12,532,326 and 20,113,588 sequence reads respectively from SM and SMT. Among the two data sets, the most abundant classes of small RNA found were 21 and 24 nt, which concurs with most reports of deep sequencing of small RNA in plants (Figure 4B) [19–21]. The sRNA were classified into eight categories (miRNA, tRNA, rRNA, snRNA, snoRNA, ta-siRNA, repeats, and others) by comparing the acquired to miRBase (http://microrna.sanger.ac.uk/sequences), Rfam (http://rfam.sanger.ac.uk/) and Genbank (http://www.ncbi.nih.gov/Genbank/) (Figure S4). Besides the unknown sRNA, miRNA made up 16.25% and 15.94% of the total acquired sequences respectively, which were represented in the main by 21 nt small RNAs. There was little variation in the different sRNA categories between SM and SMT (SD = 2.31%) indicating that sRNA categories are relatively stable in the two willow species.

Known miRNA and comparison of miRNA expression levels between the two libraries

A total of 166 known miRNAs belonging to 35 families were obtained in SM and SMT. These miRNA exhibited variable abundance. Deep sequencing analysis is a powerful strategy that allows the identification and quantification of differences in the expression of sRNA. In the 35 families, miR159 was the most accumulated miRNA with a total of 410,956 reads detected between the two libraries (87,687 miRNAs in SM and 323,269 in SMT) (Table 3). There was considerable expression level diversity between the families probably because the willow one-year-old stem is a highly differentiated organ and the genes are in the process of dynamic expression during development. For instance, miR159, miR166, miR319, and miR396 were sequenced more than one hundred thousand times while miRNAs, miR399, miR408, miR481, miR1446, and miR6459 were detected less than 10 times. Members of the same family were also not expressed equally. The two libraries, SM and SMT, had 36 miRNAs that were differentially expressed, among which 24 were up regulated, and 12 were down regulated in SM compared to SMT.

Identification of new miRNAs

Plant miRNA research has significantly progressed because of deep sequencing technology; however, stricter requirements are needed for identifying a miRNA candidate as a reliable new miRNA. Altogether, we obtained 31 predicted novel miRNAs by using two additional criteria [22]. Among these novel miRNAs, 28 and 30 were expressed in SM and SMT, respectively. The majority of the miRNAs (18 of the 31, 58.06%) have a uridine at 5′ terminal, showing a preference of AGO1 [23]. The novel miRNAs also displayed unequal expression levels as known miRNAs between the two libraries (Table S1). As none of the novel miRNAs had homologs in miRBase, we presumed that they were *S. matsudana*-specific, or were restricted to closely related species.

miRNA Targets predicting and sequences validating

Transcriptome of SM and SMT was used to identify miRNA targets. The 35 known miRNA families had 108 affiliated target genes while the 31 novel miRNAs did not target any genes. Functional annotation of the targets was found to be diverse and included the response to stimulus, signal transduction factors, transcription factors, and genes involved in other biological processes. Among the different expressed targets, 24 miRNAs-target pairs showed reverse expression changing pattern when the results from miRNA profiling and the transcriptome profiling were

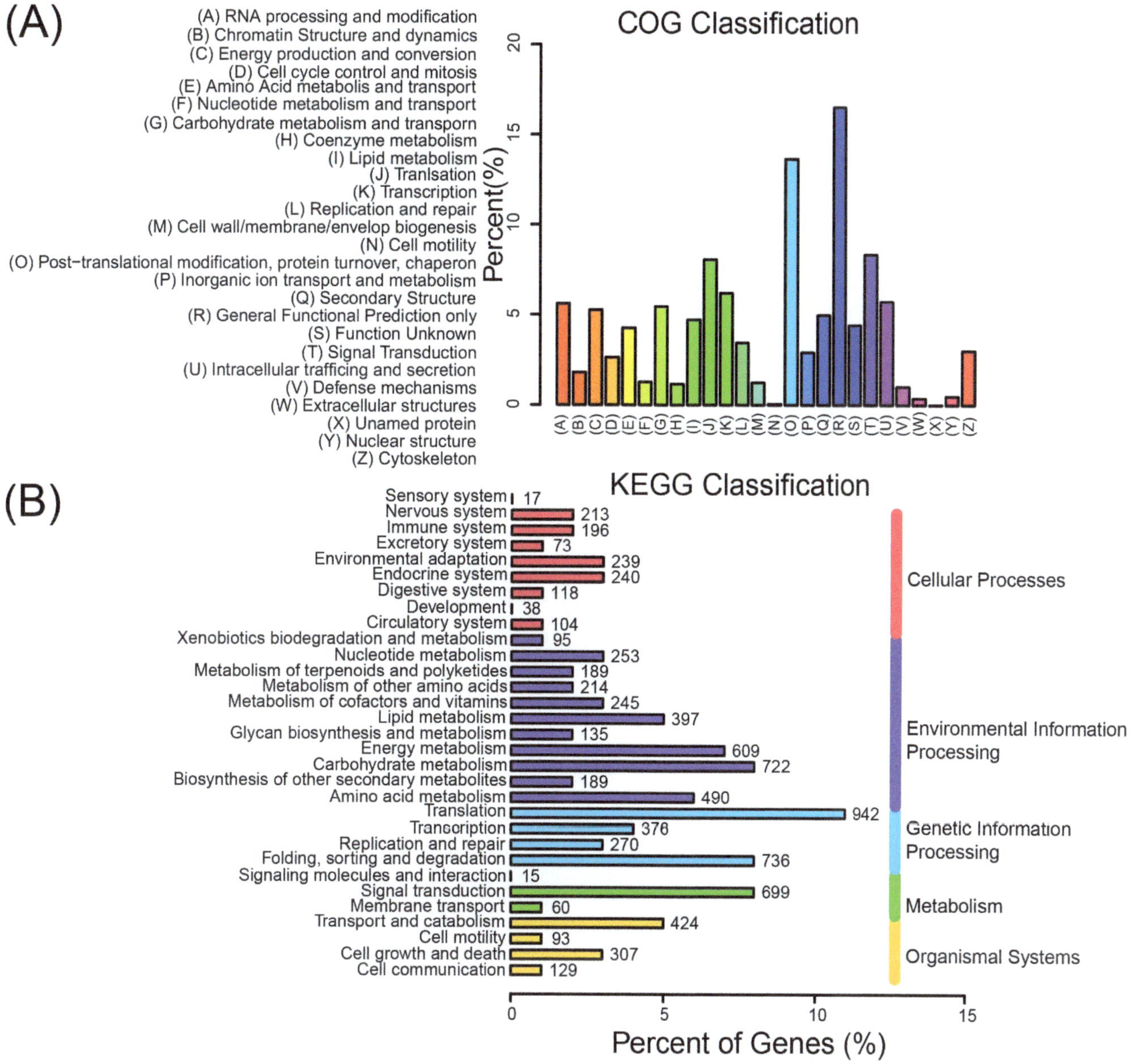

Figure 3. COG and KEGG function classification of unigenes. (**A**) COG function classification of unigenes shows 26 different functional groups. (**B**) Five subcategories (cellular processes, environmental information processes, genetic information processes, metabolism, and organismal systems) were classified by KEGG function classification.

compared (Figure 5). Functional annotations showed that the 24 miRNA-targets pairs were involved in auxin signaling, stress tolerance, signal transduction factors, and other proteins were involved in various biological processes.

To validate the reliability of RNA-seq, full-length cDNA sequences of twenty selected genes were isolated by T-A cloning and compared with the assembled sequences. The length of these genes ranged from 192 to 1,960 bp, and the sequence variation was minimal. All of the genes had the sequence pairwise identity bigger than 98% (Table S2), which validated the reliability of RNA-seq. To partially validate the miRNA and its target genes expression levels, nine miRNA-target pairs were randomly selected for a real-time qPCR experiment. The results showed that the expression of miRNA target pairs changed trends as compared with the sequencing results (Figure 6).

Discussion

Overview of the deep sequencing datasets

Although many studies have reported the transcriptome analysis of willow, here have been no studies on mRNA and miRNA expression profiles in *S. matsudana*. In this study, the RNA-seq technology was used to generate transcriptome and small RNAome data and examined the global gene expression profiling of shoot tips from *S. matsudana*. This work demonstrates that RNA-seq is a useful and effective tool for *de novo* transcriptome and small RNA assembly. Moreover, it facilitates the identification of differentially expressed miRNAs and their target genes between the two willow cultivars, even in the absence of a genomic database. We generated both mRNA and siRNA libraries from the shoot tips of the two willow species to substantially increase the

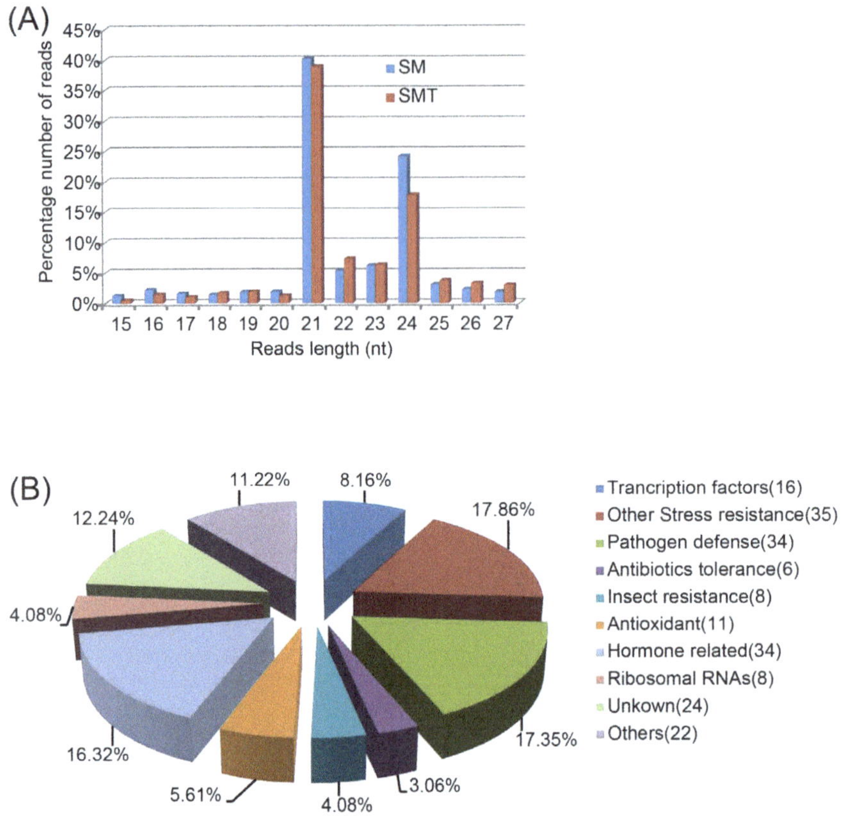

Figure 4. Classification of DEGs and sequence length distribution of small RNA libraries of SM and SMT. (**A**) Up-regulated genes in SM compared to SMT. (**B**) Sequence length distribution shows 21 and 24-nucleotides reads were of significant greater proportion than others.

available data on willow mRNAs and sRNAs to obtain genomic resources for further investigating candidate genes especially the stress resistance genes in various metabolic pathways in willow species. Comparison of transcriptome sequence data of the two libraries revealed 292 differentially expressed genes with over 67% involved in stress response. Differential expression of miRNAs also showed many miRNAs and their target genes were stress related. The real-time PCR analysis of the differentially expressed miRNAs and genes further verified the miRNA and the transcript expression level revealed by small RNA and transcriptome comparison from RNA-seq data.

Transcription factors related to stress tolerance in *S. matsudana*

Transcription factors play a key role in modulating the acclimation response of plants to various internal or external cues. They potentially control downstream gene expression in stress signal transduction pathways through activation and repression of the genes after certain stress treatments. Plant genomes contain a large number of transcription factors (TFs). It has been estimated that the Arabidopsis genome codes for at least 1,533 transcription factors and over 5.9% of its total estimated genes. Among these, about 800 genes encode AP2/EREBP, bZIP, zinc finger proteins, Myb proteins, and AUX/IAA types of transcription factors. Sixteen transcription factors have higher transcription levels in SM compared to SMT, which include C-repeat binding factors (CBF), Myb proteins, TINY proteins, AP2/EREBP, WRKY type, and other types of zinc finger proteins.

CBF proteins bind to the cold-responsive *cis*-acting elements that contain the same motif (CCGAC), are encoded by AP2/EREBP multigene families, and mediate the transcription of genes in response to stress. Overexpression of two Arabidopsis AP2/EREBP genes, CBF1 and DREB1A, resulted in stronger tolerance to drought, salt, and freezing [24–26]. Transgenic Arabidopsis plants overexpressing CBF3 under the control of cauliflower mosaic virus (CaMV) 35S promoter also showed a high tolerance to drought, high-salinity, and freezing stress [25,27,28]. Five CBF and AP2/EREBP genes were highly expressed in SM as compared to SMT. The sequenced CBF and AP2/EREBP gene reads in SM are 2 to 30 times over that in SMT indicating that the SM may have a high tolerance to abiotic stress.

WRKY family is defined by a DNA binding domain that contains the strictly conserved amino acid sequence WRKY. Members of WRKY family transcription factors have been implicated in the response to biotic and abiotic stresses. Overexpression of the GhWRKY39-1 gene enhances the resistance to pathogen infection and tolerance to high salt and oxidative stress in transgenic Nicotiana benthamiana [29]. The same results were obtained from the research of *Tamarix hispida*, overexpression of ThWRKY4 conferred tolerance to salt, oxidative and ABA treatment in transgenic plants [30]. The overexpression of a subgroup IIe WRKY transcription factor CaWRKY27 of *Capsicum annuum* also positively regulates the resistance of tobacco to the pathogen *Ralstonia solanacearum* infection [31]. In the present study of the two transcriptome libraries, three WRKY showed higher expression levels in SM as compared to SMT that may enhance the tolerance of SM to biotic and abiotic stress.

Table 3. Conserved miRNA families expression in SM and SMT.

Family	Mature sequence	Total	SM		SMT	
			Raw	TPM	Raw	TPM
miR156	UUGACAGAAGAUAGAGAGCAC	23	5	18	18	42
miR159	UUUGGAUUGAAGGGAGCUCUA	410,956	87,687	323,269	175,659	406,596
miR160	UGCCUGGCUCCCUGUAUGCCA	905	193	712	181	419
miR162	UCGAUAAACCUCUGCAUCCAG	7,920	1,690	6,230	909	2,104
miR164	UGGAGAAGCAGGGCACGUGCA	1,237	264	973	112	259
miR166	UCGGACCAGGCUUCAUUCCCC	140,711	30,024	110,687	28,821	66,712
miR167	UGAAGCUGCCAGCAUGAUCU	1,219	260	959	92	213
miR168	UCGCUUGGUGCAGGUCGGGAA	4,312	920	3,392	374	866
miR171	UGAUUGAGCCGUGCCAAUAUC	366	78	288	37	86
miR172	AGAAUCCUGAUGAUGCUGCAG	984	210	774	127	294
miR319	UUGGACUGAAGGGAGCUCCC	335,098	71,501	263,597	154,043	356,562
miR390	AAGCUCAGGAGGGAUAGCGCC	1,875	400	1,475	201	465
miR393	UCCAAAGGGAUCGCAUUGAUC	347	74	273	15	35
miR394	UUGGCAUUCUGUCCACCUCC	7,217	1,540	5,677	144	333
miR396	UUCCACAGCUUUCUUGAACUU	153,993	32,858	121,135	14,009	32,427
miR398	UUCUCAGGUCACCCCUUUGGG	951	203	748	983	2,275
miR399	UGCCAAAGGAGAUUUGCCCGG	5	1	4	1	2
miR403	UUAGAUUCACGCACAAACUCG	17,261	3,683	13,578	233	539
miR408	AUGCACUGCCUCUUCCCUGGC	5	1	4	0	0
miR472	UUUUCCCUACUCCACCCAUCCC	22,304	4,759	17,545	2,568	5,944
miR475	UUACAAUUCCAUUGAUUAAACCGU	19	4	15	11	25
miR481	AGGACCUCACCUAACAGCUUAAGC	9	2	7	0	0
miR482	UUGCCUACUCCACCCAUGCCAC	731	156	575	102	236
miR530	UGCAUUUGCACCUGCACCUUC	506	108	398	53	123
miR828	UCUUGCUCAAAUGAGUAUUCCA	37	8	29	5	12
miR1444	UCCACAUUCGGUCAAUGUUC	42	9	33	3	7
miR1446	UUCUGAACUCUCUCCCUCAA	5	1	4	0	0
miR2111	UAAUCUGCAUCCUGAGGUUUG	23	5	18	4	9
miR6421	UAGAGCAGAUUGUAAGGGAAG	28	6	22	24	56
miR6427	UCGUAAUGCUUCAUUCUCACAA	173	37	136	40	93
miR6445	UUCAUUCCUCUUCCUAAAAUGG	8,586	1,832	6,754	1,215	2,812
miR6457	UAAUCUCUCUGCAGAAUGCUG	0	0	0	28	65
miR6459	UCGAAUUUGGGCUUGAGAUUG	9	2	7	2	5
miR6474	UGUUCAGAUCAGUAGAUAGCA	80	17	63	21	49
miR6476	UCAGUGGAGAUGAAACAUGA	0	0	0	1	2

Plant zinc finger transcription factors play an important role in tolerance to various environmental stresses such as drought, cold, osmotic stress, wounding and mechanical loading. The zinc finger proteins chraracterized by zinc finger domains enable protein interaction with DNA. A large number of studies have shown that zinc finger proteins may not be specific to one particular stress, but may regulate responses to several stresses. In *A. thaliana*, the zinc finger proteins have been shown to be involved in cold, dehydration, reactive oxygen, and salt stress [32,33]. In *Populus euphratica*, the overexpression of a zinc finger protein gene *PSTZ* showed enhanced salt tolerance [34]. In rice, seven zinc finger transcription factors were involved in the response to different abiotic stresses such as cold, drought, and mechanical stress [35]. In *Capsicum annuum*, a zinc finger transcription factor gene functions as a pathogen induced early-defense gene [36]. Six zinc finger transcription factors have a higher transcription level in SM than SMT indicating that the SM has a stronger ability to tolerate abiotic stresses.

Hormones related genes involved in stress tolerance in *S. matsudana*

Plants hormones such as auxins, abscisic acid (ABA), gibberellins (GA), salicylic acid (SA), ethylene (ET), cytokinins (CK), jasmonates (JA), and brassinosteroids (BR) play crucial roles in various biotic and abiotic responses. A vast array of pathogenic microorganisms which deliver virulence factors into plant cell to promote virulence and cause disease and plant can change the

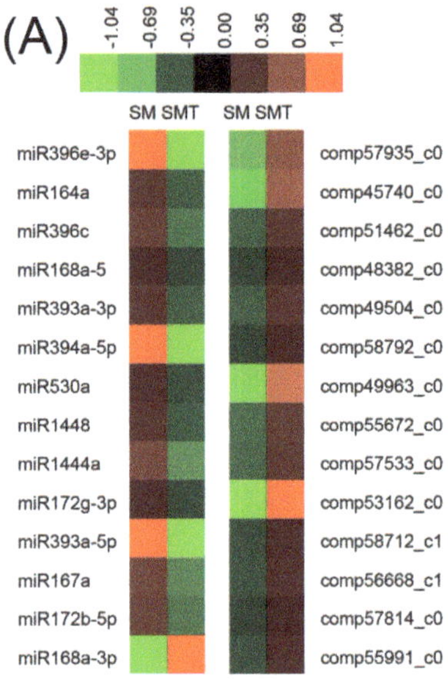

Comparing to *Populus trichocarpa*		Comparing to Arabidopsis	
Homology	Anotation	Homology	Anotation
POPTR_0018s06210	High chlorophyll fluorescence 101	AT3G24430.1	High chlorophyll fluorescence 101
POPTR_0007s03070	UDP-glucosyl transferase	AT2G43820.1	UDP-glucosyltransferase 74F2
POPTR_0007s14930	PPR repeat	AT1G30610.2	Pentatricopeptide (PPR) repeat protein
POPTR_0014s13490	Glutathione peroxidase	AT2G48150.1	Glutathione peroxidase 4
POPTR_0009s09340	Erv1/Alr family protein	AT1G49880.1	Erv1/Alr family protein
POPTR_0002s17100	DEAD box RNA helicase	AT5G08620.1	DEA(D/H)-box RNA helicase
POPTR_007G065400	Transcription activator NAC1	AT1G56010.2	NAC domain containing protein 1
POPTR_007G046200	AP2-like factor	AT4G36920.2	AP2 factor
POPTR_006G062200	Glycosyl hydrolase family 79	AT5G34940.2	Glucuronidase 3
POPTR_T101100	Glycosyl hydrolase	AT4G15210.1	Beta-amylase 5
POPTR_004G078200	Zinc finger homeobox family protein	AT1G75240.1	Homeobox protein 33
POPTR_005G227900	Auxin response factor	AT5G37020.1	Auxin response factor 8
POPTR_T033800	PCI domain	AT5G14250.1	Proteasome component domain protein
POPTR_017G095300	PCI domain	AT5G15610.1	Proteasome component domain protein

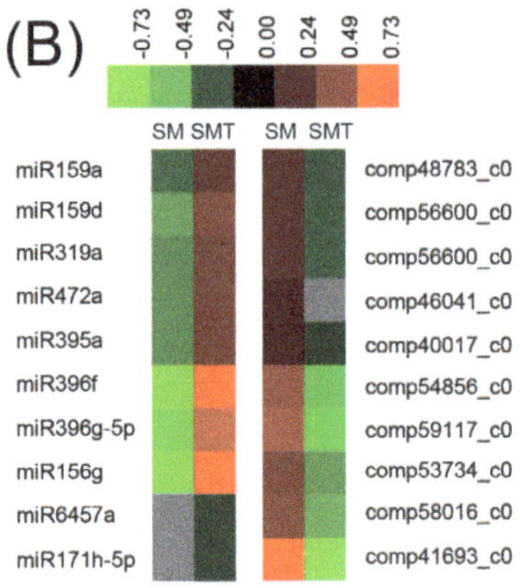

Comparing to *Populus trichocarpa*		Comparing to Arabidopsis	
Homology	Anotation	Homology	Anotation
POPTR_0005s26460	Subtilase family	AT5G59810.1	Subtilase family protein
POPTR_0018s06600	Unkonwn	AT3G25910.1	Zinc finger
POPTR_0018s06600	Unkonwn	AT3G25910.1	Zinc finger
Unkonwn	Unkonwn	Unkonwn	Unkonwn
POPTR_0009s15270	NPH3 family	AT3G44820.1	Phototropic-responsive NPH3
POPTR_0001s10070	Calcium-dependent protein kinase 1	AT4G23650.1	Calcium-dependent protein kinase 6
POPTR_0009s15270	NPH3 family	AT3G44820.1	Phototropic-responsive NPH3
POPTR_0006s12660	Unkonwn	AT2G36950.1	Heavy metal transport/detoxification superfamily prot
POPTR_0011s07050	Unkonwn	AT4G18760.1	Receptor like protein 51
POPTR_0005s06040	Unkonwn	AT5G09330.3	NAC domain containing protein 82

Figure 5. A combined view of reverse expressions between miRNA and its target. The left side of hotspot figure show miRNA expression level, and the right side show corresponding target expression levels of both SM and SMT. Up and down regulation in the expression were based on normalize data (color bar at the top) generated by Cluster 3.0 software. The tables show homologous genes of miRNA target genes by comparing to *Populus trichocarpa* and Arabidopsis. (**A**) Reverse expression of Up-regulated miRNA in SM and its targets. (**B**) Reverse expression of down-regulated miRNA in SM and its targets.

level of various phytohormones to protect the cell from infection when plants were attacked by these diverse pathogens [37,38]. Plants hormones also have the ability to adapt changing environments by mediating a wide range of adaptive responses [39,40]. It is notable that xx hormone-related transcripts have significantly higher expression levels in SM than in SMT, which include eighteen ABA related genes, ten auxin response genes, four JA, and two ET related genes.

Under abiotic stresses such as drought, cold, and salinity, ABA levels increased in vegetative tissues, which are adaptive responses essential for the plant survival and productivity [41]. In Arabidopsis, increasing the transcription of a key ABA biosynthesis gene, *NCED3*, resulted in the increased production of ABA and enhanced tolerance to drought-shock and osmotic stress [42]. Overexpression of *AtTRE1* gene that encodes the Arabidopsis trehalase conferred the ability to tolerate drought stress through increased sensitivity toward ABA-dependent stomatal closure [43]. A maize calcium-dependent protein kinase gene, *ZmCPK4*, positively regulates ABA signaling and enhanced drought stress tolerance in transgenic Arabidopsis [44]. Apart from the vital function for abiotic stress tolerance, ABA plays a modulating role in plant pathogen infections [45–47]. Many key components of ABA signal transduction have been identified including protein kinases, RNA processing factors, phosphatases, proteasome components, and chromatin remodeling proteins [48,49]. In *S. matsudana*, thirteen genes involved in the response to ABA transduction had a higher transcription level than SMT. Moreover, five miRNAs, which target five ABA transduction genes were down regulated as compared to SMT. These genes could increase the ABA levels and enhance the ability of *S. matsudana* to tolerate abiotic and biotic stresses.

Auxin responsive genes also regulate plant stress response. Overexpression of an auxin responsive gene *GH3-8* resulted in stronger pathogen resistance in rice and this resistance was shown to be independent of SA and JA signaling [50]. A rice auxin responsive gene *OsGH3.13*, encoding IAA-amido synthetase was shown to enhance the expression of genes that confer drought tolerance [51]. Ten genes related to auxin response had higher transcription levels in SM than SMT, and one miRNA had a lower expression level in SM than SMT. These genes could change the auxin levels and confer increased stress tolerance ability upon *S. matsudana*. Auxin plays a major role of in

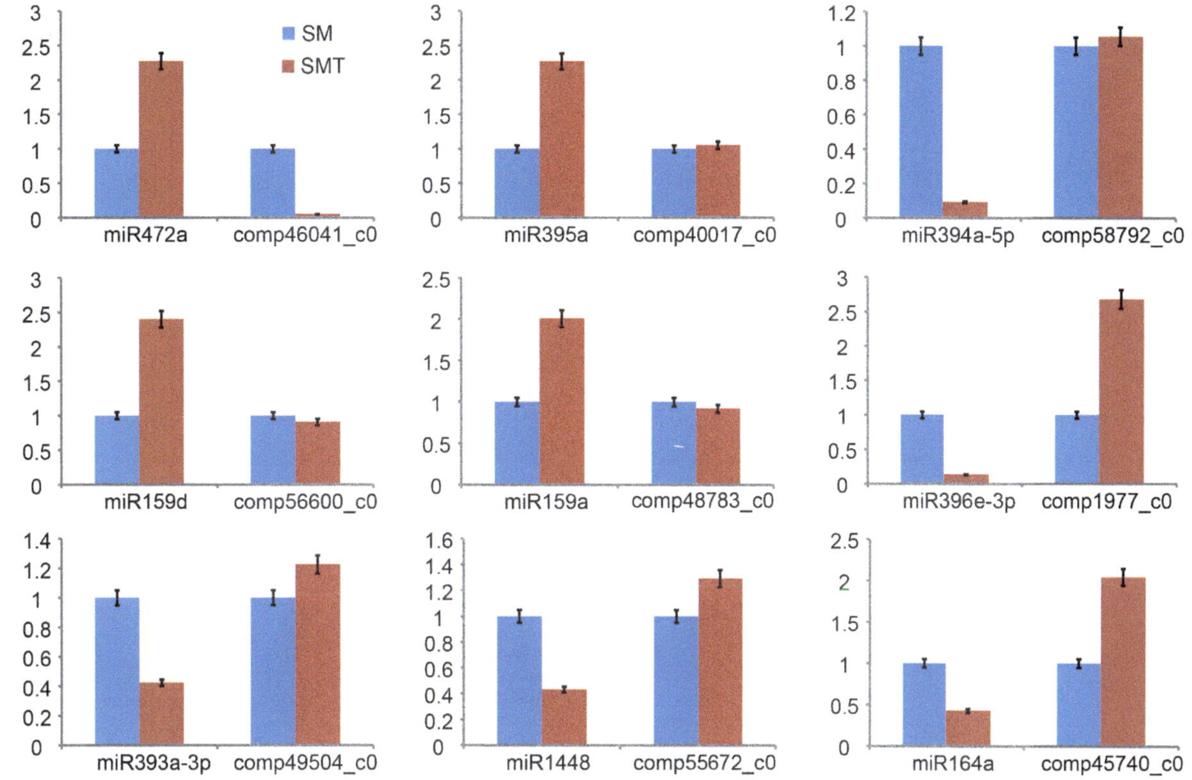

Figure 6. Real-time quantitative PCR validate expressions of 9 selected miRNA and its targets. The amount of miRNA expression was normalized to the level of 5.8rRNA, and the mRNA expression was normalized to the level of *Populus* actin gene.The normalized miRNA and its target gene levels of SM were arbitrarily set to1.

integrating the activities of multiple phytohormones to control plant growth in response to environmental signals [52]. Except ABA and auxin, JA and ET related genes were also detected that had different transcription in SM and SMT, which in turn could result in variable stress response activity. Four JA related genes and two ET related genes had higher transcript levels in SM than SMT, which could enhance the ability of *S. matsudana* to tolerance stress.

Antioxidants and detoxification genes related to stress tolerance in *S. matsudana*

Biotic and abiotic stresses such as pathogen infection, salt, drought, oxidation and heat stress are accompanied by the formation of reactive oxygen species (ROS). The ROS such as H_2O_2, O_2, and OH^- are the main damages of membranes and macromolecules [53]. Plants have developed many antioxidation strategies to scavenge ROS. Enzymes such as thioredoxin, ascorbate peroxidase (APX), superoxide dismutase (SOD), glutathione peroxidase, glutathione transferases, phospholipid-hydroperoxide glutathione peroxidase, galactose oxidase, and glutathione reductase form a part of the main ROS scavenging strategy. The increased transcript levels of these enzyme genes can result in enhanced tolerance to oxidative stress. In potato, overexpression an APX gene resulted in enhanced tolerance to methyl viologen-induced oxidative stress and high temperature in transgenic potato plants [54]. Overexpression of a tobacco glutathione S-transferase with glutathione peroxidase activity in transgenic tobacco enhanced seedling growth under a variety of stress conditions [55]. In *Solanum tuberosum*, substantial accumulation of a plant thioredoxin named CDSP 32 mRNA and protein were revealed upon oxidative treatments [56]. In *S. matsudana*, eleven genes involved in response to oxidative stress had higher transcription levels than SMT. These genes encode the abovementioned enzymes including one APX, three glutathione transferases, four thioredoxin, one galactose oxidase, and two NADP reductase. The higher transcription levels of these oxidative stress response genes in SM compared to SMT suggestion that SM has stronger stress tolerance.

Supporting Information

Figure S1 Classification of raw reads of SM and SMT trianscriptiom sequences.

Figure S2 The length distribution of the coding sequence (CDS) and predicted proteins. (A) Aligned CDS by BLASTX. (B) predicted proteins by BLASTX. (C) Aligned CDS by ESTScan. (D) predicted proteins by ESTScan.

Figure S3 Saturation test and RPKM distribution of transcriptome sequences.

Figure S4 Summary of sequence classifications of sequenced sRNA.

Table S1 Mature sequences and expression of novel miRNA.

Table S2 Primers, sequence length and sequence identity of unigenes validated by real-time PCR.

Author Contributions

Conceived and designed the experiments: GR JZ. Performed the experiments: GR JS YZ. Analyzed the data: GR CH AD. Contributed reagents/materials/analysis tools: GR CH. Wrote the paper: GR JZ.

References

1. Karp A, Hanley SJ, Trybush SO, Macalpine W, Pei M, et al. (2011) Genetic improvement of willow for bioenergy and biofuels. Journal of integrative plant biology 53: 151–165.
2. Punshon T, Dickinson NM (1997) Acclimation of *Salix* to metal stress. New Phytologist 137: 303–314.
3. Tsarouhas V, Gullberg U, Lagercrantz U (2004) Mapping of quantitative trait loci (QTLs) affecting autumn freezing resistance and phenology in *Salix*. TAG Theoretical and applied genetics Theoretische und angewandte Genetik 108: 1335–1342.
4. Hjältén J (1998) An experimental test of hybrid resistance to insects and pathogens using *Salix caprea*, *S. repens* and their F1 hybrids. Oecologia 177: 127–132.
5. Greger M, Landberg T (1999) Use of Willow in Phytoextraction. International journal of phytoremediation 1: 115–123.
6. Yang J, Yi J, Yang C, Li C (2013) Agrobacterium tumefaciens-mediated genetic transformation of *Salix matsudana* Koidz. using mature seeds. Tree physiology 33: 628–639.
7. Wong MM, Cannon CH, Wickneswari R (2011) Identification of lignin genes and regulatory sequences involved in secondary cell wall formation in *Acacia auriculiformis* and *Acacia mangium* via *de novo* transcriptome sequencing. BMC genomics 12: 342.
8. Grabherr MG, Haas BJ, Yassour M, Levin JZ, Thompson DA, et al. (2011) Full-length transcriptome assembly from RNA-Seq data without a reference genome. Nat Biotechnol 29: 644–652.
9. Li Q, Jin X, Zhu YX (2012) Identification and analyses of miRNA genes in allotetraploid Gossypium hirsutum fiber cells based on the sequenced diploid G. raimondii genome. Journal of genetics and genomics 39: 351–360.
10. Brunskill EW, Potter SS (2012) RNA-Seq defines novel genes, RNA processing patterns and enhancer maps for the early stages of nephrogenesis: Hox supergenes. Developmental biology 368: 4–17.
11. He CY, Cui K, Zhang JG, Duan AG, Zeng YF (2013) Next-generation sequencing-based mRNA and microRNA expression profiling analysis revealed pathways involved in the rapid growth of developing culms in *Moso bamboo*. BMC plant biology 13: 119.
12. Almeida R, Gonçalves S, Romano A (2005) In vitro micropropagation of endangered *Rhododendron ponticum* L. subsp. baeticum (Boissier & Reuter) Handel-Mazzetti. Biodiversity & Conservation 14: 1059–1069.
13. Rao G, Wang Y, Zhang D, Liu D, Li F, et al. (2012) Isolation and characterisation of an HpSHP gene from *Hosta plantaginea*. Molecular biology reports 39: 6887–6894.
14. Suzuki S, Li L, Sun YH, Chiang VL (2006) The cellulose synthase gene superfamily and biochemical functions of xylem-specific cellulose synthase-like genes in *Populus trichocarpa*. Plant physiology 142: 1233–1245.
15. Lv X, Song X, Rao G, Pan X, Guan L, et al. (2009) Construction vascular-specific expression bi-directional promoters in plants. Journal of biotechnology 141: 104–108.
16. Chang CH, Wang HI, Lu HC, Chen CE, Chen HH, et al. (2012) An efficient RNA interference screening strategy for gene functional analysis. BMC genomics 13: 491.
17. Muoki RC, Paul A, Kumari A, Singh K, Kumar S (2012) An improved protocol for the isolation of RNA from roots of tea (*Camellia sinensis* (L.) O. Kuntze). Molecular biotechnology 52: 82–88.
18. Zhang C, Galbraith DW (2012) RNA interference-mediated gene knockdown within specific cell types. Plant molecular biology 80: 169–176.
19. Ivanova D, Milev I, Vachev T, Baev V, Yahubyan G, et al. (2014) Small RNA analysis of Potato spindle tuber viroid infected *Phelipanche ramosa*. Plant physiology and biochemistry: PPB/Societe francaise de physiologie vegetale 74: 276–282.
20. Markakis MN, Boron AK, Van Loock B, Saini K, Cirera S, et al. (2013) Characterization of a small auxin-up RNA (SAUR)-like gene involved in *Arabidopsis thaliana* development. PloS one 8: e82596.
21. Carnavale Bottino M, Rosario S, Grativol C, Thiebaut F, Rojas CA, et al. (2013) High-throughput sequencing of small RNA transcriptome reveals salt stress regulated microRNAs in sugarcane. PloS one 8: e59423.
22. Meyers BC, Axtell MJ, Bartel B, Bartel DP, Baulcombe D, et al. (2008) Criteria for annotation of plant MicroRNAs. The Plant cell 20: 3186–3190.
23. Morgan T, Craven C, Ward K (1998) Human spiral artery renin-angiotensin system. Hypertension 32: 683–687.
24. Zhang DX, Lu HL, Liao X, St Leger RJ, Nuss DL (2013) Simple and efficient recycling of fungal selectable marker genes with the Cre-loxP recombination system via anastomosis. Fungal genetics and biology: FG & B 61: 1–8.
25. Ye ZY, Lu HL, Geng Y, Gu YZ, Xie ZY, et al. (2013) Structural, electrical, and optical properties of Ti-doped ZnO films fabricated by atomic layer deposition. Nanoscale research letters 8: 108.
26. Zheng S, Sun QQ, Yang W, Zhou P, Lu HL, et al. (2013) Modulation in current density of metal/n-SiC contact by inserting Al2O3 interfacial layer. Nanoscale research letters 8: 116.
27. Ding SJ, Chen HB, Cui XM, Chen S, Sun QQ, et al. (2013) Atomic layer deposition of high-density Pt nanodots on Al2O3 film using (MeCp)Pt(Me)3 and O2 precursors for nonvolatile memory applications. Nanoscale research letters 8: 80.
28. Liu Z, Zhang D, Liu D, Li F, Lu H (2013) Exon skipping of AGAMOUS homolog PrseAG in developing double flowers of *Prunus lannesiana* (Rosaceae). Plant cell reports 32: 227–237.
29. Shi W, Hao L, Li J, Liu D, Guo X, et al. (2014) The Gossypium hirsutum WRKY gene GhWRKY39-1 promotes pathogen infection defense responses and mediates salt stress tolerance in transgenic *Nicotiana benthamiana*. Plant cell reports 33: 483–498.
30. Zheng L, Liu G, Meng X, Liu Y, Ji X, et al. (2013) A WRKY gene from Tamarix hispida, ThWRKY4, mediates abiotic stress responses by modulating reactive oxygen species and expression of stress-responsive genes. Plant molecular biology 82: 303–320.
31. Dang F, Wang Y, She J, Lei Y, Liu Z, et al. (2014) Overexpression of CaWRKY27, a subgroup IIe WRKY transcription factor of Capsicum annuum, positively regulates tobacco resistance to *Ralstonia solanacearum* infection. Physiologia plantarum 150: 397–411.
32. Sakamoto H, Maruyama K, Sakuma Y, Meshi T, Iwabuchi M, et al. (2004) Arabidopsis Cys2/His2-type zinc-finger proteins function as transcription repressors under drought, cold, and high-salinity stress conditions. Plant physiology 136: 2734–2746.
33. Davletova S, Schlauch K, Coutu J, Mittler R (2005) The zinc-finger protein Zat12 plays a central role in reactive oxygen and abiotic stress signaling in Arabidopsis. Plant physiology 139: 847–856.
34. Wang JY, Xia XL, Wang JP, Yin WL (2008) Stress responsive zinc-finger protein gene of Populus euphratica in tobacco enhances salt tolerance. Journal of integrative plant biology 50: 56–61.
35. Figueiredo DD, Barros PM, Cordeiro AM, Serra TS, Lourenco T, et al. (2012) Seven zinc-finger transcription factors are novel regulators of the stress responsive gene *OsDREB1B*. Journal of experimental botany 63: 3643–3656.
36. Kim SH, Hong JK, Lee SC, Sohn KH, Jung HW, et al. (2004) CAZFP1, Cys2/His2-type zinc-finger transcription factor gene functions as a pathogen-induced early-defense gene in *Capsicum annuum*. Plant molecular biology 55: 883–904.
37. Adie BA, Perez-Perez J, Perez-Perez MM, Godoy M, Sanchez-Serrano JJ, et al. (2007) ABA is an essential signal for plant resistance to pathogens affecting JA biosynthesis and the activation of defenses in Arabidopsis. The Plant cell 19: 1665–1681.
38. Robert-Seilaniantz A, Navarro L, Bari R, Jones JD (2007) Pathological hormone imbalances. Current opinion in plant biology 10: 372–379.
39. Santner A, Estelle M (2009) Recent advances and emerging trends in plant hormone signalling. Nature 459: 1071–1078.
40. Argueso CT, Ferreira FJ, Kieber JJ (2009) Environmental perception avenues: the interaction of cytokinin and environmental response pathways. Plant, cell & environment 32: 1147–1160.
41. Leung J, Giraudat J (1998) Abscisic Acid Signal Transduction. Annu Rev Plant Physiol Plant Mol Biol 49: 199–222.
42. Watanabe S, Matsumoto M, Hakomori Y, Takagi H, Shimada H, et al. (2014) The purine metabolite allantoin enhances abiotic stress tolerance through synergistic activation of abscisic acid metabolism. Plant, cell & environment 37: 1022–1036.
43. Van Houtte H, Vandesteene L, Lopez-Galvis L, Lemmens L, Kissel E, et al. (2013) Overexpression of the trehalase gene AtTRE1 leads to increased drought stress tolerance in Arabidopsis and is involved in abscisic acid-induced stomatal closure. Plant physiology 161: 1158–1171.
44. Jiang S, Zhang D, Wang L, Pan J, Liu Y, et al. (2013) A maize calcium-dependent protein kinase gene, ZmCPK4, positively regulated abscisic acid signaling and enhanced drought stress tolerance in transgenic Arabidopsis. Plant physiology and biochemistry: PPB/Societe francaise de physiologie vegetale 71: 112–120.
45. Fan J, Hill L, Crooks C, Doerner P, Lamb C (2009) Abscisic acid has a key role in modulating diverse plant-pathogen interactions. Plant physiology 150: 1750–1761.
46. Oide S, Bejai S, Staal J, Guan N, Kaliff M, et al. (2013) A novel role of PR2 in abscisic acid (ABA) mediated, pathogen-induced callose deposition in *Arabidopsis thaliana*. The New phytologist 200: 1187–1199.
47. Mauch-Mani B, Mauch F (2005) The role of abscisic acid in plant-pathogen interactions. Current opinion in plant biology 8: 409–414.
48. Chinnusamy V, Zhu JK (2009) Epigenetic regulation of stress responses in plants. Current opinion in plant biology 12: 133–139.

49. Cutler SR, Rodriguez PL, Finkelstein RR, Abrams SR (2010) Abscisic acid: emergence of a core signaling network. Annual review of plant biology 61: 651–679.
50. Ding X, Cao Y, Huang L, Zhao J, Xu C, et al. (2008) Activation of the indole-3-acetic acid-amido synthetase GH3-8 suppresses expansin expression and promotes salicylate- and jasmonate-independent basal immunity in rice. The Plant cell 20: 228–240.
51. Zhang SW, Li CH, Cao J, Zhang YC, Zhang SQ, et al. (2009) Altered architecture and enhanced drought tolerance in rice via the down-regulation of indole-3-acetic acid by TLD1/OsGH3.13 activation. Plant physiology 151: 1889–1901.
52. Jaillais Y, Chory J (2010) Unraveling the paradoxes of plant hormone signaling integration. Nature structural & molecular biology 17: 642–645.
53. Mittler R (2002) Oxidative stress, antioxidants and stress tolerance. Trends in plant science 7: 405–410.
54. Kim MD, Kim YH, Kwon SY, Yun DJ, Kwak SS, et al. (2010) Enhanced tolerance to methyl viologen-induced oxidative stress and high temperature in transgenic potato plants overexpressing the *CuZnSOD*, *APX* and *NDPK2* genes. Physiologia plantarum 140: 153–162.
55. Roxas VP, Lodhi SA, Garrett DK, Mahan JR, Allen RD (2000) Stress tolerance in transgenic tobacco seedlings that overexpress glutathione S-transferase/glutathione peroxidase. Plant & cell physiology 41: 1229–1234.
56. Broin M, Cuine S, Peltier G, Rey P (2000) Involvement of CDSP 32, a drought-induced thioredoxin, in the response to oxidative stress in potato plants. FEBS letters 467: 245–248.

17

Morphological and Physiological Responses of Cotton (*Gossypium hirsutum* L.) Plants to Salinity

Lei Zhang, Huijuan Ma, Tingting Chen, Jun Pen, Shuxun Yu*, Xinhua Zhao*

State Key Laboratory of Cotton Biology, Institute of Cotton Research of CAAS, Anyang, P. R. China

Abstract

Salinization usually plays a primary role in soil degradation, which consequently reduces agricultural productivity. In this study, the effects of salinity on growth parameters, ion, chlorophyll, and proline content, photosynthesis, antioxidant enzyme activities, and lipid peroxidation of two cotton cultivars, [CCRI-79 (salt tolerant) and Simian 3 (salt sensitive)], were evaluated. Salinity was investigated at 0 mM, 80 mM, 160 mM, and 240 mM NaCl for 7 days. Salinity induced morphological and physiological changes, including a reduction in the dry weight of leaves and roots, root length, root volume, average root diameter, chlorophyll and proline contents, net photosynthesis and stomatal conductance. In addition, salinity caused ion imbalance in plants as shown by higher Na^+ and Cl^- contents and lower K^+, Ca^{2+}, and Mg^{2+} concentrations. Ion imbalance was more pronounced in CCRI-79 than in Simian3. In the leaves and roots of the salt-tolerant cultivar CCRI-79, increasing levels of salinity increased the activities of superoxide dismutase (SOD), ascorbate peroxidase (APX), and glutathione reductase (GR), but reduced catalase (CAT) activity. The activities of SOD, CAT, APX, and GR in the leaves and roots of CCRI-79 were higher than those in Simian 3. CAT and APX showed the greatest H_2O_2 scavenging activity in both leaves and roots. Moreover, CAT and APX activities in conjunction with SOD seem to play an essential protective role in the scavenging process. These results indicate that CCRI-79 has a more effective protection mechanism and mitigated oxidative stress and lipid peroxidation by maintaining higher antioxidant activities than those in Simian 3. Overall, the chlorophyll a, chlorophyll b, and Chl (a+b) contents, net photosynthetic rate and stomatal conductance, SOD, CAT, APX, and GR activities showed the most significant variation between the two cotton cultivars.

Editor: Wagner L. Araujo, Universidade Federal de Vicosa, Brazil

Funding: This work was supported by the National Natural Science Foundation of China (31301262) and China Postdoctoral Science Foundation (Grant no. 2013M540169). The funders had no role in study design, data collection and analysis, decision to publish, or preparation of the manuscript.

Competing Interests: The authors have declared that no competing interests exist.

* Email: yu@cricaas.com.cn (SY); zhaoxinhua1968@126.com (XZ)

Introduction

The proportion of agricultural land that is negatively affected by high salinity is increasing worldwide, owing to natural causes and agricultural practices [1]. This problem has been aggravated by the development of more recent agricultural practices such as irrigation. Approximately 20% of the world's cultivated lands and more than half of all irrigated lands are affected by salinity [2]. High salt concentrations in the soil cause various events that negatively impact agricultural production, such as delays in plant growth and development, inhibition of enzymatic activities and reductions in photosynthetic rates [3]. Therefore, before attempts can be made to introduce genetic and environmental factors to alleviated salt stress, it is critical to elucidate the morphological and physiological responses of particular crops and cultivars to salinity.

In general, salt stress causes an imbalance of the cellular ions resulting in ion toxicity, osmotic stress and production of reactive oxygen species (ROS) [4], thus affecting plant growth, morphology, and survival [5]. High concentrations of NaCl outside the roots reduce the water potential and make it more difficult for the root to extract water. On the other hand, high concentrations of Na^+ and Cl^- ions inside plant cells are inhibitory to many enzyme processes. Salt-tolerant plants can not only regulate ion and water movements more efficiently but should also have a better antioxidant system for effective removal of ROS [6]. Salt stress causes excessive generation of ROS such as superoxide anions (O_2^-), hydrogen peroxide (H_2O_2), and hydroxyl radicals (OH•) [7]. To mitigate the oxidative damage initiated by ROS formed under salt stress, plants possess a complex antioxidant system, including non-enzymatic antioxidants such as ascorbic acid, glutathione (GSH), tocopherols, and carotenoids; antioxidant enzymes such as superoxide dismutase (SOD, EC 1.15.1.1), catalase (CAT, EC 1.11.1.6), glutathione peroxidase (EC 1.11.1.9), and peroxidases (POD, EC 1.17.1.7); and enzymes of the so-called ascorbate-glutathione cycle, including ascorbate peroxidase (APX, EC 1.11.1.11) and glutathione reductase (GR, EC 1.6.4.2). These components of the antioxidant system act in concert to alleviate the cellular damage accumulated under conditions of oxidative stress [8], [9].

SOD is generally considered as the first line of the antioxidant defense system, as it catalyzes the dismutation of O_2^- into H_2O_2 and O_2 in the cytosol, chloroplasts, and mitochondria [10]. POD is mainly located in the apoplastic space and vacuoles, where it plays an important role in catalyzing the conversion of H_2O_2 to H_2O and O_2 [11]. H_2O_2 is scavenged by CAT and APX. CAT dismutates H_2O_2 to H_2O and O_2, whereas APX, together with monodehydroascorbate reductase, dehydroascorbate reductase, and GR, converts H_2O_2 to H_2O via the ascorbate-glutathione pathway. APX is the first enzyme in this pathway, which eliminates H_2O_2 by using ascorbate as an electron donor in an oxidation-reduction reaction [12]. GSH is the final enzyme in this pathway and functions to protect plants from oxidative stress by maintaining GSH in the reduced state [13]. Under salt stress, malondialdehyde (MDA), the decomposition product of the polyunsaturated fatty acids of biomembranes, tends to accumulate [14]. Accordingly, cell membrane stability has been widely used to differentiate between stress-tolerant and -susceptible cultivars of some crops [15], in some cases, higher membrane stability could be correlated with abiotic stress tolerance.

In most plants, higher levels of the activity of the above-mentioned antioxidant enzyme are considered as a salt tolerance mechanisms [9]. Indeed, previous studies have shown that within the same species, salt-tolerant cultivars generally have enhanced or higher constitutive antioxidant enzyme activity under salt stress when compared with sensitive-cultivars. This has been demonstrated in numerous plant species such as cotton [16], rice [17], and pea [18]. Moreover, the response of plant antioxidant enzymes to salinity has been shown to vary among plant species, tissues, and subcellular localizations [19]. Several studies have demonstrated that salt-tolerant species show increased antioxidant enzyme activities and antioxidant contents in response to salt stress, whereas salt-sensitive species fail to do so [20]. Thus, the evidence accumulated to data indicates that intrinsic antioxidant resistance mechanisms of plants may provide a strategy to enhance salt tolerance. However, to achieve efficient selection of genetically-transformed salt-tolerant plants, the mechanisms underlying the effects of salt on the morphology, physiology, growth, and antioxidative responses of plants must first be identified [21].

Salt may affect plant growth indirectly by decreasing the rate of photosynthesis. Indeed, under saline conditions, substantial reduction in photosynthesis has been associated with a decrease in total chlorophyll content and distortion in chlorophyll ultrastructures [22]. Although the factors that limit photosynthesis in salt-stressed plants have been investigated for a number of species, the mechanistic pattern of inhibition remains unclear [23].

In addition, the relationships between ion accumulation, morphological and physiological changes, salt stress, and resulting plant injury are poorly understood. A number of conflicting views have been proposed in the literature over the toxic ions and enzymatic protection and activity involved in the response to salinity [24], [25]. Oxygen radicals are generated during plant metabolism that need to be scavenged by antioxidant systems to maintain normal growth; therefore, determining any adverse effects on these antioxidant systems due to salinity is an important consideration for appropriate cultivar selection.

Cotton is one of the most economically important crops in China. Although it is classified as a salt-tolerant crop, this tolerance is not only limited but also varies according to the growth and developmental stages of the plant [26]. Several studies have been conducted to assess the effect of salinity on the germination, vegetative growth, or yield of cotton [27–29]. However, the interactions between growths rates, ionic content, enzymatic activity, and oxidation reactions are likely to be complex and perhaps vary significantly between cultivars; therefore, such interactions deserve more detailed investigation.

In this study, the effect of NaCl on the growth behavior of two cotton cultivars that differ in their tolerance to salt was investigated. Changes in growth, ion concentration, pigment contents, and photosynthesis were assessed and linked to differences in the antioxidant system observed during salt stress. Since no detailed investigations have been conducted on this responses to data, the information presented here will not only provide criteria for improving salt-tolerant cotton specifically but also for selecting other salt-tolerant species and cultivars.

Materials and Methods

Experimental design

Seeds of two cotton cultivars, CCRI-79 (salt tolerant) and Simian 3 (salt sensitive), were obtained from the National Medium-Term Gene Bank of Cotton in China and soaked in sterile deionized water at 28°C for 6 h. They were then transferred to two sheets of sterile filter paper moistened with deionized water and placed in plastic trays for germination at 28°C for 72 h in the dark. The seeds were then sown into pots filled with perlite and grown under controlled conditions (light/dark regime of 16/8 h at 23°C, relative humidity of 60%–70%, photosynthetic photon flux density of 350 $\mu mol \cdot m^{-2} \cdot s^{-1}$). Germinated seeds were sown into holes in Styrofoam boards that were placed in deionized water and grown hydroponically in a growth room for 3 weeks under fluorescent and incandescent lights.

After 3 weeks, healthy and uniform seedlings were transplanted to 4-L plastic pots (10 plants per pot) filled with an aerated Hoagland nutrition solution (pH 5.2). The nutrition solution was aerated constantly and replaced twice a week throughout the experiment. Plants were cultured under non-saline conditions for 10 d to ensure full establishment before starting the salinity treatments. Salt stress treatment was initiated by providing the plants with full-strength Hoagland's solutions containing 0, 80, 160, or 240 mM NaCl. To avoid osmotic shock, salt concentrations were increased daily by 40 mM NaCl, until reaching the required concentration. After 1 week, the plants were harvested, cleaned and their fresh weights were measured.

Leaf photosynthesis measurements

Net photosynthetic rate (Pn) and stomatal conductance (gs) of leaves were measured in three plants per cultivar per treatment with a photosynthesis system (Li-6400, Li-COR Inc., NE, USA) under 1500 $\mu mol \cdot m^2 \cdot s^2$ light intensity, 65%±5% relative humidity, 32°C±2°C leaf temperature, and 380 $\mu mol \cdot mol^{-1}$ CO_2 concentrations at 9:30–11:00 AM.

Growth parameter measurements

From each treatment group, 10 plants were randomly selected and separated into leaf and root fractions. The leaf area of the youngest fully developed leaf of each plant was measured. Root samples were placed in a rectangular glass dish with a thin layer of water (4–5 mm depth) to allow all roots to spread appropriately. Entire roots were scanned with an EPSON Transparency unit (Epson Perfection V700 Photo; Indonesia), and then analyzed with WinRHIZO software version 5.0 (Regent Instruments, Inc.; Canada) to calculate the total root length, total root surface area, total root volume, and average root diameter. Leaves and roots were washed with deionized water and dried at 70°C for 48 h to determine dry weight before being ground to determine the ion contents.

In addition, another six plants per replicate of each of the four treatments were harvested. Fresh roots of seedlings were separated and frozen immediately in liquid nitrogen before being stored at −80°C pending further analysis.

Chlorophyll and carotenoid content measurements

To determine chlorophyll a (Chl a), chlorophyll b (Chl b), and carotenoids levles, 3–5 discs (0.8-cm diameter) were cut from the upper-most fully expanded leaves randomly selected from five plants per replicate. Discs were homogenized with 2 mL of acetone (80%) and washed twice with an additional 2 mL of acetone. The absorbance of the pooled extracts was measured using a spectrophotometer at 480 nm, 645 nm, and 663 nm. Contents of Chl a, Chl b, and carotenoids in the extracts was determined using MacKinney equations [30]:

$$Chla(mg\ g^{-1}DW) = (12.72 \times A663) - (2.58 \times A645)$$

$$Chlb(mg\ g^{-1}DW) = (22.87 \times A645) - (4.67 \times A663)$$

$$Carotenoids(mg\ g^{-1}DW) = (0.114 \times A663) + A480 - (0.638 \times A645)$$

Ion analyses

The concentration of calcium (Ca), magnesium (Mg), potassium (K), and sodium (Na) were analyzed in subsamples of dried plant materials, which were finely ground in a mill grinder. Approximately 0.5 g of finely ground plant samples were placed into digesting tubes, to which 10 mL of concentrated nitric acid and 3 mL perchlorate acid were added. All the samples were soaked for 12 h and then burned at 300°C for 3 h. The residue was transferred to a 50-mL volumetric flask, which was topped up to 50 mL with distilled water. The cation content was then measured using an atomic absorption spectrophotometer (TAS-986; Persee; China) [31]. For the determination of Cl^- content, leaf samples (0.1 g) were extracted in 10 mL of distilled water by heating at 80°C for 3 h [32]. The Cl^- content in the extracts was analyzed by ion chromatography (DX-300; Sunnyvale, CA, USA) [33].

Determination of lipid peroxidation

Frozen leaf and root segments (0.5 g) were homogenized in a 0.1% (w/v) trichloroacetic acid (TCA) solution. The homogenate was centrifuged at 15,000× *g* for 10 min and 1 mL of the supernatant was added to 4 mL 0.5% (w/v) 2-Thiobarbituric acid (TBA) in 20% (w/v) TCA. The mixture was incubated at 90°C for 30 min, and the reaction was stopped by placing the reaction tubes in an ice water bath. Samples were centrifuged at 10,000× *g* for 5 min and the absorbance of the supernatant was read at 532 nm. The value for non-specific absorption at 600 nm was subtracted from the measured values. The concentration of MDA was calculated using an extinction coefficient of 155 $mM^{-1}\cdot cm^{-1}$.

H_2O_2 determination

H_2O_2 content was estimated according to the methods of Bernt and Bergmeyer [34]. Approximately 0.5 g of root and leaf samples from control and treatment groups were homogenized with liquid nitrogen, and the powders were suspended in 1.5 mL of 100 mM potassium phosphate buffer (pH 6.8). The suspensions were then centrifuged at 18,000× *g* for 20 min at 4°C. The enzymatic reaction was initiated with 0.25 mL supernatant and 1.25 mL peroxidase reagent, consisting of 83 mM potassium phosphate buffer (pH 7.0), 0.005% (w/v) *O*-dianizidine, and 40 μg peroxidase/mL at 30°C. The reaction was stopped after 10 min by adding 0.25 mL of 1 N perchloric acid and the reaction mixture was centrifuged at 5000× *g* for 5 min. The absorbance of the supernatant was read at 436 nm, and the amount of H_2O_2 was determined using an extinction coefficient of 39.4 $mM^{-1}\cdot cm^{-1}$.

Extraction and assay of antioxidant enzyme assay

For enzyme extractions, frozen root and leaf samples (0.3 g) were ground into a fine powder by using a mortar that was placed in an ice bath and a pestle that was pre-cooled with liquid nitrogen, and homogenized in 50 mM potassium phosphate buffer (pH 7.8) containing 1 mM ascorbate and 2% (w/v) polyvinylpolypyrrolidone. Homogenates were then centrifuged at 20,000× *g* for 30 min at 4°C.

SOD activity was determined according to the methods of Foster and Hess [35]. The reaction was performed in a total volume of 1 mL containing 50 mM potassium phosphate buffer (pH 7.8, containing 0.1 mM ethylenediaminetetraacetic acid [EDTA]), 0.1 mM cytochrome c, 0.1 mM xanthine, enzyme extract, and 0.3 U/mL of xanthine oxidase. The reaction was initiated by the addition of xanthine oxidase and absorbance was measured at 560 nm. One unit of SOD activity was defined as the amount of enzyme that inhibits the rate of cytochrome c reduction by 50%.

Total CAT activity was measured according to the method reported by Beers and Sizer [36], with minor modifications. The reaction mixture (1.5 mL) consisted of 100 mM phosphate buffer (pH 7.0), 0.1 mM EDTA, 20 mM H_2O_2, and 50 μL enzyme extract. The reaction was initiated by the addition of the enzyme extract. The decrease in H_2O_2 was monitored at 240 nm and was quantified by its molar extinction coefficient (36 $M^{-1}\cdot cm^{-1}$).

Peroxidase activity was analyzed in 2.9 mL of 0.05 M phosphate buffer, containing 1.0 mL of 0.05 M guaiacol and 1.0 mL of 2% H_2O_2 [37]. Increases in absorbance at 470 nm were recorded after adding 2.0 mL of 20% chloroacetic acid.

APX activity was determined according to Nakano and Asada [38] by following the decline in absorbance at 290 nm as ascorbate was oxidized. The oxidation rate of ascorbate was estimated between 1 and 60 s after starting the reaction with the addition of H_2O_2. The 1-mL reaction mixture contained 50 mM HEPES-NaOH (pH 7.6), 0.22 mM ascorbate, 1.0 mM H_2O_2, and an enzyme extract. Corrections were made for the low, non-enzymatic oxidation of ascorbate in the absence of H_2O_2.

GR activity was measured as described by Foyer et al. [39]. The assay medium contained 1 mM EDTA in 50 mM potassium phosphate buffer (pH 7.8), 0.1 mM NADPH, enzyme extract, and 0.1 mM glutathiol (GSSG) in a total volume of 1 mL. The reaction was initiated by adding GSSG and the NADPH oxidation rate was monitored at 340 nm. GR activity was calculated using an extinction coefficient of 6.2 $mM^{-1}\cdot cm^{-1}$ for NADPH, and one unit of enzyme was defined as the amount of enzyme required to oxidize 1 μmol NADPH per minute. The specific enzyme activity for all enzymes was expressed as units/mg protein.

Statistical analysis

The experiments were set up as a completely randomized design, including two cotton cultivars and four salinity levels. All data obtained were subjected to ANOVA, and the mean difference was compared by the LSD test at 95% or 99% levels of probability. In all figures, the spread of values is shown as error bars representing standard errors of the means.

Results and Discussion

Growth parameters

Salinity exposure can lead to various physiological and biochemical changes within plant cells causing numerous changes in their structure and function [40]. In the present study, salt-induced changes in the growth and antioxidant profile of cotton plants were evaluated. Increasing NaCl concentration, up to 240 mM, gradually reduced leaf and root growth (Fig. 1). In general, both cultivars showed decreased growth rates of the roots and leaves with increasing salt concentration, but there was no significant variation in the leaf dry weight of Simian 3 when

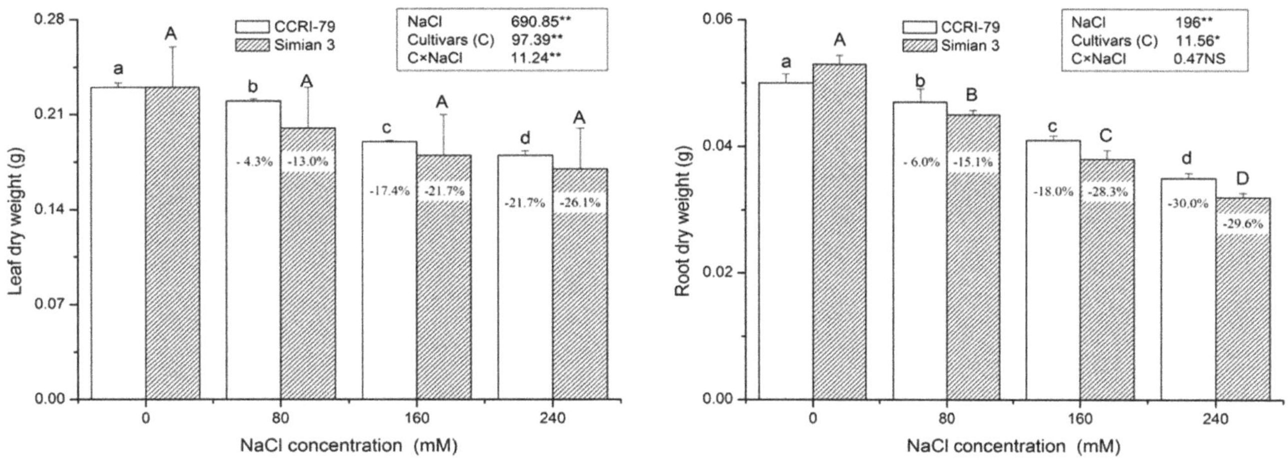

Figure 1. Effect of NaCl salinity on the leaf and root dry weight of two cotton cultivars. Vertical bars represent ± standard error (n = 3). Bars labeled with the different lowercase letters on open square bars or uppercase letters on closed square bars are significant difference (P<0.05). *, ** Significant at P<0.05 and P<0.01, respectively. NS, not significant. Figures in column indicate± increase/decrease under salinity stress over control.

subjected to salt stress. However, the percentage reduction in leaf and root dry weights due to salinity over control was lower in CCRI-79 as compared with Simian 3, indicating that CCRI-79 is a more salt-tolerant cultivar. Inhibition of growth due to NaCl stress in CCRI-79 is comparable to the observations of Takemura et al. [41]. This reduction in growth may be due to osmotic injury or specific ion toxicity caused by the uptake of salt [42]. However, the differential response of growth to salinity observed between CCRI-79 and Simian 3 could be due to genotypic differences, which have also been reported by Qidar and Shams [26]. An increase in the tissue maintenance process (through respiration) is believed to be the primary cause of growth decline during salinity stress, and could represent a mechanism of adaptation to salinity [25]. The sacrifice of leaf photosynthetic tissue during salt adaptation may serve to conserve energy that can then be redirected to maintaining leaf multiplication and growth, indicating successful use of the tissue maintenance process. Thus, the fact that the leaf dry weight of the salt-sensitive cultivar Simian 3 was maintained indicates that it is not reallocating its energy reserves when faced with high-salinity conditions as compared to the more salt-tolerant cultivar CCRI-79.

When plants are grown under conditions of salt stress, the immediate response is a cessation in the expansion of the leaf surface [43]. Several authors have reported the phenomenon of leaf expansion in response to salinity in halophytes as well as in glycophytes [44]. Similarly, in our study, leaf area was highest in the control group (0 mM NaCl), whereas it decreased continuously as salinity increased (Table 1). One possible reason for this decline might be related to salt osmotic effects, which affect cell turgor and expansion [45].

Compared with the no-salt control treatment, salt stress significantly (P<0.01) reduced the length, surface area, volume,

Table 1. Morphological parameters of cotton at different NaCl concentrations and results of ANOVA (F-ratios).

Cultivar	NaCl (mM)	Leaf area (cm^2)	Root length (cm)	Surface area (cm^2)	Volume (cm^3)	Average diameter (mm)
CCRI-79	0	36.19±0.91a	370.86±3.45a	52.07±0.94a	0.57±0.014a	0.54±0.016a
	80	35.18±1.06a	320.33±2.53b	44.04±1.29b	0.48±0.016b	0.45±0.014b
	160	34.43±0.46a	192.72±4.38c	32.45±1.67c	0.42±0.009c	0.43±0.009b
	240	34.07±0.12a	161.28±4.88d	28.03±0.71d	0.34±0.014d	0.35±0.007c
Simian 3	0	35.13±0.85a	302.58±3.02a	49.07±0.78a	0.62±0.041a	0.52±0.005a
	80	33.99±1.07a	260.86±8.67b	35.71±1.28b	0.39±0.043b	0.49±0.018a
	160	33.46±1.36a	188.32±14.26c	31.12±0.98c	0.38±0.023b	0.40±0.014b
	240	32.98±0.63a	138.58±1.05d	24.19±0.60d	0.30±0.008c	0.31±0.004c
NaCl		4.31NS	673.99**	338.36**	75.49**	182.17**
Cultivars(C)		5.73*	134.92**	50.99**	4.60 NS	3.12 NS
C×NaCl		0.01 NS	20.54**	6.70*	5.15*	9.41**

Values are the mean of three replicates ± S.E. Means followed by a different letter within a column for each cotton cultivar are significantly different at P<0.05 according to the Student's LSD test.
*, ** Significant at P<0.05 and P<0.01, respectively.
NS, not significant.

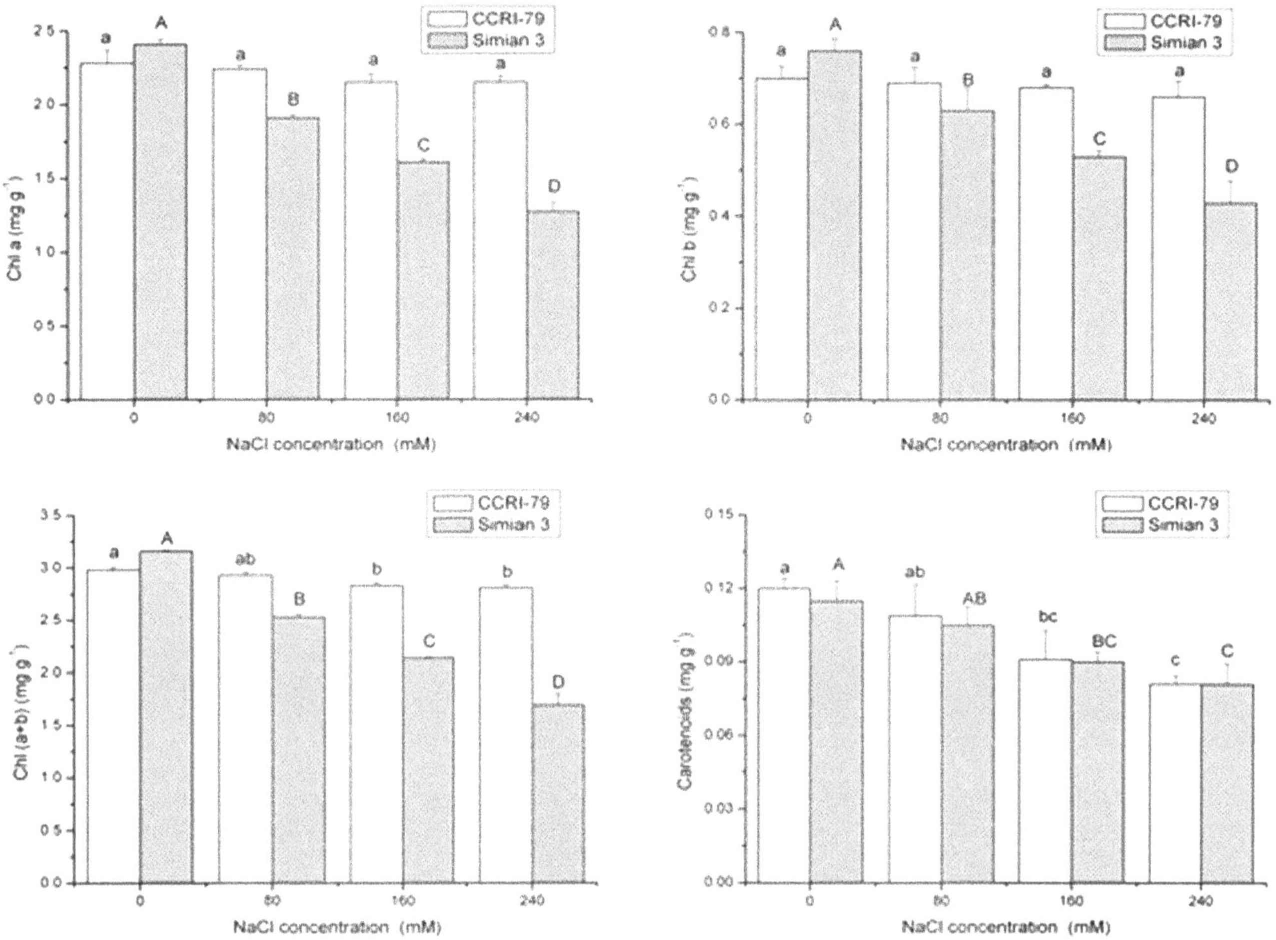

Figure 2. Concentrations of chlorophylls and carotenoids in cotton grown at different NaCl concentrations. Vertical bars represent ± standard error (n = 3). Bars labeled with the different lowercase letters on open square bars or uppercase letters on closed square bars are significant difference (P<0.05).

and average diameter of the roots in both cotton cultivars (Table 1). Significant differences were observed when NaCl concentration was increased from 80 to 240 mM. There are several reasons for the reduced root length, including cell growth restriction due to low water potential of the external medium, interference caused by ions, or the toxicity of accumulated ions [46]. Our findings are consistent with the results of Siroka et al. [47], who reported that salinity suppressed the development of maize roots cell. The inhibition of root growth in terms of root length, surface area, volume, and average diameter can be attributed to the inhibition of mitosis, reduced synthesis of cell wall components, damage to the Golgi apparatus, and changes in polysaccharide metabolism [48]. However, this decrease was more

Table 2. *F* values of ANOVA of the effects of NaCl, cultivars, and their interaction for chlorophylls, carotenoids content, net photosynthetic rate (Pn), and stomatal conductance (gs).

Item	NaCl	Cultivars(C)	C×NaCl
Chl a	108.52**	245.65**	66.29**
Chl b	45.49**	61.62**	26.86**
Chl (a+b)	186.62**	380.12**	113.08**
Carotenoids	13.62**	0.28 NS	0.10 NS
Pn	98.15**	200.15**	51.5**
gs	56.31**	77.17**	23.44**

*, ** Significant at P<0.05 and P<0.01, respectively.
NS, not significant.

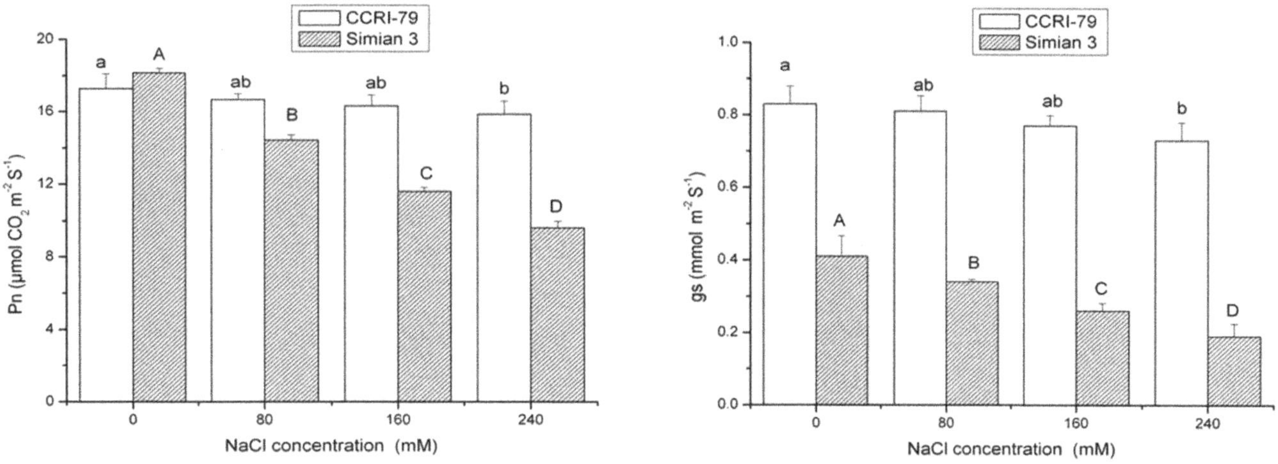

Figure 3. Net photosynthetic rate (Pn) and stomatal conductance (gs) of two cotton cultivars as affected by different NaCl concentrations. Vertical bars represent ± standard error (n = 3). Bars labeled with the different lowercase letters on open square bars or uppercase letters on closed square bars are significant difference (P<0.05).

predominant in Simian 3 than in CCRI-79, indicating that CCRI-79 was more tolerant to salinity than Simian 3.

Chlorophyll and carotenoid contents and photosynthesis

In general, the reduction in growth and productivity when plants are grown under conditions of salt stress is accompanied by as strong reduction in the rate of photosynthesis owing to severe impairments in photosynthetic activities and the photosynthetic apparatus, the degree of which depends on the varieties of species considered [49]. As shown in Fig. 2, the contents of Chl a, Chl b, and Chl (a+b) in the plants significantly ($P<0.01$) decreased as salinity increased in Simian 3, but only marginal changes were observed in CCRI-79 (Table 2). Maintenance of chlorophyll content has been reported in other salt-tolerant crops such as durum wheat and legume species [50]. One possible mechanisms of salt tolerance in these species may be the possession of a salt exclusion and/or sequestration trait that prevents leaf injury, thus maintaining chlorophyll content. Since chlorophyll content is directly correlated with growth and development of the plant [51], the decrease in chlorophyll content in Simian 3 suggested substantial damage to the photosynthetic mechanism, as reported previously in salt-treated rice and sorghum plants [52], [53]. The inhibitory effects of salt on chlorophylls could be due to the suppression of specific enzymes responsible for the synthesis of chlorophyll. Our findings regarding total chlorophyll content are comparable with the observations of Meloni et al. [54], who demonstrated that Guazuncho, another salt-sensitive cotton cultivar, showed a 35% reduction in total chlorophyll content after 21 days of salinity treatment.

Carotenoids are reported to play an important role in ROS scavenging, thereby protecting membranes from salt stress [55]. However, we did not observe changes in carotenoid contents in response to salinity treatments in either cultivar, suggesting that these pigments are not involved in the response to salt stress in cotton. Although the rate of change was slower in carotenoids than in chlorophylls, carotenoid content also showed a decreasing trend with increasing salt stress, indicating that this trait could also serve as a useful indicator of NaCl stress in cotton.

Since plant growth is dependent on photosynthesis, environmental stresses affecting growth will also affect photosynthesis [56]. In the present study, the net photosynthetic rate (Pn) and stomatal conductance (gs) of both cotton cultivars were inhibited by salinity due to NaCl (Fig. 3, Table 2). However, the net photosynthetic rate and stomatal conductance were significantly lower for Simian 3 than for CCRI-79 under conditions of salt stress. Compared with the control treatment, the net photosynthetic rate and stomatal conductance of Simian 3 significantly decreased with increasing salinity, whereas there was no significant difference in either trait in CCRI-79.

Photosynthesis was reduced in both cotton cultivars in response to salt stress, which was likely caused by the reduction in stomatal conductance. Indeed, parallel decreases in stomatal conductance and net photosynthesis due to NaCl salinity have previously been reported for cotton [57]. Our results suggest that stomatal closure limited the leaves' photosynthetic capacity in the NaCl-treated plants of both cultivars, Although only Simian 3 showed a significant decline in the leaf chlorophyll contents due to NaCl stress for 7 days. Similarly, Delfine et al. reported no changes in the chlorophyll content in spinach plants (*Spinacia oleracea* L.) exposed to salt stress for 20 days [58]. Our results suggest that the greater reduction in stomatal conductance accompanied by decreased leaf chlorophyll content could have contributed to the higher reduction in the leaf photosynthetic rate of Simian 3 when compared with that of CCRI-79.

Further examination showed that the decrease in the Chl a, Chl b, Chl (a+b), net photosynthetic rate, and stomatal conductance levels with increasing salt concentrations occurred more rapidly compared with the rate of decrease in carotenoid content; this trend was more conspicuous in Simian 3 than in CCRI-79.

Ion concentrations

High external salt concentration causes an ion imbalance or disturbance in ion homeostasis [43]. In our experiments, the leaves and roots of both cultivars had higher levels of Na^+ and Cl^- ions under salt stress due to nonspecific ion uptake and/or membrane leakage. However, as the NaCl concentration increased, the levels of Na^+ and Cl^- also increased further, suggesting that these cultivars may not differ in terms of Na^+ uptake and its transportation to leaves, and thus the increase observed can not be explained by an ionic exclusion mechanism (Fig. 4). In addition, Na^+ ion concentrations in leaves were higher than those in the roots of both cultivars at all salinity levels, which indicated

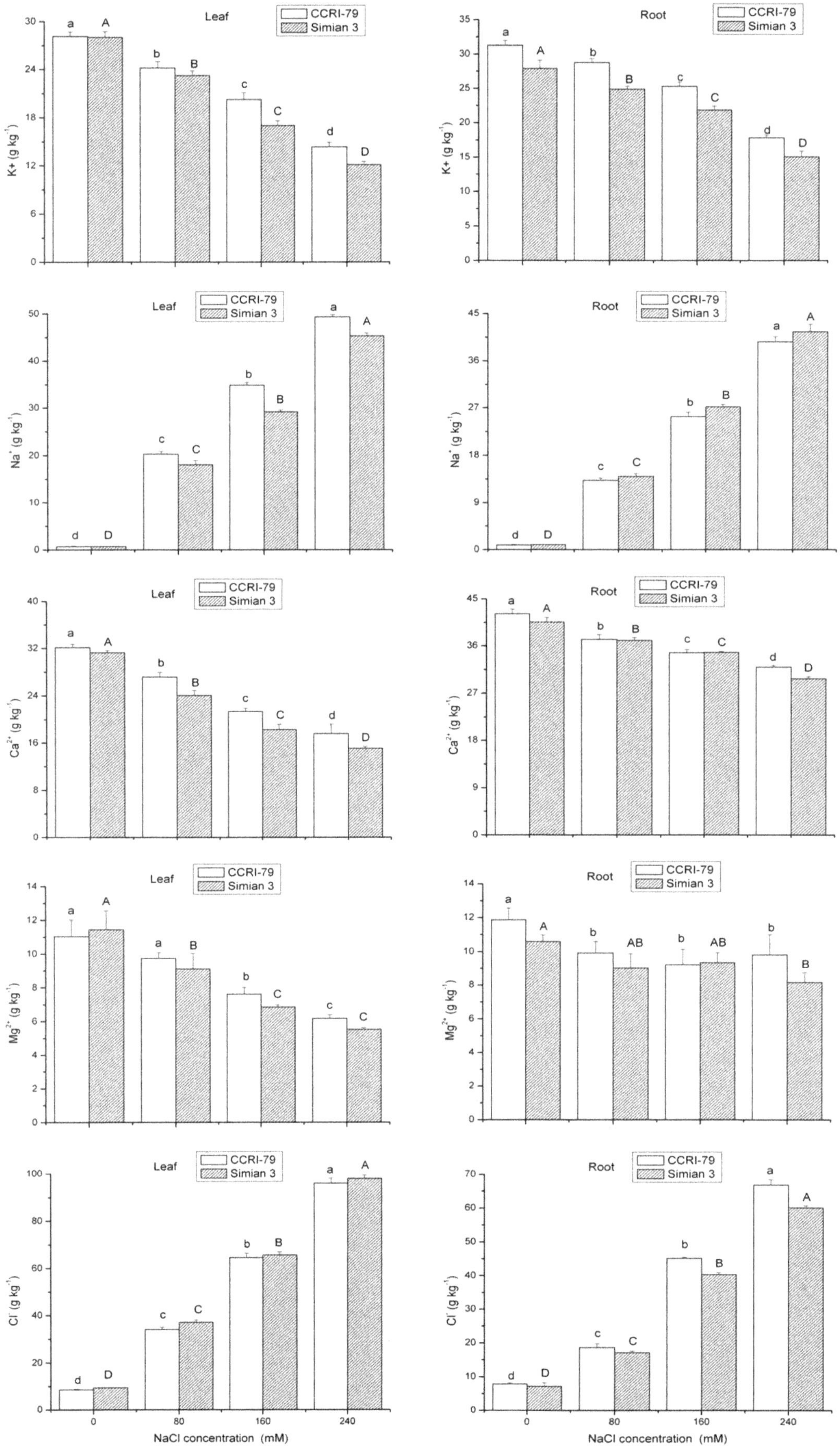

Leaf
CCRI-79
Simian 3
K+ (g kg-1)
Root
CCRI-79
Simian 3
K+ (g kg-1)
Leaf
CCRI-79
Simian 3
Na+ (g kg-1)
Root
CCRI-79
Simian 3
Na+ (g kg-1)
Leaf
CCRI-79
Simian 3
Ca2+ (g kg-1)
Root
CCRI-79
Simian 3
Ca2+ (g kg-1)
Leaf
CCRI-79
Simian 3
Mg2+ (g kg-1)
Root
CCRI-79
Simian 3
Mg2+ (g kg-1)
Leaf
CCRI-79
Simian 3
Cl- (g kg-1)
Root
CCRI-79
Simian 3
Cl- (g kg-1)
0
80
160
240
NaCl concentration (mM)

Figure 4. Effect of NaCl salinity on the concentrations of K^+, Na^+, Ca^{2+}, Mg^{2+} and Cl^- in the roots and leaves of cotton. Vertical bars represent ± standard error (n = 3). Bars labeled with the different lowercase letters on open square bars or uppercase letters on closed square bars are significant difference ($P<0.05$).

the inability of these cultivars to prevent Na^+ ion transportation from the roots to leaves. Chlorine ions showed a similar distribution pattern to Na^+ ions, despite being at higher salinity levels. Na^+ and Cl^- are highly water-soluble and are readily taken up by plants and transported into leaves; these ions most likely act as osmotica, but only moderate concentrations can be tolerated before growth and photosynthesis are reduced.

High Na^+ and Cl^- absorption competes with the uptake of other nutrient ions such as K^+, Mg^{2+}, and Ca^{2+}, leading to a deficiency of these ions and an imbalance among cations [59]. During the same period, K^+ and Ca^{2+} content in the leaves and roots of both cultivars were significantly reduced, and a further decrease in these ions was observed with increasing NaCl concentration (Fig. 4, Table 3). Salt tolerance involves not only adaptation to Na^+ influx but also acquisition of K^+, the uptake of which is adversely affected by high external Na^+ concentration due to the chemical similarity of these two ions [60], [61]. Indeed, selective uptake of K^+ as opposed to Na^+ is considered to be one of the key physiological mechanisms contributing to salt tolerance in many plant species [62]. K^+ efflux has already been used as an indicator of cellular toxicity for a range of toxic compounds, and losses of K^+ and Ca^{2+} have already been documented during salinity stress [63], [64]. A large and permanent efflux of K^+ and Ca^{2+} usually indicates damage to the limiting membranes.

On the other hand, salt treatments induced a significant decrease in Mg^{2+} concentrations in cotton leaves in the present study. The significantly lower levels of Mg^{2+} in the leaves under salinity conditions are probably related to the lower levels of chlorophylls present in the NaCl-treated leaves. However, in the roots of NaCl-treated plants, Mg^{2+} concentrations were close to or lower than those observed in control plants, even at the highest salt dose (240 mM NaCl). This was in contrast to the Ca^{2+} content, which was significantly lower in salt-treated plants. Our findings are similar to those of Khan [65], who reported that NaCl treatment induced a decline in Ca^{2+} and Mg^{2+} levels in *Ceriops tagal* plants.

In Addition, there were no significant differences between the two cotton cultivars in the variation of ion content (including K^+, Na^+, Mg^{2+}, Ca^{2+} and Cl^-) under NaCl stress. This similarity in ionic levels may be a consequence of the shoot culture method used, which would not involve a selective mechanisms of ion transport. For example, there was likely to be no regulation between the xylem parenchyma and xylem interface, in which ion selection and reabsorption from the medium may be regulated. Previous results reported for other species are consistent with these observations [66]. However, the restriction of entry of ions into metabolically active areas of cells in the more-tolerant CCRI-79 cultivar can not be ruled out as a mechanism to maintain ionic equilibrium when ions are highly concentrated in the external environment [67].

Lipid peroxidation and proline content

Salt stress is known to result in extensive lipid peroxidation, which has often been used as an indicator of salt-induced oxidative damage in membranes [68]. The MDA content increased with increasing salinity in the leaves and roots of both cotton cultivars (Fig. 5, Table 4), indicating cell membrane damage in both cotton cultivars. However, as the salinity increased, the accumulation of MDA was higher in Simian 3 as compared to CCRI-79, indicating a higher degree of lipid peroxidation in Simian 3 due to salt stress. Lipid peroxidation could be a result of light-dependent formation of singlet oxygen during stress conditions [69]. The low values of MDA content obtained with CCRI-79 might account for the lower lipid peroxidation levels observed and the reduced effect on membrane permeability. Low levels of lipid peroxidation may have contribute to the observed tolerance of CCRI-79 plants exposed to high levels of salinity. Similar results for lipid peroxidation have been reported by other researchers in barley [70].

Many plants accumulate proline as a nontoxic and protective osmoylte under saline conditions [55]. In this study, the levels of proline continued to increase in both of the cultivars as the NaCl concentration increased (Fig. 5). The proline concentration of

Table 3. *F* values of ANOVA of the effect of NaCl, cultivars, and their interaction for K^+, Na^+, Ca^{2+}, Mg^{2+}, and Cl^- contents.

Tissue	Item	Cultivars (C)	NaCl	C×NaCl
Leaf	K^+	22.47**	349.67**	4.05 NS
	Na^+	131.7**	5755.89**	22.16**
	Ca^{2+}	27.27**	220.03**	1.28 NS
	Mg^{2+}	2.09 NS	69.77**	0.94 NS
	Cl^-	7.03*	3175.89**	0.58 NS
Root	K^+	72.66**	204.40**	0.34 NS
	Na^+	21.19**	4546.89**	3.12 NS
	Ca^{2+}	15.30**	311.67**	4.96*
	Mg^{2+}	5.19NS	6.21*	0.91 NS
	Cl^-	65.95**	3497.79**	11.22**

*, ** Significant at $P<0.05$ and $P<0.01$, respectively.
NS, not significant.

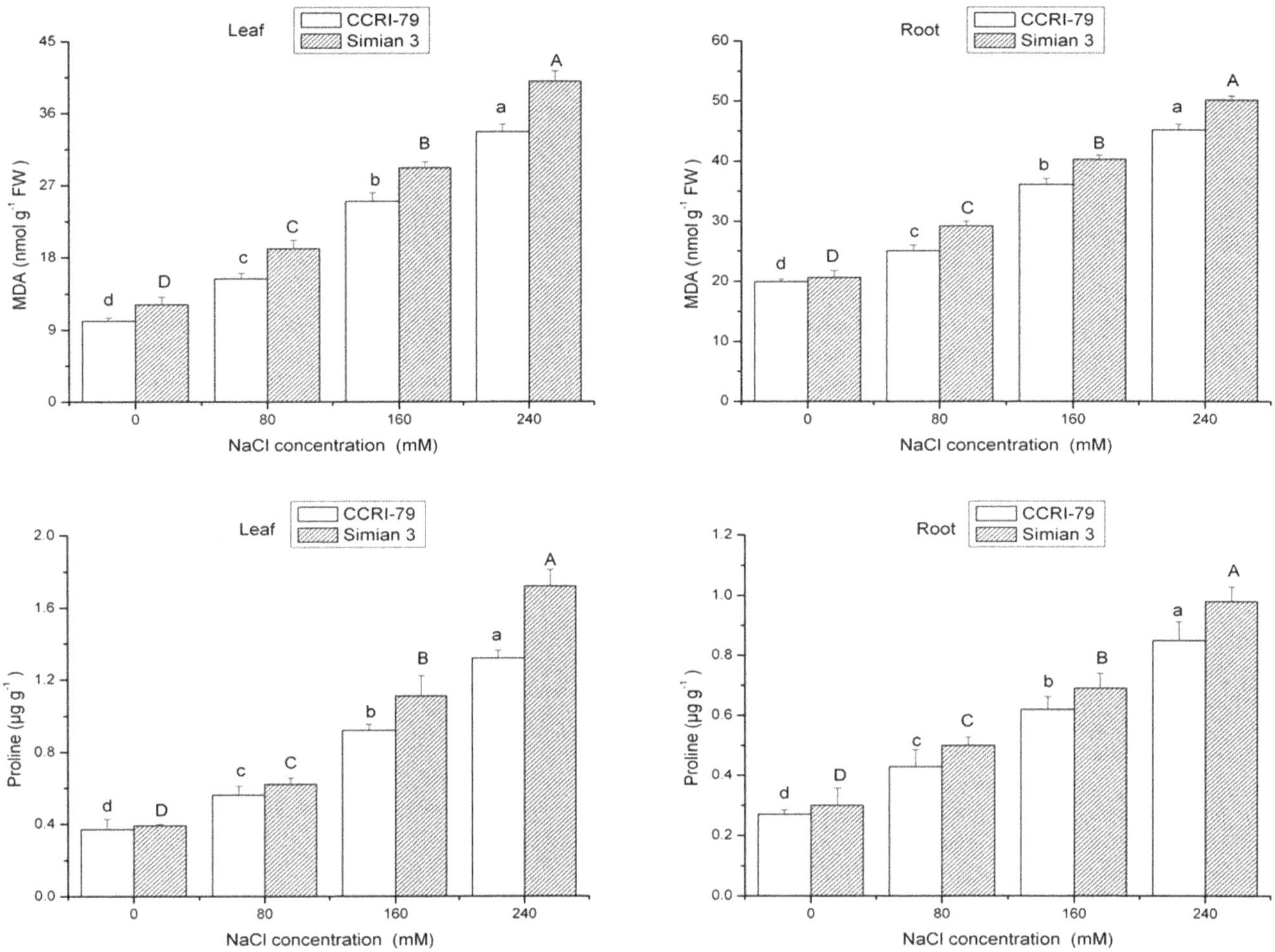

Figure 5. Effect of NaCl salinity on the concentrations of malondialdehyde (MDA) and proline in the roots and leaves of cotton. Vertical bars represent ± standard error (n = 3). Bars labeled with the different lowercase letters on open square bars or uppercase letters on closed square bars are significant difference ($P<0.05$).

CCRI-79 plants was lower than that of Simian 3 plants, especially at the highest salinity level, which could be attributable to the greater salt resistance of the CCRI-79 cultivar, i.e., less injury was induced by the salt [71]. Similar trends were observed by Rabie and Almadini in broad bean plants [72]. Moreover, in our study, proline accumulation in the leaves was higher than that in the roots, which is similar to the findings of Sharma and Dietz [73], indicating that proline plays a more important protective role in the leaves of cotton seedlings than in the roots under salinity stress.

H_2O_2 content

Stress conditions enhance H_2O_2 production in different compartments of plants cells through both enzymatic and non-enzymatic processes [74]. In our study, both genotypes had similar levels of H_2O_2 levels at 0 mM NaCl treatment. However, salinity treatments caused a marked increase in H_2O_2 content, and the Simian 3 cultivar had a higher H_2O_2 content than did CCRI-79 after NaCl treatment (Fig. 6, Table 4), which could be mainly due to the decreased H_2O_2-scavenging activity in the salt-sensitive cultivar. These results are comparable to those reported in a previous study [75], where accumulation of H_2O_2 in the roots of the salt-tolerant rice cultivar FL478 was significantly higher than that of the salt-sensitive rice cultivar IR-29 in response to moderate salt stress applied for 12 days. Simultaneously, the H_2O_2 content were markedly lower in the leaves than in the roots, regardless of NaCl dose, which contrasts with the findings of Lee et al. [76] in rice. This discrepancy in the effects of salt on H_2O_2 content between studies may be related to technical difficulties, and therefore these results should be evaluated with caution [77]. Furthermore, H_2O_2 has been shown to induce cytosolic APX [78]; therefore, accumulation of H_2O_2 under high salinity conditions may be a signal to initiate an adaptive response to the stress [79]. Although differences in salt tolerance among different cultivars are not necessarily related to differences in the ability to detoxify ROS, many comparative studies using salt-tolerant and salt-sensitive genotypes have shown a correlation between salt tolerance and increased activity of antioxidant enzymes [80].

Activities of antioxidant enzymes

Environmental stresses that limit photosynthesis can increase oxygen-induced cellular damage due to increased ROS generation [81]. Therefore, salt stress resistance may depend, at least in part, on the enhancement of the antioxidative defense system, which involves antioxidant compounds and several antioxidant enzymes. In the present study, the responses of SOD, POD, CAT, APX, and GR enzyme activities suggested that oxidative stress is an important component of salt stress in cotton plants.

Table 4. *F* values of ANOVA of the effect of NaCl, cultivars, and their interaction for malondialdehyde (MDA), proline, H_2O_2 content, and antioxidant enzyme activities.

Tissue	Item	Cultivars (C)	NaCl	C×NaCl
Leaf	MDA	78.15**	608.62**	3.65 NS
	Proline	25.01**	230.23**	6.58*
	H_2O_2	74.86**	1357.74**	11.54**
	SOD	315.65**	213.58**	63.01**
	CAT	874.75**	445.12**	43.46**
	POD	0.47 NS	5.52*	0.49 NS
	APX	1.36 NS	5.93*	0.08 NS
	GR	28.88**	194.18**	15.77**
Root	MDA	60.16**	708.42**	4.47*
	Proline	9.10*	119.51**	0.72 NS
	H_2O_2	174.18**	4622.30**	40.92**
	SOD	8559.93**	1510.48**	1357.86**
	CAT	111.07**	1.92 NS	28.60**
	POD	50.42**	322.16**	2.08 NS
	APX	37.10**	208.76**	5.83*
	GR	17.47**	82.69**	1.45 NS

*, ** significant at P = 0.05 and P = 0.01 levels, respectively.
NS, not significant.
H_2O_2, hydrogen peroxide; SOD, superoxide dismutase; CAT, catalase; POD, peroxidase; APX, ascorbate peroxidase; GR, glutathione reductase.

Because SOD can catalyze the dismutation of superoxide to molecular oxygen and H_2O_2, this enzyme is considered the most effective intracellular enzymatic antioxidant. Indeed, it has been suggested that SOD plays an important role in plant stress tolerance and provides the first line of defense against the toxic effects of elevated levels of ROS [82]. In this study, salt stress increased SOD activity in the leaves of both cultivars and in the roots of CCRI-79 only (Fig. 7, Table 4). However, increased SOD activity in both the leaves and roots was more conspicuous in the salt tolerant cultivar CCRI-79 than in the salt-sensitive cultivar Simian 3, suggesting that the salt-tolerant genotype has a more efficient $O_2^{\bullet -}$ radical-scavenging ability. Similar results have also been shown in both the leaves and roots of cotton and pea plants [14], [83]. In plants, high induction of SOD activity can lead to H_2O_2 accumulation as well as lipid peroxidation [84], which could contribute to the increased H_2O_2 content observed in the roots than in the leaves of cotton seedlings exposed to salinity.

POD is the primary enzyme that detoxifies H_2O_2 in the chloroplasts and cytosol of plant cells [85]. CAT plays an important role in the antioxidant system because it converts H_2O_2 into oxygen and water [86]. These two enzymes constitute the main H_2O_2-scavenging systems in cells. The present data showed that the roots had higher POD activity compared to the leaves in both cultivars; however, the enzyme activity in the roots and leaves responded differently to incremental levels of salinity. In the roots of both cultivars, there was a significant decline in POD activity with an increase in salinity levels, whereas there was no significant difference in the leaf of Simian 3 across salt treatments. However, POD activity in the leaf of CCRI-79 showed no significant difference when subjected to 0 to 160 mM NaCl concentrations. Conversely, the 240 mM NaCl concentration induced a significant decrease in the leaf POD activity of CCRI-79 when compared to the control (0 mM NaCl concentration). Amor et al. reported that H_2O_2 accumulation under salinity stress was related to a decrease in CAT activity in *C. maritima* [87]. In our study, CAT activity in the leaves and roots of Simian 3 declined with an increase in NaCl concentration (Fig. 7, Table 4). However, although the CAT activity in CCRI-79 increased with salinity in the roots, it decreased with increasing salinity in the leaves. CAT activity in the leaves and roots of CCRI-79 was significantly higher than that in Simian 3 leaves and roots for all salinity treatments. This indicates that the scavenging of H_2O_2 is more effective in CCRI-79 than in Simian 3. The roots of both the CCRI-79 and Simian 3 cultivars showed higher CAT activities than did the leaves. The stimulation of POD and CAT suggests that these enzymes are important in the detoxification of H_2O_2 in plant seedlings under salinity stress [88]. In contrast, POD and CAT activities were found to be reduce in response to excess salinity in the leaves and roots of various other plants species [52], [89]. Thus, the effect of salinity on antioxidant enzyme activities varies among plant species, organs, and even treatment concentrations.

H_2O_2 scavenging is also accomplished by the glutathione-ascorbate cycle, a series of coupled redox reactions involving three enzymes, APX, GR and monodehydroascorbate reductase [90]. APX plays a key role in protecting the plant against oxidative stress by scavenging H_2O_2 in different cell compartments. It also has a higher affinity for H_2O_2 than POD and CAT and, as such, may play a more crucial role in the management of ROS during stress [82]. As shown in Fig. 7, there was a significant difference in the effect of salt on APX activity between the leaves and roots of the two cultivars. As salinity increased, APX activity in the root increased in CCRI-79 but decreased in Simian 3. However, there was no significant difference in the leaf APX activity of plants subjected to 0 to 160 mM NaCl concentrations in both cultivars. Conversely, the 240 mM NaCl concentration induced a significant

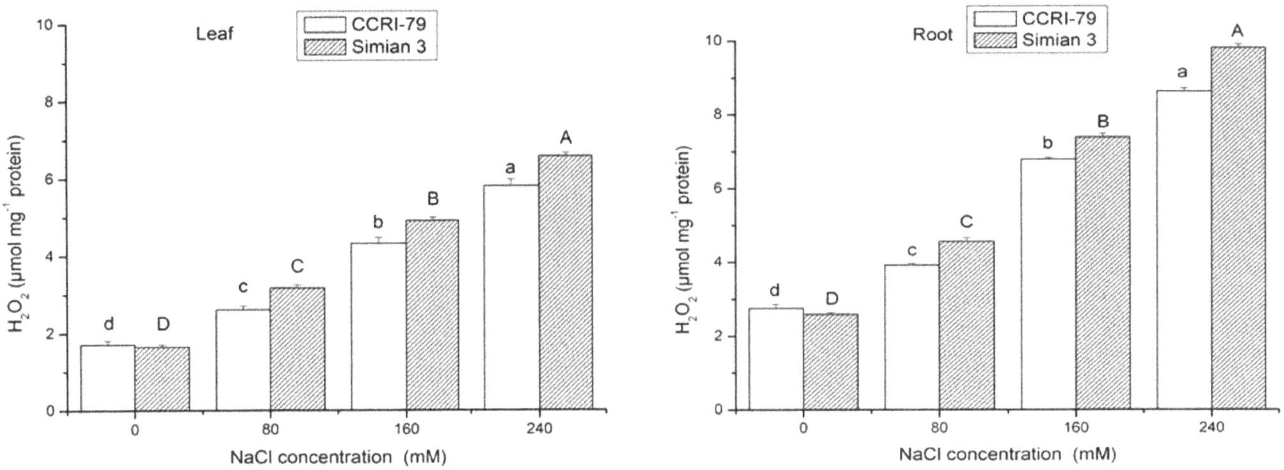

Figure 6. Effect of NaCl salinity on the concentrations of H_2O_2 in the roots and leaves of cotton. Vertical bars represent ± standard error (n = 3). Bars labeled with the different lowercase letters on open square bars or uppercase letters on closed square bars are significant difference ($P < 0.05$).

decrease in the leaf APX activity of both cultivars when compared to the control (0 mM NaCl concentration). We also found that higher APX activities were accompanied by higher CAT activities in CCRI-79, implying that CCRI-79 has a more effective H_2O_2-scavenging mechanism than Simian 3.

The role of GSH and GR in H_2O_2 scavenging in plant cells has been well established in the Halliwell-Asada enzyme pathway. GR catalyzes the rate-limiting and last step of the ascorbate-glutathione pathway [91]. In our study, GR activities increased in the leaves of both cotton cultivars and in the roots of CCRI-79 only in response to salt stress. On the other hand, salinity significantly reduced GR activity in the roots of Simian 3. Several studies investigating salt-tolerant and salt-sensitive cultivars have suggested that the salt tolerance trait is related to increased GR activity in salt-tolerant cultivars [83], [92], which is similar to our results. The elevated levels of GR activity may increase the rate of NADPH oxidation to $NADP^+$, thereby ensuring the availability of $NADP^+$ to accept electrons from the photosynthetic electron transport chain. Under such conditions, the flow of electrons to O_2^-, and therefore the formation of O_2^- can be minimized. However, in the roots of Simian 3, the reduction in GR activity suggests a decrease in the GSH turnover rate. Considering that salinity also reduced APX activity in the roots of Simian 3, these results suggests that salt-sensitive cultivars exhibit a less-active ascorbate-glutathione cycle in the roots, which may be a key enzyme for the development of salt-tolerance in cotton plants.

Furthermore, the variation in SOD, CAT, APX, and GR activities differed significantly between CCRI-79 and Simian 3 under the NaCl concentrations tested. The increased salinity resistance of CCRI-79 was associated with its ability to maintain higher activity of these antioxidant enzymes, which resulted in lower H_2O_2 production, lipid peroxidation, and higher membrane stability. This provides further evidence that the H_2O_2-scavenging mechanisms were more effective in CCRI-79. By contrast, the relatively lower CAT, APX, and GR activities in salt-stressed Simian 3 compared to control plants indicated that H_2O_2 scavenging was less effective in this cultivar. This excess of H_2O_2 may be the main contributor to the extensive lipid peroxidation and growth inhibition observed in Simian 3.

When compared with the other scavenging enzymes tested, CAT and APX had much higher H_2O_2-scavenging activities in the leaves and roots of both cotton cultivars in our study. The importance of these two enzymes in H_2O_2 scavenging has been demonstrated in several previous studies, in which increased CAT and APX activities were correlated with tolerance to salt and other environmental stresses [93]. These enzymes were also shown to be important in salt tolerance of barley and mulberry [94], [95]. Therefore, the results suggest that the coordination of CAT and APX with SOD activity could comprise an additional constituent in the enzymatic antioxidant mechanism of cotton plants against oxidative stress.

Conclusions

In this study, we compared the response of two cotton cultivars that differ with respect to salt tolerance to increasing NaCl concentrations. Overall, salinity significantly reduced the leaf and root dry weights, root volume, root length, root surface area, root average diameter, chlorophyll content and photosynthesis in the cotton plants of both cultivars. In contrast, antioxidant enzyme activity and proline and MDA contents increased in response to salinity. The salt-tolerant cultivar CCRI-79 showed evidence of possessing a more efficient antioxidant defense system against oxidative stress and lipid peroxidation by maintaining higher SOD, CAT, APX, and GR activities than those in Simian 3 during salt stress. The differences in the antioxidant enzyme activity of the leaves and roots may, at least in part, explain the greater tolerance to salt stress exhibited by CCRI-79 compared to that exhibited by Simian 3. Besides differences in antioxidant enzyme activities, the two cotton cultivars also showed marked variation in Chl a, Chl b, and Chl (a+b) contents, net photosynthetic rate, and stomatal conductance in response to NaCl stress. Therefore, acquisition of tolerance to salt may not only involve improved resistance to oxidative stress owing to enzymes that primarily function to protect membranes and tissues from such damage, but might also involve improvement in the biosynthetic pathway of photosynthetic pigments to maintain higher rates of photosynthesis in the face of stress. However, it should be noted that salt stress was only assessed at concentrations of 0, 80, 160, and 240 mM NaCl; therefore, further studies should be conducted to verify and screen the selection criteria for salt-tolerant species and cultivars. Nonetheless, the data presented

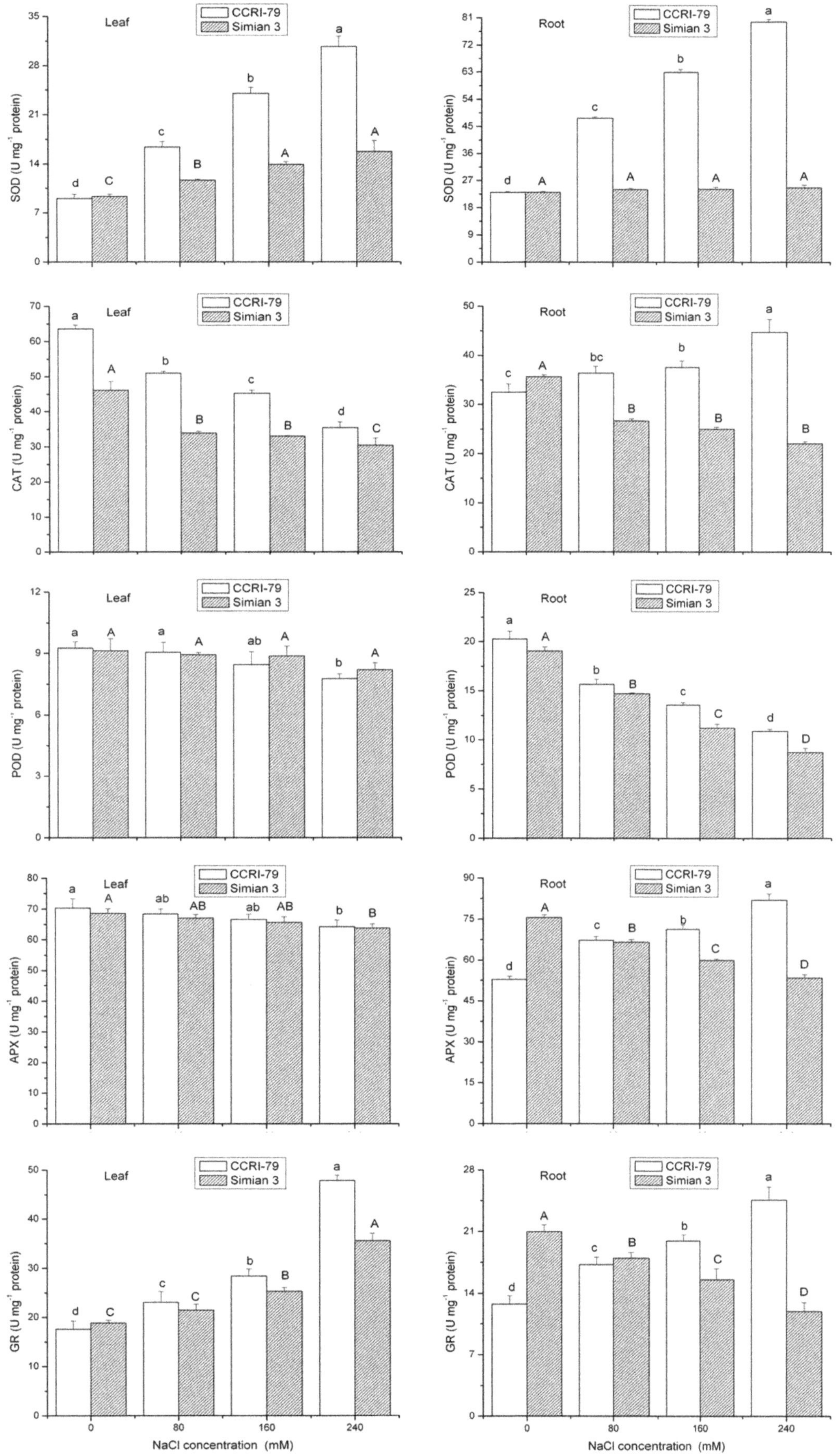
Leaf
Root
CCRI-79
Simian 3
SOD (U mg⁻¹ protein)
CAT (U mg⁻¹ protein)
POD (U mg⁻¹ protein)
APX (U mg⁻¹ protein)
GR (U mg⁻¹ protein)
0
80
160
240
NaCl concentration (mM)

Figure 7. Effect of NaCl salinity on the antioxidant enzyme activities in the roots and leaves of cotton. Vertical bars represent ± standard error (n = 3). Bars labeled with the different lowercase letters on open square bars or uppercase letters on closed square bars are significant difference (P<0.05). SOD, superoxide dismutase; CAT, catalase; POD, peroxidase; APX, ascorbate peroxidase; GR, glutathione reductase.

herein provide novel information on the mechanisms and traits involved in salt tolerance, which could be exploited for cultivar selection and breeding to increase crop production in the face of increased salinity stress.

Acknowledgments

The authors thank Dr. Zhiguo Zhou for critically reading this manuscript and technical assistance.

Author Contributions

Conceived and designed the experiments: SXY XHZ. Performed the experiments: LZ JP HJM TTC. Analyzed the data: LZ. Contributed reagents/materials/analysis tools: LZ JP TTC. Wrote the paper: LZ. Final approval of the version to be published: XHZ.

References

1. Munns R, Tester M (2008) Mechanisms of salinity tolerance. Annu Rev Plant Biol 59: 651–681.
2. Arzani A (2008) Improving salinity tolerance in crop plants: a biotechnological view. In Vitro Cell Dev Biol Anim 44: 373–383.
3. Gaber MA (2010) Antioxidative defense under salt stress. Plant Signaling &Behaviour 5: 369–374.
4. Khan MA, Ungar IA, Showalter AM (2000) Effects of salinity on growth, water relations and ion accumulation in the subtropical perennial halophyte, *Atriplex griffithii var. stocksil*. Ann Bot 85: 225–232.
5. Locy RD, Chang CC, Neilson BL, Singh NK (1996) Photosynthesis in salt-adapted heterotrophic tobacco cells and regenerated plants. Plant Physiol 110: 321–328.
6. Rout NP, Shaw BP (2001) Salt tolerance in aquatic macrophytes: possible involvement of the antioxidative enzymes. Plant Sci 160: 415–423.
7. Zheng C, Jiang D, Liu F, Dai T, Jing Q, et al. (2009) Effects of salt and water logging stresses and their combination on leaf photosynthesis, chloroplast ATP synthesis, and antioxidant capacity in wheat. Plant Sci 176: 575–582.
8. Foyer CH, Halliwell B (2000) Oxygen processing in photosynthesis: regulation and signaling. New Phytol 146: 359–388.
9. Ashraf M (2009) Biotechnological approach of improving plant salt tolerance using antioxidants as markers. Biotechnol Adv 27: 84–93.
10. Sigaud-Kutner TSC, Pinto E, Okamoto OK, Latorre LR, Colepicolo P (2002) Changes in superoxide dismutases activity and photosynthetic pigment content during growth of marine phytoplankters in batch-cultures. Plant Physiol 114: 566–571.
11. Gratao PL, Polle A, Lea PJ, Azevedo RA (2005) Making the life of heavy metal-stressed plants a little easier. Funct Plant Biol 32: 481–494.
12. Noctor G, Foyer CH (1998) Ascorbate and glutathione: keeping active oxygen under control. Annu Rev Plant Phys 49: 249–279.
13. Blokhina O, Virolainen E, Fagestedt KV (2002) Antioxidants, oxidative damage, and oxygen deprivation stress: A review. Ann Bot 91:179–194.
14. Gosset DR, Millhollon EP, Lucas MC (1994) Antioxidant response to NaCl stress in salt-tolerant and salt-sensitive cultivar of cotton. Crop Sci. 34: 706–714.
15. Blum A, Ebercon A (1981) Cell membrane stability as a measure of drought and heat tolerance in wheat. Crop Sci 21: 43–47.
16. Gosset DR, Millhollon EP, Lucas MC (1994) Antioxidant response to NaCl stress in salt-tolerant and salt-sensitive cultivars of cotton. Crop Sci 34: 706–714.
17. Dionisio-Sese ML, Tobita S (1998) Antioxidant response of rice seedlings to salinity stress. Plant Sci 135: 1–9.
18. Hernandez JA, Jimenez A, Mullineaux P, Sevilla F (2000) Tolerance of pea (*Pisum sativum* L.) to long-term salt stress is associated with induction of antioxidant defenses. Plant Cell Environ 23: 853–862.
19. Mittova V, Tal M, Volokita M, Guy M (2003) Up-regulation of the leaf mitochondrail and peroxisomal antioxidative systems in response to salt-induced oxidative stress in the wild salt-tolerant tomato species *Lycopersicon pennellii*. Plant Cell Environ 26: 845–856.
20. Meneguzzo S, Navario-Izzo F, Izzo R (1999) Antioxidative responses of leaves and roots of wheat to increasing NaCl concentrations. J Plant Physiol 155: 274–280.
21. Xiong L, Zhu JK (2002) Molecular and genetic aspects of plant responses to osmotic stress. Plant Cell Environ 25: 131–139.
22. Meng HB, Jiang SS, Hua SJ, Lin XY, Li YL, et al. (2011) Comparison between a tetraploid turnip and its diploid progenitor (*Brassica rapa* L.): the adaptation to salinity stress. ASC 10: 363–375.
23. Steduto P, Albrizio R, Giorio P (2000) Gas exchange response and stomatal and non-stomatal limitations to carbon assimilation of sunflower under salinity. Environ Exp Bot 44: 243–255.
24. Fedina IS, Tsoner T, Guleva EI (1993) The effect of pretreatment with proline on the response of *Pisum sativum* to salt stress. Photosynthetica 29: 521–527.
25. Flowers TJ, Yeo AR (1995) Breeding for salinity resistance in crop plants: Where next? Aust J Plant Physiol 22: 875–884.
26. Qidar M, Shams M (1997) Some agronomic and physiological aspects of salt tolerance in cotton (*Gossypium hirsutum* L.). Journal of Agronomy and Crop Science 179: 101–106.
27. Guo WX, Mass SJ, Bronson KF (2012) Relationship between cotton yield and soil electrical conductivity, topography, and Landsat imagery. Precision Agronomy 13(2): 678–692.
28. Zhang L, Zhang GW, Wang YH, Zhou ZG, Meng YL, et al. (2013) Effect of soil salinity on physiological characteristics of functional leaves of cotton plants. J Plant Res 126(2): 293–304.
29. Ahmads S, Khan N, Iqbal MZ (2002) Salt tolerance of cotton (*Gossypium hirsutum* L.). Asian Journal of Plant Science 1(6): 715–719.
30. Sestak K, Catsky J, Jarvis PG (1971) Plant photosynthetic production. Dr. W Junk N.V publishers The Hague.
31. Zheng Y, Wang Z, Sun X, Jia A, Jiang G, et al. (2008) Higher salinity tolerance cultivars of winter wheat relieved senescence at reproductive stage. Environ Exp Bot 62(2): 129–138.
32. Ashraf M, Orooj A (2006) Salt stress effects on growth, ion accumulation and seed oil concentration in an arid zone traditional medicinal plant ajwain (*Trachyspermum ammi* [L.] Sprague). J Arid Environ 64(2): 209–220.
33. Liu J, Guo WQ, Shi DC (2010) Seed germination, seedling survival, and physiological response of sunflowers under saline and alkaline conditions. Photosynthetica 48(2): 278–286.
34. Bernt E, Bergmeyer HU (1974) Inorganic peroxidases. In: Bergmeyer HU (ed) methods of enzymatic analysis, vol 4. Academic press, New York: 2246–2248.
35. Foster JG, Hess JL (1980) Responses of superoxide dismutase and glutathione reductase activities in cotton leaf tissue exposed to an atmosphere enriched in oxygen. Plant Physiol 66: 482–487.
36. Beers RF, Sizer IW (1952) A spectrophotometric method of measuring the breakdown of hydrogen peroxide by catalase. J Biol Chem 195: 133–140.
37. Tan W, Liu J, Dai T, Jing Q, Cao W, et al. (2008) Alterations in photosynthesis and antioxidant enzyme activity in winter wheat subjected to post-anthesis waterlogging. Photosynthetica 46: 21–27.
38. Nakano Y, Asada K (1981) Hydrogen peroxide is scavenged by ascorbate-specific peroxidase in spinach chloroplasts. PCPhy 22(5): 867–880.
39. Foyer CH, Halliwell B (1976) The presence of glutathione and glutathione reductase in chloroplasts: a proposed role in ascorbic acid metabolism. Planta 133: 21–25.
40. Takemura T, Hanagata N, Sugihara K, Baba S, Karube I, et al. (2000) Physiological and biochemical responses to salt stress in the mangrove, *Bruguiera gymnorrhiza*. Aquat Bot 68: 15–28.
41. Garratt LC, Janagoudar BS, Lowe KC, Anthony P, Power JB, et al. (2002) Salinity tolerance and antioxidant status in cotton cultures. Free Radical Biol Med 33(4): 502–511.
42. Meloni DA, Oliva MA, Ruiz HA, Martinez CA (2001) Contribution of proline and inorganic solutes to osmotic adjustment in cotton under salt stress. J Plant Nutr 24: 599–612.
43. Parida AK, Das AB (2005) Salt tolerance and salinity effects on plants: a review. Ecotoxicol Environ Saf 60: 324–349.
44. Curtis PS, Läuchli A (1986) The role of the leaf area development and photosynthetic capacity in determining growth of Kenaf under moderate salt stress. Aust J Plant Physiol 18: 553–565.
45. Thiel G, Lynch J, Läuchli A (1988) Short-term effects of salinity stress on the turgor and elongation of growing barley leaves. J Plant Physiol 132: 38–44.
46. Cuartero J, Fernandez-Munoz R (1999) Tomato and salinity. HortScience 78: 83–125.
47. Siroka B, Huttova J, Tamas L, Simonoviva M, Mistrik I (2004) Effect of cadmium on hydrolytic enzymes in maize root and celeopptile. Biologia 59: 513–517.
48. Berkelaar E, Beverley H (2000) The relationship between morphology and cadmium accumulation in seedlings of two durum wheat cultivars. Canadian Journal of Botany 78: 381–387.

49. Singh AK, Dubey RS (1995) Changes in chlorophyll a and b contents and activities of photosystems I and II in rice seedlings induced by NaCl. Photosynthetica 31: 489–499.
50. Munns R, James RA, Lauchli A (2006) Approaches to increasing the salt tolerance of wheat and other cereals. J Exp Bot 57: 1025–1043.
51. Zhang Q, Alfarra MR, Worsnop D, Allan JD, Coe H, et al. (2005) Deconvolution and quantification of hydrocarbon-like and oxygenated organic aerosols based on aerosol mass spectrometry. Environment Science and Technology 39: 4938–4952.
52. Lee MH, Cho EJ, Wi SG, Bae H, Kim JE, et al. (2013) Divergences in morphological changes and antioxidant responses in salt-tolerant and salt-sensitive rice seedlings after salt stress. Plant Physiol Biochem 70: 325–335.
53. de Lacerda CF, Cambraia J, Oliva MA, Ruiz HA, Prisco JT (2003) Solute accumulation and distribution during shoot and leaf development in two sorghum genotypes under salt stress. Environ Exp Bot 49(2): 107–120.
54. Meloni DA, Oliva MA, Martinez CA, Cambraia J (2003) Photosynthesis and activity of superoxide dismutase, peroxidase and glutathione reductase in cotton under salt stress. Environ Exp Bot 49(1): 69–76.
55. Asharf M, Foolad MR (2005) Pre-sowing seed treatment- a shotgun approach to improve generation, plant growth, and crop yield under saline and non-saline conditions. Advances in Agronomy 88: 223–271.
56. Dubey RS (1997) Photosynthesis in plants under stressful conditions. In: Pessarakli, M. (Ed.), Handbook of Photosynthesis. Marcel Dekker, New York,: 859–975.
57. Brugnoli E, Lauteri M (1991) Effects of salinity on stomatal conductance, photosynthetic capacity and capacity and carbon isotope discrimination of salt-tolerant (*Gossypium hirsutum* L.) and salt-sensitive bean (*Phaseilus vulgaris* L.) C3 non-halophytes. Plant Physiol 95: 658–635.
58. Delfine S, Alvino A, Villani MC, Loreto F (1999) Restrictions to carbon dioxide conductance and photosynthesis in spainach leaves recovering from salt stress. Plant Physiol 119: 1101–1106.
59. Parida A, Das AB, Mittra B (2004) Effects of salt on growth, ion accumulation, photosynthesis and leaf anatomy of the mangrove, *Bruguiera parviflora*. Trees: Structure and Function 18: 167–174.
60. Kaya C, Tuna AL, Asharf M, Altunlu H (2007) Improved salt tolerance of molon (*Cucumis melo* L.) by the addition of proline and potassium nitrate. Environ Exp Bot 60: 397–403.
61. Borsani O, Valpuesta V, Botella MA (2003) Developing salt tolerant plants in a new century: a molecular biology approach. Plant Cell Tissue and Organ Culture 73: 101–115.
62. Asgari HR, Cornelis W, Damme PV (2012) Salt stress on wheat (*Triticum aestivum* L.) growth and leaf ion concentrations. International Journal of Plant Production 6: 195–208.
63. Taleisnik E, Grunberg K (1994) Ion balance in tomato cultivars differing in salt tolerance. I. Soidium and potassium accumulation and fluxes under moderate salinity. Plant Physiol 92: 528–534.
64. Graifenberg A, Giustiniani L, Temperini O, Paola ML (1995) Allocation of Na, Cl, K and Ca within plant tissues in globe artichoke (*Cynara scolimus* L.) under saline-sodic conditions. Scientia Horticulturae 63: 1–10.
65. Khan MA (2001) Experimental assessment of salinity tolerance of *Ceriops tagal* seedlings and saplings from the *Indus delta*. Pakistan Journal of Acquatic Botany 70: 259–268.
66. Munns R, Schachtman DP, Condon AG (1995) The significance of a two-phase growth response to salinity in wheat and barely. Aust J Plant Physiol 22: 561–569.
67. Niu X, Bressan RA, Hasegawa PM, Pardo JM (1995) Ion homeostasis in NaCl stress environments. Plant Physiol 109: 735–742.
68. Hernandez JA, Jimenez A, Mullineaux P, Sevilla F (2002) Tolerance of pea (*Pisum sativum* L.) to long-term salt stress is associated with induction of antioxidant defenses. Plant Cell Environ 23: 853–862.
69. Boo YC, Jung J (1999) Water deficit-induced oxidative stress and antioxidative defense in rice plants. Plant Physiol 155: 255–261.
70. Seckin B, Turkan I, Sekmen AH, Ozfidan C (2010) The role of antioxidant defense systems at differential salt tolerance of *Hordeum marinum* Hds. (sea barlygrass) and *Hordeum vulgare* L. (cultivated barley). Environ Exp Bot 69: 76–85.
71. Ruiz-Lozano JM, Azcon R (1997) Effect of calcium application on the tolerance of mycorrhizal lettuce plants to polyethylene glycol-induced water stress. Symbiosis 23: 9–21.
72. Rabie GH, Almadini AM (2005) Role of bioinoculants in development of salt-tolerance of *Vicia faba* plants under salinity stress. Africa Journal of Biotechnology 4: 210–220.
73. Sharma SS, Dietz KJ (2006) The significance of amino acids and amino acid derived molecules in plant responses and adaptation to heavy metal stress. J Exp Bot 57: 711–726.
74. Foyer CH, Lopez-Delgado H, Dat JF, Scott IM (1997) Hydrogen peroxide and glutathione-associated mechanism of acclimatory stress tolerance and signaling. Plant Physiol 100: 241–254.
75. Senadheera P, Tirimanne TLS, Maathuis FJM (2012) Long term salinity stress reveals variety specific differences in root oxidative stress response. Rice Sci 19: 36–43.
76. Lee DH, Kim YS, Lee CB (2001) The inductive responses of the antioxidant enzymes by salt stress in the rice (*Oryza sativa* L.). J Plant Physiol 158(6): 737–745.
77. Queval G, Hager J, Gakiere B, Noctor G (2008) Why are literature data for H_2O_2 contents so variable? A discussion of potential difficulties in the quantitative assay of leaf extracts. J Exp Bot 59: 135–146.
78. Morita S, Kaminaka H, Masumura K, Tanaka K (1999) Induction of rice cytosolic ascorbate peroxidase mRNA by oxidative stress: the involvement of hydrogen peroxide in oxidative stress signaling. Plant Cell Physiol 40: 417–422.
79. Breusegem FV, Vranova E, Dat JF, iNZE D (2001) The role of active oxygen species in plant signal transduction. Plant Sci 161: 405–414.
80. Sekmen AH, Turkan I, Takio S (2007) Differential responses of antioxidative enzymes and lipid peroxidation to salt stress in salt-tolerant Plantago maritima and salt-sensitive Plantago media. Plant Physiol 131: 399–411.
81. Mittler R (2002) Oxidative stress, antioxidants and stress tolerance. Trends in Plant Science 7(9): 405–410.
82. Gill SS, Tuteja N (2010) Reactive oxygen species and antioxidant machinery in abiotic stress tolerance in crop plants. Plant Physiol Biochem 48: 909–930.
83. Henandez JA, Jimenez A, Mullineaux P, Sevilla F (2000) Tolerance of pea to long-term salt stress in associated with induction of antioxidant defenses. Plant Cell Environ 23: 853–862.
84. Laspina NV, Groppa MD, Tomaro ML, Benavides MP (2005) Nitric oxide protects sunflower leaves against Cd-induced oxidative stress. Plant Sci 169: 323–330.
85. Zhang Q, Zhang JZ, Chow WS, Sun LL, Chen JW, et al. (2011) The influence of low temperature on phtosynthesis and antioxidant enzymes in sensitive banana and tolerant plantain (*Musa sp.*) cultivars. Photosynthetica 49: 201–208.
86. Asada K (2006) Production and scavenging of reactive oxygen species in chloroplasts and their functions. Plant Physiol 141: 391–396.
87. Amor NB, Jimenez A, Megdiche W, Lundqvist M, Sevilla M, et al. (2007) Kinetics of the antioxidant response to salinity to the halophyte *Cakile maritime*. J Integrative Plant Biol 126: 446–457.
88. Ben Amor N, Ben Hamed K, Debez A, Grignon C, Abdelly C (2005) Physiological and antioxidant responses of the perennial halophyte *Crithmum maritimum* to salinity. Plant Sci 168(4): 889–899.
89. Xu R, Yamada M, Fujiyama H (2013) Lipid peroxidation and antioxidative enzymes of two turgrass species under salinity stress. Pedosphere 23(2): 213–222.
90. Nakano Y, Asada K (1981) Hydrogen peroxide is scavenged by ascorbate-specific peroxidase in spinsth chloroplasts. PCPhy 22: 867–880.
91. Bower C, Montagu MV, Inze D (1992) Superoxide dismutase and stress tolerance. Annu Rev Plant Physiol Plant Mol Biol 43: 81–116.
92. Demiral T, Turkan I (2005) Comparative lipid peroxidation, antioxidant defense systems and proline content in roots of two rice cultivars differing in salt tolerance. Environ Exp Bot 53: 247–257.
93. Locato V, Gadaleta C, Gara I, Pinto MC (2008) Production of reactive species and modulation of antoxidant network in response to heat shock: a critical balance for cell fate. PCPhy 31: 1606–1619.
94. Liang Y, Chen Q, Liu Q, Zhang W, Ding R (2003) Exogenous silicon (Si) increases antioxidant enzyme activity and reduces lipid peroxidation in roots of salt-stressed barley (*Hordeum vulgare* L.). J Plant Physiol 160: 1157–1164.
95. Sudhakar C, Lakshmi A, Giridarakumar S (2001) Changes in the antioxidant enzyme efficiency in two high yielding genotypes of mulberry (*Morus alba* L.) under NaCl salinity. Plant Sci 161: 613–619.

18

Transcriptome-Wide Profiling and Expression Analysis of Diploid and Autotetraploid *Paulownia tomentosa* × *Paulownia fortunei* under Drought Stress

Enkai Xu[1,2], Guoqiang Fan[1,2]*, Suyan Niu[1,2], Zhenli Zhao[1,2], Minjie Deng[1,2], Yanpeng Dong[1,2]

1 Institute of Paulownia, Henan Agricultural University, Jinshui Area, Zhengzhou, Henan, P.R. China, **2** College of Forestry, Henan Agricultural University, Jinshui Area, Zhengzhou, Henan, P.R. China

Abstract

Paulownia is a fast-growing deciduous hardwood species native to China, which has high ecological and economic value. In an earlier study, we reported ploidy-dependent differences in Paulownia drought tolerance by the microscopic observations of the leaves. Autotetraploid Paulownia has a higher resistance to drought stress than their diploid relatives. In order to obtain genetic information on molecular mechanisms responses of Paulownia plants to drought, Illumina/Solexa Genome sequencing platform was used to *de novo* assemble the transcriptomes of leaves from diploid and autotetraploid *Paulownia tomentosa* × *Paulownia fortunei* seedlings (PTF2 and PTF4 respectively) grown under control conditions and under drought stress and obtained 98,671 nonredundant unigenes. A comparative transcriptome analysis revealed that hundreds of unigenes were predicted to be involved mainly in ROS-scavenging system, amino acid and carbohydrate metabolism, plant hormone biosynthesis and signal transduction, while these unigenes exhibited differential transcript alteration of the two accessions. This study provides a comprehensive map of how *P. tomentosa* × *P. fortunei* responds to drought stress at physiological and molecular levels, which may help in understanding the mechanisms involve in water-deficit response and will be useful for further study of drought tolerance in woody plants.

Editor: Keqiang Wu, National Taiwan University, Taiwan

Funding: This work was supported by GQF, National Natural Science Foundation of China (Grant no. 30271082, 30571496,U1204309, http://www.nsfc.gov.cn/), GQF, Outstanding Talents Project of Henan Province (Grant no. 122101110700), GQF, Transformation Project of the National Agricultural Scientific and Technological Achievement of China (Grant no. 2012GB2D000271), and GQF, Science and Technology Innovation Team Project of Zhengzhou City (Grant no. 121PCXTD515). The funders had no role in study design, data collection and analysis, decision to publish, or preparation of the manuscript.

Competing Interests: The authors have declared that no competing interests exist.

* Email: zlxx64@126.com

Introduction

Plants are constantly exposed to a wide range of environmental stresses such as drought, high salt, heat, and extremes of temperature. Water scarcity is one of the major environmental stresses that causes substantial losses in many plants [1]. It affects every aspect of plant growth, modifying anatomy, morphology, physiology, and biochemistry, which negatively affects plant growth and productivity [2].

Global climate change will likely reduce water availability further increasing the need for drought tolerant plants [3]. To adapt to drought stress, trees have evolved complex mechanisms based on modifications in metabolites, gene expression, and proteins. Clearly, drought tolerance is a complex trait that is regulated by multiple genes. Based on their putative biological functions in the drought response process, drought-responsive genes have been classified into two groups: (1) genes encoding proteins that defend the cells from the effects of water-deficit, for example, proteins that combat the accumulation of solutes (e.g., enzymes for biosynthesis of osmolytes like proline, betaine, sugars), passive and active transport systems across membranes (e.g., water channel proteins and membrane transporters), proteins that protect and stabilize cell structures from damage by reactive oxygen species (e.g., detoxification enzymes such as catalase and superoxide dismutase), and other proteins that protect the macromolecules (e.g., late embryogenesis-abundant (LEA) proteins and osmotin) [4–6]; and (2) genes encoding proteins that regulate stress signal transduction and alter gene expression (e.g., transcription factors, protein kinases, protein phosphatases, proteinases, and phytohormones) [7,8].

The underlying mechanisms of the drought stress response in plants have been investigated using various genomic and genetic tools. Cohen et al. [9] assessed the genotype-dependent component of a drought-induced transcriptome response in two poplar genotypes with different drought tolerances using a multi-species designed microarray. Zheng et al. [10] compared genome-wide gene expression profiles between drought-tolerant and drought-sensitive lines of maize during drought stress by microarray.

High-throughput sequencing is a powerful and cost-efficient tool that has been used effectively in many areas, including the study of the transcriptomes of plants under biotic and abiotic stresses. Over 24,000 drought-responsive genes were identified from *Populus euphratica* leaves using a Roche/454 platform [11]. The transcriptomes of 22 different developmental stage tissues from

Cicer arietinum L. were sequenced using a Illumina/Solexa system, which revealed thousands of drought-responsive genes [12]. Using Illumina/Solexa sequencing, Peng et al. [13] reported differences in the transcriptome levels between male and female *Populus yunnanensis* in response to drought stress.

Paulownia is a species of tree that is native to China, where it has been cultivated for over 2000 years. This species has also been planted in many other countries and can now be found throughout the world, except Antarctica [14]. Paulownia is an adaptable and fast-growing deciduous tree, which provides raw material for manufacturing furniture, plywood, and musical instruments. It has a unique biological characteristic of late leaf emergence and fall, a deep root system, and sparse branching; therefore, exemplary areas of agroforestry have been built on the broad plains in North-Central China [15]. Paulownia has a very large leaf area with a high transpiration rate; therefore, sufficient moisture is very important for Paulownia growth. *Paulownia tomentosa* × *Paulownia fortunei* is grown mainly in hilly loess areas or in salinized soil [16] where the annual precipitation is below 500 mm. *P. tomentosa* × *P. fortunei* grow quite well without artificial irrigation because of their inherent ability to withstand drought. Therefore, *P. tomentosa* × *P. fortunei* is an ideal species in which to study the mechanisms of drought tolerance.

Polyploidization of chromosomes is thought to be one of the most important mechanisms for species evolution [17]. Polyploidy usually changes plant morphological and anatomical characteristics introducing, for example, superior breast-height diameter, volume, leaf, fruit, and stoma [18–20]. Polyploidy also changes the physiology, biochemistry, and gene expression in plants [17,21,22]. Several studies have reported that polyploids exhibit higher stress resistance compared with their corresponding diploid relatives [23–25].

To benefit from the possible value of polyploidy, in 2010, an autotetraploid *P. tomentosa* × *P. fortunei* ($2n = 4x = 80$) was induced from the leaves of diploid *P. tomentosa* × *P. fortunei* ($2n = 2x = 40$) using colchicines (an established system of *in vitro* plantlet regeneration) and a detailed physiological study of diploid and autotetraploid plants was conducted [26].

Drought tolerance varies considerably between diploid and autotetraploid Paulownia. In an earlier paper, we studied diploid and autotetraploid Paulownia plants for their drought and cold tolerance by microscopic observations of the leaves [27]. However, little is known about the molecular mechanisms of diploid and autotetraploid Paulownia under drought conditions.

In the present study, to reveal the transcriptome-wide response to drought stress in *P. tomentosa* × *P. fortunei* and to investigate diploid and autotetraploid differences in leaf transcriptome remodeling during drought stress, we analyzed the gene expression profiles of diploid and autotetraploid seedlings using an Illumina/Solexa Genome Analyzer IIx (GAIIx) platform. To the best of our knowledge, this is the first comprehensive analysis of differentially expressed genes in drought-stressed *P. tomentosa* × *P. fortunei*. A comparison of the diploid and autotetraploid transcriptomes revealed differences and similarities, which helped identify stress-related candidate genes and gene regulatory networks for future screening. The acquired information will provide new insights into the molecular mechanisms involved in the *P. tomentosa* × *P. fortunei* response to drought stress.

Materials and Methods

Plant material

Uniformly grown tissue culture seedlings of diploid *P. tomentosa* × *P. fortunei* (PTF2) and tetraploid *P. tomentosa* × *P. fortunei* (PTF4) were collected from the Institute of Paulownia, Henan Agricultural University, Zhengzhou, Henan Province, China and cultivated in nutrition blocks containing ordinary garden soil for 30 days. Samples with the same crown size and height were then planted in plastic pots (20 cm in diameter at the bottom and 20 cm deep) containing ordinary garden soil, one for each plant. The plants were housed randomly in an outdoor nursery for 50 days before being subjected to stress treatment. Twelve plants from each accession (PTF2 and PTF4) were obtained, including three controls and nine plants for water-deficit treatment (divided into three groups, each group contained three individual biological replicates).

The seedlings were subjected to drought conditions in a water-controlled experiment according to the method of Moumeni et al. [28]; the controls were watered to field capacity every day. For drought treatment, the seedlings of the two accessions were subjected to mild (5 days without irrigation, 55% relative soil water content), moderate (15 days without irrigation, 40% relative soil water content), and severe (20 days without irrigation, 25% relative soil water content) drought stress. The mature, fully expanded leaves (second leaf from the apex) were sampled and mixed in equal numbers from each plant in each of the replicate groups. The leaf tissue was frozen in liquid nitrogen and then stored at −80°C for later use.

Physiological measurements

Phenotypic and physiological characteristics of the PTF2 and PTF4 seedlings were measured under the control, mild, moderate, and severe drought stresses.

Physiological measurements were made between 8.30 am and 11.00 am on single, fully expanded leaves (third and fourth leaves from the terminal bud of a twig) immediately after excision. The measurements were taken on three replicates of the leaves from the same plants.

The percent relative water content of the leaf was calculated as (fresh weight - dry weight)/(turgid weight - dry weight) × 100 [29]. Chlorophyll content was measured according to a described method previously [30] and calculated as described by Arnon [31]. The malondialdehyde concentration was measured using the thiobarbituric acid method as described by Hodges et al. [32]. Relative electrical conductivity was measured as described previously [33]. Superoxide dismutase activity was determined using the nitroblue tetrazolium method reported by Dhindsa et al. [34]. Values were expressed in units/gmHb [35]. Total soluble protein content was measured following the procedure of Guy et al. [36], and the protein concentration was measured as described by Bradford [37] using bovine serum albumin (Sigma, USA) as the standard. Proline content was measured using the acid ninhydrin method of Bates et al. [38]. Soluble sugar content was measured using the anthrone colorimetry method described by Irigoyen et al. [39].

Statistical analysis

Analyses of variance (ANOVA) of the parameters were performed using the MSTAT-C computer program (MSTAT Development Team, Michigan State University, 1989) and SPSS (v. 16.5). The mean values of each of the measured traits were compared using the least significance difference (LSD) test at a 5% level of probability according to Snedecor and Cochran [40].

RNA extraction and double-stranded cDNA synthesis

The PTF2 and PTF4 seedlings from the severe drought-stressed group were chosen for sequencing based on the physiological measurements (see the results section for details).

Approximately 8 mg of leaves from the diploid control (well-watered PTF2) and the severe drought-treated PTF2, and 8 mg of leaves from the tetraploid control (well-watered PTF4 control) and the severe drought-treated PTF4 in three replications were used for total RNA extraction.

Total RNA was extracted from the four samples with TRIzol reagent (Invitrogen, Carlsbad, CA). RNA was purified using RNeasy MiniElute Cleanup Kit (Qiagen, Valencia, CA) according to the manufacturer's protocol. Magnetic beads with Oligo (dT) were used to isolate poly-(A) mRNA from the total RNA. The mRNA was mixed with fragmentation buffer and fragmented into short fragments. The first-strand cDNA was synthesized with SuperScript II reverse transcriptase (Life Technologies, Carlsbad, CA), and DNA polymerase I and RNase H were then used to synthesize the second-strand cDNA. The short cDNA fragments were purified and dissolved in EB buffer for end reparation and addition of a single nucleotide A (adenine). The short fragments were then connected to adapter sequences. Suitable fragments were selected as templates for PCR amplification. The cDNA libraries were qualified using a 2100 Bioanalyzer (Agilent Technologies, Inc., Santa Clara, CA), then sequenced on a Illumina Genome Analyzer IIx (GAIIx) platform (Illumina, San Diego, CA) by Beijing Genomics Institute (BGI)-Shenzhen, Shenzhen, China, following the manufacturer's standard cBot and sequencing protocols.

Assembly and annotation

The raw image data were transformed by base calling into sequence data (raw reads). Raw reads that contained adaptor sequences, ambiguous reads, or low-quality reads were removed using DynamicTrim in the SolexaQA package (http://solexaqa.sourceforge.net/) to create clean reads. The clean reads used in this publication have been deposited in the NIH Short Read Archive (SRA) database (http://www.ncbi.nlm.nih.gov/sra) under accession number SRP041407. The quality sequences were assembled into unigenes using the Trinity program [41]. Unigenes from multiple samples from the same accession were further processed by removing redundancy and splicing, and then non-redundant unigenes without gaps were clustered using the TIGR Gene Indices clustering tools (TGICL) [42]. The unigene sequences were annotated by aligning them against the NCBI non-redundant protein sequence database (Nr) (http://www.ncbi.nlm.nih.gov/), Swiss-Prot (http://www.expasy.ch/sprot/), Cluster of Orthologous Groups (COG) (http://www.ncbi.nlm.nih.gov/cog/), and the Kyoto Encyclopedia of Genes and Genomes (KEGG) (http://www.genome.jp/kegg/) using BLASTX (E-value <1.0E-5), and then against the NCBI nucleotide sequence database (Nt) using BLASTN (E-value <1.0E-5).

The best alignments were used to annotate the unigenes and to decide the coding region and sequence direction of the unigenes. Non-annotated unigenes with no matches in any of the searched databases were scanned using ESTScan [43] to predict the sequence (5′–3′) direction and to predict the coding region to obtain the translated amino sequence. Gene Ontology (GO) functional classification of the unigenes was performed using Blast2GO [44] and WEGO software [45]. Pathway annotations were obtained by aligning the unigenes with the KEGG database [46].

Unigene expression difference analysis

Unigene expression in each library was calculated using the FPKM (fragments per kb per million fragments) method [47] as:

$$FPKM = \frac{10^6 C}{NL/10^3}$$

where *FPKM* is the expression of unigene A, C is the number of fragments that aligned uniquely to unigene A, *N* is the total number of fragments that aligned uniquely to all the unigenes, and *L* is the number of bases in the coding sequence of unigene A.

Differentially expressed unigenes (DEUs) between the PTF2 and PTF4 transcriptomes were identified using the rigorous algorithm described by Audic et al. [48].

To screen the DEUs, the p-value threshold in multiple hypothesis testing and analyses was determined by manipulating the false discovery rate (FDR) value [49]. A FDR ≤ 0.001 and absolute value of the |log2Ratio| >2 were used as the threshold to determine the significance of the differences in gene expression [49]. The DEUs were then subjected to GO functional and KEGG Pathway analyses.

GO and pathway enrichment analysis

GO functional enrichment analysis was performed to classify the GO functional annotations for the DEUs. First, all the DEUs were mapped to terms in the GO database (http://www.geneontology.org/) and the gene numbers for each GO term were calculated to obtain a gene list and the gene numbers for every mapped GO term. The GO enrichment analysis was performed by applying the hypergeometric test to find significantly enriched GO terms in the DEUs compared with the whole transcriptome background of *P. tomentosa* × *P. fortunei*. The p-value was calculated as:

$$p = 1 - \sum_{i=0}^{m-1} \frac{\binom{M}{i}\binom{N-M}{n-i}}{\binom{N}{n}}$$

where N is the total number of genes with GO annotation; n is the number of DEUs in N; M is the total number of all genes that were annotated to a certain GO term; and m is the number of DEUs in M. After applying the Bonferroni correction to the calculated p-value we chose a corrected p-value of ≤ 0.05 as the threshold, and GO terms with $p \leq 0.05$ were defined as significantly enriched GO terms for the DEUs.

Similarly, a pathway enrichment analysis to identify significantly enriched biochemical or signal transduction pathways associated with the DEUs was performed by comparing the pathway annotations with the whole transcriptome background of *P. tomentosa* × *P. fortunei*. A Q-value, defined as the FDR, was used to correct the p-value. After multiple test corrections, we chose a Q-value of ≤ 0.05 to identify significantly enriched pathways for the DEUs.

Real-time quantitative PCR analysis of potential drought response DEUs

A real-time quantitative PCR (qRT-PCR) analysis was conducted on selected DEUs to validate the Illumina paired-end sequencing results. *P. tomentosa* × *P. fortunei* leaves collected from the severe drought and control PTF2 and PTF4 plants were used to extract total RNA. The first-strand cDNA was synthesized from 2 mg total RNA with the iScriptc DNA synthesis kit (Bio-Rad, Hercules, CA) according to the manufacturer's instructions.

The specific primers used for the real-time PCR assay were designed using Beacon Designer, version 7.7 (Premier Biosoft International, Ltd., Palo Alto, CA) and are listed in Table S1. The cDNA was amplified using a CFX96TM Real-Time System (Bio-Rad) with So Fast Eva Green Supermix (Bio-Rad) according to the manufacturer's instructions. PCR mixtures with a final volume of 20 μL contained 10 μL SYBR Green PCR mix, 1 μL cDNA, 0.4 μM forward primer, 0.4 μM reverse primer, and 7.0 μL SteriLe ddH$_2$O. The PCR cycles were as follows: 95°C for 1 min, followed by 40 cycles of 95°C for 10 s, and 55°C for 15 s. Relative expression levels of the genes were calculated using the $2^{-\Delta\Delta Ct}$ method and normalized with 18S rRNA from *P. tomentosa* × *P. fortunei*.

Results

Physiological characterization of P. tomentosa × P. fortunei under drought stress

The phenotypic and physiological responses of PTF2 and PTF4 to water deficit were assessed using seedlings exposed to progressive water loss for about three weeks.

There were no significant differences in stress phenotype between the PTF2 and PTF4 under the well-watered conditions (Figure 1a and 1b). The leaves of the PTF2 seedlings drooped after five days of drought stress (Figure 1c), whereas the leaves of the PTF4 seedlings remained green and turgid (Figure 1d). After 15 days, the PTF2 plants showed stunted growth and had leaves that were paler and more wilted compared with the PTF4 plants (Figure 1e and 1f); after 20 days, the PTF2 and PTF4 plants both showed distinct effects of drought stress. However, the PTF4 plants showed much better development, less wilting, and higher biomass compared with the PTF2 plants (Figure 1 g and 1 h).

In general, the PTF4 seedlings maintained markedly higher growth vigor under the drought stress conditions than the PTF2 seedlings.

Changes in physiological parameters in PTF2 and PTF4 plants in response to drought

Drought stress significantly inhibited the relative water content (RWC) and chlorophyll content of the plants. The RWC of PTF2 was similar to that of PTF4 under well-watered conditions. The RWC of both accessions declined with the increase of the drought, as characterized by their significantly lower RWC compared with the corresponding controls (well-watered), particularly under the severe drought treatment. However, after 20 days of drought, the PTF2 plants showed a sharp decrease in RWC (13.25%) compared with the RWC (12.23%) of the PTF4 plants (Figure 2a). PTF2 and PTF4 were recorded the highest chlorophyll content under well-watered conditions while, during the drought treatment, the chlorophyll content significantly decreased in both accessions. The chlorophyll content also obviously showed difference between PTF2 and PTF4 plants, and the PTF4 seedlings always contained higher levels of total chlorophyll (chlorophyll a and b) than the PTF2 seedlings under drought stress (Figure 2b).

Relative electrical conductivity (REC) and malondialdehyde (MDA) content showed overall increased tendencies in the PTF2 and PTF4 plants under drought stress. However, after 20 days of drought stress the REC and MDA values increased significantly in both accessions, whereas the REC and MDA in the PTF2 plants were always at much higher levels than in the PTF4 plants over the entire treatment process (Figure 2c and 2d).

Compared to the control conditions, both the PTF2 and PTF4 plants under mild and moderate drought conditions showed a substantial increase in superoxide dismutase (SOD) activity and soluble proteins content, and then decreased dramatically under severe drought stress, whereas the values of these two traits were higher in both PTF2 and PTF4 plants under severe drought than under well-watered conditions. Besides, SOD activity and soluble proteins content measurements showed striking differences between the two accessions under the same drought treatments. In particular, PTF4 showed higher SOD activity and soluble proteins content than PTF2 at the end of the 20-day experiment (Figure 2e and 2f).

Additionally, the proline and soluble sugar contents tended to increase as the drought strengthened in the two accessions. During the different drought treatments, the proline and soluble sugar contents in PTF4 increased significantly, but in PTF2 they showed relatively little changes. Overall, the PTF4 plants showed much higher levels of the proline and soluble sugar contents compared with the PTF2 plants (Figure 2 g and 2 h).

Paired-end sequencing and de novo assembly of the P. tomentosa × P. fortunei reads

To understand the molecular mechanisms of drought tolerance in *P. tomentosa* × *P. fortunei*, four leaf cDNA libraries, two for the well-watered controls (PTF2W, PTF4W) and two for the severe drought-treated plants (PTF2T, PTF4T), were constructed and sequenced on the GAIIx platform. Over 291 million raw reads with a mean length of 110 nt, were generated from the four libraries; approximately 73 million reads from each sample. After removing adapter sequences, duplication sequences, and ambiguous and low-quality reads, 264 (23.78 Gb) million clean reads

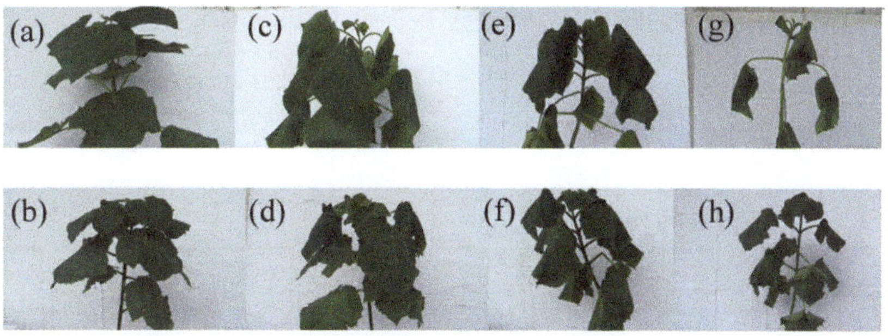

Figure 1. Physiological feature of diploids and autotetraploid in response to drought. (a) PTF2W, well-watered diploid, 75% relative soil water content; (b) PTF4W, well-watered tetraploid, 75% relative soil water content; (c) PTF2T-1, 5-day drought treated diploid, 55% relative soil water content; (d) PTF4T-1, 5-day drought treated tetraploid, 55% relative soil water content; (e) PTF2T-2, 15-day drought treated diploid, 40% relative soil water content; (f) PTF4T-2, 15-day drought treated tetraploid, 40% relative soil water content; (g) PTF2T-3, 20-day drought treated diploid, 25% relative soil water content; (h) PTF4T-3, 20-day drought treated tetraploid, 25% relative soil water content.

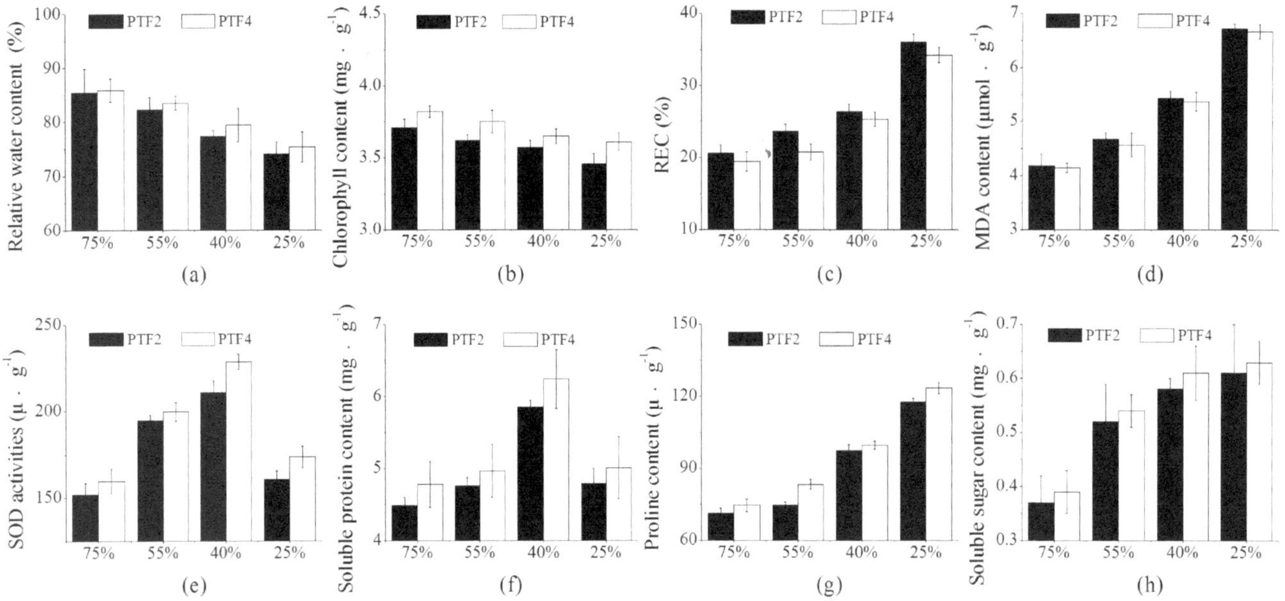

Figure 2. Effects of drought stress on *P. tomentosa* × *P. fortunei* physiology. PTF2 represents diploid *P. tomentosa* × *P. fortunei*, PTF4represents autotetraploid *P. tomentosa* × *P. fortunei*, 75%, 55%, 40% and 25% relative soil water contents were used. (a) Effect of drought stress on leaf relative water content;(b) Effect of drought stress on leaf chlorophyll content; (c) Effect of drought stress on leaf relative electrical conductivity; (d) Effect of drought stress on leaf malondialdehyde (MDA) content; (e) Effect of drought stress on leaf superoxide dismutase (SOD) activities; (f) Effect of drought stress on leaf soluble protein content; (g) Effect of drought stress on leaf proline content; (h) Effect of drought stress on leaf soluble sugar content.

remained. The Q20 (sequencing error rate, 1%) percentage, N percentage, and GC content were 97.43%, 0%, and 49.04%, respectively (Table 1). The *de novo* assembly of the clean reads using Trinity generated 599,908 contigs without gaps, with an average length of 304 nt and an N50 of 471 nt. After clustering using the TGICL software, a total of 98,671 nonredundant unigenes were produced, including 47,582 clusters and 51,089 singletons. The total length of the unigenes was 96,940,229 nt, with a mean length of 982 nt and an N50 of 1626 nt (Table 2); 63,015 (63.86%) unigenes were longer than 500 nt and 38,134 (38.65%) were longer than 1,000 nt. The length distribution of the unigenes is shown in Figure S1. None of the unigenes had gaps, indicating that the sequences were of high quality. All the assembled unigenes were aligned against the Nr, Swiss-Prot, KEGG, and COG protein databases, and a total of 70,524 coding sequences (CDSs) were predicted; 67,605 CDSs were inferred using BLASTX hits and 2,919 were assigned using ESTScan.

All the statistics showed that the sequencing of the *P. tomentosa* × *P. fortunei* transcriptomes was of high depth and that high quality reads were produced.

Annotation and functional analysis of the unigenes

The BLAST searches against the various databases yielded a total of 67,326 (68.23%) unigenes that found matches against the Nr database, followed by Nt (58,508, 59.30%), GO (55,413, 56.16%), Swiss-Prot (43,851, 44.44%), KEGG (41,206, 41.76%), and COG (28,165, 28.55%) (Table S2).

Overall, 70,100 (71.04%) unigenes shared similarities with entries in one or more of the six databases. No matches were found for 28,571 of the unigene sequences in any of these databases (E-value >10-5), suggesting that some of these unigenes may serve as a source for novel gene discovery.

Based on the Nr annotations, we determined that 34.0% of the mapped unigenes had very significant homology to known sequences (E-value <10–100), 35.9% showed significant homology (10–100<E-value <10–30), and 30.1% showed weak homology (E-value 10–30 to 10–5) (Figure 3A); 25.1% of unigenes shared >80% similarity with known sequences, 67.8% shared between 40% and 80% similarity, and 7.2% shared 19% to 40% similarity (Figure 3B). Further, the unigenes shared strong homology with sequences from *Vitis vinifera* (46.0%), *Ricinus communis* (12.8%), *Populus trichocarpa* (10.8%), *Glycine max* (5.9%), and *Nicotiana tabacum* (2.0%) (Figure 3C).

Table 1. Summary of the sequence assembly after Illumina sequencing.

Samples	Total Raw Reads	Total Clean Reads	Total Clean Nucleotides (nt)	Q20 percentage	N percentage	GC percentage
PTF2W	72,147,152	65,657,050	5,909,134,500	97.50%	0.00%	46.42%
PTF2 T	72,176,810	65,227,994	5,870,519,460	97.42%	0.00%	46.07%
PTF4W	76,705,742	69,169,960	6,225,296,400	97.43%	0.00%	46.90%
PTF4 T	70,571,228	64,168,586	5,775,172,740	97.37%	0.00%	47.29%

Table 2. Summary statistics of contig and unigene in four libraries from *P. tomentosa* × *P. fortune*.

	Sample	Total Number	Total Length(nt)	Mean Length(nt)	N50	Total Consensus Sequences	Distinct Clusters	Distinct Singletons
Contig	PTF2W	144,755	45,259,057	313	508	-	-	-
	PTF2 T	161,799	47,608,475	294	441	-	-	-
	PTF4W	143,358	45,001,926	314	494	-	-	-
	PTF4 T	149,996	44,356,767	296	442	-	-	-
Unigene	PTF2W	78,883	59,470,771	754	1332	78,883	32,443	46,440
	PTF2 T	89,241	69,693,869	781	1429	89,241	36,185	53,056
	PTF4W	81,816	57,290,184	700	1209	81,816	31,763	50,053
	PTF4 T	79,025	55,450,884	702	1240	79,025	30,515	48,510
	All	98,671	96,940,229	982	1626	98,671	47,582	51,089

COG functional categories

When the translated protein sequences of the unigenes were searched against the COG database, 28,165 unigenes (28.55%) were assigned to 25 function categories. The largest category was "General function prediction only" (9,316, 9.44%), followed by "Transcription" (5,232, 5.30%) and "Replication, recombination and repair" (4,299, 4.35%), while "Extracellular structures" (44, 0.045%) and "Nuclear structure" (8, 0.0081%) were the smallest categories (Figure 4).

GO annotations

The GO classification system was used to assign possible functions to the unigenes based on the annotations in Nr. A total of 55,413 unigenes (56.16%) were categorized into the three main GO categories (biological process, cellular component, and molecular function) and 44 functional groups. The largest groups under biological process were "cellular process" (36,028, 36.51%), "metabolic process" (34,472, 34.94%), and "response to stimulus" (19,059, 19.32%). Under cellular component, "cell" and "cell part" were the two largest groups, both containing 44,582 unigenes (45.18%). Under molecular function, the top three groups were "catalytic activity" (27,131, 27.50%), "binding" (26,550, 26.90%), and "transporter activity" (4,071, 4.12%) (Figure 5).

KEGG pathways

To identify the different biochemical pathway that the annotated unigenes were associated with, we assigned the Enzyme Commission numbers (EC numbers) in the KEGG pathways. A total of 41,206 unigenes (approximately 41.76%) were mapped to 128 KEGG pathways, the top four of which were "Metabolic pathways"(ko01100, 10,041 unigenes, 24.37%), "Biosynthesis of secondary metabolites" (ko01110, 4,681 unigenes, 11.36%), "Plant-pathogen interaction" (ko04626, 2,284 unigenes, 5.54%), and "Plant hormone signal transduction" (ko04075, 2,089 unigenes, 5.07%), while "Betalain biosynthesis" (ko00965, 3 unigenes, 0.01%) was the least represented (Table S3).

Transcriptional profiling of PTF2 and PTF4 under drought stress

To identify transcriptional changes under drought stress, we performed four pairwise comparisons of the transcriptomes as follows: drought-treated PTF2 plants versus well-watered PTF2 plants (PTF2T vs. PTF2W), and drought-treated PTF4 versus well-watered PTF4 plants (PTF4T vs. PTF4W), well-watered PTF4 plants versus well-watered PTF2 plants (PTF4W vs. PTF2W), and drought-treated PTF4 versus drought-treated PTF2 plants (PTF4T vs. PTF2T). Large numbers of DEUs were detected (Table S4, S5, S6 and S7). The results of the pairwise comparisons of the DEUs are shown in Figure 6. Co-regulated DEUs between the PTF2T vs. PTF2W and PTF4T vs. PTF4W (the drought-treated vs. the well-watered) libraries were identified; 1,616 DEUs were up-regulated and 3,457 DEUs were down-regulated in the two comparisons. Similarly, co-regulated DEUs between the PTF4W vs. PTF2W and PTF4T vs. PTF2T (the tetraploid vs. the diploid) libraries were also identified; 173 DEUs were up-regulated and 98 DEUs were down-regulated in the two comparisons. Cell wall genes, photosynthesis-related genes, and genes encoding drought stress-related kinases, reactive oxygen species (ROS) scavenging enzymes, and osmotic adjustment proteins were among the most abundant transcripts. Some of the DEUs had no matches to known proteins; however, they may

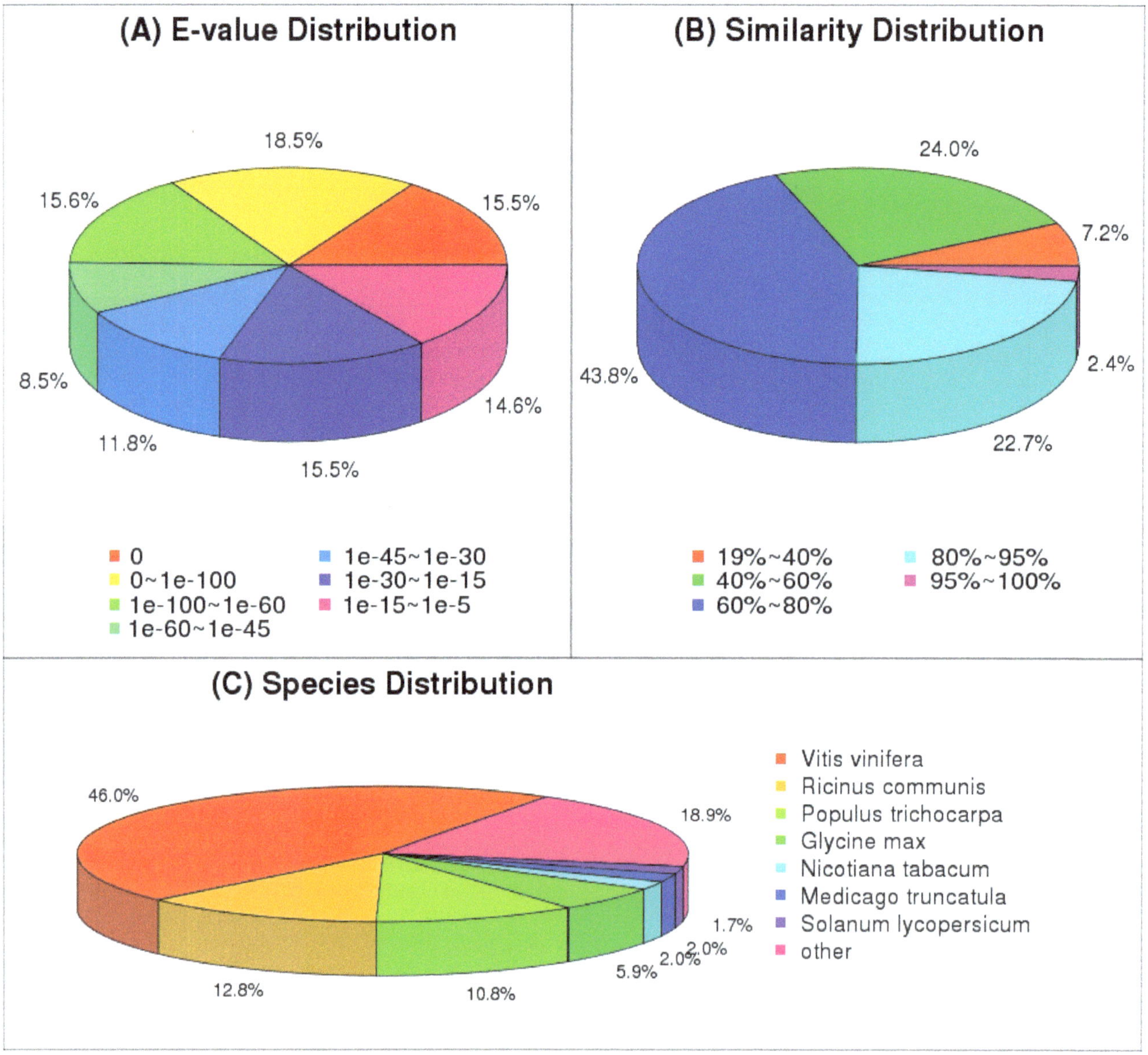

Figure 3. E-value Distribution. (A) similarity Distribution (B) and species distribution (C) of the BLAST matches of the transcriptome unigenes. This figure shows the distributions of unigenes BLASTX matches against the nr protein database (cutoff value E <1.0E-5) and the proportions for each species.

be found to contribute to drought tolerance when studied further (Table S8 and S9).

GO assignments allowed us to determine the major biological process, cellular component, and molecular function with which DEUs were associated (Figure 7 and 8). We then performed a GO enrichment analysis to identify the functional categories that were significantly enriched among the DEUs. Under biological process, stress-responsive processes (e.g., response to water deprivation, response to biotic stimulus, response to hormone stimulus, response to abiotic stimulus, and) were more represented in the up-regulated than in down-regulated DEUs, suggesting an increase of the related activities during drought treatment. Additionally, as reported previously in drought-stressed *P. euphratica* [50], GO terms like signal transduction, response to osmotic stress, small molecule metabolic process, carbohydrate catabolic process, arginine and proline metabolism, maltose metabolic process, disaccharide metabolic process, glucose catabolic process, hexose catabolic process, monosaccharide catabolic process, oligosaccharide metabolic process, and the related MAP kinase kinase activity were enriched. These findings may suggest that the osmotic stress-activated MAPK pathway and osmotic adjustment were crucial in the responses of PTF2 and PTF4 plants to water deprivation. Under cellular component, cell wall and plant-type cell wall were significantly enriched. Under molecular function, catalytic activity and binding were highly enriched for the drought-treated vs. the well-watered and the tetraploid vs. the diploid expression profiles, suggesting that the activities of many genes may be modulated in response to drought stress [51].

The co-regulated DEUs were also mapped to the KEGG metabolic pathways. Thirty-nine and seven metabolic pathways were significantly enriched (Q-value ≤0.05) in the drought-treated

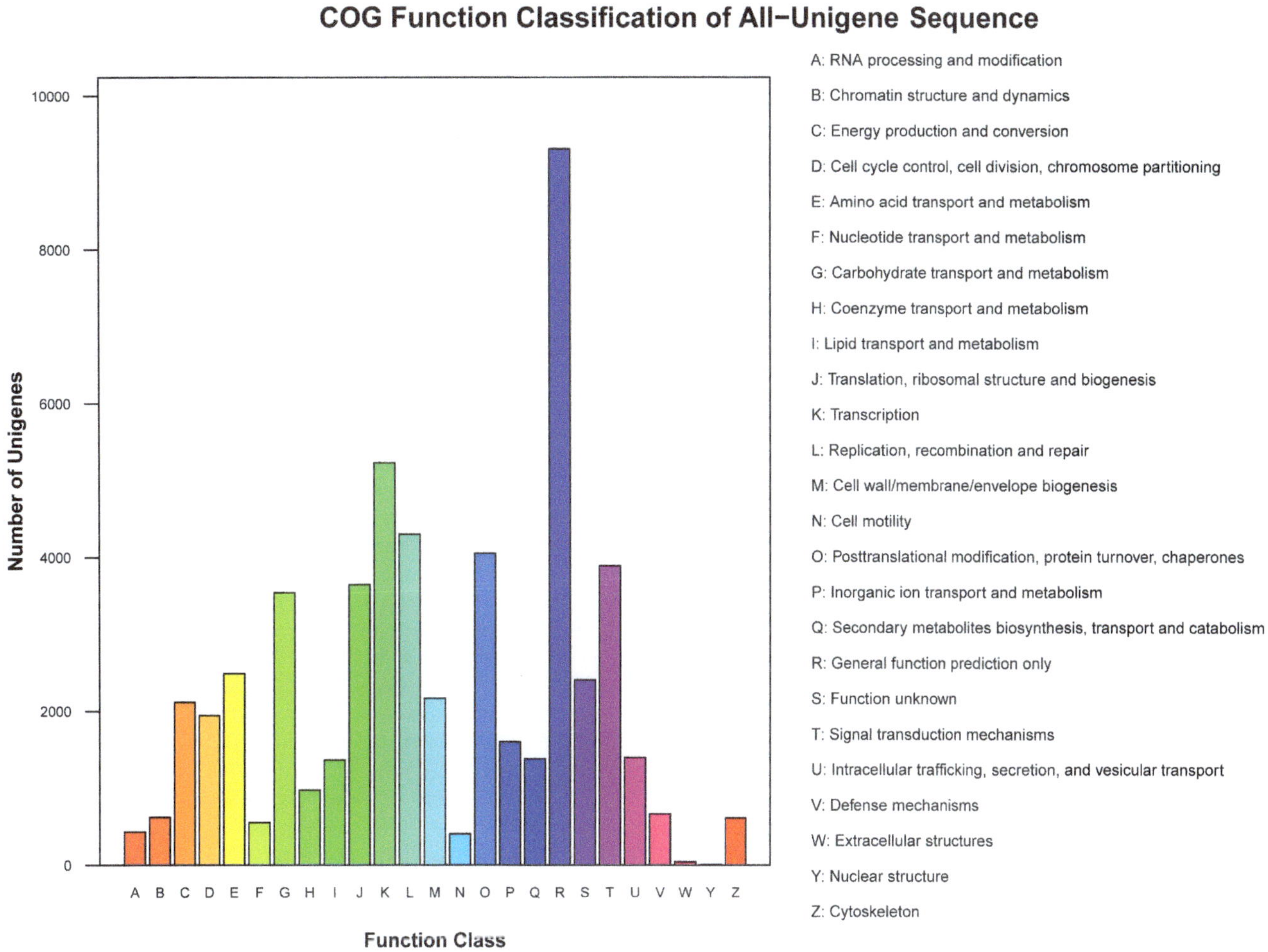

Figure 4. Classification of the clusters of orthologous groups (COG) for the transcriptome of *P. tomentosa* × *P. fortune.*

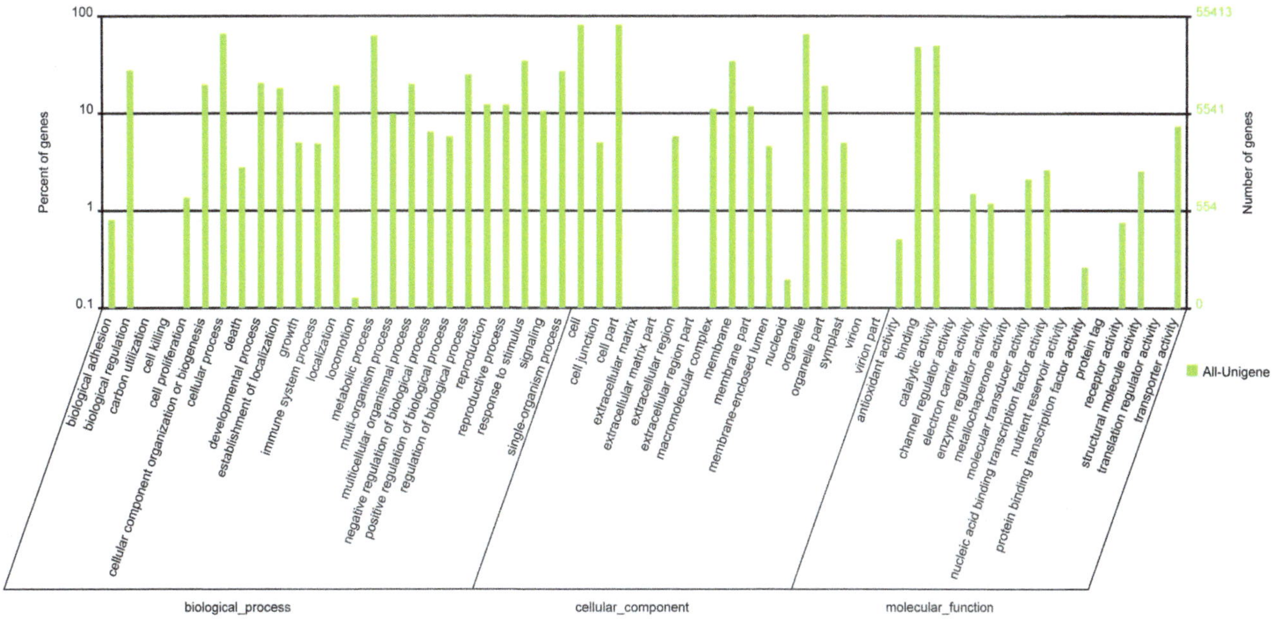

Figure 5. Classification of the gene ontology (GO) for the transcriptome of *P. tomentosa* × *P. fortune.*

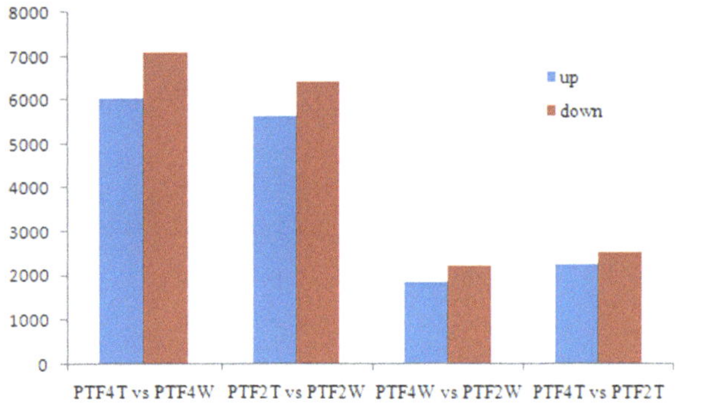

Figure 6. Statistics of differentially expressed genes in each pairwise comparison. Red bars represent the up-regulated genes while blue bars represent the down-regulated ones. PTF4T, 20 days drought treated tetraploid. PTF4W, well-watered tetraploid. PF2FT, 20 days drought treated diploid. PTF2W, well-watered diploid.

vs. the well-watered and the tetraploid vs. the diploid, respectively (Table S10 and S11).

qRT-PCR verification of genes related to drought response in Paulownia leaves

The expression profiles of 13 DEUs from leaves of the drought-treated and the well-watered plants were assessed by qRT-PCR. The results showed that except for the mannose-1-phosphate guanyl transferase beta gene in the PTF2 plants, the expressions of all the other genes were consistent with the expressions from the transcriptome analyses (Figure 9).

Discussion

Water deficiency significantly decreased plant growth in terms of height and basal stem diameter in the PTF2 and PTF4 plants. Paulownia is regarded as a relatively drought-tolerant species, and *P. tomentosa* × *P. fortune* plants, which were established in an ecosystem characterized by drought, have evolved special mechanisms that allow them to deal with insufficient water supply. In this study, the PTF4 plants exhibited more significant growth than the PTF2 plants in both the drought-treated and the well-watered samples. We also observed that PTF2 and PTF4 plants showed substantial differences in several physiological parameters in response to drought stress. The PTF4 seedlings exhibited higher relative water content (RWC), chlorophyll content, superoxide dismutase (SOD) activity, proline content and soluble protein content than the PTF2 seedlings under drought stress, this may indicate that PTF4 has stronger water absorption and cell turgor, and a more efficient enzymatic antioxidant system under water deficit conditions. Additionally, PTF4 seedlings showed lower malondialdehyde (MDA) content and relative electrical conductivity (REC) than the PTF2 seedlings in response to drought stress, suggesting more damage to membranes in PTF2 plants. These physiological parameters imply that PTF4 plants may be better capable of surviving drought conditions, which is consistent with the findings of our previous study [27]. In the present study, the transcriptome-wide gene expression profiling revealed differences in the abundance of drought-responsive genes between the PTF2 and PTF4 plants (Table S4, S5, S6, and S7). Under control conditions, transcript abundances of the majority of DEUs co-up-regulated in PTF2 plants were higher than those in PTF4 plants. However, under water-stress conditions, transcript abundances of co-up-regulated DEUs significantly increased in PTF4, but they maintained a relatively stable level in PTF2 plants, especially DEUs involved in ROS-scavenging system, amino acid and carbohydrate metabolism, and plant hormone biosynthesis and transduction (Table S8 and S9). This implies that PTF4 plants had a greater capability to tackle drought stress by increasing abundance of metabolism and defense-related unigenes. GO terms and pathways analysis showed that diverse metabolic processes were altered and some DEUs indicated the possibility

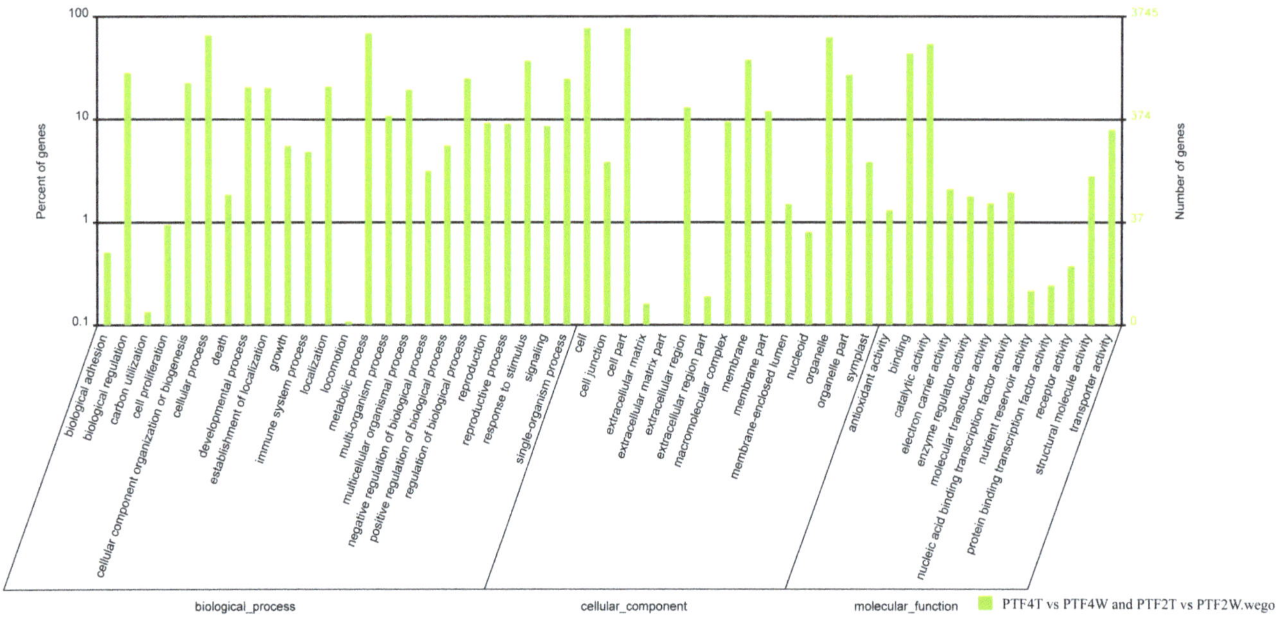

Figure 7. GO function analysis results for the consistently differentially expressed genes in both the PTF4T vs. PT4W and PTF2T vs. PT2W comparisons. PTF4T, 20 days drought treated tetraploid. PTF4W, well-watered tetraploid. PF2FT, 20 days drought treated diploid. PTF2W, well-watered diploid.

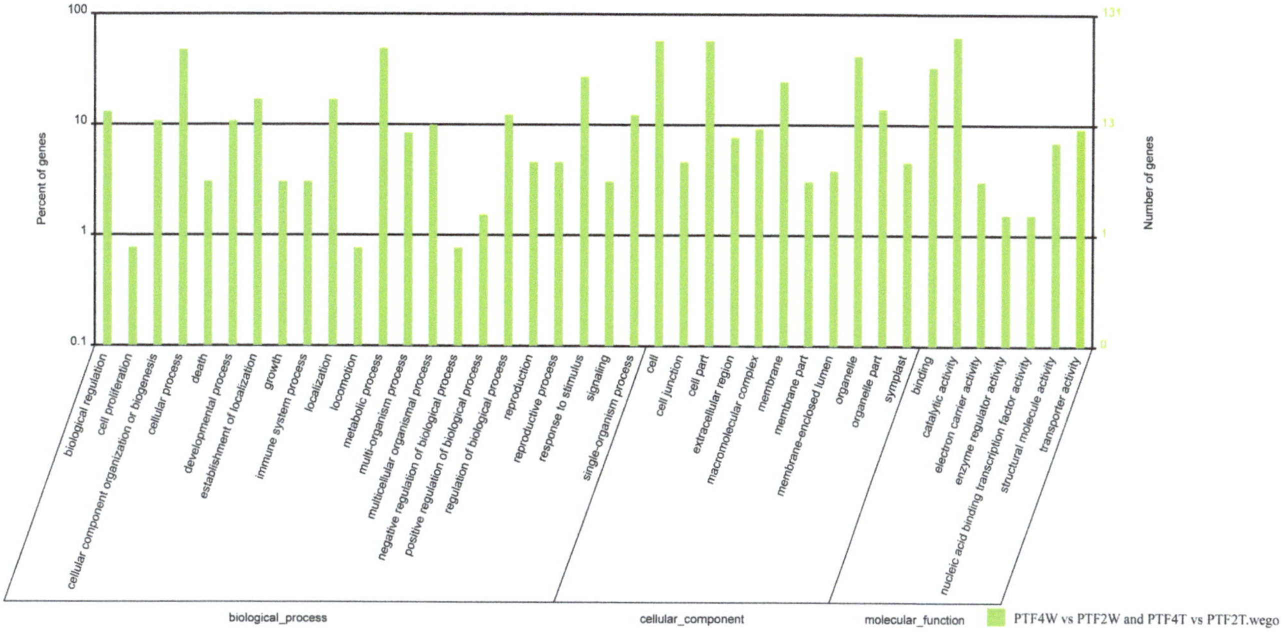

Figure 8. GO function analysis results for the consistently differentially expressed genes in both the PTF4W vs. PTF2W and PTF4T vs. PTF2T comparisons. PTF4W, well-watered tetraploid. PTF2W, well-watered diploid. PTF4T, 20 days drought treated tetraploid. PF2FT, 20 days drought treated diploid.

of crosstalk between pathways in the plants during drought stress. These changes might help the PTF2 and PTF4 plants survive under drought conditions.

Protection from oxidative damage

ROS accumulate in chloroplasts and produce a leakage of electrons towards oxygen under water deficit conditions. ROS serve as integral cellular signal molecules in stress sensing/signal transduction processes; however, excessive levels of ROS may induce oxidative damage to proteins, DNA, and lipids [52]. To protect tissues from oxidative damage, plants have developed an enzymatic antioxidant defense system to remove excess ROS. The ROS-scavenging system of plants includes several important antioxidant enzymes such as superoxide dismutase and ascorbate peroxidase [53]. In the present study, major ROS-scavenging enzymes and antioxidants including superoxide dismutase, ascorbate peroxidase, glutathione peroxidase, and lactoylglutathione lyase were up-regulated by drought stress (Table S8). However, differences in the expressions of these antioxidant genes were also observed between the two accessions, and the fold changes of these unigenes in the SOD, APX, and CAT families were almost always greater in the PTF4 plants than that in the PTF2 plants (Table S9). This is supported by the observation that both PTF2 and PTF4 plants showed increasing SOD activity, and the values of this trait in the PTF4 plants were always at much higher levels than in the PTF2 plants over the entire treatment process.

A senescence-associated protein was up-regulated in the PTF2 and PTF4 plants under drought stress, and its homologs, senescence associated protein 1 and senescence-associated gene 21, were over-expression in Arabidopsis to protect the plant from oxidative damage [54]. The Unigene22025 transcript was strongly enriched and up-regulated in the PTF2 and PTF4 plants during drought stress; interestingly, this transcript was also reported to be significantly differentially expressed in a previous study [55]. Additionally, the PTF2 and PTF4 plants were found to accumulate of high amounts of galactinol and raffinose during drought stress (discussed below), which may also help improve ROS scavenging in the drought-stressed plants [56].

Amino acid and carbohydrate metabolism

Low-molecular weight compatible osmolytes (osmoprotectants), such as proline, glycine, betaine, and sugar alcohols have been reported to accumulate in a variety of plant species in response to environmental stresses. This accumulation allows plants to maintain their water absorption and cell turgor under water deficit conditions by allowing rapid water intake from rather dry soil, thereby, mitigating the effects of water stress [57–59]. In the present study, various genes that may be involved in proline and carbohydrate metabolism were found.

Proline is a common osmoprotectant in water-stressed plants and its dramatic accumulation has been shown to confer osmotic tolerance to plants during drought stress [60]. In our study, unigenes encoding aldehyde dehydrogenases and glutamate synthase, which play important roles in catalyzing the conversion of Δ^1-pyrroline 5-carboxylate synthetase (P5C) to glutamate, were found to be down-regulated in the PTF2T and PTF4T plants; however, the expression levels of these DEUs were higher in PTF2T than in PTF4T. Conversely, a unigene encoding proline dehydrogenase (ProDH) was up-regulated in PTF2T, while its expression was unaltered in PTF4T plants (Table S8 and S9).

In plants, glutamate is catalyzed to P5C by Δ^1-pyrroline 5-carboxylate synthetase (P5CS) and P5C is then converted to proline by P5C reductase (P5CR). However, ProDH catalyzes the conversion of proline to P5C [61]. Therefore, the accumulation of proline in dehydrated plants has been linked to the up-regulation of P5CS and the down-regulation of ProDH [62]. In our study, unigenes encoding P5CS and P5CR were not among the selected DEUs (Table S8 and S9). However, the values of physiological parameters showed that the proline content in the PTF4 plants increased significantly with the increase of the drought stress, we

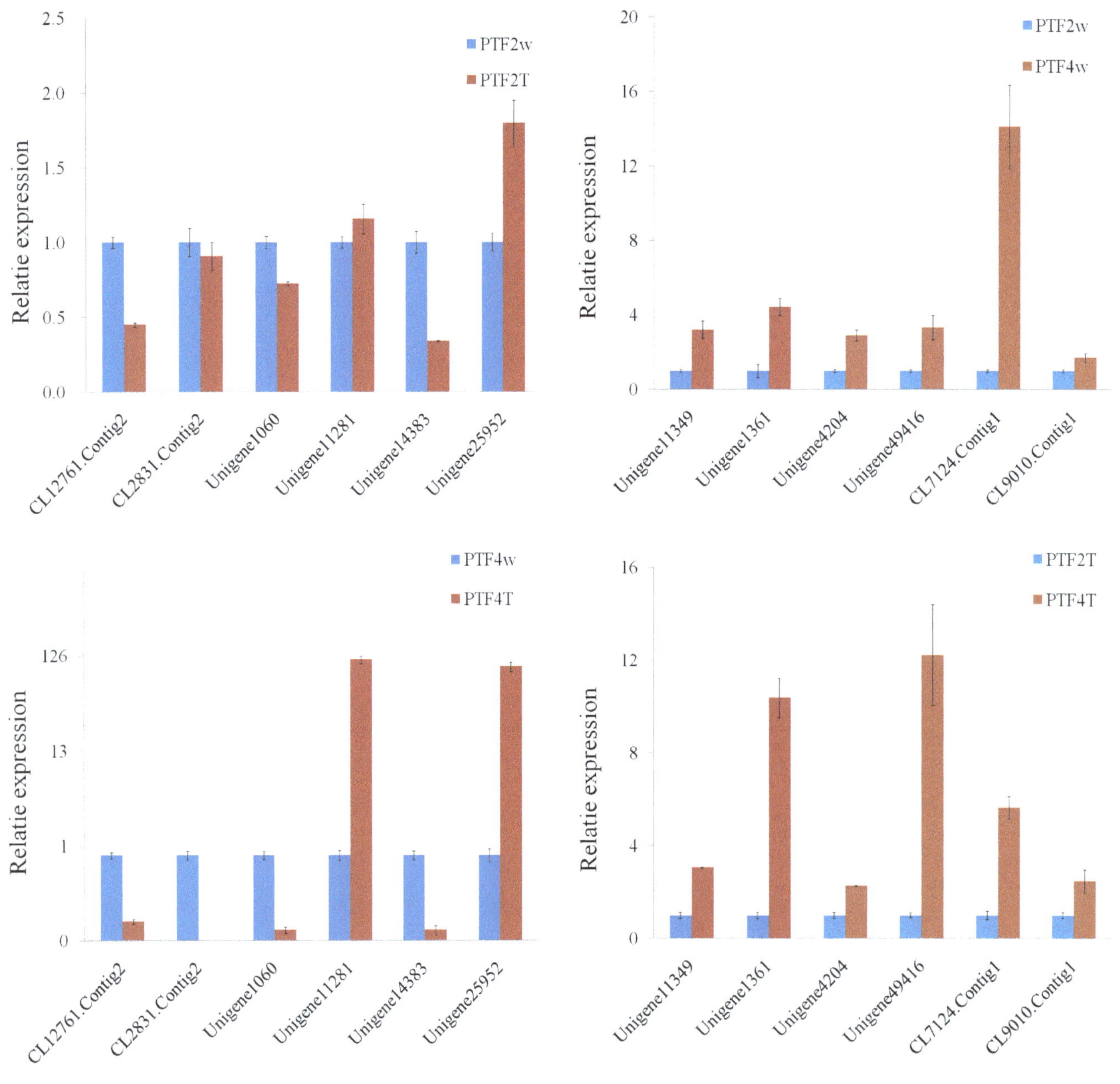

Figure 9. qRT-PCR analysis of candidate drought response genes. A, PTF2W, well-watered diploid; PTF2T, drought treated diploid. B, PTF4W, well-watered diploid; PTF4T, drought treated diploid.

therefore speculate that processes other than transcriptional regulation might play a role in influencing the accumulation of proline in PTF4 plants.

Unigenes encoding proteins involved in carbohydrate metabolism (e.g., galactinol, sucrose, fructose, and mannose metabolism) accounted for a large number of the DEUs. It has been reported that the accumulation of soluble carbohydrates is strongly correlated with stress tolerance in plants [63]. Under drought stress, we observed that most of these DEUs involved in carbohydrate metabolism were co-up-regulated in the PTF4 and PTF2 plants; however, the fold changes of these unigenes were greater in PTF4 than in PTF2 plants (Table S8 and S9). The expression of the unigene encoding galactinol synthase (GolS), which is involved in galactinol and raffinose family oligosaccharides (RFO) metabolism, was increased in PTF2T and PTF4T plants. In Arabidopsis, the homologs of *GolS*, *AtGolS1* and/or *AtGolS2*, were reported to be over-expressed, which increased galactinol and raffinose accumulation and enhanced the plant's tolerance to drought stress and oxidative damage [56]. The GolS gene in *Populus balsamifera* was found to be significantly differentially expressed in response to drought stress [64]. GolS catalyzes the first step in the biosynthetic pathway of galactinol and raffinose family oligosaccharides using galactose and myo-inositol as substrates. Several unigenes encoding aldose 1-epimerase, which are involved in galactose and myo-inositol synthesis, were found to be elevated in both PTF4T and PTF2T plants (Table S8 and S9). The expressions of several trehalose synthase encoding unigenes were found to be significant enhanced in response to drought stress in both accessions. The trehalose-6-phosphate synthase 1 (TPS1) encoding gene, which was homologous to *AtTPS1*, was of particular interest because *AtTPS1* plays an important role in abscisic acid (ABA) and sugar signaling and its over-expression resulted in improved drought tolerance in Arabidopsis [65]. Further, the over-expression of *AtTPS1* in rice was reported to elevate the sucrose, fructose, and glucose content and enhance the drought stress response [66]. Several DEUs with homology to glucose-6-phosphate/phosphate translocator 2 (GPT2), which was found to be involved in glucose 6-phosphate transport in Arabidopsis [67], were up-regulated in PTF4T and PTF2T, implying that the sugar signal triggered by GPT2 may be important in regulating drought stress tolerance in *P. tomentosa* × *P. fortunei* plants. We also found a very robust increase in soluble sugar contents that was more pronounced in PTF4 than PTF2 plants in our physiological measurements. The accumulation of these osmoprotectants in the leaves could help enhance the osmotic stress tolerance of *P. tomentosa* × *P. fortunei*.

Hormone biosynthesis

Plant hormones are closely related to plant growth and stress tolerance [68,69]. ABA plays crucial roles in various stress responses and the accumulation of endogenous ABA is known to be associated with drought tolerance [70,71]. Recently, several studies have revealed the molecular basis of ABA biosynthesis and its response mechanisms [72–74].

The over-expression of 9-cis-epoxycarotenoid dioxygenase (NCED), Zeaxanthin epoxidase (ABA2), and phytoene synthase were reported to be associated with the conversion of zeaxanthin to ABA, which can be metabolized and catalyzed by the ABA hydroxylase CYP707A1 [75]. In transgenic Arabidopsis plants, *AtNCED3* and the phytoene synthase gene were induced by drought and their up-regulation was reported to enhance drought tolerance by increasing the endogenous ABA level [76,77]. Furthermore, ABA2 was found to be a negative regulator of ABA in Arabidopsis [78,79]. Therefore, ABA biosynthesis can be enhanced by positive regulators (NCED3, phytoene synthase) and inhibited by negative regulators (ABA2). In our study, transcript abundances of hormone biosynthesis-related DEUs were significantly different between PTF2 and PTF4. The unigenes that were most similar to the Arabidopsis *NCED3* and phytoene synthase gene sequences were up-regulated, and the unigenes encoding the ABA2 enzymes were down-regulated in the PTF2T and PTF4T plants, transcript abundances of these co-up-regulated and co-down-regulated DEUs have higher fold change in PTF4 than in PTF2 (Table S8 and S9). Our data suggest that these genes may play important roles in regulating the biosynthesis of ABA in the two accessions.

ABA signaling pathways have been linked to phospholipase D (PLD), protein phosphatase 2C (PP2C), SNF1-related kinases (SnPKs), and CBL-interacting protein kinases (CIPKs), and unigenes encoding these enzymes were significantly up- or down-regulation more in the PTF4T than that in the PTF2T plants under drought stress (Table S8 and S9). PP2C was found previously to negatively regulate ABA signaling in Arabidopsis [80] and links between PP2C and SnRK2 have been reported [81–83]. Several members of the SnRK2 family in Arabidopsis and rice were shown to be activated in response to ABA or osmotic stress [84,85], and in fava bean guard cells SnRK2 was shown to play an important role in ABA-dependent stomatal closure [86]. Together these results suggest that SnRK2 is an important regulator of ABA signal transduction during drought stress [83]. CIPKs and their interacting partner calcineurin B-like (CBL) (also known as SnRK3) form a complex that was reported to be a central hub of the ABA-mediated signaling pathway under drought, cold, or salt stress [87,88]. Although several genes encoding potential factors of the ABA signaling pathway were identified in the PTF2T and PTF4T plants, the molecular mechanisms that control this complex signaling pathway in *P. tomentosa* × *P. fortunei* need to be investigated further.

Recent studied have revealed that the plant hormone indol-3yl-acetic acid (IAA) plays a critical role in regulating plant growth and cell enlargement [89]. The auxin transporter protein AUX1 has been found to be a major influx carrier [90]. Aux/IAA proteins and the ARF transcription factors are important regulators of IAA-modulated gene expression, and auxin-responsive promoter (GH3) and auxin-responsive protein (SAUR) have been reported to regulate plant growth and cell enlargement [89,91]. In the present study, several increased transcript abundance of the DEUs in both PTF2T and PTF4T were annotated as being involved in crosstalk among the IAA encoding unigenes, whereas we observed that the expression level of these DEUs were higher in PTF4 under drought stress (Table S8 and S9), suggesting that different levels of IAA accumulation might explain the cell enlargement and plant growth differences observed between the PTF2 and PTF4 plants.

Multi-stress responsive genes in *P. tomentosa* × *P. fortunei*

Transcription factors (TFs) have been shown to be critical upstream regulators of plant responses to various biotic and abiotic stresses [92]. In the present study, a large number of DEUs encoded TFs belonging to the MYB/MYC, ABI3VP1, AP2-EREBP, DREB, ARF, bzip, ABA responsive, NAC, WRKY, bHLH, ZF-HD, AREB/ABF, and ZFP families. Under control conditions, most of the DEUs encoded TFs had higher transcript abundance in PTF2 than in PTF4, while under drought stress, expression changes of these DEUs were significantly higher in PTF4 than in PTF2 plants (Table S8 and S9). These TFs were divided into ABA-dependent signaling systems and ABA-independent signaling pathways. Over-expression of the MYB/MYC, bHLH, NAC, WRKY, ZFP, bzip, and ZF-HD families has been shown to play significant roles in controlling the expression of specific stress-related genes and plant stress signal transduction, thereby greatly enhancing the tolerance of plants to drought and high salinity in Arabidopsis and rice [93–97]. Recently, transcriptional activators that up-regulate stress-responsive genes have been used widely to impart drought tolerance to transgenic plants [98]. The over-expression of the AP2-type TF and ethylene-responsive transcription factors (SHN1-3) in transgenic plants were found to regulate several stress inducible target genes and increase drought tolerance [99]. Increased expression of the NAC transcription factor 45 (OsNAC045) in transgenic rice plants was found to enhance drought tolerance, which may serve as a useful tool for improving drought stress response in genetically engineered rice [94]. Recent studies have shown that ABRE-binding bZIP and OsABF2 were detectable in transgenic rice where they positively regulated drought tolerance and ABA signaling [100]. Transcriptional repressors that have the ability to repress the expression of genes under drought stress have been shown to be important in engineering drought tolerance in plants [101]. AtMYB60 was reported to adjust stomatal movement and function as a transcriptional repressor in the plants response to dehydration; specifically, the expression of this gene in Arabidopsis guard cells was negatively regulated during drought stress [102]. A null T-DNA insertion mutation in the AtMYB60 gene led to reduced stomatal opening and minimized wilting during drought stress, this mutation resulted in defective guard cell but had no significant effect on other physiological processes and agricultural traits [102]. In our study, several signal transduction pathways that independently respond to drought stress (in both an ABA dependent- and independent- manner) were identified in the PTF2T and PTF4T plants. ABA-dependent signaling pathway have been illustrated that they could mediate stress adaptation by the action of ABA-inducible TFs controlling the expression of genes containing cis-acting ABA response elements (e.g., TFs such as MYB/MYC and AREB/ABF), while ABA-independent stress signaling pathway leading to gene expression involve different TFs such as (1) NAC and ZF-HD, and (2) CBF/DREB (Figure S2) [103–106], suggesting that drought stress tolerance at the transcriptional level was controlled by an extremely intricate gene regulatory network. Thus, functional analysis of drought inducible TFs will help in the development of drought tolerant plants that will survive well in acute field condition.

Different sets of genes are known to be associated with specific stress-responsive signal transduction. Mitogen-activated protein kinases (MAPKs) are a highly conserved family of serine/threonine protein kinases that are involved in a variety of fundamental cellular processes such as proliferation, differentiation, stress response, and survival. The MAPK cascade has been found to lead to accumulated production of osmolytes by activating, for example, the protein tyrosine kinases and G-protein receptors [107,108]. The MAPK pathway phosphorylates a three-tiered kinase cascade: MAPKKK→MAPKK→MAPK, which propagates downstream stress signals [109]. Activated MAPK can migrate to the nucleus to activate the TFs directly. These findings indicate that modulating MAPK may lead to altered expression of specific regulatory TFs and protein kinases towards enhanced drought tolerance in plants. In this study, many DEUs in the PTF2T and PTF4T plants were identified as MAPKs including MAPK8, MAPK9, MAPK14, MAPKK2, MAPKK6, MAPKKK1, and MAPKKK2, and a relatively large proportion of DEUs encoding MAPKs were up-regulated both in the PTF2 and PTF4 plants under drought stress, but the fold changes of these DEUs were greater in PTF4 than in PTF2. We also identified a great number of calcium-dependent protein kinases (CDPKs) that were activated and the expression levels of these unigenes were induced strongly in PTF4 plants under drought treatments (Table S8 and S9). Previous studies have indicated that CDPKs play a crucial role in Ca^{2+}-mediated signal transduction in plants in response to drought and salinity stresses, suggesting that signal transduction may play a critical role in the *P. tomentosa* × *P. fortunei* drought response [107,110].

Plants have been found to survive different stresses using multifunctional genes that enable crosstalk among the various abiotic stress resistance systems [111]. We also found that in PTF4, several stress-regulated unigenes like lipid transfer protein, late embryogenesis abundant protein (LEA), osmotin, small heat shock protein, HSP70, HSP90, antiporter, and dehydrins were expressed at higher transcript abundance than in PTF2 plants under drought stress. Several study reported that these genes were generally involved in cell rescue and defense mechanisms and it was suggested that their overexpression may play an important role in conferring dehydration and drought stress by causing stomatal closure and osmotic adjustment [112–116].

Conclusions

PTF4 trees have high ecological and economic value. In the present work, we performed large-scale transcriptome sequencing of diploid and autotetraploid *P. tomentosa* × *P. fortune* plants under drought stress using an Illumina/Solexa sequencing platform. More than 291 million reads were generated from four libraries and *de novo* assembled into 98,671 nonredundant unigenes that will provide a useful resource for this species. GO terms enrichment and pathways analyses of the annotated DEUs revealed changes in several metabolic processes as a result of drought stress. These results may shed some light on the biology of the drought response and tolerance in Paulownia species. In addition to the previously reported drought stress related genes detected in the libraries, a large number of unigene sequences could not be mapped to known sequences in the protein databases. Some of the unidentified DEU sequences may be novel and/or unknown Paulownia genes associated with drought stress. Functional analysis of these genes may contribute to a deeper understanding of the response of Paulownia plants to drought. Our findings will help enrich our understanding of the drought response in diploid and autotetraploid *P. tomentosa* × *P. fortune* and may provide new insights into the responses of woody plants to drought stress.

Supporting Information

Figure S1 Distribution of unigene lengths in the transcriptome of *P. tomentosa* × *P. fortunei*. The sizes of all unigenes were calculated.

Figure S2 A schematic representation of transcriptional regulatory networks of cis-acting elements and transcription factors involved in abiotic-stress-responses. Transcription factors are shown in ellipses; cis-acting elements are shown in boxes.

Table S1 Primers of quantitative RT-PCR analysis of candidate drought response genes. -f represents forward primers and -r represents reverse primers.

Table S2 Unigene annotations based on their BLASTX hits in the nr database.

Table S3 KEGG annotation of unigenes.

Table S4 Differetially expressed genes in the PTF2T vs. PTF2W comparison. PTF2T, 20 days drought-treated diploid; PTF2W, well-watered diploid.

Table S5 Differetially expressed genes in the PTF4T vs. PTF4W comparison. PTF4T, 12 days drought-treated tetraploid; PTF4W, well-watered tetraploid.

Table S6 Differetially expressed genes in the PTF4W vs. PTF2W comparison. PTF4W, well-watered tetraploid; PTF2W, well-watered diploid.

Table S7 Differetially expressed genes in the PTF4T vs. PTF2T comparison. PTF4T, 20 days drought-treated tetraploid; PTF2T, 20 days drought-treated diploid.

Table S8 Consistently differentially expressed genes in both the PTF2T vs. PTF2W and PTF4T vs. PTF4W comparisons. PTF2T, 20 days drought-treated diploid; PTF2W, well-watered diploid; PTF4T, 20 days drought-treated tetraploid; PTF4W, well-watered tetraploid.

Table S9 Consistently differentially expressed genes in both the PTF4W vs. PTF2W and PTF4T vs. PTF2T comparisons. PTF4W, well-watered tetraploid; PTF2W, well-watered diploid; PTF4T, 20 days drought-treated tetraploid; PTF2T, 20 days drought-treated diploid.

Table S10 KEGG pathway analysis results for the consistently differentially expressed genes in both the PTF2T vs. PTF2W and PTF4T vs. PTF4W comparisons. PTF2T, 20 days drought-treated diploid; PTF2W, well-watered diploid; PTF4T, 20 days drought-treated tetraploid; PTF4W, well-watered tetraploid.

Table S11 KEGG pathway analysis results for the consistently differentially expressed genes in both the PTF4W vs. PTF2W and PTF4T vs. PTF2T comparisons. PTF4W, well-watered tetraploid; PTF2W, well-watered diploid; PTF4T, 20 days drought-treated tetraploid; PTF2T, 20 days drought-treated diploid.

Author Contributions

Conceived and designed the experiments: EKX GQF SYN ZLZ MJD YPD. Performed the experiments: EKX GQF SYN ZLZ MJD YPD. Analyzed the data: EKX GQF SYN ZLZ MJD YPD. Contributed reagents/materials/analysis tools: EKX GQF SYN ZLZ MJD YPD. Contributed to the writing of the manuscript: EKX GQF SYN.

References

1. Valliyodan B, Nguyen HT (2006) Understanding regulatory networks and engineering for enhanced drought tolerance in plants. Current opinion in plant biology 9: 189–195.
2. Wang W, Vinocur B, Shoseyov O, Altman A (2000) Biotechnology of plant osmotic stress tolerance physiological and molecular considerations. pp. 285–292.
3. Hamdy A, Ragab R, Scarascia-Mugnozza E (2003) Coping with water scarcity: water saving and increasing water productivity. Irrigation and drainage 52: 3–20.
4. Bohnert HJ, Ayoubi P, Borchert C, Bressan RA, Burnap RL, et al. (2001) A genomics approach towards salt stress tolerance. Plant Physiology and Biochemistry 39: 295–311.
5. Seki M, Narusaka M, Ishida J, Nanjo T, Fujita M, et al. (2002) Monitoring the expression profiles of 7000 Arabidopsis genes under drought, cold and high-salinity stresses using a full-length cDNA microarray. The Plant Journal 31: 279–292.
6. Fowler S, Thomashow MF (2002) Arabidopsis transcriptome profiling indicates that multiple regulatory pathways are activated during cold acclimation in addition to the CBF cold response pathway. The Plant Cell Online 14: 1675–1690.
7. Seki M, Kamei A, Satou M, Sakurai T, Fujita M, et al. (2004) Transcriptome analysis in abiotic stress conditions in higher plants. Plant Responses to Abiotic Stress: Springer. pp. 271–308.
8. Shinozaki K, Yamaguchi-Shinozaki K (2007) Gene networks involved in drought stress response and tolerance. Journal of experimental botany 58: 221–227.
9. Cohen D, Bogeat-Triboulot M-B, Tisserant E, Balzergue S, Martin-Magniette M-L, et al. (2010) Comparative transcriptomics of drought responses in Populus: a meta-analysis of genome-wide expression profiling in mature leaves and root apices across two genotypes. BMC genomics 11: 630.
10. Zheng J, Fu J, Gou M, Huai J, Liu Y, et al. (2010) Genome-wide transcriptome analysis of two maize inbred lines under drought stress. Plant molecular biology 72: 407–421.
11. Tang S, Liang H, Yan D, Zhao Y, Han X, et al. (2013) Populus euphratica: the transcriptomic response to drought stress. Plant molecular biology 83: 539–557.
12. Hiremath PJ, Farmer A, Cannon SB, Woodward J, Kudapa H, et al. (2011) Large-scale transcriptome analysis in chickpea (Cicer arietinum L.), an orphan legume crop of the semi-arid tropics of Asia and Africa. Plant biotechnology journal 9: 922–931.
13. Peng S, Jiang H, Zhang S, Chen L, Li X, et al. (2012) Transcriptional profiling reveals sexual differences of the leaf transcriptomes in response to drought stress in Populus yunnanensis. Tree physiology 32: 1541–1555.
14. Hall T (2008) Paulownia: An Agroforestry Gem. Trees for Life Journal 3: 3.
15. Zhu Z, Chao C, Lu X, Xiong Y (1986) Paulownia in China: cultivation and utilization: International Development Research Centre, beijing, china.
16. Ipekci Z, Gozukirmizi N (2003) Direct somatic embryogenesis and synthetic seed production from Paulownia elongata. Plant Cell Reports 22: 16–24.
17. Masterson J (1994) Stomatal size in fossil plants: evidence for polyploidy in majority of angiosperms. Science (New York, NY) 264: 421–424.
18. Dwivedi N, Suryanarayana N, Sikdar A, Susheelamma B, Jolly M (1989) Cytomorphological studies in triploid mulberry evolved by diploidization of female gamete cells. Cytologia 54: 13–19.
19. Leitch A, Leitch I (2008) Genomic plasticity and the diversity of polyploid plants. Science 320: 481–483.
20. Zhang ZY, Yu XS, Zhu ZT (1997) Advances in the triploid breeding programme of Populus tomentosa in China. Journal of Beijing Forestry University, English edition 6: 1–8.
21. Stupar RM, Bhaskar PB, Yandell BS, Rensink WA, Hart AL, et al. (2007) Phenotypic and transcriptomic changes associated with potato autopolyploidization. Genetics 176: 2055–2067.
22. Riddle NC, Jiang H, An L, Doerge R, Birchler JA (2010) Gene expression analysis at the intersection of ploidy and hybridity in maize. Theoretical and applied genetics 120: 341–353.
23. Fock II, Collonnier C, Purwito A, Luisetti J, Souvannavong VV, et al. (2000) Resistance to bacterial wilt in somatic hybrids between Solanum tuberosum and Solanum phureja. Plant Sci 160: 165–176.
24. Huang S, Sirikhachornkit A, Su X, Faris J, Gill B, et al. (2002) Genes encoding plastid acetyl-CoA carboxylase and 3-phosphoglycerate kinase of the Triticum/Aegilops complex and the evolutionary history of polyploid wheat. Proc Natl Acad Sci U S A 99: 8133–8138.
25. Zhang XY, Hu CG, Yao JL (2010) Tetraploidization of diploid Dioscorea results in activation of the antioxidant defense system and increased heat tolerance. J Plant Physiol 167: 88–94.
26. Fan GQ, Zhai XQ, Wei ZZ, Yang ZQ (2010) Induction of autotetraploid from the somatic cell of *Paulownia tomentosa* × *P. fortunei* and its in vitro plantlet regeneration. Journal of Northwest Forestry University 38: 22–26.
27. Zhang XS, Zhai XQ, Fan GQ, Deng MJ, Zhao ZL (2013) Observation on microstructure of leaves and stress tolerance analysis of different tetraploid Paulownia. Journal of Henan Agricultural University 46: 646–650.
28. Moumeni A, Satoh K, Kondoh H, Asano T, Hosaka A, et al. (2011) Comparative analysis of root transcriptome profiles of two pairs of drought-tolerant and susceptible rice near-isogenic lines under different drought stress. BMC plant biology 11: 174.
29. Barrs HD WP (1962) A re-examination of the relative turgidity technique for estimating water deficit in leaves. Aust J Biol Sci 15: 413–428.
30. Bojović BM, Stojanović J (2005) Chlorophyll and carotenoid content in wheat cultivars as a function of mineral nutrition. Archives of Biological Sciences 57: 283–290.
31. Arnon DI (1949) Copper enzymes in isolated chloroplasts. Polyphenoloxidase in Beta vulgaris. Plant physiology 24: 1–15.
32. Hodges DM, DeLong JM, Forney CF, Prange RK (1999) Improving the thiobarbituric acid-reactive-substances assay for estimating lipid peroxidation in plant tissues containing anthocyanin and other interfering compounds. Planta 207: 604–611.
33. Liu FH, Guo Y, Gu DM, Xiao G, Chen ZH, Chen SY (1996) Salt tolerance of transgenic plants with BADH cDNA. Acta Genetica Sinica 24: 54–58.
34. Dhindsa RS, Plumb-Dhindsa P, Thorpe TA (1981) Leaf senescence: correlated with increased levels of membrane permeability and lipid peroxidation, and decreased levels of superoxide dismutase and catalase. Journal of Experimental botany 32: 93–101.
35. Sun Y, Oberley LW, Li Y (1988) A simple method for clinical assay of superoxide dismutase. Clinical Chemistry 34: 497–500.
36. Guy C, Haskell D, Neven L, Klein P, Smelser C (1992) Hydration-state-responsive proteins link cold and drought stress in spinach. Planta 188: 265–270.
37. Bradford MM (1976) A rapid and sensitive method for the quantitation of microgram quantities of protein utilizing the principle of protein-dye binding. Analytical biochemistry 72: 248–254.
38. Bates L, Waldren R, Teare I (1973) Rapid determination of free proline for water-stress studies. Plant and soil 39: 205–207.
39. Irigoyen J, Einerich D, Sánchez-Díaz M (1992) Water stress induced changes in concentrations of proline and total soluble sugars in nodulated alfalfa (Medicago sativd) plants. Physiologia Plantarum 84: 55–60.
40. SnedecorGW., WG C (1980). Statistical methods7th Edition: Iowa State University Press, Ames, IOWA. pp. 507.
41. Grabherr MG, Haas BJ, Yassour M, Levin JZ, Thompson DA, et al. (2011) Full-length transcriptome assembly from RNA-Seq data without a reference genome. Nat Biotechnol 29: 644–652.
42. Pertea G, Huang X, Liang F, Antonescu V, Sultana R, et al. (2003) TIGR Gene Indices clustering tools (TGICL): a software system for fast clustering of large EST datasets. Bioinformatics 19: 651–652.

43. Iseli C, Jongeneel CV, Bucher P (1999) ESTScan: a program for detecting, evaluating, and reconstructing potential coding regions in EST sequences. Proc Int Conf Intell Syst Mol Biol: 138–148.
44. Conesa A, Gotz S, Garcia-Gomez JM, Terol J, Talon M, et al. (2005) Blast2GO: a universal tool for annotation, visualization and analysis in functional genomics research. Bioinformatics 21: 3674–3676.
45. Ye J, Fang L, Zheng H, Zhang Y, Chen J, et al. (2006) WEGO: a web tool for plotting GO annotations. Nucleic Acids Res 34: W293–297.
46. Kanehisa M, Araki M, Goto S, Hattori M, Hirakawa M, et al. (2008) KEGG for linking genomes to life and the environment. Nucleic Acids Res 36: D480–484.
47. Ali Mortazavi BAW, Kenneth McCue, et al. (2008) Lorian Schaeffer and Barbara Wold Mapping and quantifying mammalian transcriptomes by RNA-Seq. Nature Methods 5: 621–628.
48. Audic S, Claverie JM (1997) The significance of digital gene expression profiles. Genome Res 7: 986–995.
49. Benjamini Y, Yekutieli D (2001) The control of the false discovery rate in multiple testing under dependency. Annals of statistics: 1165–1188.
50. Tang S, Liang H, Yan D, Zhao Y, Han X, et al. (2013) Populus euphratica: the transcriptomic response to drought stress. Plant Mol Biol 83: 539–557.
51. Cohen D, Bogeat-Triboulot MB, Tisserant E, Balzergue S, Martin-Magniette ML, et al. (2010) Comparative transcriptomics of drought responses in Populus: a meta-analysis of genome-wide expression profiling in mature leaves and root apices across two genotypes. BMC Genomics 11: 630.
52. Mittler R, Zilinskas BA (1994) Regulation of pea cytosolic ascorbate peroxidase and other antioxidant enzymes during the progression of drought stress and following recovery from drought. Plant J 5: 397–405.
53. Bowler C, Slooten L, Vandenbranden S, De Rycke R, Botterman J, et al. (1991) Manganese superoxide dismutase can reduce cellular damage mediated by oxygen radicals in transgenic plants. The EMBO journal 10: 1723.
54. Woo HR, Kim JH, Nam HG, Lim PO (2004) The delayed leaf senescence mutants of Arabidopsis, ore1, ore3, and ore9 are tolerant to oxidative stress. Plant Cell Physiol 45: 923–932.
55. Yan DH, Fenning T, Tang S, Xia X, Yin W (2012) Genome-wide transcriptional response of Populus euphratica to long-term drought stress. Plant Sci 195: 24–35.
56. Nishizawa A, Yabuta Y, Shigeoka S (2008) Galactinol and raffinose constitute a novel function to protect plants from oxidative damage. Plant Physiol 147: 1251–1263.
57. Ashraf M, Foolad M (2007) Roles of glycine betaine and proline in improving plant abiotic stress resistance. Environmental and Experimental Botany 59: 206–216.
58. Csonka L, Gelvin S, Goodner B, Orser C, Siemieniak D, et al. (1988) Nucleotide sequence of a mutation in the proB gene of Escherichia coli that confers proline overproduction and enhanced tolerance to osmotic stress. Gene 64: 199–205.
59. Delauney AJ, Verma DPS (1993) Proline biosynthesis and osmoregulation in plants. The plant journal 4: 215–223.
60. Verma DPS (1999) Osmotic Stress Tolerance in Plants: Role of Proline and Sulfur Metabolisms. Molecular Responses to Cold, Drought, Heat and Salt Stress in Higher Plants: Landes, Austin, pp. 153–168.
61. Kishor PK, Sangam S, Amrutha R, Laxmi PS, Naidu K, et al. (2005) Regulation of proline biosynthesis, degradation, uptake and transport in higher plants: its implications in plant growth and abiotic stress tolerance. Curr Sci 88: 424–438.
62. Evers D, Lefèvre I, Legay S, Lamoureux D, Hausman J-F, et al. (2010) Identification of drought-responsive compounds in potato through a combined transcriptomic and targeted metabolite approach. Journal of experimental botany 61: 2327–2343.
63. Oliver MJ, Cushman JC, Koster KL (2010) Dehydration tolerance in plants. Methods Mol Biol 639: 3–24.
64. Wilkins O, Waldron L, Nahal H, Provart NJ, Campbell MM (2009) Genotype and time of day shape the Populus drought response. Plant J 60: 703–715.
65. Kolukisaoglu U, Weinl S, Blazevic D, Batistic O, Kudla J (2004) Calcium sensors and their interacting protein kinases: genomics of the Arabidopsis and rice CBL-CIPK signaling networks. Plant Physiol 134: 43–58.
66. Yoshida R, Umezawa T, Mizoguchi T, Takahashi S, Takahashi F, et al. (2006) The regulatory domain of SRK2E/OST1/SnRK2.6 interacts with ABI1 and integrates abscisic acid (ABA) and osmotic stress signals controlling stomatal closure in Arabidopsis. J Biol Chem 281: 5310–5318.
67. Niewiadomski P, Knappe S, Geimer S, Fischer K, Schulz B, et al. (2005) The Arabidopsis plastidic glucose 6-phosphate/phosphate translocator GPT1 is essential for pollen maturation and embryo sac development. Plant Cell 17: 760–775.
68. He XJ, Mu RL, Cao WH, Zhang ZG, Zhang JS, et al. (2005) AtNAC2, a transcription factor downstream of ethylene and auxin signaling pathways, is involved in salt stress response and lateral root development. The Plant Journal 44: 903–916.
69. Wolters H, Jürgens G (2009) Survival of the flexible: hormonal growth control and adaptation in plant development. Nature Reviews Genetics 10: 305–317.
70. Wasilewska A, Vlad F, Sirichandra C, Redko Y, Jammes F, et al. (2008) An update on abscisic acid signaling in plants and more.... Molecular Plant 1: 198–217.
71. Zeevaart J, Creelman R (1988) Metabolism and physiology of abscisic acid. Annual review of plant physiology and plant molecular biology 39: 439–473.
72. Shinozaki K, Yamaguchi-Shinozaki K (1997) Gene expression and signal transduction in water-stress response. Plant Physiology 115: 327–334.
73. Zhu JK (2002) Salt and drought stress signal transduction in plants. Annual review of plant biology 53: 247–273.
74. Nambara E, Marion-Poll A (2005) Abscisic acid biosynthesis and catabolism. Annu Rev Plant Biol 56: 165–185.
75. Nambara E, Marion-Poll A (2005) Abscisic acid biosynthesis and catabolism. Annu Rev Plant Biol 56: 165–185.
76. Iuchi S, Kobayashi M, Taji T, Naramoto M, Seki M, et al. (2001) Regulation of drought tolerance by gene manipulation of 9-cis-epoxycarotenoid dioxygenase, a key enzyme in abscisic acid biosynthesis in Arabidopsis. Plant J 27: 325–333.
77. Thompson AJ, Mulholland BJ, Jackson AC, McKee JM, Hilton HW, et al. (2007) Regulation and manipulation of ABA biosynthesis in roots. Plant Cell Environ 30: 67–78.
78. Cheng WH, Endo A, Zhou L, Penney J, Chen HC, et al. (2002) A unique short-chain dehydrogenase/reductase in Arabidopsis glucose signaling and abscisic acid biosynthesis and functions. Plant Cell 14: 2723–2743.
79. Hoth S, Morgante M, Sanchez JP, Hanafey MK, Tingey SV, et al. (2002) Genome-wide gene expression profiling in Arabidopsis thaliana reveals new targets of abscisic acid and largely impaired gene regulation in the abi1-1 mutant. J Cell Sci 115: 4891–4900.
80. Bertauche N, Leung J, Giraudat J (1996) Protein phosphatase activity of abscisic acid insensitive 1 (ABI1) protein from Arabidopsis thaliana. Eur J Biochem 241: 193–200.
81. Mustilli AC, Merlot S, Vavasseur A, Fenzi F, Giraudat J (2002) Arabidopsis OST1 protein kinase mediates the regulation of stomatal aperture by abscisic acid and acts upstream of reactive oxygen species production. Plant Cell 14: 3089–3099.
82. Yoshida R, Hobo T, Ichimura K, Mizoguchi T, Takahashi F, et al. (2002) ABA-activated SnRK2 protein kinase is required for dehydration stress signaling in Arabidopsis. Plant Cell Physiol 43: 1473–1483.
83. Fujii H, Verslues PE, Zhu JK (2007) Identification of two protein kinases required for abscisic acid regulation of seed germination, root growth, and gene expression in Arabidopsis. Plant Cell 19: 485–494.
84. Wang Y, Ying J, Kuzma M, Chalifoux M, Sample A, et al. (2005) Molecular tailoring of farnesylation for plant drought tolerance and yield protection. Plant J 43: 413–424.
85. Mikolajczyk M, Awotunde OS, Muszynska G, Klessig DF, Dobrowolska G (2000) Osmotic stress induces rapid activation of a salicylic acid-induced protein kinase and a homolog of protein kinase ASK1 in tobacco cells. Plant Cell 12: 165–178.
86. Li J, Wang XQ, Watson MB, Assmann SM (2000) Regulation of abscisic acid-induced stomatal closure and anion channels by guard cell AAPK kinase. Science 287: 300–303.
87. Cheong YH, Kim KN, Pandey GK, Gupta R, Grant JJ, et al. (2003) CBL1, a calcium sensor that differentially regulates salt, drought, and cold responses in Arabidopsis. Plant Cell 15: 1833–1845.
88. Hrabak EM, Chan CW, Gribskov M, Harper JF, Choi JH, et al. (2003) The Arabidopsis CDPK-SnRK superfamily of protein kinases. Plant Physiol 132: 666–680.
89. Jain M, Kaur N, Tyagi AK, Khurana JP (2006) The auxin-responsive GH3 gene family in rice (Oryza sativa). Functional & integrative genomics 6: 36–46.
90. Vandenbussche F, Petrášek J, Žádníková P, Hoyerová K, Pešek B, et al. (2010) The auxin influx carriers AUX1 and LAX3 are involved in auxin-ethylene interactions during apical hook development in Arabidopsis thaliana seedlings. Development 137: 597–606.
91. Liscum E, Reed J (2002) Genetics of Aux/IAA and ARF action in plant growth and development. Auxin Molecular Biology: Springer. pp. 387–400.
92. Baloglu M OM, Oktem H, Yucel M (2012) Expression analysis of TaNAC69-1 and TtNAMB-2, wheat NAC family transcription factor genes under abiotic stress conditions in durum wheat (Triticum turgidum). Plant Mol Biol Report 30: 1246–1252.
93. Abe H, Urao T, Ito T, Seki M, Shinozaki K, et al. (2003) Arabidopsis AtMYC2 (bHLH) and AtMYB2 (MYB) function as transcriptional activators in abscisic acid signaling. Plant Cell 15: 63–78.
94. Zheng X, Chen B, Lu G, Han B (2009) Overexpression of a NAC transcription factor enhances rice drought and salt tolerance. Biochem Biophys Res Commun 379: 985–989.
95. Sugano S, Kaminaka H, Rybka Z, Catala R, Salinas J, et al. (2003) Stress-responsive zinc finger gene ZPT2-3 plays a role in drought tolerance in petunia. Plant J 36: 830–841.
96. Choi H, Hong J, Ha J, Kang J, Kim SY (2000) ABFs, a family of ABA-responsive element binding factors. J Biol Chem 275: 1723–1730.
97. Tran LS, Nakashima K, Sakuma Y, Osakabe Y, Qin F, et al. (2007) Co-expression of the stress-inducible zinc finger homeodomain ZFHD1 and NAC transcription factors enhances expression of the ERD1 gene in Arabidopsis. Plant J 49: 46–63.
98. Muoma J, Ombori O, Jesse M (2011) Signal Transduction One of the Current Molecular Approaches hi the Management of Drought Stress hi the Sub-Saharan Region. Biotechnology 10: 469–476.

99. Aharoni A, Dixit S, Jetter R, Thoenes E, van Arkel G, et al. (2004) The SHINE clade of AP2 domain transcription factors activates wax biosynthesis, alters cuticle properties, and confers drought tolerance when overexpressed in Arabidopsis. Plant Cell 16: 2463–2480.
100. Hossain MA, Cho JI, Han M, Ahn CH, Jeon JS, et al. (2010) The ABRE-binding bZIP transcription factor OsABF2 is a positive regulator of abiotic stress and ABA signaling in rice. J Plant Physiol 167: 1512–1520.
101. Zhang JZ (2003) Overexpression analysis of plant transcription factors. Curr Opin Plant Biol 6: 430–440.
102. Cominelli E, Galbiati M, Vavasseur A, Conti L, Sala T, et al. (2005) A guard-cell-specific MYB transcription factor regulates stomatal movements and plant drought tolerance. Curr Biol 15: 1196–1200.
103. Chen JH, Song YP, Zhang H, Zhang DQ (2013) Genome-Wide Analysis of Gene Expression in Response to Drought Stress in Populus simonii. Plant Molecular Biology Reporter 31: 946–962.
104. Tardieu F, Tuberosa R (2010) Dissection and modelling of abiotic stress tolerance in plants. Curr Opin Plant Biol 13: 206–212.
105. Hadiarto T, Tran LSP (2011) Progress studies of drought-responsive genes in rice. Plant Cell Reports 30: 297–310.
106. Umezawa T, Fujita M, Fujita Y, Yamaguchi-Shinozaki K, Shinozaki K (2006) Engineering drought tolerance in plants: discovering and tailoring genes to unlock the future. Curr Opin Biotechnol 17: 113–122.
107. Tuteja N, Mahajan S (2007) Calcium signaling network in plants: an overview. Plant signaling & behavior 2: 79–85.
108. Hasegawa PM, Bressan RA, Zhu J-K, Bohnert HJ (2000) Plant cellular and molecular responses to high salinity. Annual review of plant biology 51: 463–499.
109. Nadarajah K, Sidek HM (2010) The Green MAPKs. Asian Journal of Plant Sciences 9: 1–10.
110. Wan B, Lin Y, Mou T (2007) Expression of rice Ca2+-dependent protein kinases(CDPKs) genes under different environmental stresses. FEBS letters 581: 1179–1189.
111. Hu X, Zhang Z, Xu P, Fu Z, Hu S, et al. (2010) Multifunctional genes: the cross-talk among the regulation networks of abiotic stress responses. Biologia plantarum 54: 213–223.
112. Watkinson JI, Sioson AA, Vasquez-Robinet C, Shukla M, Kumar D, et al. (2003) Photosynthetic acclimation is reflected in specific patterns of gene expression in drought-stressed loblolly pine. Plant Physiology 133: 1702–1716.
113. Queitsch C, Sangster TA, Lindquist S (2002) Hsp90 as a capacitor of phenotypic variation. Nature 417: 618–624.
114. Alvim FC, Carolino SM, Cascardo JC, Nunes CC, Martinez CA, et al. (2001) Enhanced accumulation of BiP in transgenic plants confers tolerance to water stress. Plant Physiology 126: 1042–1054.
115. Cho EK, Hong CB (2006) Over-expression of tobacco NtHSP70-1 contributes to drought-stress tolerance in plants. Plant cell reports 25: 349–358.
116. Wittenmayer L, Merbach W (2005) Plant responses to drought and phosphorus deficiency: contribution of phytohormones in root-related processes. Journal of Plant Nutrition and Soil Science 168: 531–540.

19

Genome-Wide Investigation and Expression Profiling of AP2/ERF Transcription Factor Superfamily in Foxtail Millet (*Setaria italica L.*)

Charu Lata[1,2], Awdhesh Kumar Mishra[3], Mehanathan Muthamilarasan[3], Venkata Suresh Bonthala[3], Yusuf Khan[3], Manoj Prasad[3]*

1 National Research Centre on Plant Biotechnology, New Delhi, India, **2** CSIR-National Botanical Research Institute, Lucknow, Uttar Pradesh, India, **3** National Institute of Plant Genome Research, New Delhi, India

Abstract

The APETALA2/ethylene-responsive element binding factor (AP2/ERF) family is one of the largest transcription factor (TF) families in plants that includes four major sub-families, namely AP2, DREB (dehydration responsive element binding), ERF (ethylene responsive factors) and RAV (Related to ABI3/VP). AP2/ERFs are known to play significant roles in various plant processes including growth and development and biotic and abiotic stress responses. Considering this, a comprehensive genome-wide study was conducted in foxtail millet (*Setaria italica* L.). A total of 171 *AP2/ERF* genes were identified by systematic sequence analysis and were physically mapped onto nine chromosomes. Phylogenetic analysis grouped *AP2/ERF* genes into six classes (I to VI). Duplication analysis revealed that 12 (~7%) *SiAP2/ERF* genes were tandem repeated and 22 (~13%) were segmentally duplicated. Comparative physical mapping between foxtail millet *AP2/ERF* genes and its orthologs of sorghum (18 genes), maize (14 genes), rice (9 genes) and *Brachypodium* (6 genes) showed the evolutionary insights of *AP2/ERF* gene family and also the decrease in orthology with increase in phylogenetic distance. The evolutionary significance in terms of gene-duplication and divergence was analyzed by estimating synonymous and non-synonymous substitution rates. Expression profiling of candidate *AP2/ERF* genes against drought, salt and phytohormones revealed insights into their precise and/or overlapping expression patterns which could be responsible for their functional divergence in foxtail millet. The study showed that the genes *SiAP2/ERF-069, SiAP2/ERF-103* and *SiAP2/ERF-120* may be considered as potential candidate genes for further functional validation as well for utilization in crop improvement programs for stress resistance since these genes were up-regulated under drought and salinity stresses in ABA dependent manner. Altogether the present study provides new insights into evolution, divergence and systematic functional analysis of *AP2/ERF* gene family at genome level in foxtail millet which may be utilized for improving stress adaptation and tolerance in millets, cereals and bioenergy grasses.

Editor: Swarup K. Parida, National Institute of Plant Genome Research (NIPGR), India

Funding: This work was financially supported by core grant of National Institute of Plant Genome Research, New Delhi, India, and Department of Science & Technology, Government of India through INSPIRE Faculty Award [IFA-11LSPA-01]. CL is the recipient of INSPIRE Faculty Award from Department of Science & Technology, while AKM and MM are the recipients of Research Fellowships from Council of Scientific and Industrial Research and University Grants Commission, New Delhi, respectively. The funders had no role in study design, data collection and analysis, decision to publish, or preparation of the manuscript.

Competing Interests: The corresponding author is serving as Academic Editor of PLOS ONE and this does not alter the authors' adherence to all the PLOS ONE policies on sharing data and materials.

* Email: manoj_prasad@nipgr.ac.in

Introduction

Plants frequently confront numerous environmental stresses which ultimately affect their growth and productivity. Therefore, in order to cope with these recurrent challenges, a plant species must acquire stress responsive and adaptive mechanisms that may assist in better survival and yield. Grass species of *Setaria* genus especially *S. italica* (foxtail millet) and *S. viridis* (green foxtail) prove to be excellent examples of stress adaptation and tolerance among graminaceous species [1]. Foxtail millet is a stress tolerant crop having a small genome (~515 Mb; 2n = 2x = 18) with relatively lower repetitive DNA, short life-cycle and inbreeding nature which makes it a perfect model for understanding various biological aspects including architecture, phylogeny and physiology of related Panicoid crops, particularly potential bioenergy grasses which have closely related but relatively composite genomes [2]. Considering its significance as a model system for evolution and biological studies, its genome has recently been sequenced by Beijing Genomics Institute (BGI), China [3] and Joint Genome Institute (JGI), Department of Energy, USA [4] independently [5]. The availability of foxtail millet genome sequence has consequently encouraged plant biology researchers to work towards deciphering its structural and functional genomics that may give new insights for its stress response and adaptation mechanisms and eventually support crop improvement programmes to ensure sustainable food security [6]. However, stress response and adaptation is a complex process as stress may possibly occur at different stages of plant development with different intensities (moderate to severe) and often several stresses may act together, thus increasing the effects manifold. It is hypothesized that plants have evolved an intricate signaling network to survive stress conditions that begins with stress perception, initiation of signal transduction, modulation of stress responsive gene(s) expression and finally its manifestation at cellular and physiological levels. Stress response and adaptation entails differential gene expression which is controlled by specific transcription factors (TFs) that directly regulate majority of downstream multiple stress responsive gene expression in a synchronized manner. Hence TFs are attractive targets for application in plant molecular biology for gene manipulation and crop improvement. Among various TF families, the ethylene responsive TF (ERF) family plays an important role in plant growth and development and also enables them to adapt to changing environmental conditions, and therefore it is important to understand molecular functions of these genes in order to improve plant adaptability and productivity under varied ambiance/environmental changes.

The APETALA2/ethylene-responsive element binding factor (AP2/ERF) superfamily is a large group of TFs which is distinguished by the number of repetitions and the sequence of AP2/ERF DNA-binding domain based on which it is categorized into AP2, ERF and RAV families [7,8]. The AP2/ERF domain was first reported in the *Arabidopsis* homeotic *AP2* gene implicated in floral development [9]. This conserved DNA-binding domain usually consists of 60–70 amino acid residues and is known to interact directly with *cis*-acting elements namely GCC box and/or dehydration responsive element (DRE)/C-repeat element (CRT) present in the promoter regions of downstream target genes [10,11]. The homologous sequences of this domain have been found in homing endonucleases (HNH-endonucleases) of the cyanobacterium *Trichodesmium erythraeum*, the ciliate *Tetrahymena thermophila*, and the viruses Enterobacteria phage RB49 and Bacteriophage Felix 01 and hence it has been postulated that a horizontal transfer of an HNH-AP2 endonuclease from prokaryotes into plants resulted in evolution of the AP2/ERF superfamily [12,13]. Among the threeAP2/ERF families, the members of AP2 family contain two AP2/ERF domains connected by a 25 amino acid linker, whereas the members of ERF subfamily contain a single AP2/ERF domain. The RAV family members contain a single AP2/ERF domain and an additional B3 DNA-binding motif [14]. In addition, the AP2 family is again categorized into AP2 and AINTEGUMENTA (ANT) monophyletic groups in seed plants [15], while the ERF family is further subdivided into ERF and DREB subfamilies [6,7]. The ERF subfamily is characterized by the presence of conserved alanine and aspartic acid at 14th and 19th position respectively in the DNA-binding domain, while the DREB subfamily has valine and glutamine at respective positions [16].

A large number of AP2/ERF TFs have been identified and studied in various plants including *Arabidopsis*, rice, wheat, poplar, barley, castor bean, grape, cucumber, soybean, *Brassica* and *Malus* [8,17–25]. The genome-wide analyses of AP2/ERF TF superfamily have been performed in these crops, both to categorize each family member in an ordered nomenclature system as well as to investigate their expression profiles and chromosomal positions. As already mentioned, two full genome sequences of *Setaria italica* cv. Zhang gu and inbred Yugu 1 are available which have not only provided a useful genomic platform but have also paved pathway for researchers to carry out advanced genetic and genomic studies in this model crop. As AP2/ERF TFs show wide diversity of functions including regulation of several developmental processes such as vegetative and reproductive development, cell proliferation, and responses to various abiotic and biotic stresses and plant hormones [8,10], their superfamily stands as one of the best candidates to examine important traits in foxtail millet. With this aim, a genome wide investigation of foxtail millet AP2/ERF TF superfamily and their expression profiling has been taken up in this study. Hence, this is the first comprehensive report on genome-wide survey, expression profiling and evolutionary analysis of AP2/ERF proteins in foxtail millet (named as *Setaria italica* AP2/ERF; 'SiAP2/ERF').

Materials and Methods

Sequence retrieval and identification of AP2/ERF proteins from *Setaria italica*

The Hidden Markov Model (HMM) profile of the AP2/ERF domain (PF00847) was obtained from Pfam v27.0 database (http://Pfam.sanger.ac.uk/) [26] and searched against the PHYTOZOME database of *Setaria italica* (www.phytozome.net/). All hits with expected (E) values less than 1.0 were retrieved and the non-redundant sequences were examined for the presence of conserved AP2/ERF domain by executing HMMSCAN (http://hmmer.janelia.org/search/hmmscan).

Chromosomal location, gene structure and genomic distribution of *AP2/ERF* genes

The identified AP2/ERF domain-containing proteins were BLASTP searched against *S. italica* genome of PHYTOZOME database with default settings, and physical map was constructed using MapChart [27]. Segmental duplications were calculated based on the method of Plant Genome Duplication Database [28] using MCScan v0.8 [29] and visualized using Circos v0.55 [30]. Tandem duplications were identified manually as described elsewhere [31,32] and marked on the physical map. The exon-intron organizations of the genes were ascertained using Gene Structure Display Server (http://gsds.cbi.pku.edu.cn/) [33].

Phylogenetic analysis, Gene Ontology (GO) annotation, promoter analysis and identification of miRNAs targeting *SiAP2/ERFs*

The amino acid sequences of AP2/ERF proteins were imported into MEGA5 and an unrooted phylogenetic tree based on the Neighbor-joining method was generated after 1000 bootstrap replications [34]. The GO annotation of AP2/ERF protein sequences was performed using Blast2GO [35] and *cis*-regulatory elements were identified using PLACE (http://www.dna.affrc.go.jp/PLACE/) database. Further. the *S. italica* miRNAs reported by Khan et al. [36] were retrieved and searched for their targets in 171 *SiAP2/ERF* transcripts using psRNATarget tool (http://plantgrn.noble.org/psRNATarget/).

Comparative mapping and evolutionary analysis of paralogs and orthologs

The amino acid sequences of SiAP2/ERF proteins that were physically mapped onto foxtail millet genome were BLASTP searched against protein sequences of sorghum, maize, rice and *Brachypodium* (http://gramene.org/; www.phytozome.net). Reciprocal BLAST was also carried out to establish unique relationship between the orthologous genes. Hits with E-value≥ 1e-05 and at least 80% homology were considered significant. The comparative orthologous relationships of *AP2/ERF* genes among foxtail millet, sorghum, maize, rice and *Brachypodium* were finally

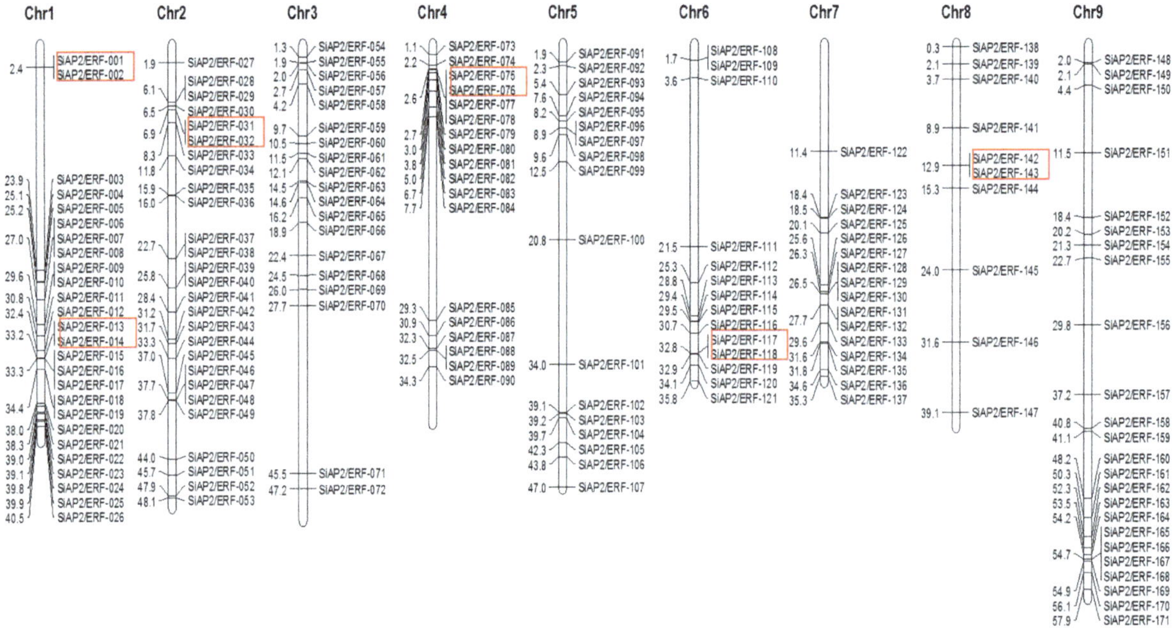

Figure 1. Physical map of 171 *SiAP2/ERF* genes. The bars represent the chromosomes with numbers at the left indicating the physical position (in Mb). The tandemly duplicated gene pairs are indicated within boxes.

illustrated using Circos [30]. For estimating the synonymous (Ks) and non-synonymous (Ka) substitution rates, the corresponding amino-acid as well as cDNA sequences of paralogous and orthologous SiAP2/ERF proteins were analyzed using PAL2NAL (http://www.bork.embl.de/pal2nal/) [37]. Time (million years ago, Mya) of duplication and divergence was calculated using a synonymous mutation rate of l substitutions per synonymous site per year as $T = Ks/2\lambda$ ($\lambda = 6.5\times10^{-9}$) [38,39].

Tissue-specific expression profiling using RNA-seq data

S. italica Illumina RNA-HiSeq reads from 4 tissues namely spica, stem, leaf and root, retrieved from European Nucleotide Archive [SRX128226 (spica); SRX128225 (stem); SRX128224 (leaf); SRX128223 (root)] [40] and were mapped onto the gene sequences of *Setaria italica* using CLC Genomics Workbench v.4.7.1 (http://www.clcbio.com/genomics). Normalization of the mapped reads was done using RPKM (reads per kilobase per million) method. The heat map for tissue-specific expression profile was generated based on the RPKM values for each gene in all the tissue samples using TIGR MultiExperiment Viewer (MeV4) software package [41,42].

Plant materials, growth conditions and stress treatments

Seeds of drought tolerant foxtail millet cultivar IC-403579 [8] were obtained from National Bureau of Plant Genetic Resources (NBPGR), Hyderabad, India. The seeds were sown in composite soil (peat compost: vermiculite: sand, 2:2:1) in glass house at National Phytotron Facility, Indian Agricultural Research Institute (IARI), New Delhi, India at 28±1°C day/23±1°C night temperature with 70±5% relative humidity and natural sunlight during June–July, 2013. For stress treatments, two week old seedlings were exposed to 20% polyethylene glycol (PEG 6000) (drought), 250 mM NaCl (salt), 100 μM abscisic acid (ABA), 100 μM salicylic acid (SA), 100 μM methyl jasmonate (MeJA) and 100 μM ethephon (Et) for 1 h (early) and 24 h (late) based on our previous studies (Lata et al. 2010; Lata et al. 2011a; Lata et al. 2011b). The plants were supplemented with water and Hoagland solution on alternate days. Unstressed plants were maintained as control. After stress treatments, whole seedlings were carefully harvested and immediately frozen in liquid nitrogen and stored at −80°C until RNA isolation. Three independent experiments were conducted for precision and reproducibility, and for each experiment, ~100 mg seedling samples were collected by random sampling.

RNA extraction and expression analysis using qRT-PCR

Total RNA was isolated from the 14-day old unstressed and stressed (1 h and 24 h) foxtail millet cv. IC-403579 seedlings using TRIzol Reagent (Sigma, USA) following manufacturer's instructions. DNA contamination was removed from the RNA samples using RNase-free DNaseI (1 U μl^{-1}, Fermentas). The quality and purity of the RNA preparations were determined by measuring the OD_{260}/OD_{280} absorption ratio (1.9–2.0), and the integrity of the preparations was determined by electrophoresis in a 1.2% agarose gel containing formaldehyde as described in previous studies [43,44]. RNA concentrations were measured by a spectrophotometer (Nanodrop, USA). About 1 μg of total RNA was used to synthesize first strand cDNA primed with OligodT in a 20 μl reaction mix using 200 U/μl of PrimeScript M-MuLV reverse transcriptase (Takara Bio Inc., USA) following manufacturer's instructions. Quantitative real time (qRT) PCR was performed using SYBR Premix ExTaq II (Tli RNaseH Plus) (Takara Bio Inc., USA) on Mastercycler ep realplex system (Eppendorf) in triplicate. The constitutive gene RNA Polymerase II (RNA POL II; Accession No Si033113m) from foxtail millet was used as endogenous control which gave an amplification product of 146 bp [45]. The qRT-PCR primers of the *SiAP2/ERF* genes were designed from non-conserved regions of the corresponding genes using GenScript real-time PCR (TaqMan) Primer Design tool (www.genscript.com) using default parameters (Table S1). The PCR mixtures and reactions were used as detailed previously [45]. Melting curve analysis (60 to 95°C after 40 cycles) and

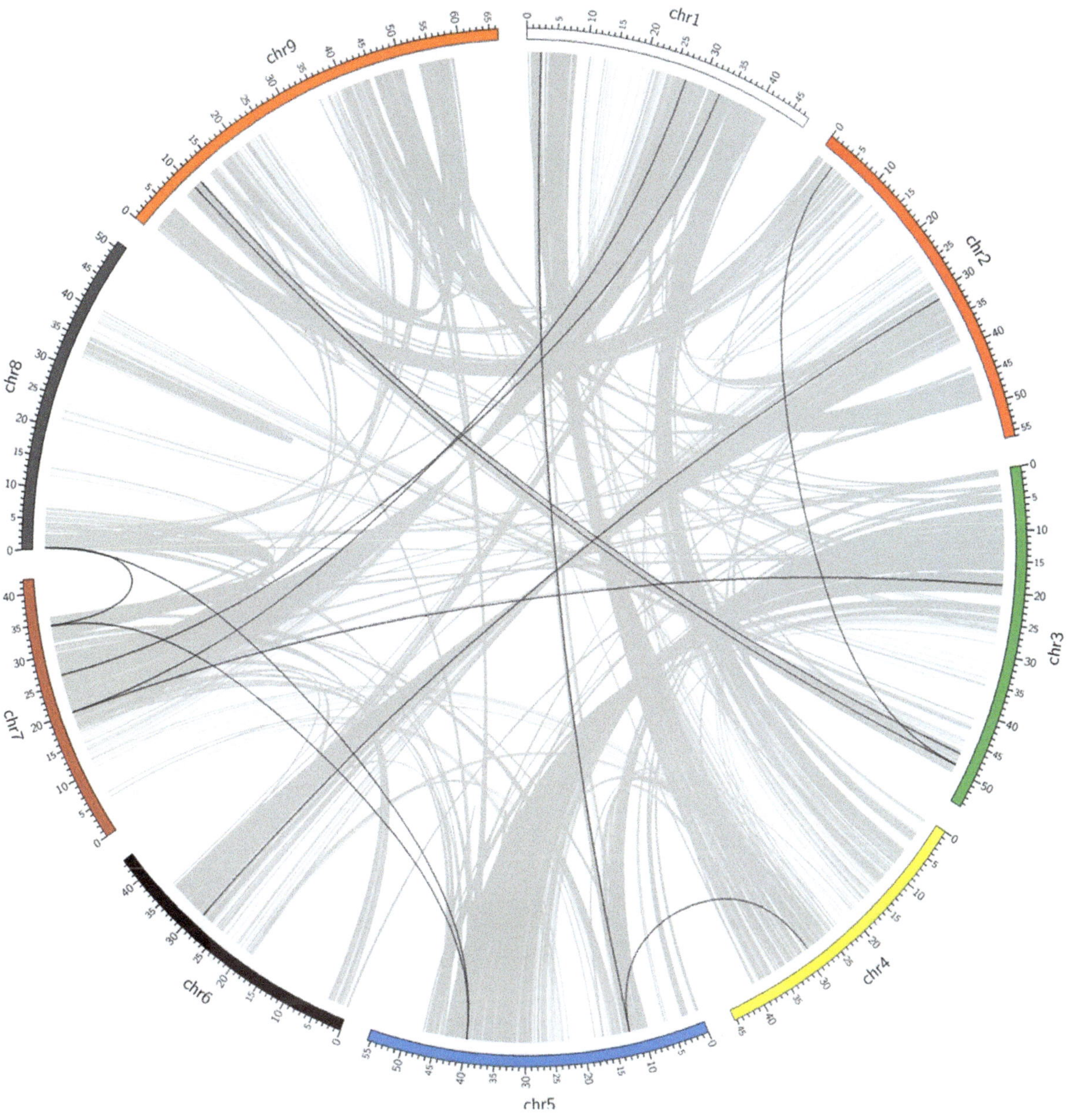

Figure 2. Distribution of segmentally duplicated *SiAP2/ERF* genes on foxtail millet chromosomes. Grey lines indicate collinear blocks in whole foxtail millet genome, and black lines indicate duplicated *SiAP2/ERF* gene pairs.

agarose gel electrophoresis were performed to check the amplification specificity of *AP2/ERF* genes normalized to the internal control RNA POL II and were analyzed using $2^{-\Delta\Delta Ct}$ method [46]. qRT-PCR data analysis was done according to previous studies [45,46]. The PCR cycling conditions were: initial denaturation at 95°C for 2 min, 95°C for 15 s, and 60°C for 1 min for 40 cycles followed by melting curve analysis using default parameters i.e. 95°C for 15 s, 60°C for 15 s, 95°C for 15 s with ramp time of 20 min.

Identification of molecular markers and homology modeling of SiAP2/ERF proteins

The presence of various types of DNA-based markers including simple sequence repeats (SSRs) [47], EST-derived SSRs (eSSRs) [48] and intron length polymorphic (ILPs) markers [49] retrived from FmMDb (http://www.nipgr.res.in/foxtail.html) [50] were searched in the *SiAP2/ERF* genes using in-house perl script. For homology modeling, all the SiAP2/ERF proteins were queried against the Protein Data Bank (PDB) [51] to identify the best template with similar amino acid sequence and known 3D structure. The data was fed in Phyre2 server (Protein Homology/AnalogY Recognition Engine; http://www.sbg.bio.ic.ac.uk/phyre2) for predicting the three-dimensional structure of proteins by homology modeling under 'normal' mode [52]. Active site was predicted using COACH server (http://zhanglab.ccmb.med.umich.edu/COACH/) and highlighted using UCSF Chimera 1.8.

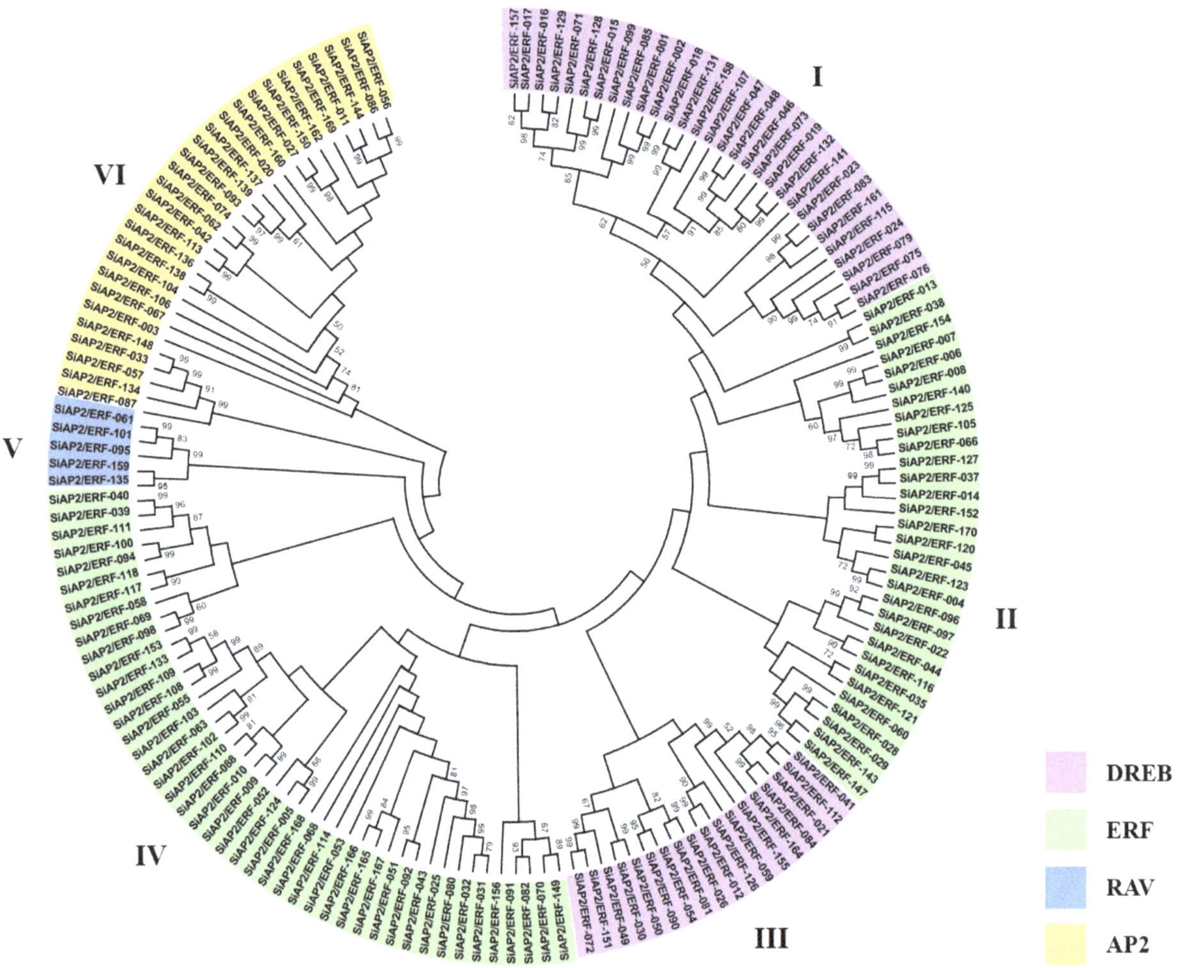

Figure 3. Phylogenetic tree of 171 *SiAP2/ERF* genes. Bootstrap confidence values from 1000 replicates are indicated at each branch. The different classes of SiAP2/ERF proteins are highlighted in different colors.

Results and Discussion

Identification of the AP2/ERF family transcription factors in foxtail millet genome

The HMM BLAST identified a total of 186 AP2/ERF protein sequences from foxtail millet. Fifteen proteins were found to be splice variants of primary transcripts, removal of which led to the identification of a total of 171 putative SiAP2/ERF proteins (Table S2) which represents approximately 0.4407% of all annotated genes (38801 genes total) in the *Setaria* genome [3] which is very similar to those present in poplar (0.4390%) and rice (0.4315%) however approximately 0.10% smaller than that of Arabidopsis (0.5481%) [7,8,18,53,54]. Among splice variants, Si022619m gene was found to encode a maximum of 8 alternate

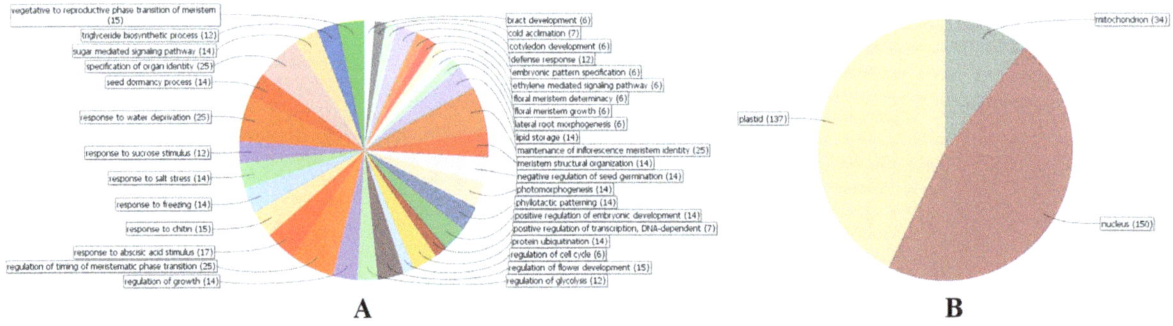

Figure 4. Gene Ontology annotation of SiAP2/ERF proteins. The Blast2GO output defining the (**A**) biological processes and (**B**) cellular localization of 171 SiAP2/ERF proteins.

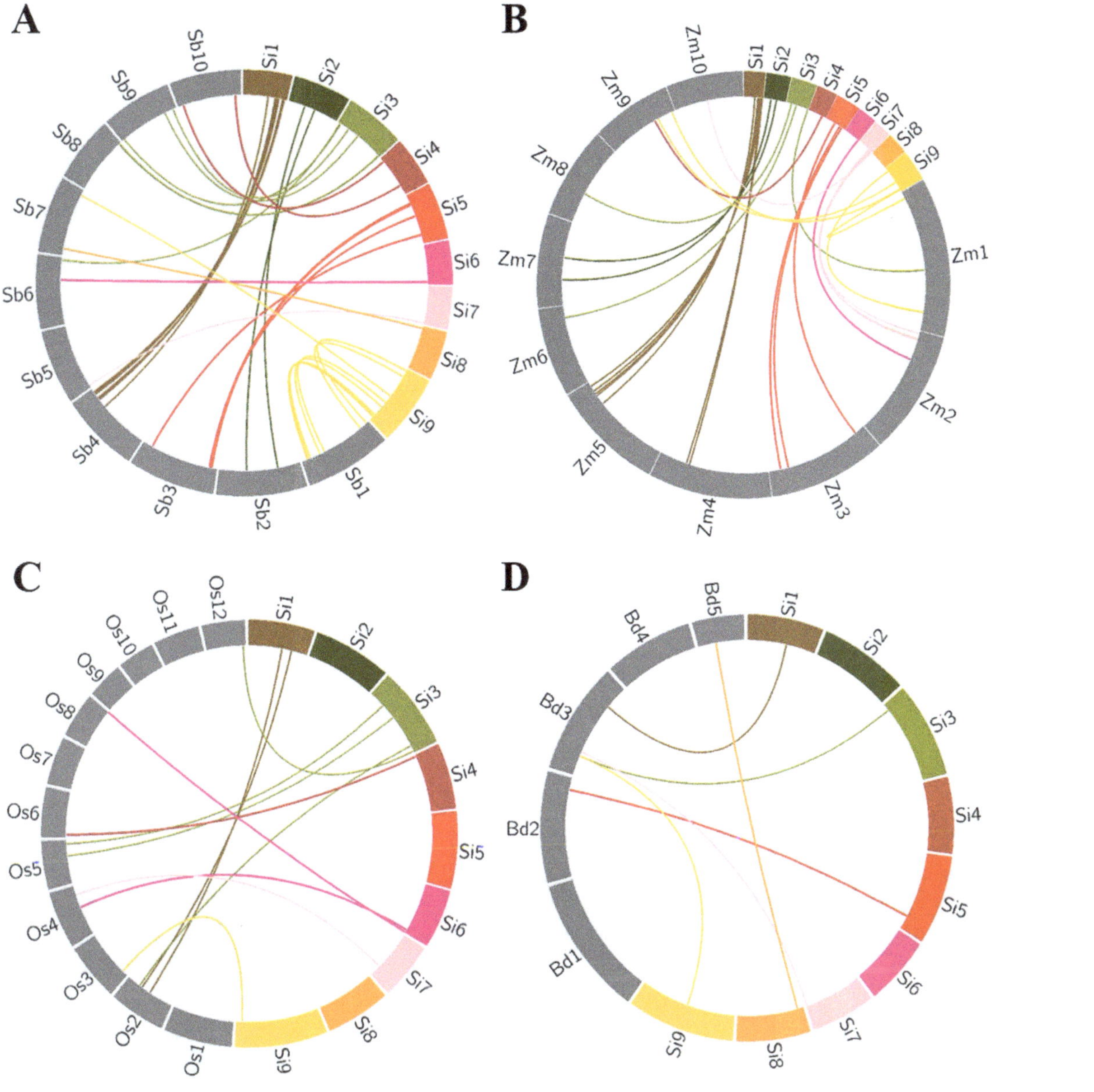

Figure 5. Comparative physical mapping showing the degree of orthologous relationships of *SiAP2/ERF* genes located on nine chromosomes of foxtail millet with (A) sorghum, (B) maize, (C) rice and (D) *Brachypodium*.

transcripts (Si022998m, Si022997m, Si022990m, Si022995m, Si022996m, Si022991m, Si022621m, Si022989m), followed by Si006802m and Si010289m which has 2 splice variants each (Si006941m, Si006905m in Si006802m and Si010292m, Si010301m in Si010289m) (Table S2). Genes Si030514m, Si036615m and Si036647m comprised of one alternate transcript Si030506m, Si036606m and Si036938m, respectively. The number conforms with the number of AP2/ERFs reported in Foxtail millet Transcription Factor Database [55]. Of note, the identification of pseudogenes among these 171 AP2/ERFs require further experimentations. In addition, the respective gene sequences encoding these proteins were retrieved and the presence of AP2/ERF domain was ascertained (Table S3). Due to lack of proper annotation, the existing identities of the genes were highly disordered and therefore for convenience, all 171 genes were assigned consecutive numbers from SiAP2/ERF-001 to SiAP2/ERF-171 in the order of their chromosomal locations. All *SiAP2/ERF* genes varied greatly in the size and sequence of their encoded proteins as well as in their physico-chemical properties (Table S2). Additionally all *SiAP2/ERF* genes were characterized by the presence of one or two highly conserved AP2/ERF DNA-binding domains and a B3 domain in case of RAV proteins (Table S3). The lengths of the identified proteins vary from 88 to 691 amino acids. ExPASy analysis revealed large variation in iso-electric point (pI) values ranging from 4.26 to 11.7 and molecular weight ranging from 9.92 to 72.31 kDa. Interestingly, this wide variation in pI and molecular weight revealed the presence of putative novel variants of SiAP2/ERFs and this is in accordance to previous genome-wide reports on DCL, AGO, RDR, C_2H_2 zinc finger and MYB genes in foxtail millet [56–58]. The characteristic features of SiAP2/ERF protein sequences are summarized in Table S2. Dual targeting is a term used to infer the ability of proteins to localize

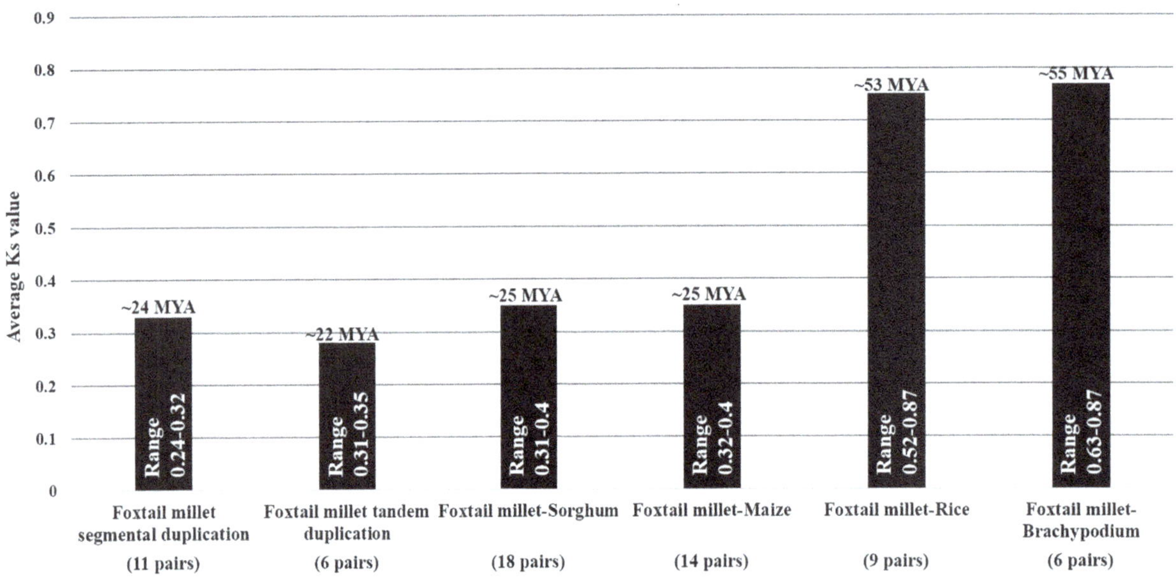

Figure 6. Time of duplication and divergence (MYA) based on synonymous substitution rate (Ks) estimated using paralogous and orthologous *SiAP2/ERF* gene pairs.

into more than one cellular compartment and it can also be viewed as a post translational regulatory mechanism [59]. Localization of 171 AP2/ERF proteins was determined using Blast2GO. Majority of the SiAP2/ERF proteins were predicted to be dual targeted or localized to nucleus, plastid and/or mitochondrion except SiAP2/ERF-025, SiAP2/ERF-032,

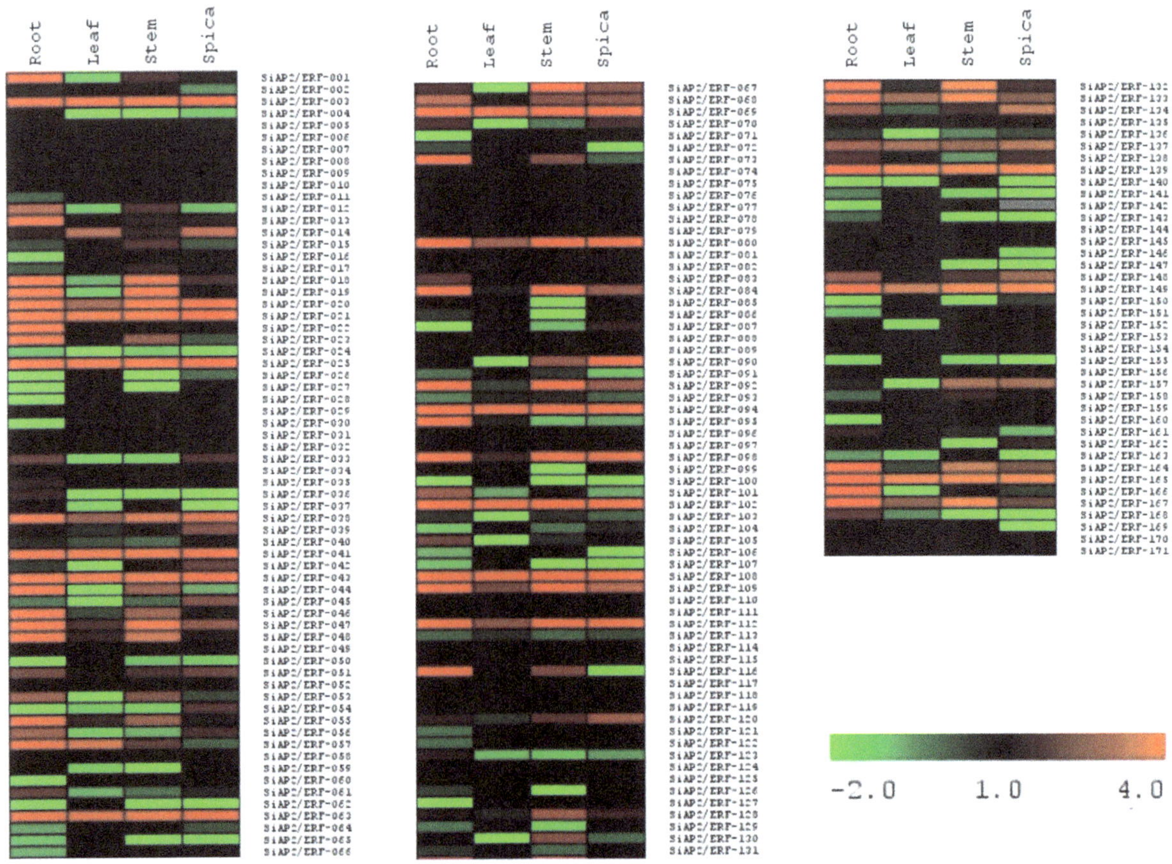

Figure 7. Heat-map showing the expression pattern of *SiAP2/ERF* genes in four tissues namely leaf, root, stem and spica. The color scales for fold-change values are shown at the bottom right.

SiAP2/ERF-035, SiAP2/ERF-040, SiAP2/ERF-051, SiAP2/ERF-055, SiAP2/ERF-063, SiAP2/ERF-065, SiAP2/ERF-091, SiAP2/ERF-100, SiAP2/ERF-108, SiAP2/ERF-121, SiAP2/ERF-122, SiAP2/ERF-153, SiAP2/ERF-165, SiAP2/ERF-166 (nucleus); SiAP2/ERF-024, SiAP2/ERF-059, SiAP2/ERF-066, SiAP2/ERF-078, SiAP2/ERF083 and SiAP2/ERF096 (plastid); and SiAP2/ERF-034, SiAP2/ERF-075, SiAP2/ERF-077, SiAP2/ERF-079, SiAP2/ERF-111 and SiAP2/ERF-159 (intracellular membrane-bound organelle) localized. Further, AP2/ERF superfamily was divided into four major families on the basis of nature and number of DNA-binding domains, namely AP2, ERF, DREB and RAV. The AP2/ERF proteins of *Setaria italica* were also classified into these families. Out of 171 genes, 28 belong to AP2, 90 to ERF, 48 to DREB and 5 to RAV (Table S3) indicating that foxtail millet genome supports large ERF and DREB subfamilies similar to Chinese cabbage genome [60].

Chromosomal distribution and structure of SiAP2/ERF proteins

Physical mapping of *SiAP2/ERFs* on all 9 chromosomes of foxtail millet revealed an uneven distribution of *SiAP2/ERF* genes in the genome (Figure 1). Among all chromosomes, chromosome 2 contained highest number of *AP2/ERF* genes (27; ~16%) followed by chromosome 1 (26; ~15%), while minimum number of genes were assigned on chromosome 8 (10; ~6%). The precise position (in bp) of each *SiAP2/ERF* on foxtail millet chromosomes is detailed in Table S2. Distribution pattern of these genes on individual chromosomes also pointed certain physical regions with a relatively higher accumulation of *AP2/ERF* gene clusters. As for example, *SiAP2/ERF* genes located on chromosomes 1, 6, 7 and 9 appear to congregate at the lower end of the arms as compared to chromosome 3 and 4 where these genes appear to cluster together at the upper end of the arm (Figure 1). It has recently been reported that foxtail millet genome underwent whole-genome duplication similar to other grass species about 70 million years ago [3] and hence occurrence of such large number of *SiAP2/ERF* genes in foxtail millet genome suggests huge amplification of this gene family during the course of evolution. Hence, duplication of these genes was studied and found that 12 (~7%) *SiAP2/ERF* genes were tandem repeated (Figure 1) and 22 (11 pairs; ~13%) were segmentally duplicated (Figure 2). The tandem duplicated genes included six clusters (2 genes each) including two clusters on chromosome 1 and one each on chromosomes 2, 4, 6 and 8. The distance between these genes ranged from 6.2 kb to 32.2 kb. Among the segmentally duplicated gene pairs, three genes namely, SiAP2/ERF-001 (chromosome 1), SiAP2/ERF-013 (chromosome 1) and SiAP2/ERF-072 (chromosome 5) were duplicated twice in the genome forming six paralogs in chromosomes 2, 4, 5, 7 and 9 (Figure 2). Chromosomal localization study of *SiAP2/ERF* genes thus indicates that tandem- and segmental-duplication may be one of the contributing factors in evolution of new genes in foxtail millet genome. Moreover, analysis of *SiAP2/ERF* gene structures indicated highly diverse distribution of intronic regions (from 1 to 10 in numbers) among the exonic sequences suggesting significant evolutionary changes in the foxtail millet genome. Interestingly, 89 (~52%) *SiAP2/ERF* genes were found to be intronless (Figure S1). Similar results were also observed in case of *Arabidopsis* [7] and *Lotus corniculatus* [61]. Further, the shortest *SiAP2/ERF* gene was merely 266 bp (SiAP2/ERF-156), whereas the longest one was identified as SiAP2/ERF-150 with 3.7 kb genomic sequence.

Phylogenetic analysis of SiAP2/ERF proteins

Phylogenetic analysis is essential for understanding the evolutionary history of crop species. Therefore, to understand the evolutionary significance of domain architecture, a phylogenetic tree was constructed with 171 SiAP2/ERF proteins. The phylogenetic analysis clustered all the SiAP2/ERFs into distinct clades (AP2, ERF, DREB and RAV) comprising of 28, 90, 48, and 5 proteins, respectively, according to their domain composition (Figure 3). Interestingly, the DREB formed two clades intervened

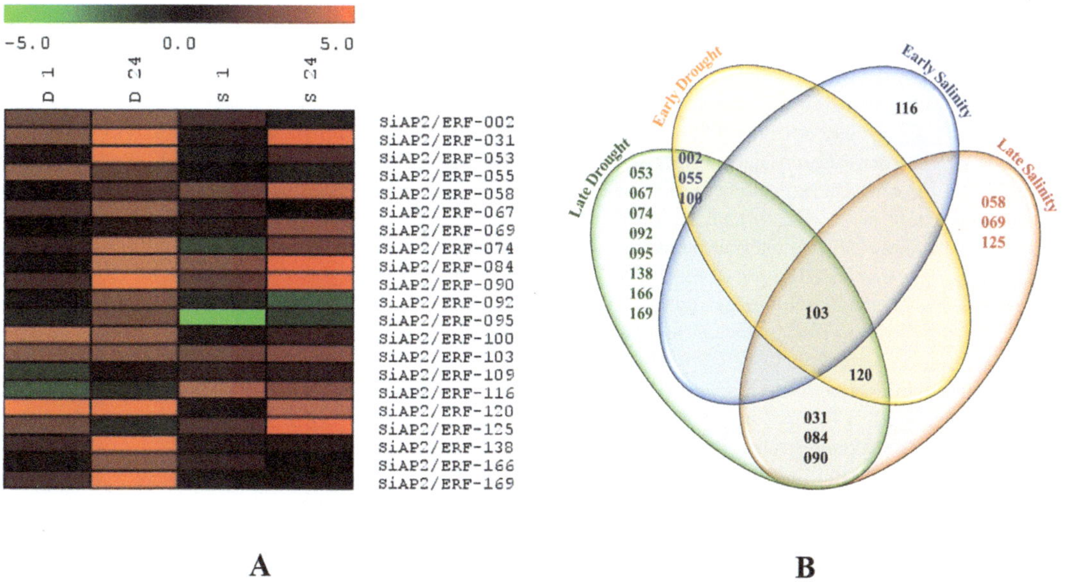

Figure 8. Expression profile of 21 *SiAP2/ERF* genes in response to dehydration and salinity stresses. (**A**) Heat map showing differential gene expression in response to dehydration (D) and salinity (S) stress across two time points (1 h and 24 h). The heat-map has been generated based on the fold-change values in the treated sample when compared with its unstressed control sample. The color scale for fold-change values is shown at the top. (**B**) Venn diagram showing stress-specific higher-expression of *SiAP2/ERF* genes during early and late stresses. The common subset of genes regulated by two stresses is marked by the overlapping circle. The numbers provided in the venn diagram corresponds to the *SiAP2/ERF* ID listed in Table S2.

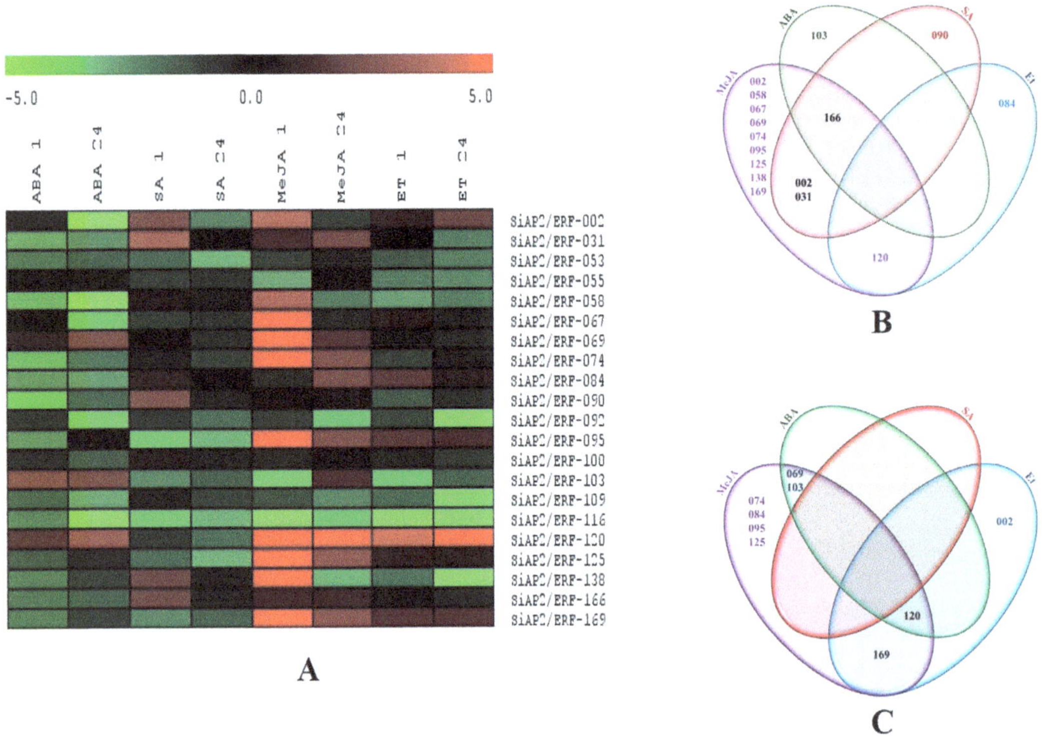

Figure 9. Expression profile of 21 *SiAP2/ERF* genes in response to Abscisic acid (ABA), salicylic acid (SA), methyl jasmonate (MeJA) and Ethephone (Et) treatments. (**A**) Heat map showing differential gene expression in response to ABA, SA, MeJA, Et treatments across two time points (1 h and 24 h). The heat-map has been generated based on the fold-change values in the treated sample when compared with its treated control sample. The color scale for fold-change values is shown at the top. (**B**) Venn diagram showing stress-specific higher-expression of *SiAP2/ERF* genes during early hormonal treatment. (**C**) Venn diagram showing stress-specific higher-expression of *SiAP2/ERF* genes during early hormonal treatment. The common subset of genes regulated by the hormonal treatments is marked by the overlapping circle. The numbers provided in the venn diagram corresponds to the *SiAP2/ERF* ID listed in Table S2.

by ERFs. Although similar observation was not reported in genome-wide studies of AP2/ERF conducted in other plants, in Chinese cabbage it has been found that the DREB clade was intervened by AP2 [60]. Further in-depth *in silico* analysis is requisite for finding the possible reason for such observation. The tree was divided into six groups based on the distribution of AP2, ERF, DREB and RAV. The derivation of statistically reliable pairs of possible homologous proteins sharing similar functions from a common ancestor was confirmed owing to high bootstrap values observed for a good number of the internal branches of the unrooted phylogenetic tree. Close association of SiAP2/ERF sub-families with their corresponding counterparts in other plant systems in terms of expression and/or biological and regulatory functions may be an implication of sequence conservation and also evidence to their similar *in planta* roles. Such phylogeny-based function prediction is obviously a rational systematic approach to facilitate identification of orthologous genes and has near-perfectly been applied for prediction of AP2/ERF proteins in other plant species such as rice, grapes and *Brassica* [20,21,60]. Thus, members of the sub-families of AP2/ERF superfamily in foxtail millet also have similar regulatory roles as those of their orthologs in other crop species.

Gene Ontology annotation

The GO analysis performed using rice protein sequences as reference showed the putative participation of SiAP2/ERF proteins in diverse biological, cellular and molecular processes (Figure 4; Table S4). The analysis of biological processes mediated by SiAP2/ERF depicted that a predominant of SiAP2/ERF proteins were involved in stress responses, such as response to water deprivation, salt stress and freezing. In addition, SiAP2/ERF proteins were also evidenced to participate in regulation of timing of meristematic phase transition, specification of organ identity and maintenance of inflorescence meristem identity. The molecular processes of SiAP2/ERF proteins clearly showed that all the 171 proteins possess sequence-specific DNA binding transcription factor activity (Table S4). Further, cellular component analysis revealed the localization of SiAP2/ERF proteins in nucleus, plastids, mitochondria and other intracellular membrane-bound organelles. These are in concordance with the experimental findings reported earlier [8,10,61,62].

Promoter analysis and miRNA targets of *SiAP2/ERF* genes

Cis-regulatory elements are DNA sequences that are situated upstream of genes in the promoter region and act as TF-binding sites. These are known to play crucial roles in determining tissue-specific as well as stress-responsive gene expression [63]. There have also been reports that multi-stimuli genes are closely correlated with *cis*-regulatory elements in their promoter sequences [64]. Therefore, a comprehensive promoter analysis of all the 171 *SiAP2/ERF* genes was conducted in order to further understand transcriptional regulation and support functional prediction of the respective proteins (Table S5). A total of 300 cis-regulatory elements were found to be present in one or the other *SiAP2/ERF* gene. The cis-regulatory elements CACTFTPPCA1, CAATBOX1, EBOXBNNAPA, MYCCON-

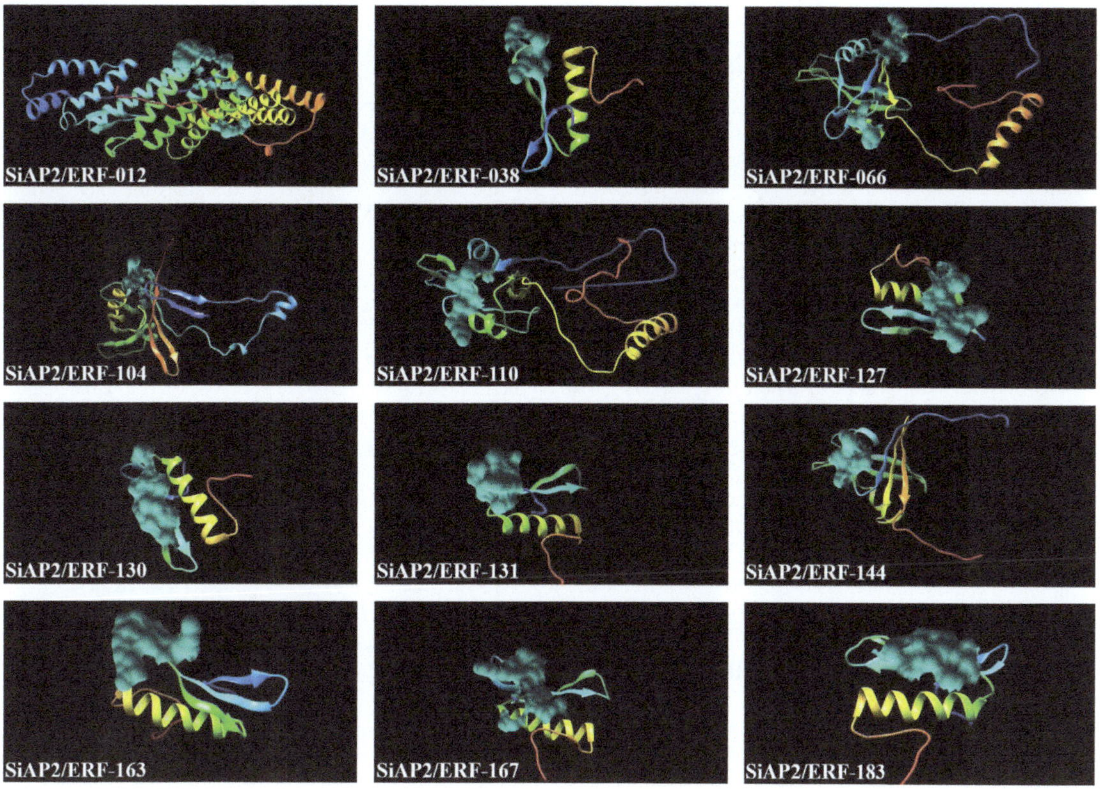

Figure 10. Predicated structures of SiAP2/ERF proteins. The structures of 12 SiAP2/ERF proteins with greater than 90% confidence level were shown along with its potential active site.

SENSUSAT, WRKY71OS, GT1CONSENSUS, ARR1AT, DOFCOREZM, GTGANTG10, RAV1AAT and GATABOX were present in all the 171 genes whereas HSE, VSF1PVGRP18, GMHDLGMVSPB, ABREAZMRAB28, PALBOXLPC, DR5GMGH3, SITEIIAOSPCNA, ABREBNNAPA, ABASEED1, ABAREG2, O2F3BE2S1, OPAQUE2ZM22Z, RGATAOS, CEREGLUBOX3PSLEGA, CPRFPCCHS, LREBOXIIPCCHS1, AGL2ATCONSENSUS, TOPOISOM, NONAMERATH4, PALBOXPPC, WRECSAA01, SORLIP4AT, NONAMERMOTIFTAH3H4, ABREMOTIFIOSRAB16B, ACIIIPVPAL2, ABRE3OSRAB16, B2GMAUX28, AT1BOX, SPHZMC1, D3GMAUX28, INTRONUPPER, WINPSTPIIIK, JASE1ATOPR1, OCETYPEIINTHISTONE and MSACRCYM were unique to *SiAP2/ERF-020, SiAP2/ERF-023, SiAP2/ERF-024, SiAP2/ERF-026, SiAP2/ERF-029, SiAP2/ERF-036, SiAP2/ERF-038, SiAP2/ERF-059, SiAP2/ERF-063, SiAP2/ERF-066, SiAP2/ERF-067, SiAP2/ERF-069, SiAP2/ERF-071, SiAP2/ERF-072, SiAP2/ERF-073, SiAP2/ERF-073, SiAP2/ERF-082, SiAP2/ERF-083, SiAP2/ERF-091, SiAP2/ERF-091, SiAP2/ERF-091, SiAP2/ERF-093, SiAP2/ERF-094, SiAP2/ERF-099, SiAP2/ERF-110, SiAP2/ERF-112, SiAP2/ERF-128, SiAP2/ERF-132, SiAP2/ERF-140, SiAP2/ERF-152, SiAP2/ERF-157, SiAP2/ERF-159, SiAP2/ERF-166, SiAP2/ERF-169 and SiAP2/ERF-170*, respectively (Table S5). In addition, putative microRNAs (miRNAs) targeting the *SiAP2/ERF* genes were also detected using psRNATarget server. The analysis showed five *SiAP2/ERF* genes to be targeted by *Setaria italica* miRNAs (Sit-miRs) (Table S6; Figure S2). The miRNAs identified in this study would be helpful in interpreting the post-transcriptional control of gene regulation during various physiological and stress-induced cellular responses in this otherwise naturally stress tolerant crop.

Orthologous relationships of *SiAP2/ERF* genes between foxtail millet and other grass species

To derive orthologous relationships of SiAP2/ERFs, comparative mapping approach was followed wherein the physically mapped *AP2/ERF* genes of foxtail millet were compared with those in the chromosomes of related grass genomes namely sorghum, maize, rice and *Brachypodium* (Figure 5). Maximum orthology of genes annotated on the foxtail millet chromosomes was exhibited with sorghum (18; ~11%) followed by maize (14; ~8%), rice (9; 5%) and least with *Brachypodium* (6; 4%). The extensive synteny among foxtail millet, sorghum and maize at gene level supports their close evolutionary relationships [3,4]. Intriguingly, most of *SiAP2/ERF* genes showed syntenic bias towards particular chromosomes of sorghum, maize, rice and *Brachypodium* and this suggests that the chromosomal rearrangement events like duplication and inversion predominantly shaped the distribution and organization of *AP2/ERF* genes in these grass genomes. The comparative mapping information thus offers a useful preface for understanding the evolution of *AP2/ERF* genes among grasses including foxtail millet. In addition, this study would be helpful in selecting candidate *SiAP2/ERF* genes and utilize them in genetic improvement of related grass family members. As for example, AP2-like ethylene-responsive transcription factors PLETHORA 1 and 2 are essential for QC specification and stem cell activity in roots of Arabidopsis [65]. It is thus likely that its orthologous foxtail millet gene (SiAP2/ERF-011; Phytozome ID: Si016558m) and rice gene (LOC_Os02g40070.1) may also be involved in similar function.

Duplication and divergence rate of the *SiAP2/ERF* genes

Whole genome duplications such as tandem and segmental duplications usually give rise to multiple copies of genes in a gene family. Such gene duplication events have been reported in various plant TF families including MYB, NAC and AP2/ERF [21,56–58,66–68]. Association of Darwinian positive selection in duplication and divergence of *AP2/ERF* in foxtail millet was explored by estimating the ratios of non-synonymous (Ka) versus synonymous (Ks) substitution rate (Ka/Ks) for 6 tandem and 11 segmentally duplicated gene-pairs as well as between orthologous gene-pairs of *SiAP2/ERF* with those of sorghum (18 pairs), maize (14 pairs), rice (9 pairs) and *Brachypodium* (6 pairs) (Figure 6) (Tables S7–S12). The Ka and Ks are a measure to examine the course of divergence after duplication, and the Ka/Ks ratio is a measure of the selection pressure to which a gene pair is subjected wherein Ka/Ks <1 means purifying selection, Ka/Ks = 1 stands for neutral selection, and Ka/Ks >1 signifies accelerated evolution with positive selection [38]. The Ka/Ks ratio for tandem duplicated gene-pairs in foxtail millet AP2/ERF genes ranged from 0.10 to 0.15 with an average of 0.13, whereas Ka/Ks for segmentally duplicated gene-pairs ranged from 0.03 to 0.14 with an average of 0.09 (Tables S7–S8). The data indicated that the duplicated *SiAP2/ERF* genes were under strong purifying selection pressure and had gone through substitution elimination and enormous selective constraint by natural selection during the course of evolution since their Ka/Ks ratios estimated as <1. Further, the duplication event of these *SiAP2/ERF* tandemly and segmentally duplicated genes may be estimated to have occurred around ~22 and ~24 Mya, respectively (Figure 6). Among the orthologous gene-pairs of *SiAP2/ERF* with those of other grass species, the average Ka/Ks value was maximum between *Brachypodium* and foxtail millet (0.4) and rice and foxtail millet (0.4), and least for sorghum-foxtail millet and maize-foxtail millet gene-pairs (0.2) (Tables S9–S12). The relatively higher rate of synonymous substitution between *Brachypodium*-foxtail millet and rice-foxtail millet *AP2/ERF* genes pointed their earlier divergence around 53–55 Mya from foxtail millet as compared to sorghum and maize *AP2/ERF* genes (Figure 6). Remarkably, the *AP2/ERF* gene-pairs between sorghum and foxtail millet, and maize and foxtail millet (average Ka/Ks = 0.2) appear to have undergone extensive intense purifying selection in comparison to foxtail millet-rice and foxtail millet-*Brachypodium* (average Ka/Ks = 0.4 for both) *AP2/ERF* genes. This is in agreement to their recent time of divergence, around 25 Mya. The estimation of tandem and segmental duplication time (average of 23 Mya) of foxtail millet *AP2/ERF* genes in between the divergence time of foxtail millet-rice (53 Mya), foxtail millet-*Brachypodium* (55 Mya) and foxtail millet-maize and foxtail millet-sorghum (both 25 Mya) orthologous *AP2/ERF* gene-pairs are comparable to evolutionary studies involving the protein-coding genes annotated from the recently released draft genome sequences of foxtail millet [3,4]. Interestingly, though the *SiAP2/ERF* gene-pairs showing segmental (Ka/Ks = 0.09) and tandem duplication (Ka/Ks = 0.13) events are not under similar evolutionary pressure, both set of gene pairs revealed that these events took place almost at similar time (22 Mya for tandem and 24 Mya for segmentally duplicated gene pairs). Therefore, overall, it can be concluded that the segmental and tandem duplication events including the divergence events of *SiAP2/ERF* genes with other grass species have played a major role in evolution of this gene family in foxtail millet. This is also in agreement with earlier genome-wide studies conducted for important gene families in foxtail millet [56–58,67,68].

In silico tissue-specific expression profiling of *SiAP2/ERF* genes

Tissue-specific expression data at a given developmental stage is helpful in identifying genes involved in defining precise nature of individual tissues. Therefore, in order to examine tissue-specific expression profiles of 171 *SiAP2/ERF* genes, a heat map was generated based on the RPKM values for each gene in all tissue samples using RNA-Seq data. A differential expression for all 171 transcripts was observed in 4 tissue samples namely root, leaf, stem and spica (Figure 7). A relative comparison of expression profiles of all 171 SiAP2/ERF showed a relatively higher expression of SiAP2/ERF-020, SiAP2/ERF-021, SiAP2/ERF-025, SiAP2/ERF-041, SiAP2/ERF-043, SiAP2/ERF-063, SiAP2/ERF-069, SiAP2/ERF-094, SiAP2/ERF-108, SiAP2/ERF-139 and SiAP2/ERF-165 in all the four tissues suggesting their importance as potential targets for further functional characterization. In general, majority of the SiAP2/ERFs exhibited root-specific expression (56; ~33%) followed by expression in stem (47; ~27%), then spica (43; ~25%) and least in leaves (26; ~15%). The results indicated that *AP2/ERF* genes in foxtail millet are mostly expressed in roots as confirmed by earlier studies [25,60]. The tissue-specific expression profiling of *SiAP2/ERF*s would further aid the combinatorial involvement of these genes in transcriptional regulation of various tissues, while ubiquitously expressed *SiAP2/ERF*s might control a broad set of genes at transcriptional level. The heat map data also facilitates the overexpression studies of *SiAP2/ERF*s across the tissues to impart stress tolerance to both foxtail millet as well as related grass species.

Expression profiling of *SiAP2/ERFs* during abiotic stresses and phytohormone treatments

Gene expression studies can provide essential indications regarding functions of a gene. In order to analyze the role of *AP2/ERF* genes in foxtail millet, we examined the expression profiles of 21 selected genes representing different sub-families using quantitative real-time (qRT) PCR analysis in response to drought (20% PEG 6000), salt (250 mM NaCl), 100 µM ABA, 100 µM SA, 100 µM MeJA and 100 µM Et during early (1 h) and late (24 h) durations of treatments. The heat map illustration of expression profiles of 21 selected *SiAP2/ERF* genes under drought and salinity is shown in Figure 8. The qRT-PCR analysis demonstrated an overall differential expression patterns to one or more stresses for the genes under study (Figs. 8, 9). The *SiAP2/ERF* genes, in general, were up-regulated by drought and salt treatments except SiAP2/ERF-116 which was down-regulated under drought stress and SiAP2/ERF-092 and SiAP2/ERF-095 which were down-regulated under salt stress at both time points. Only SiAP2/ERF-103 was co-regulated as it was induced by both stresses at all time points. However, 8 *SiAP2/ERF* genes were activated exclusively at late drought stress and 3 at late salinity stress suggesting their role in stress adaptation (Figure 8). The variability in gene expression patterns observed in this study indicated that *SiAP2/ERF*s might play an important role in regulating a complex web of stress responsive pathways for stress adaptation and tolerance towards multiple abiotic stresses.

Phytohormones or plant growth regulators not only play a crucial role in regulation of various plant processes including growth and development but also in signaling and gene expression during environmental stresses both abiotic and biotic. Therefore it was attempted to analyze the expression patterns of the selected 21 *SiAP2/ERF* genes under various hormone treatments. A hierarchical clustering demonstrated overlapping and specific gene expression patterns in response to various phytohormones

(Figure 9). No single gene was exclusively induced in all hormone treatments indicating their treatment-specific roles. However, several genes were exclusively repressed (SiAP2/ERF053, SiAP2/ERF-055, SiAP2/ERF-092, SiAP2/ERF-109 and SiAP2/ERF-116) in all hormone treatments indicating that these genes may be a part of general hormone response. Overall majority of the *SiAP2/ERF* genes were down-regulated in response to ABA except SiAP2/ERF-069, SiAP2/ERF-103 and SiAP2/ERF-120 confirming the previous reports that *AP2/ERF* genes (mostly ERFs and DREBs) are generally regulated in an ABA-independent manner with a few exceptions [11,43]. The regulation of certain *AP2/ERF* genes by SA, MeJA or Et suggests their potential roles in biotic stress responses. Several genes were found to be regulated exclusively by a specific or more than one hormone treatments (Figure 9). As for example, as many as 5 genes were specifically up regulated by MeJA at both early and late time points, while SiAP2/ERF-120 was induced by MeJA and Et at all time points. Phytohormones generally act synergistically or antagonistically to each-other thus influencing signaling response for maintaining cellular homeostasis [69]. Thus *SiAP2/ERF* TFs also act as important mediators of this signaling process. The differential expression patterns of *SiAP2/ERF* genes in this investigation again underlines the intimidating task of understanding the global milieu associated with any stress response. However, as an outcome of this study, we are able to compare their expression profiles during several environmental stress stimuli at early and late time points for precise identification of potential candidate genes for crop improvement programmes. In this regard, *SiAP2/ERF-069, SiAP2/ERF-103* and *SiAP2/ERF-120* may be considered as potential candidate genes for further functional validation as well for utilization in crop improvement programs for stress resistance since these genes were up-regulated under drought and salinity stresses in ABA dependent manner. It can thus be concluded that certain members of *AP2/ERF* gene family in foxtail millet exhibit stimulus-specific and temporal responses and hence expanding current knowledge on molecular basis of stress tolerance and adaptation conferred on plants by them.

Identification of markers in *SiAP2/ERF* genes

Marker-assisted selection (MAS) is a combination of conventional breeding and molecular biology and offers a methodology for accelerating the procedure of crop improvement. The tagging of useful genes, such as those involved in plant hormone synthesis, and those responsible for conferring stress resistance to plants, namely drought and salinity, has been a major target for improving crop growth and productivity [70,71]. With the use of molecular markers, it is now easy to trace important alleles either in segregating or natural populations. Some of the recent studies have shown the importance of *AP2/ERF* TFs, especially DREB TFs, in marker-aided breeding and crop-improvement strategies [16,71]. Considering this, the presence of previously reported DNA-based molecular markers such as SSRs [47], eSSRs [48] and ILPs [49] were searched for their presence in all the 171 *SiAP2/ERF* genes. The analysis identified 54 SSRs and 1 ILP marker in *SiAP2/ERF* genes (Table S13). These markers would be useful in genotyping and MAS for crop improvement.

Homology modeling of *SiAP2/ERF* proteins

Three dimensional protein models of twelve proteins were constructed by sequence similarity searching against the PDB database using BLASTP. These 12 proteins were selected owing to their higher homology to the known protein sequences in the PDB and Phyre2 was used for homology modeling of their predicted structures. The protein structure of all the 12 SiAP2/ERFs were modelled at 90% confidence and the potential active sites were identified (Figure 10). The 3D structure revealed the presence of conserved AP2/ERF domain of about 60–70 amino acids in all the SiAP2/ERF proteins with a typical three-dimensional conformation ordered into a layer of three antiparallel β-sheets followed by a parallel α-helix. Further examination of the AP2/ERF domain showed the presence of two regions namely YRG and RAYD. The YRG region was 20-amino acid long N-terminal stretch rich in basic and hydrophilic residues and was reported to play a crucial role in establishing direct contact with the DNA [72]. Conversely, RAYD region comprises about 40 amino acids and this region was reported to mediate protein-protein interactions through α-helix. Moreover, reports also indicate that RAYD region is involved in DNA binding through interactions of hydrophobic face of the α-helix with the major groove of DNA [72]. The AP2 sub-family members possess two AP2/ERF domains separated by a linker sequence of 25 amino acids which is responsible for positioning of the DNA-binding domains [73]. The molecular modeling thus proved that all the predicted protein structures were highly consistent and this data would offer a preliminary foundation for comprehending the molecular functions of SiAP2/ERF proteins.

Conclusion

The AP2/ERF TFs are important regulators of various plant processes including growth, development and stress responses and thus have been subjected to intensive investigations in various crop plants (Figure S3). However, to the best of our knowledge, no such study has been taken up in otherwise naturally stress tolerant model panicoid C_4 crop *Setaria italica*. The present study identified 171 AP2/ERF TFs in the foxtail millet genome. Isolation and identification of these functional TF genes are expected to aid knowledge towards understanding the molecular genetic basis for foxtail millet stress adaptation and genetic improvement, and may also provide functional gene resources for genetic engineering approaches. To date, only one gene representing this TF superfamily has been characterized from foxtail millet [10]. Hence the present comprehensive study would assist in explicating AP2/ERF family gene function in regulations of stress signaling pathways, and defense responses as well as in providing new opportunities to discover foxtail millet stress tolerance and adaptation mechanisms. The *in silico* structure prediction might provide basic resources to study the molecular regulation of foxtail millet development and stress tolerance. However, extensive *in planta* characterization of putative candidate *SiAP2/ERF* genes is must to further explore its biological roles.

Supporting Information

Figure S1 Gene structures of 171 SiAP2/ERF proteins. Exons and introns are represented by green boxes and black lines, respectively.

Figure S2 Diagrammatic representation of alignment between the miRNA and the SiAP2/ERF targets.

Figure S3 Distribution of AP2/ERFs in sequenced plant genomes.

Table S1 List of primers used in quantitative real time-PCR expression analysis of *SiAP2/ERF* genes.

Table S2 Characteristic features of *SiAP2/ERF* Transcription factor gene family members identified in *Setaria italica*.

Table S3 Summary of functional domains present in the SiAP2/ERF proteins.

Table S4 Blast2GO annotation details of SiAP2/ERF protein sequences.

Table S5 Characteristics of the promoter region of *SiAP2/ERF* genes.

Table S6 List of putative *Setaria italica* miRNAs targeting *SiAP2/ERF* transcripts.

Table S7 The Ka/Ks ratios and estimated divergence time for tandemly duplicated *SiAP2/ERF* genes.

Table S8 The Ka/Ks ratios and estimated divergence time for segmentally duplicated *SiAP2/ERF* genes.

Table S9 The Ka/Ks ratios and estimated divergence time for orthologous SiAP2/ERF proteins between foxtail millet and sorghum.

Table S10 The Ka/Ks ratios and estimated divergence time for orthologous SiAP2/ERF proteins between foxtail millet and maize.

Table S11 The Ka/Ks ratios and estimated divergence time for orthologous SiAP2/ERF proteins between foxtail millet and rice.

Table S12 The Ka/Ks ratios and estimated divergence time for orthologous SiAP2/ERF proteins between foxtail millet and *Brachypodium*.

Table S13 Details of SiAP2/ERF transcription factor-based markers.

Acknowledgments

Timely assistance from Mr. Rohit Khandelwal is appreciated.

Author Contributions

Conceived and designed the experiments: MP CL. Performed the experiments: CL AKM MM VSB YK. Analyzed the data: MP CL. Contributed to the writing of the manuscript: CL MM MP.

References

1. Muthamilarasan M, Prasad M (2014) Advances in *Setaria* genomics for genetic improvement of cereals and bioenergy grasses. Theor Appl Genet DOI:10.1007/s00122-014-2399-3.
2. Lata C, Gupta S, Prasad M (2013) Foxtail millet: a model crop for genetic and genomic studies in bioenergy grasses, Crit Rev Biotechnol 33: 328–343.
3. Zhang G, Liu X, Quan Z, Cheng S, Xu X, et al. (2012) Genome sequence of foxtail millet (*Setaria italica*) provides insights into grass evolution and biofuel potential. Nature Biotechnol 30: 549–554.
4. Bennetzen JL, Schmutz J, Wang H, Percifield R, Hawkins J, et al. (2012) Reference genome sequence of the model plant *Setaria*. Nature Biotechnol 30: 555–561.
5. Lata C, Prasad M (2013) *Setaria* genome sequencing: An overview. J Plant Biochem Biotechnol 22: 257–260.
6. Muthamilarasan M, Theriappan P, Prasad M (2013) Recent advances in crop genomics for ensuring food security. Curr Sci 105: 155–158.
7. Sakuma Y, Liu Q, Dubouzet JG, Abe H, Shinozaki K, et al. (2002) DNA-binding specificity of the ERF/AP2 domain of Arabidopsis DREBs, transcription factors involved in dehydration- and cold-inducible gene expression, Biochem Biophys Res Commun. 290: 998–1009.
8. Nakano T, Suzuki K, Fujimura T, Shinshi H (2006) Genome-wide analysis of the ERF gene family in Arabidopsis and rice. Plant Physiol 140: 411–432.
9. Jofuku KD, Boer BGW, Montagu MV, Okamuro JK (1994) Control of Arabidopsis flower and seed development by the homeotic gene APETALA2. Plant Cell 6: 1211–25.
10. Lata C, Yadav A, Prasad M (2011) Role of plant transcription factors in abiotic stress tolerance. In: Shanker A. and Venkateshwarulu B. (eds) Abiotic Stress Response in Plants, INTECH Open Access Publishers, 269–296.
11. Mizoi J, Shinozaki K, Yamaguchi-Shinozaki K (2012) AP2/ERF family transcription factors in plant abiotic stress responses. Biochimica et Biophysica Acta 1819: 86–96.
12. Magnani E, Sjölander K, Hake S (2004) From endonucleases to transcription factors: evolution of the AP2 DNA binding domain in plants, Plant Cell 16: 2265–2277.
13. Shigyo M, Hasebe M, Ito M (2006) Molecular evolution of the AP2 subfamily, Gene 366: 256–265.
14. Saleh A, Pagés M (2003) Plant AP2/ERF transcription factors. Genetika 35: 37–50.
15. Shigyo M, Ito M (2004) Analysis of gymnosperm two-AP2-domain-containing genes. Development Genes Evoln 214: 105–14.
16. Lata C, Prasad M (2011) Role of DREBs in regulation of abiotic stress responses in plants. J Exp Bot 62: 4731–4748.
17. Riechmann JL, Meyerowitz EM (1998) The AP2/EREBP family of plant transcription factors. Biol Chem, 379: 633–646.
18. Zhuang J, Cai B, Peng RH, Zhu B, Jin XF, et al. (2008) Genome-wide analysis of the AP2/ERF gene family in *Populus trichocarpa*. Biochemical Biophysical Res Comm 371: 468–474.
19. Gil-Humanes J, Piston F, Martin A, Barro F (2009) Comparative genomic analysis and expression of the APETALA2-like genes from barley, wheat, and barley-wheat amphiploids. BMC Plant Biol 9: 66.
20. Dietz KJ, Vogel MO, Viehhauser A (2010) AP2/EREBP transcription factors are part of gene regulatory networks and integrate metabolic, hormonal and environmental signals in stress acclimation and retrograde signalling. Protoplasma 245: 3–14.
21. Licausi F, Giorgi FM, Zenoni S, Osti F, Pezzotti M, et al. (2010) Genomic and transcriptomic analysis of the AP2/ERF superfamily in *Vitis vinifera*. BMC Genomics 11: 719.
22. Zhuang J, Chen J-M, Yao Q-H, Xiong F, Sun C-C, et al. (2011) Discovery and expression profile analysis of AP2/ERF family genes from *Triticum aestivum*. Mol Biol Rep 38: 745–753.
23. Zhuang J, Yao Q-H, Xiong A-S, Zhang J (2011) Isolation, Phylogeny and Expression Patterns of AP2-Like Genes in Apple (Malus × domestica Borkh). Plant Mol Biol Rep 29: 209–216.
24. Zhang G, Chen M, Chen X, Xu Z, Guan S, et al. (2008) Phylogeny, gene structures, and expression patterns of the ERF gene family in soybean (*Glycine max* L.). J Exp Bot 59: 4095–4107.
25. Xu W, Li F, Ling L, Liu A (2013) Genome-wide survey and expression profiles of the AP2/ERF family in castor bean (*Ricinus communis* L.). BMC Genomics 14: 785.
26. Punta M, Coggill PC, Eberhardt RY, Mistry J, Tate J, et al. (2012) The Pfam protein families database. Nucleic Acids Res 40: D290–D301.
27. Voorrips RE (2002) MapChart: software for the graphical presentation of linkage maps and QTLs. J Hered 93: 77–78.
28. Lee TH, Tang H, Wang X, Paterson AH (2012) PGDD: a database of gene and genome duplication in plants. Nucleic Acids Res 41: D1152–D1158.
29. Tang H, Bowers JE, Wang X, Ming R, Alam M, et al. (2008) Synteny and Collinearity in Plant Genomes, Science 320: 486–488.
30. Krzywinski M, Schein J, Birol I, Connors J, Gascoyne R, et al. (2009) Circos: an information aesthetic for comparative genomics. Genome Res 19: 1639–1645.
31. Shiu S-H, Bleecker AB (2003) Expansion of the Receptor-Like Kinase/Pelle Gene Family and Receptor-Like Proteins in Arabidopsis. Plant Physiol 132: 530–543.

32. Du D, Zhang Q, Cheng T, Pan H, Yang W, et al. (2012) Genome-wide identification and analysis of late embryogenesis abundant (LEA) genes in *Prunus mume*. Mol Biol Rep 40: 1937–1946.
33. Guo AY, Zhu QH, Chen X, Luo JC (2007) GSDS: a gene structure display server. Yi Chuan 29: 1023–1026.
34. Tamura K, Peterson D, Peterson N, Stecher G, Nei M, et al. (2011) MEGA5: Molecular evolutionary genetics analysis using maximum likelihood, evolutionary distance, and maximum parsimony methods. Mol Biol Evol 28: 2731–2739.
35. Conesa A, Gotz S (2008) Blast2GO: a comprehensive suite for functional analysis in plant genomics. Int J Plant Genomics 2008: 619832.
36. Khan Y, Yadav A, Suresh BV, Muthamilarasan M, Yadav CB, et al. (2014) Comprehensive genome-wide identification and expression profiling of foxtail millet [*Setaria italica* (L.)] miRNAs in response to abiotic stress and development of miRNA database. Plant Cell Tiss Organ Cult 118: 279–292.
37. Suyama M, Torrents D, Bork P (2006) PAL2NAL: robust conversion of protein sequence alignments into the corresponding codon alignments. Nucleic Acids Res 34: W609–W612.
38. Lynch M, Conery JS (2000) The evolutionary fate and consequences of duplicate genes. Science 290: 1151–1155.
39. Yang Z, Gu S, Wang X, Li W, Tang Z, et al. (2008) Molecular evolution of the cpp-like gene family in plants: insights from comparative genomics of Arabidopsis and rice. J Mol Evol 67: 266–277.
40. Cochrane G, Alako B, Amid C, Bower L, Cerdeño-Tárraga A, et al. (2013) Facing growth in the European Nucleotide Archive. Nucleic Acids Res 41: D30–D35.
41. Saeed AI, Bhagabati NK, Braisted JC, Liang W, Sharov V, et al. (2006) TM4 microarray software suite. Methods Enzymol 411: 134–193.
42. Saeed AI, Sharov V, White J, Li J, Liang W, et al. (2003) TM4: a free, open-source system for microarray data management and analysis. Biotechniques 34: 374–378.
43. Lata C, Sahu PP, Prasad M (2010) Comparative transcriptome analysis of differentially expressed genes in foxtail millet (*Setaria italica* L.) during dehydration stress. Biochem Biophys Res Commun 393: 720–727.
44. Lata C, Bhutty S, Bahadur RP, Majee M, Prasad M (2011) Association of a SNP in a novel DREB2-like gene *SiDREB2* with stress tolerance in foxtail millet [*Setaria italica* (L.)]. J Exp Bot 62: 3387–3401.
45. Kumar K, Muthamilarasan M, Prasad M (2013) Reference genes for quantitative Real-time PCR analysis in the model plant foxtail millet (*Setaria italica* L.) subjected to abiotic stress conditions. Plant Cell Tiss Organ Cult 115: 13–22.
46. Livak KJ, Schmittgen TD (2001) Analysis of relative gene expression data using realtime quantitative PCR and the $2^{-\Delta\Delta Ct}$ method. Methods 25: 402–408.
47. Pandey G, Misra G, Kumari K, Gupta S, Parida SK, et al. (2013) Genome-wide development and use of microsatellite markers for large-scale genotyping applications in foxtail millet [*Setaria italica* (L.)]. DNA Res 20: 197–207.
48. Kumari K, Muthamilarasan M, Misra G, Gupta S, Subramanian A, et al. (2013) Development of eSSR-markers in *Setaria italica* and their applicability in studying genetic diversity, cross-transferability and comparative mapping in millet and non-millet species. PLoS ONE 8: e67742.
49. Muthamilarasan M, Venkata Suresh B, Pandey G, Kumari K, Parida SK, et al. (2014) Development of 5123 intron-length polymorphic markers for large-scale genotyping applications in foxtail millet. DNA Res 21: 41–52.
50. Suresh BV, Muthamilarasan M, Misra G, Prasad M (2013) FmMDb: a versatile database of foxtail millet markers for millets and bioenergy grasses research. PLoS ONE 8: e71418.
51. Berman HM, Westbrook J, Feng Z, Gilliland G, Bhat TN, et al. (2000) The protein data bank. Nucleic Acids Res 28: 235–242.
52. Kelley LA, Sternberg MJE (2009) Protein structure prediction on the Web: a case study using the Phyre server. Nature Protocols 4: 363–371.
53. Zhuang J, Peng R-H, Cheng Z-M, Zhang J, Cai B, et al. (2009) Genome-wide analysis of the putative AP2/ERF family genes in *Vitis vinifera*. Scientia Horticul. 123: 73–81.
54. Rashid M, Guangyuan H, Guangxiao Y, Hussain J, Xu Y (2012) AP2/ERF transcription factor in rice: genome-wide canvas and syntenic relationships between monocots and eudicots. Evol Bioinform Online 8: 321–355.
55. Bonthala VS, Muthamilarasan M, Roy R, Prasad M (2014) FmTFDb: a foxtail millet transcription factors database for expediting functional genomics in millets. Mol Biol Rep 41: 6343–6348.
56. Yadav CB, Muthamilarasan M, Pandey G, Prasad M (2014) Identification, characterization and expression profiling of Dicer-like, Argonaute and RNA-dependent RNA polymerase gene families in foxtail millet. Plant Mol Biol Rep DOI:10.1007/s11105-014-0736-y.
57. Muthamilarasan M, Bonthala VS, Mishra AK, Khandelwal R, Khan Y, et al. (2014) C_2H_2 type of zinc finger transcription factors in foxtail millet define response to abiotic stresses. Funct Integr Genomics 14: 531–543.
58. Muthamilarasan M, Khandelwal R, Yadav CB (2014) Identification and molecular characterization of MYB transcription factor superfamily in C_4 model plant foxtail millet (*Setaria italica* L.). PLOS ONE 9: e109920.
59. Karniely S, Pines O (2005) Single translation-dual destination: mechanisms of dual protein targeting in eukaryotes. Embo Reports 6: 420–425.
60. Song X, Li Y, Hou X (2013) Genome-wide analysis of the AP2/ERF transcription factor superfamily in Chinese cabbage (*Brassica rapa* ssp. pekinensis). BMC Genomics 14: 573.
61. Sun Z-M, Zhou M-L, Xiao X-G, Tang Y-X, Wu Y-M (2014) Genome-wide analysis of AP2/ERF family genes from *Lotus corniculatus* shows LcERF054 enhances salt tolerance. Funct Integr Genomics 14: 453–466.
62. Chen J, Xia X, Yin W (2009) Expression profiling and functional characterization of a DREB2-type gene from *Populus euphratica*. Biochem Biophys Res Commun 378: 483–487.
63. Le DT, Nishiyama R, Watanabe Y, Vankova R, Tanaka M, et al. (2012) Identification and expression analysis of cytokinin metabolic genes in soybean under normal and drought conditions in relation to cytokinin levels. PLoS ONE 7(8): e42411.
64. Fang YJ, You J, Xie KB, Xie WB, Xiong LZ (2008) Systematic sequence analysis and identification of tissue-specific or stress-responsive genes of NAC transcription factor family in rice. Mol Genet Genomics 280: 547–563.
65. Aida M, Beis D, Heidstra R, Willemsen V, Blilou I, et al. (2004) The PLETHORA genes mediate patterning of the Arabidopsis root stem cell niche. Cell 119: 109–120.
66. Cannon SB, Mitra A, Baumgarten A, Young ND, May G (2004) The roles of segmental and tandem gene duplication in the evolution of large gene families in *Arabidopsis thaliana*. BMC Plant Biol 4: 10.
67. Puranik S, Sahu PP, Mandal SN, B VS, Parida SK, et al. (2013) Comprehensive genome-wide survey, genomic constitution and expression profiling of the NAC transcription factor family in foxtail millet (*Setaria italica* L.). PLoS ONE 8: e64594.
68. Mishra AK, Muthamilarasan M, Khan Y, Parida SK, Prasad M (2014) Genome-wide investigation and expression analyses of WD40 protein family in the model plant foxtail millet (*Setaria italica* L.) PLoS ONE 9: e86852.
69. Zeller G, Henz SR, Widmer CK, Sachsenberg T, Ratsch G, et al. (2009) Stress induced changes in the *Arabidopsis thaliana* transcriptome analyzed using whole genome tiling arrays. Plant J 58: 1068–1082.
70. Lopez CG, Banowetz GM, Peterson CJ, Kronstad WE (2003) Dehydrin expression and drought tolerance in seven wheat cultivars. Crop Science 43: 577–582.
71. Lata C, Prasad M (2014) Association of an allele-specific marker with dehydration stress tolerance in foxtail millet suggests SiDREB2 to be an important QTL. J Plant Biochem Biotechnol 23: 119–122.
72. Okamuro JK, Caster B, Villarroel R, van Montagu M, Jofuku KD (1997) The AP2 domain of APETALA2 defines a large new family of DNA binding proteins in Arabidopsis. Proc Natl Acad Sci USA 94: 7076–7081.
73. Wolfe SA, Nekludova L, Pabo CO (2000) DNA recognition by Cys2His2 zinc finger proteins. Annu Rev Biophys Biomol Struct 29: 183–212.

Permissions

All chapters in this book were first published in PLOS ONE, by The Public Library of Science; hereby published with permission under the Creative Commons Attribution License or equivalent. Every chapter published in this book has been scrutinized by our experts. Their significance has been extensively debated. The topics covered herein carry significant findings which will fuel the growth of the discipline. They may even be implemented as practical applications or may be referred to as a beginning point for another development.

The contributors of this book come from diverse backgrounds, making this book a truly international effort. This book will bring forth new frontiers with its revolutionizing research information and detailed analysis of the nascent developments around the world.

We would like to thank all the contributing authors for lending their expertise to make the book truly unique. They have played a crucial role in the development of this book. Without their invaluable contributions this book wouldn't have been possible. They have made vital efforts to compile up to date information on the varied aspects of this subject to make this book a valuable addition to the collection of many professionals and students.

This book was conceptualized with the vision of imparting up-to-date information and advanced data in this field. To ensure the same, a matchless editorial board was set up. Every individual on the board went through rigorous rounds of assessment to prove their worth. After which they invested a large part of their time researching and compiling the most relevant data for our readers.

The editorial board has been involved in producing this book since its inception. They have spent rigorous hours researching and exploring the diverse topics which have resulted in the successful publishing of this book. They have passed on their knowledge of decades through this book. To expedite this challenging task, the publisher supported the team at every step. A small team of assistant editors was also appointed to further simplify the editing procedure and attain best results for the readers.

Apart from the editorial board, the designing team has also invested a significant amount of their time in understanding the subject and creating the most relevant covers. They scrutinized every image to scout for the most suitable representation of the subject and create an appropriate cover for the book.

The publishing team has been an ardent support to the editorial, designing and production team. Their endless efforts to recruit the best for this project, has resulted in the accomplishment of this book. They are a veteran in the field of academics and their pool of knowledge is as vast as their experience in printing. Their expertise and guidance has proved useful at every step. Their uncompromising quality standards have made this book an exceptional effort. Their encouragement from time to time has been an inspiration for everyone.

The publisher and the editorial board hope that this book will prove to be a valuable piece of knowledge for researchers, students, practitioners and scholars across the globe.

List of Contributors

Sagarika Mishra and Lingaraj Sahoo
Department of Biotechnology, Indian Institute of Technology Guwahati, Guwahati, India

Hemasundar Alavilli and Byeong-ha Lee
Department of Life Science, Sogang University, Mapo-gu, Seoul, Korea

Sanjib Kumar Panda
Department of Life Sciences and Bioinformatics, Assam University, Silchar, India
Department of Biochemistry & Molecular Biology, Noble Research Centre, Oklahoma
State University, Stillwater, OK, United States of America

Sudhir P. Singh, Raja Jeet, Jitendra Kumar, Vishnu Shukla, Shrikant S. Mantri and Rakesh Tuli
National Agri-Food Biotechnology Institute, Department of Biotechnology (DBT)

Rakesh Srivastava
School of Biotechnology, Banaras Hindu University, Varanasi, India

Gabriel J. Bellante, Scott L. Powell and Rick L. Lawrence
Department of Land Resources and Environmental Sciences, Montana State University, Bozeman, Montana, United States of America

Kevin S. Repasky
Department of Electrical and Computer Engineering, Montana State University, Bozeman, Montana, United States of America

Tracy Dougher
Department of Plant Sciences and Plant Pathology, Montana State University, Bozeman, Montana, United States of America

Li-Fei Yang
College of Horticulture, Nanjing Agricultural University, Nanjing, China

Yi Chen
College of Horticulture, Nanjing Agricultural University, Nanjing, China
Institute of Food Quality and Safety, Jiangsu Academy of Agricultural Sciences, Nanjing, China

You-Qin Li and Jian Chen
Institute of Food Quality and Safety, Jiangsu Academy of Agricultural Sciences, Nanjing, China

Hai-Zhen Mo and Liang-Bin Hu
Department of Food Science, Henan Institute of Science and Technology, Xinxiang, Henan Province, China

José Tomás Matus
Center for Research in Agricultural Genomics CSIC-IRTA-UAB-UB, Bellaterra, Barcelona, Spain
Departamento de Genética Molecular y Microbiología, Pontificia Universidad Católica de Chile, Santiago, Chile

Felipe Aquea, Carmen Espinoza, Paola Cañón, Amparo Rodríguez-Hoces de la Guardia, Jennifer Serrano and Patricio Arce-Johnson
Departamento de Gené tica Molecular y Microbiología, Pontificia
Universidad Cató lica de Chile, Santiago, Chile

Andrea Vega
Departamento de Ciencias Vegetales, Facultad de Agronomía e Ingeniería Forestal, Pontificia
Universidad Cató lica de Chile, Santiago, Chile

Erika Cavallini, Silvia Dal Santo and Giovanni Battista Tornielli
Department of Biotechnology, University of Verona, Verona, Italy

Guimei Jiang, Peitao Lü, Jitao Liu, Junping Gao and Changqing Zhang
Department of Ornamental Horticulture, College of Agriculture and Biotechnology, China Agricultural University, Beijing, PR China

Xinqiang Jiang
College of Landscape Architecture and Forestry, Qingdao Agricultural University, Qingdao, PR China

Eric L. Sparks and Just Cebrian
Dauphin Island Sea Lab, Dauphin Island, Alabama, United States of America
Marine Sciences, University of South Alabama, Mobile, Alabama, United States of America

Sara M. Smith
Dauphin Island Sea Lab, Dauphin Island, Alabama, United States of America

Lipan Hu, Lixia Xiang, Li Zhang, Xiaoting Zhou, Zhirong Zou and Xiaohui Hu
College of Horticulture, Northwest Agricultural & Forest University, Yangling, China
Key Laboratory of Protected Horticultural Engineering in Northwest, Ministry of Agriculture, Shaanxi Yangling, China

Srivathsa Nallanchakravarthula, Shahid Mahmood, Sadhna Alströ m and Roger D. Finlay
Uppsala BioCenter, Department of Forest Mycology and Plant Pathology, Swedish University of Agricultural Sciences, Uppsala, Sweden

Li Xu
College of Resources and Environmental Sciences, Nanjing Agricultural University, Nanjing, China
Zhongyuan Lin, Qing Tao, Mingxiang Liang, Gengmao Zhao, Xiangzhen Yin and

Ruixin Fu
College of Resources and Environmental Sciences, Nanjing Agricultural University, Nanjing, China
Jiangsu Key Lab of Marine Biology, Nanjing, China

Vivek Dogra and Tejpal Gill
Division of Biotechnology, CSIR-Institute of Himalayan Bioresource Technology, Palampur, Himachal Pradesh, India

Amrina Shafi, Paramvir Singh Ahuja and Yelam Sreenivasulu
Division of Biotechnology, CSIR-Institute of Himalayan Bioresource Technology, Palampur, Himachal Pradesh, India
Academy of Scientific and Innovative Research (AcSIR), CSIR- Institute of Himalayan Bioresource Technology, Palampur, Himachal Pradesh, India

Amit Kumar Chaturvedi, Manish Kumar Patel, Avinash Mishra, Vivekanand Tiwari and Bhavanath Jha
Discipline of Marine Biotechnology and Ecology, CSIR-Central Salt and Marine Chemicals Research Institute, Bhavnagar, Gujarat, India

Xiaolian Zhang, Ning Wang, Pei Chen, Mengmeng Gao, Juge Liu, Yufeng Wang, Tuanjie Zhao, Yan Li and Junyi Gai
National Key Laboratory of Crop Genetics and Germplasm Enhancement, National Center for Soybean Improvement, Key Laboratory for Biology and Genetic Improvement of Soybean (General, Ministry of Agriculture), Nanjing Agricultural University, Nanjing, Jiangsu, China

Guo-Hong Yu, Lin-Lin Jiang, Meng-Meng Liu, Xian-Guo Cheng and Shu-Guang Shan
Key Lab. of Plant Nutrition and Fertilizer, Ministry of Agriculture, Institute of Agricultural Resources and Regional Planning, Chinese Academy of Agricultural Sciences, Beijing, China

Xue-Feng Ma
Key Lab. of Plant Nutrition and Fertilizer, Ministry of Agriculture, Institute of Agricultural Resources and Regional Planning, Chinese Academy of Agricultural Sciences, Beijing, China
Institute of Agro-Products Processing Science and Technology, Chinese Academy of Agricultural Sciences, Beijing, China

Zhao-Shi Xu
Institute of Crop Science Chinese Academy of Agricultural Sciences (CAAS)/National Key Facility for Crop Gene Resources and Genetic Improvement, Key Laboratory of Biology and Genetic Improvement of Triticeae Crops, Ministry of Agriculture, Beijing, China

Dan Zhu
College of Life Science, Qingdao Agricultural University, Qingdao, P.R. China
Plant Bioengineering Laboratory, Northeast Agricultural University, Harbin, P.R. China

Rongtian Li
Key Laboratory of Molecular Biology, College of Heilongjiang Province, Heilongjiang University, Harbin, P.R. China

Xin Liu
College of Life Science, Qingdao Agricultural University, Qingdao, P.R. China

Mingzhe Sun, Jing Wu, Ning Zhang and Yanming Zhu
College of Life Science, Qingdao Agricultural University, Qingdao, P.R. China
Plant Bioengineering Laboratory, Northeast Agricultural University, Harbin, P.R. China

Guodong Rao and Jinkai Sui
Key Laboratory of Tree Breeding and Cultivation, State Forestry Administration, Research Institute of Forestry, Chinese Academy of Forestry, Beijing, Republic of China

Yanfei Zeng, Caiyun He, Aiguo Duan and Jianguo Zhang
State Key Laboratory of Tree Genetics and Breeding, Research Institute of Forestry, Chinese Academy of Forestry, Beijing, Republic of China

Key Laboratory of TreeBreeding and Cultivation, State Forestry Administration, Research Institute of Forestry, Chinese Academy of Forestry, Beijing, Republic of China

Lei Zhang, Huijuan Ma, Tingting Chen, Jun Pen, Shuxun Yu and Xinhua Zhao
State Key Laboratory of Cotton Biology, Institute of Cotton Research of CAAS, Anyang, P. R. China

Enkai Xu, Guoqiang Fan, Suyan Niu, Zhenli Zhao, Minjie Deng and Yanpeng Dong
Institute of Paulownia, Henan Agricultural University, Jinshui Area, Zhengzhou, Henan, P.R. China
College of Forestry, Henan Agricultural University, Jinshui Area, Zhengzhou, Henan, P.R. China

Charu Lata
National Research Centre on Plant Biotechnology, New Delhi, India
CSIR-National Botanical Research Institute, Lucknow, Uttar Pradesh, India

Awdhesh Kumar Mishra, Mehanathan Muthamilarasan, Venkata Suresh Bonthala, Yusuf Khan and Manoj Prasad
National Institute of Plant Genome Research, New Delhi, India

Index

N

O

P

R

S

T

W

Y

Z